T0207313

Designing and Conducting Business Surveys

WILEY SERIES IN SURVEY METHODOLOGY
Established in Part by WALTER A. SHEWHART AND SAMUEL S. WILKS

A complete list of the titles in this series appears at the end of this volume.

Designing and Conducting Business Surveys

GER SNIJKERS
Statistics Netherlands, Heerlen, The Netherlands

GUSTAV HARALDSEN
Statistics Norway, Kongsvinger, Norway

JACQUI JONES
Office for National Statistics, Newport, Gwent, United Kingdom

DIANE K. WILLIMACK
United States Census Bureau, Washington, DC, USA

Disclaimer

Any views expressed on methodological, technical, or operational issues are those of the authors and not necessarily those of Statistics Netherlands, Statistics Norway, the UK Office for National Statistics, or the US Census Bureau.

Cover design: John Wiley & Sons, Inc.
Cover illustration: © Colourbox/Courtesy of Siri Boquist

Published by John Wiley & Sons, Inc., Hoboken, New Jersey.

Published simultaneously in Canada.

Library of Congress Cataloging-in-Publication Data:

Snijkers, Ger, 1963-
Designing and conducting business surveys / Ger Snijkers, Gustav Haraldsen, Jacqui Jones, Diane K. Willimack.
pages cm.–(Wiley series in survey methodology ; 568)
Includes bibliographical references and index.
ISBN 978-0-470-90304-9
1. Survey methodology. 2. Establishment/Organizational surveys 3. Social surveys
4. Research – Official Statistics – Statistical methods
HB137.S63 2013
338.0072′3–dc23

2012020962

Printed in the United States of America

10 9 8 7 6 5 4 3 2 1

We dedicate this book in memory of Seymour Sudman

We add our personal dedications:
From Ger, for Sjaak Snijkers (my late father) and Ingrid;
From Gustav, for Ellinor and Katarina;
From Jacqui, for Jonah, Owain, Rhiannon, and my girls
(Delila and Sophie); and
From Diane, for Steven, my parents Edwin and Betty Willimack, and
my late sister Jeanne Paulsen.

Contents

Preface

The book you are about to read has a history of about ten years. For the last decade we have been thinking about preparing a comprehensive book on business surveys. In the literature on social survey methodology, many books about designing and conducting surveys are available [e.g., Groves et al. (2004); Czaja and Blair, 2005; de Leeuw et al. (2008); Dillman et al. (2009)]. This also holds for business research methods focusing on business market and management research. There is, however, no textbook devoted specifically to business surveys, that is, surveys that collect data from businesses for statistical, research, or policy reasons, although some books contain a chapter devoted to business surveys [e.g., Dillman et al. (2009)]. The one exception is the monograph entitled *Business Survey Methods* edited by Cox et al. (1995). This book was published following the first International Conference on Establishment Surveys (ICES) held in 1993. Although highly regarded as a valuable and scholarly overview of business survey issues, it is a compendium of different aspects of business surveys, rather than focusing on designing and conducting business surveys. Since publication of the Cox book, the business survey literature has continued to remain primarily dispersed in journals and conference papers.

Our mutual interest in developing a book devoted to business survey methodology became apparent when we all attended the Workshop on Questionnaire Pretesting Methods (QUEST) at Statistics Netherlands in April 2005. Here we discussed starting a similar workshop on Business Data Collection Methodology (BDCM). We felt that formal conferences on business surveys, such as the International Conference on Establishment Surveys, held once every 7 years, were too infrequent and that we needed an informal platform for researchers and practitioners in business surveys to discuss design issues and practices. We agreed that the dispersed literature on business surveys continued to be "a stumbling block for progress in solving business surveys' unique problems" (Cox et al. 1995, p. xiii). Indeed, this very issue had been the catalyst for Brenda G. Cox and B. Hanjamma Chinkappa to set up the first ICES event. To fill this void, the first BDCM Workshop was held in April 2006 in London, hosted by the UK Office for National Statistics (ONS). The goal of the workshop was to exchange information and learn from each other; the content included research papers, research in progress, unanswered questions, hands-on experience, and errors and pitfalls. Interest was evident in that the first BDCM Workshop was attended by twenty nine business survey researchers and practitioners representing twelve countries, and led to two more workshops held in September 2008 (Ottawa, hosted by Statistics Canada) and April 2010 [Wiesbaden, hosted by the German Federal Statistical Office (FSO)].

The first BDCM Workshop was set up as a satellite workshop to the European Conference on Quality in Official Statistics (hosted by UK ONS) in Cardiff, Wales. At this conference we discussed the need for a book devoted to designing and conducting business surveys with the leading survey methodologist Robert Groves. He agreed that the survey methodology field needed a comprehensive book on this topic, and encouraged us to write it.

We continued to discuss the book when organizing the second BDCM Workshop, and in February 2009, we convened in Amsterdam to draft an outline of the book. Next we contacted Wiley with our proposal. Following submission of a sample chapter, in January 2010, Wiley's publishing committee provided positive feedback on our proposed book. This served as the green light for us to start writing. We next convened in April 2010, at the third BDCM Workshop to finalize the outline for the book. We pooled our resources and divided initial writing tasks among ourselves, utilizing our various areas of expertise. We also contributed varying perspectives in preparing individual chapters based on the diversity of our international and organizational experiences. In addition, we gratefully acknowledge the contributions of Paul Smith (UK ONS), who wrote Chapter 5, on sampling and estimation; Deirdre Giesen (Statistics Netherlands) and Li-Chun Zhang (Statistics Norway), who collaborated with us on Chapter 6 on response burden; and Mike Hidiroglou (Statistics Canada), who coauthored Chapter 11 on capturing, coding, and cleaning survey data. It has taken us 2 years to write this book (to April 2012, when the manuscript was completed).

In the period that we have been preparing and writing the book, we have been supported and inspired by many people. First, we are very grateful to our respective institutions (Statistics Netherlands and Utrecht University, Statistics Norway, UK Office for National Statistics, and the US Census Bureau), which have been very supportive and enabled us to organize the BDCM Workshops and write this book. We are very grateful to Lars Lyberg for his inspiration and support during the writing process. By hosting the three BDCM Workshops, UK ONS, Statistics Canada, and the German FSO have contributed to the preparation of this book; once again, we'd like to thank them for their hospitality and contributions to the book by facilitating the international discussion on business survey data collection methodology. But above all, we would like to thank the participants of the BDCM Workshops and colleagues around the world (who continually asked about the progress of the book, telling us that they are eager to read it) for their contributions and support. The inspiring discussions we have had at the BDCM Workshops contributed to many ideas for this book. We have been writing this book with these colleagues in mind, as the audience of this book: practitioners, researchers, analysts, students, and others involved in business surveys from national statistical institutes, central banks, universities, and survey organizations conducting business surveys at large.

Although the focus is on business surveys, we feel that this book may be of interest to survey methodologists in general.

Last but not least, we would like to thank our spouses and families, who have been very patient with us for three years while we focused our energy on writing and finalizing this book.

We may have made errors; and we are sure that we have missed relevant studies. Still, to our best efforts, we have tried to integrate theories, evidence-based knowledge, and good practices into this book from a multidisciplinary process-quality-oriented perspective. Furthermore, we have tried to provide guidelines to enable readers to make well-informed decisions about tradeoffs that need to be taken when designing and conducting a business survey. The chapters are structured around this dichotomy; theoretical frameworks and high-level principles, as well as practices and guidelines, are discussed. The result is a how-to book.

When writing the book, we found many issues that still require additional research. The field of business surveys needs more documentation describing practices (practices that both do and do not work), case studies, pilots and experiments to identify and isolate best practices. So, we encourage the readers to do more research, and to share the results with colleagues around the world. Also, we believe that our field would benefit from new multidisciplinary perspectives for building relevant theoretical models that provide a basis for understanding and improving the processes in business surveys, in order to reduce response burden, improve data quality, and generate relevant, reliable statistical products.

We invite the readers to contact us regarding any relevant information on designing and conducting business surveys; comments, suggestions for improvement, and research reports are welcome at `WileyBookBusinessSurveys@gmail.com`. Additionally, information about future BDCM Workshops can also be acquired at this email address.

We look forward to hearing from you. Happy reading!

Statistics Netherlands, Heerlen GER SNIJKERS

Statistics Norway, Kongsvinger GUSTAV HARALDSEN

UK Office for National Statistics, Newport, Gwent JACQUI JONES

US Census Bureau, Washington, DC DIANE K. WILLIMACK

May 2013

Contributors

Dierdre Giesen, Statistics Netherlands, Heerlen/The Hague, The Netherlands

Gustav Haraldsen, Statistics Norway, Kongsvinger/Oslo, Norway

Mike Hidiroglou, Statistics Canada, Ottawa, Ontario, Canada

Jacqui Jones, UK Office for National Statistics, Newport, Gwent, United Kingdom

Paul Smith, UK Office for National Statistics, Newport, Gwent, United Kingdom

Ger Snijkers, Statistics Netherlands, Heerlen/The Hague, and Utrecht University (2006–2011), The Netherlands

Diane K. Willimack, United States Census Bureau, Washington, DC, USA

Li-Chun Zhang, Statistics Norway, Kongsvinger/Oslo, Norway

Abbreviations

AAPOR	American Association for Public Opinion Research
ABS	Australian Bureau of Statistics
ACTR	automatic coding by text recognition
AIC/BIC	Akaike/Bayes information criterion
ANCOVA	analysis of covariance
ANOVA	analysis of variance
ARIMA	autoregressive integrated moving average
ARTS	annual retail trade survey
ASHE	annual survey of hours and earnings
AUM	animal unit month (found in Fig. 8.9)
BDCM	business data collection methodology
BLS	Bureau of Labor Statistics (US)
BLUE	best linear unbiased estimator
BRDIS	Business Research and Development Innovation Survey (US)
CAPI	computer-assisted personal interview
CATI	computer-assisted telephone interview
CBA	cost/benefit analysis
COLI	cost-of-living index
CPA	classification of products by activity
CPC	central product classification
CPI	consumer price index
CRA	Canada Revenue Agency
CSAQ	computerized self-administered questionnaire
CSO	Central Statistical Office (UK)
CSPI	corporate services price index
CV	coefficient of variation
DBT	design–build–test
DDI	data documentation initative
DP	data provider
EDI	electronic data interchange
EIN	employer identification number

ESA	European system of (national) accounts
ESS	European Statistical System
ETS	enterprise and trade statistics
FPC	Financial Policy Committee (of Bank of England)
FTE	full-time equivalent
GDP	gross domestic product
GSBPM	generic statistical business process model
GSS	Government Statistical Service (UK)
GST	goods and services tax (Australia)
GVA	gross value-added (account, etc.)
HLG	high-level group
HT	Horvitz–Thompson
ICES	International Conference on Establishment Surveys
ICR	intelligent character recognition
IGEN	Interagency Group on Establishment Nonresponse
IMF	International Monetary Fund
I/O	input/output
ISIC	International Standard Industrial Classification (of all economic activities)
ISI	International Statistical Institute
ISO	International Organization for Standardization
IT	information technology
IVR	Interactive voice recognition (response)
KAU	kind of activity unit
MIBSR	multidimensional integral business survey response
MPC	Monetary Policy Committee (of Bank of England)
MSE	mean-squared error
NACE	Classification of Economic Activities in the European Community (actual title in French)
NASS	National Agricultural Statistics Service (US)
n.e.c.	not elsewhere classified
NSI	national statistical institute
OCR	optical character recognition
OECD	Organization for Economic Cooperation and Development
OIG	Office of Inspector General (US)
OMR	optical mark recognition
ONS	Office for National Statistics (UK)
OQRM	object-oriented quality risk management
PA	personal assistant
PAYE	pay as you earn (tax)
PDCA	plan–do–check–act (cycle)
PDI	proactive dependent interviewing
PEI	principal economic analysis (PEEI — principal European economic analysis)
PMI	purchasing managers index

PPI	producer price index
PPS	probability proportional to size
PRaM	postrandomization method
PRODCOM	Products of the European Community
PV	punched and verified
RAS	response analysis survey
RC	response coordinator
RDI	reactive dependent interviewing
RPI	retail price index
SBS	structural business statistics (or survey)
SCM	standard cost model
SDC	socially distributed cognition
SDMX	statistical data and metadata exchange
SIC	Standard Industrial Classification
SIRD	survey of industrial research and development
SITC	Standard international Trade Classification
SNA	system of national accounts
SOC	Standard Occupation Classification
SPPI	services producer price index
SRS	simple random sampling
STC	short-term statistics
TDE	telephone (touchtone) data entry
TDM	tailored design method
UNECE	United Nations Economic Commission for Europe
VAT	value-added tax
VDU	visual display unit
WCAG	Web content accessibility guidelines
WPI	wholesale price index

CHAPTER 1

Surveys and Business Surveys

Jacqui Jones, Ger Snijkers, and Gustav Haraldsen

1.1 THE OBJECTIVE OF THIS BOOK

The objective of this book is to provide a coherent overview of designing and conducting business surveys. Using the generic statistical business process model (GSBPM) as a high-level framework, the book brings together what we currently know about planning, designing, and conducting business surveys, right through to producing and disseminating statistics or other research results from the collected data. This knowledge is accumulated from various disciplines such as survey methodology, organizational sciences, sociology, and psychology. The result is a multidisciplinary process-quality approach. The contents of the book reflect the existing literature (books, journal articles, conference papers, etc.) and the experiences and observations of the authors. The book is intended to help anyone involved in designing and/or conducting business surveys, producing statistics or other research results from business surveys, or using statistics produced from business surveys; and open up new areas for future business survey research.

Businesses are central to a country's economy in terms of both economic growth and maintaining the nation's infrastructure. Because of their importance, data from businesses are needed for national, regional, local, and individual business monitoring, and policymaking and decisionmaking, as well as decisionmaking for individual members of society. The required data may relate, for example, to the economic performance of businesses, business perceptions of current and future performance, or working practices, policies, or conditions.

Business surveys are one method for collecting data from businesses, which are organizations involved in the production and/or trade of goods and/or services. In some respects the methods used to design, build, collect, and process a business survey are the same as those used in surveys of individuals and households. A sample frame is

Designing and Conducting Business Surveys, First Edition.
Ger Snijkers, Gustav Haraldsen, Jacqui Jones, and Diane K. Willimack.

required; a sample is designed and selected; data are collected; various communication methods are used to elicit contact and cooperation with respondents; the collected data need to be captured, coded, and cleaned; and the survey methods need to focus on minimizing errors, response burden, and building trust. Yet, underlying these high-level processes are unique features relevant only to business surveys.

Cox and Chinnappa (1995) note that there is no generally accepted definition of a business and they use an expanded approach including businesses (e.g., manufacturing, construction, retail, financial organizations), farms (e.g., crop, livestock, vineyards) and institutions (e.g., schools, hospitals, prisons) (ibid., p. 3). Although this book focuses on surveying businesses, the methodologies, procedures, and practices described can be applied to surveying farms and institutions.

A business can also be defined from a national accounts perspective, where the focus is on institutional units that must be capable of "owning assets, incurring liabilities and engaging with economic activities and in transactions with other entities" (United Nations 2008, p. 61). Businesses can be privately owned profit or not-for-profit organizations, state-owned not-for-profit, or a hybrid of private and state-owned.

This book is intended as a "how to" handbook covering the end-to-end business survey process set, in the context of the statistical production process. It is not, however, a cookbook type treatise; there is no standard recipe for designing and conducting a business survey. Every survey has its own special features that are relevant to the overall statistical production process. For every phase in the business survey process this book provides guidelines that facilitate educated tradeoff decisions in order to minimize survey errors, costs, and response burden.

This chapter is divided into three main parts: Sections 1.2 and 1.3, Sections 1.4–1.7, and Section 1.8. These are followed by an appendix outlining the history of UK official business statistics. In Section 1.2, we discuss the statistical production process; in Section 1.3, surveys in general and the survey process, with a brief historical overview of surveys in Section 1.3.3; in Section 1.4, types of business data outputs; in Section 1.5, how business data outputs are used; in Section 1.6, obtaining business data; and in Section 1.7, business surveys. In Section 1.7.2.1, we discuss business survey constraints and considerations, with resulting unique features of business surveys. These features are discussed further throughout the book. Section 1.8, outlines the organization of the book and summarizes the topics discussed in each chapter (see Section 1.8.3).

1.2 THE STATISTICAL PRODUCTION PROCESS

The generic statistical business process model (GSBPM) (Vale 2009) (Figure 1.1) provides a useful overview of the required phases and subprocesses for producing statistical outputs from any type of source; it can also be applied to the development and maintenance of statistical registers. Having a model helps define and describe statistical processes in a coherent way. It should also assist with standardizing terminology and comparing, benchmarking, and identifying synergies within and

Quality Management / Metadata Management

1 Specify Needs	2 Design	3 Build	4 Collect	5 Process	6 Analyze	7 Disseminate	8 Archive	9 Evaluate
1.1 Determine needs for information	2.1 Design output	3.1 Build data collection instrument	4.1 Select sample	5.1 Integrate data	6.1 Prepare draft output	7.1 Update output systems	8.1 Define archive rules	9.1 Gather evaluation inputs
1.2 Consult and confirm needs	2.2 Design variable descriptions	3.2 Build or enhance process components	4.2 Set up collection	5.2 Classify and code	6.2 Validate output	7.2 Produce dissemination products	8.2 Manage archive repository	9.2 Conduct evaluation
1.3 Establish output objectives	2.3 Design data collection methodology	3.3 Configure workflows	4.3 Run collection	5.3 Review, validate and edit	6.3 Scrutinize and explain	7.3 Manage release of dissemination products	8.3 Preserve data and associated metadata	9.3 Agree on action plan
1.4 Identify concepts	2.4 Design frame and sample methodology	3.4 Test production system	4.4 Finalize collection	5.4 Impute	6.4 Apply disclosure control	7.4 Promote dissemination products	8.4 Dispose of data and associated metadata	
1.5 Check data availability	2.5 Design statistical processing methodology	3.5 Test statistical business process		5.5 Derive new variables and statistical units	6.5 Finalize outputs	7.5 Manage user support		
1.6 Prepare business case	2.6 Design production systems and workflow	3.6 Finalize production system		5.6 Calculate weights				
				5.7 Calculate aggregates				
				5.8 Finalize data files				

Figure 1.1 The generic statistical business process model.

between organizations. The GSBPM was originally based on the business process model developed by Statistics New Zealand, and two additional phases (archive and evaluate) were added from reviewing models used in other national statistical institutes (NSIs). Originally it was seen as a vehicle for developing statistical metadata and processes (Vale 2010). The GSBPM identifies the phases in producing statistics independent of data sources (e.g., surveys, censuses, administrative sources, register-based statistics, mixed sources), and acknowledges that the order of phases will vary between the production of individual statistical outputs. There are four levels in the GSBPM (Vale 2009):

Level 0: the statistical business process
Level 1: the nine phases of the statistical business process
Level 2: the subprocesses within each phase
Level 3: a description of those subprocesses

In addition to the GSBPM phases and processes highlighted in Figure 1.1, there are 12 overarching themes (see Section 4.6), although only the first two themes are represented in Figure 1.1 (quality management and metadata management). These 12 themes need to be considered and strategies implemented to manage the numerous aspects of the statistical production process. What is evident here is the importance of planning and managing the whole statistical production process (see Chapters 4 and 10).

In this book the GSBPM is used as the high-level framework for the organization of the chapters, but with additional focus on the survey process phases required to collect and process business data.

1.3 SURVEYS

1.3.1 Definition of Surveys

Surveys typically collect data from a subset of the population of interest; if the population is appropriately sampled, the results are generalizable to the population. This is not as easy as it seems and is challenging from a number of perspectives, for example, selecting a sample representative of the population of interest, selecting appropriate data collection modes, designing questions and questionnaires that will collect the required data and that respondents can successfully complete, developing and implementing strategies for improving response, and quality-assuring and cleaning the collected data.

To meet known data needs, surveys can provide timely data using specific survey questions. Surveys can collect a variety of different types of data, for example, factual data (e.g., date that the business started), monitoring data (e.g., the amount of money invested in capital), attitudinal data (e.g., attitudes toward government's economic policy), and/or perception data (e.g., perception of the business' performance). However, the type of data collected can impact on key characteristics of the survey design. Such as design of the data collection instrument, the mode of data collection, and how the data are validated. Surveys can also provide better control of processing and coverage errors (Lavallée 2002).

A survey is a method of collecting quantitative data from a sample of the population of interest. In some cases qualitative data, from responses to open-ended questions, may be collected but will be coded during the processing phase to produce quantitative results. Groves (1989) acknowledges that the term *survey* is not well defined, as does De Leeuw et al. (2008), who found differing definitions, with some definitions defining surveys by the "major components of surveys and survey error." From a review of the literature, De Leeuw et al. (ibid, p.2) identified the common characteristics of existing survey definitions and description stating that "a survey can be seen as a research strategy in which quantitative information is systematically collected from a relatively large sample taken from a population." The survey outputs are *statistics*; "quantitative descriptors" (Groves et al. 2004, p. 2).

In Section 1.2 we described the GSBPM as providing an overview of the statistical production process regardless of source (e.g., survey or existing source); what is not evident are the key components of the surveys process. What distinguishes the production of statistical outputs from survey data, in comparison to, say, using administrative data, is essentially the fact that the survey components (sample frame; sample; questions and questionnaire; survey communication; and data capture, coding, and estimation) need to be designed, built, and tested, and then the data collected and processed, within the context of predefined statistical research purposes. This is in contrast to existing data sources where the data have already been requested, collected and processed, aimed at administrative use of the data, and 'organic' data that are generated automatically within processes (like traffic loop data, mobile phone and GPS data, scanner data on purchases (Groves, 2011)). Table 1.1 shows the distinguishing characteristics of the survey process set in the context of the GSBPM.

Once collected and processed the survey data are used to produce statistical outputs. The survey process and statistical production process therefore comprise several phases, which all need to be planned, designed, and tested to run a successful survey and produce fit for purpose outputs.

1.3.2 Survey Constraints, Considerations, and Tradeoffs

When designing and conducting a survey, you will be faced with constraints and consideration relating to: (1) the survey project, including financial resources, relevant expertise, time, and the production environment (see Chapter 4); and (2) the design and conduct of the survey, including survey errors (see Chapter 3), response burden (see Chapter 6), and trust. Undoubtedly these constraints and considerations will lead to tradeoff decisions (i.e., the tradeoff between, e.g., timeliness and accuracy).

1.3.2.1 Survey Errors and the Process Quality Approach

Groves (1989, p. 35) states that "survey estimates are subject to many sources of error" and categorizes these errors as nonobservation (coverage, sample, and

Table 1.1 Distinguishing Characteristics of the Survey Process Set in the Context of the GSBPM

Design, Build, and Test[a]	Collect: Running the Survey	Process: Processing the Survey Data
Whom you need to request data from (sample frame and sample).	Select sample.	
How you will request and collect the data (survey communication, questions and questionnaire, and mode).	Request and collect the data.	
For recurring surveys with panel or overlapping sample designs, how you will build and maintain respondent relations.	Implement questionnaires, and survey communication, taking recurring contacts into account.	
How you will process the data (data capture, coding, editing and imputation).	Initial data capture into the data collection instrument. Data transfer to the survey organization.	Data capture processing, coding and cleaning.
How you will use the survey data to provide estimates of the population of interest.		Estimate to the population.
How you will minimize, measure, and monitor sampling and nonsampling errors, costs, and actual and perceived response burden; trust in the survey organization and the produced outputs are also linked to this characteristic.	Implement quality and process management to minimize and measure sampling and nonsampling errors, costs, and actual and perceivd response burden.	Implement quality and process management.

[a]Please note that the survey processes listed are not linear.

nonresponse) and observations errors (construct and measurement). The ultimate concern is that potential errors be recognized and factored into the design and conduct of surveys, as they ultimately can affect the quality of the survey outputs. For example, if the sampling frame has missing units, then this will affect the quality of the statistics; if the questions are not comprehended as intended, different data will be collected. In 2004 (Groves et al. 2004, p. 48) related these errors to the survey process: "good survey estimates require simultaneous and coordinated attention to the different steps in the survey process." In this book we will take the same process quality perspective. Chapter 3 provides a detailed discussion of total survey error in the context of the business survey process, integrating quality perspectives related to the survey design, the survey organization, and users.

1.3.2.2 Response Burden

Response burden is multidimensional and includes actual and perceived response burden. Traditionally response burden has been regarded as the time spent responding to the survey (actual burden). But this economic cost is not the only consideration. Perceptions of burden are also significant, as these perceptions can impact the quality of collected data. For example, respondents who perceive the topic of the survey as uninteresting and/or irrelevant may be less likely to carefully consider the questions asked and their responses. Perceived response burden addresses this wider quality issue and includes respondents' perceptions of the usefulness of the statistics, interest in the survey topic(s), and knowledge of the survey organization (Dale et al. 2007). Chapter 6 discusses response burden in more detail.

1.3.2.3 Trust

In relation to any statistics, there are trust considerations from the user perspective. If the statistics are produced from survey data, trust will also be relevant from the respondent perspective. From the user perspective the key determinants of trust are: (1) perceived independence in the production of the statistics (e.g., produced without political interference), (2) perceived accuracy of the statistics, (3) personal knowledge of the trustworthiness of the statistics, and (4) personal experience of the trustworthiness of the statistics. Users will use these four determinants to derive an "overall level of trust in specific statistics" (Wilmot et al. 2005, p. 5). The qualitative study carried out by Wilmot et al. found that the more information available on the independence and accuracy of the statistics, the less weight was placed on personal knowledge and experience. From the opposite perspective, then, the less information available on the independence and accuracy of the statistics, the more weight placed on personal knowledge and experience (Wilmot et al. 2005, p. 25): Figure 1.2 provides an overview of the determinants of trust from the user perspective.

From the respondent perspective, the key determinants of trust are intrinsically linked to data confidentiality. Respondents need to believe with confidence that the data that they provide to the survey organization will not be: (1) lost (e,g., left on a train), (2) given to people outside the survey or statistical production area without their permission, (3) accessed by unauthorized people, and (4) identifiable in the statistics that are released (disclosure control). Chapters 9 and 12 provide further information on these issues.

$$\begin{array}{c}\text{Level}\\\text{of}\\\text{trust}\end{array} = \left(\begin{array}{c}\text{perceived}\\\text{independence}\end{array} + \begin{array}{c}\text{perceived}\\\text{accuracy}\end{array}\right) + \left(\begin{array}{c}\text{self-validation}\\\text{using knowledge}\end{array} + \begin{array}{c}\text{self-validation}\\\text{using experience}\end{array}\right)$$

Figure 1.2 User determinants of trust. [*Source:* Wilmot et al. (2005, p. 25). © Crown copyright 2005.]

1.3.3 A Brief Historical Look at Surveys

From the earliest days of collecting survey data, there existed nonsampling errors (e.g., coverage, measurement, processing) that impacted the quality of the collected data and ultimately the statistical outputs; measurement error was explicitly identified as an issue in the 19th century by Quetelet, the Dutch/Belgian mathematician-astronomer (Bethlehem 2009). Collections of data using samples were seen in historical population studies in the 1600s and 1700s [e.g., Webster, 1755 (Anderson 2011)], but these samples were not scientifically chosen, instead relying on convenience sampling. In the 1700s there was growing pressure to collect data using surveys, as surveys began to gain acceptance as vehicles for obtaining information to make decisions. For example, in 1767 Sir James Steuart, one of the earliest writers on modern economic science, "recommended local surveys as the only safe basis of political and financial regulations" (Sinclair 1837, p. 2). The development of modern sampling theory started around 1895, when the Norwegian Anders Kiaer (founder and first director of Statistics Norway) published his "representative method", which consisted in selecting and interviewing a large number of people. This method stressed the relevance of sample representativeness, the selected group should mirror the population (Bethlehem 2009).

It was not until the 1906 Bowley study that sample surveys became accepted as a scientific method for producing statistics, as Bowley made the first steps toward statistical inference. The first scientific survey was conducted by Bowley in 1912, surveying working class conditions in five British cities. Between the 1930s and 1950s sample surveys gained popularity and standardized instruments to measure attitudes and opinions were developed [see e.g., Bogardus (1925), Thurstone (1928), Likert (1932), and Guttman (1950)].

During the 1930s–1950s there was an increasing demand for survey research; World War II saw an increased focus on the need to measure the attitudes and opinions of society. By the 1960s low cost collection methods such as paper self-completion further increased the demand for survey outputs (Groves et al. 2004). The 1970s saw increased use of the telephone as a general data collection mode. Prior to then there was concern about the representativeness of telephone ownership, following the failure of the 1936 *Literary Digest* poll failure (in which telephone directories were employed to help create the sample frame) (Squire 1988). With increased telephone ownership and the need to reduce survey costs, the telephone became accepted as a general mode of data collection (Massey 1988). Moving onto the 1980s and 1990s the shift in emphasis was toward development of alternative modes of data collection and new technologies, such as computer assisted data collection and the web (Couper et al. 1998). More recently the focus has been on measurement and data quality (De Heer et al. 1999).

The first recorded successful mail survey (achieving 100% response) commenced in 1790 and took 7.5 years to complete (see Figure 1.3). The survey faced many of the challenges that we still face today in conducting surveys: balancing the need for data with the number of data requests that can reasonably be expected for people to provide in one survey, motivating respondents to respond through respondent communication, using different modes of data collection as the survey period progresses, and safeguarding the data collected (some of the survey data was lost in a fire). To achieve the high response rate, the surveyors used various incentives, eventually sending out data collectors to collect data from nonresponders. The output from this survey was the first Statistical Account of Scotland, which became regarded as the "model book of the nation" for every country in Europe and initiated similar activities in other countries.

John Sinclair, a former pupil of Adam Smith, and a lay member of the General Assembly of the Church of Scotland, undertook the task of collecting information on the state of every Scottish parish. The original intention of this initiative was to "close his History of the Revenue with a general view of the political circumstances of the country, but had been obliged to abandon the attempt from the scantiness of existing information" (Sinclair 1837, p. 8). So in May 1790, having abandoned his previous attempt, he attempted to collect this information himself and publish it in volumes. He developed a circular letter, which he sent to the minister of every parish in Scotland, containing 160 queries under the five headings of: geography, natural history, population, productions, and miscellaneous subjects. To quality assure the queries, Sir John sent them to Bishop Watson who responded that "your statistical queries are all good but they are too numerous to be answered with precision by a country clergyman" (Sinclair 1837, p. 9).

By 1791 Sir John had received some parish returns and published a specimen volume containing the accounts of four parishes. He sent the specimen volume to ministers as an incentive for them to respond, along with a second copy of the circular. Yet for many respondents there remained resistance to responding, even though Sir John was well known to them. Resistance originated from a variety of factors (Sinclair 1837, p. 11):

- "Boldness of an individual expecting that a whole nation would 'consider him a fit centre for general co-operation'"
- Concern that their responses would be criticized
- Inability to respond because of ill health
- For large parishes the requested information was too burdensome

By the middle of 1791 he had received enough information to compile and publish the first volume of the "Statistical Account of Scotland." Yet, nonresponse remained an issue, with many parishes suspecting that the requested information would be used to increase rents or introduce new government taxes. By mid-1792, 413 parishes had still not responded. So, as a further incentive to respond, Sir John arranged that the profits from the Statistical Account would be given to a newly established society for the "Benefit of the Sons of the Clergy." At the time this society could not distribute any of its funds until their capital had doubled. Sir John persuaded the societies' directors to apply for a royal grant—the grant application was successful and given "as a reward to the clergy for their statistical exertions" (Sinclair 1837, p. 16). Following this, Sir John sent out a third circular requesting the same information, followed shortly after by a fourth circular.

Sir John then sought and achieved endorsement from the General Assembly of the Church for his inquiry. This was then followed by dispatch of another circular "entreating compliance with the recommendation of the supreme ecclesiastical court." Despite all these efforts, nonresponse remained an issue. In total, Sir John sent out 23 circulars as reminders. The last circular was written in red ink.

Figure 1.3 The first recorded successful mail survey.

In a last attempt to receive outstanding responses, Sir John employed people, at his own expense, to go out as "statistical missionaries" and visit nonresponding parishes and collect the information themselves. Eventually responses were received for 100% of parishes.

Now, with all parish data collected, John Sinclair began the process of organizing, classifying, and editing the data. How to present the enormous amount of data was an issue, and Sir John sought advice from his friends, including the historian Dr. Adam Ferguson, who suggested the use of tables and a full index. Repeat requests for the information from some parishes had to be made as the 14th volume was destroyed in a fire at the printers and 12 parish returns were lost. Finally, however, the 21 volumes of the "Statistical Account of Scotland" were published in January 1798, "seven years, seven months and seven days" after starting. This represented "the contributions of about nine hundred individuals" (Sinclair 1837, p. 22). Following publication, it became regarded as the model book of the nation for every country in Europe and initiated similar activities in other countries, for example: The 1800 Census Act of Great Britain and the first census in 1801, Cesar Moreau's Statistical Works on France, and Dr. Seybert's enquiries as to the United States of America.

First use of the word *statistics* in the English language

Sir John Sinclair was also the first person to use the word *statistics* in the English language. In his 1791 volume, containing the accounts of four parishes, the terms *statistics* and *statistical*, which occurred continually in this volume, were such novelties in the British *nomenclature* of economic science, that Sir John thought it necessary to apologize for their introduction. He explained that he had derived the term from the German, although he employed it in a sense somewhat different from its foreign acceptance. In Germany, a statistical enquiry related to the *political strength* of the country, or to questions of state policy, whereas he employed the word to express an enquiry into the state of a country, for the purpose of ascertaining the account of *happiness* enjoyed by its inhabitants, and the means of its future improvement (Sinclair 1837, pp. 9–10).

Figure 1.3 *(Continued)*

1.4 TYPES OF BUSINESS DATA OUTPUTS

There are various types of business data outputs, produced by a variety of organizations (e.g., NSIs, central banks, universities, research organizations). This section provides an overview of some of them.

1.4.1 Official Statistics

Business data collected by NSIs are used to produce official business statistics. They provide a statistical picture of the economy of a country. Official business statistics can be broadly categorized into national accounts, structural business statistics and short term statistics.

1.4.1.1 National Accounts

The *system of national accounts* (SNA) is an international framework used to measure the economic activity of a country; the most recent manual is *SNA 2008*. There is also a European version of SNA called the European *system of national accounts* (ESA); the most recent manual is *ESA 2010*, which is approved under regulation by the European Council. Both SNA and ESA focus on economic activities and market transactions, and are periodically reviewed to ensure that they align with economic changes (Giovannini 2008).

National accounts employ an accounting technique to measure the economic activity of a country. Central to the SNA framework are accounts that are comprehensive, consistent, and integrated, showing the economic flows and stocks in the production of goods and services in a particular country. SNA focuses on institutional units and their activities related to production, consumption, and assets. These are recorded as transactions between institutional units. SNA also distinguishes between two kinds of institutional units: households and legal entities (profit and not-for-profit businesses, government units). These institutional units are then grouped together into the following sectors, which make up the whole economy (United Nations 2008, p. 2):

- Nonfinancial corporations
- Financial corporations
- Government units, including social security funds
- Nonprofit institutions serving households
- Households

Three different accounts are included, each representing a different economic function: the production account, consumption account, and the adding-to-wealth account. The sequence of accounts begins with the production accounts and moves to the primary distribution of income accounts, the secondary distribution of income accounts, the use of income accounts, the capital account, the financial account, and finally the balance sheet (United Nations 2003, p. 7); there is also a rest-of-the-world account.

The focus of national accounts is on producing a balanced and comprehensive picture of the flow of products in the economy and the relationships between producers and consumers of goods and services; this is achieved by construction of supply and use tables [Mahajan (2007) provides a useful overview of the development, compilation, and methodology for UK supply and use tables]. These macroeconomic accounts are then used for a variety of analytical and policy purposes such as identifying potential gaps in supply and demand.

National accounts provide an overall picture of a country's economy by presenting the expenditure, income, and production activities of corporations, government, and households in the country's economy, and economic relations with the rest of the world. Data used to compile national accounts come primarily from business surveys and administrative data. National accounts are published both quarterly and annually. The UK annual national accounts include data from annual business surveys, generally available 15 months after the relevant year, and quarterly data, which are used to estimate the latest period (Mahajan 2007). Lequiller and Blades (2006) provide a useful overview of national accounts.

The aggregates derived from national accounts are also used as key economic indicators in their own right. For example, quarterly and annual gross domestic product (GDP) provide an indication of whether the economy is growing or contracting, and gross value-added (GVA) is used to analyze the importance of different industries. The GVA metric is the difference between output and intermediate consumption (goods and services consumed or used as inputs in production).

1.4.1.2 Structural Business Statistics

Structural business statistics (SBS) by definition collect information on the structure and development of businesses in industry, construction, and distribution trade and services. The data collected are more detailed than national accounts data and can be broken down to a very detailed sectoral level. The information collected includes monetary values (current prices) or counts (e.g., number of people employed) and can be used for point-in-time and change-over-time comparisons. In contrast, national accounts are regarded as providing more comprehensive coverage of economic activities (OECD 2006). It is challenging to collect data that reflects up-to-date changes in business demography, and SBS data are generally compiled by NSIs from a combination of annual business survey data and their business register, which results in a time lapse between data collection and statistical production.

- In the United Kingdom information is collected via the annual business survey, with updates from the statistical business register (number of enterprises).
- In the United States information is collected via the economic census, the annual survey of manufacturers and services, with updates from the statistical business register (number of enterprises and establishments).
- In the Netherlands information is collected via the annual structural business survey and administrative data, with updates from the statistical business register (number of enterprises).
- In Norway information is collected via numerous sources, including annual surveys, annual company accounts, the value-added tax (VAT) register, and the statistical business register (OECD 2006).

In Europe SBS are produced under European regulation. SBS provide information on wealth creation, investment, and labor input of different economic activities. The statistics are disseminated by enterprise size, which allow for comparison of structures, activities, competitiveness, and performance of businesses.

Table 1.2 European STS Indicators and Variables

Production	Construction	Retail Trade and Repair	Other Services
Turnover	Production	Turnover	Turnover
Number of people employed	Number of people employed	Number of people employed	Number of people employed
Hours worked	Hours worked		
Gross wages and salaries	Gross wages and salaries		
Output prices	Construction costs	Deflator of sales	Services producer price index
Import prices	Construction permits		

1.4.1.3 Short-Term Statistics

Short-term statistics (STS) cover the domains of industry, construction, retail trade, and other services (e.g., transport, information and communication, business services but not financial services). The data are generally published monthly and presented as seasonally and workday-adjusted indices. Indicators classified as STS include production, turnover, prices, number of people employed, and gross wages. STS generally include monetary values (current prices) or counts (e.g., number of people employed). The statistics are used to identify changes between industries in terms of how productive they are, and can be used for point-in-time comparisons and changes over time (OECD 2006).

In Europe comparability of short-term statistics across member states is made possible by harmonized: definitions, level of detail, and reference periods laid down in the Short Term Statistics Regulation 1165/98. Table 1.2 summarizes the indicators and the variables compiled for STS.

1.4.1.4 International and European Comparisons

The Organization for Economic Cooperation and Development (OECD) collates and publishes on a monthly basis a wide range of short-term indicators from 34 OECD countries and a number of nonmember countries. Types of indicators included are gross domestic product, private and government consumption, gross fixed-capital formation, imports and exports, business confidence, consumer confidence, and inflation rate (OECD 2009). The collation of indicators allows for country comparisons on the economic status of countries.

In Europe, principal European economic indicators (PEEIs) are produced to ensure a reliable and regular supply of statistics. They consist of statistical indicators from national accounts, external trade, balance of payments, prices, business statistics, and monetary and financial statistics, as well as the labor market (Eurostat 2009c, pp. 9–10). Their dissemination provides information on the economic status of individual European member states and can be used to produce aggregated macroeconomic models.

1.4.2 Other Types of Business Data Outputs

Economic business data are not the only data collected from businesses. From a policy perspective, business surveys are carried out to review, for example, the status of company policies. For example, the European Foundation for the Improvement of Living and Working Conditions (Eurofound) periodically carries out a European company survey that focuses on company policies in relation to flexibility practices and employee participation in the workplace (Eurofound 2010). The survey is carried out across European member states and candidate countries, and in 2009 included more than 27,000 establishments in the private and public sectors that had 10 or more employees. Via computer-assisted telephone interviews (CATI), survey data are collected from both management and employee representatives in the businesses. The findings from the survey assist in monitoring the European objectives stated in the Lisbon strategy: making the European economy competitive, dynamic, and knowledge based, with more and better jobs (Eurofound 2010).

At a European Union (EU) level, the European Commission funds the joint harmonized EU program of business and consumer surveys (European Commission 2007) to complement official statistics produced by EU NSIs. The program began in 1961, and its benefits are seen to be the production of timely and harmonized indicators that can identify economic turning points. Originally the program focused on the manufacturing industry and then expanded to include the construction sector (1966), consumers (1972), retail trade (1984), the service sector (1996), and more recently the financial services sector. Every 3–4 years the contracts for carrying out the surveys in each member state go out to tender, which result, across member states, in a variety of different organizations running the surveys, such as, NSIs, central banks, research organizations, business associations, or private companies (European Commission 2007). The surveys all include harmonized questions and a harmonized timetable; the size of the samples is dependent on the heterogeneity of the member states economy and population size. Nearly all the questions ask about perceptions and the majority of response categories are based on a three-option ordinal scale, such as increase, remain unchanged, or decrease; more than sufficient, sufficient, or not sufficient; or too large, adequate, or too small (European Commission 2007). Table 1.3 provides some examples of the different indicators, with examples of harmonized questions and response categories. The indicators are used as an index to allow comparisons over time. For example, the economic sentiment indicator is referenced from 1990 to 2010 as the long-term average, therefore equaling 100. Thus, if the indicator is above 100, the economic sentiment is improving; if below, it is decreasing.

Similar economic indicators are produced around the world. For example, in the US the Institute for Supply Management produces the *purchasing managers index* (PMI), which measures the economic health of the US manufacturing sector on the basis of five indicators: new orders, inventory levels, production, supplier deliveries, and the employment environment. An example of the PMI output is presented in Table 1.4. To compile the index, a monthly survey is dispatched to nationwide purchasing and supply respondents, who are asked about changes, in relation to a

Table 1.3 Examples of EU Business Indicators with Example Questions and Response Categories

Indicator	Examples of Questions Asked	Response Categories
Industrial confidence indicators	Do you consider your current overall order books to be. . . ?	More than sufficient (above normal) Sufficient (normal for the season) Not sufficient (below normal)
	Do you consider your current stock of finished products to be. . . ?	Too large (above normal) Adequate (normal for the season) Too small (below normal)
	How do you expect your production to develop over the next 3 months? It will . . .	Increase Remain unchanged Decrease
Services confidence indicator	How has your business situation developed over the past 3 months? It has . . .	Improved Remained unchanged Deteriorated
	How has demand (turnover) for your company's services changed over the past 3 months? It has . . .	Increased Remained unchanged Decreased
	How do you expect the demand (turnover for your company's services to change over the next 3 months? It will . . .	Increase Remain unchanged Decrease
Economic sentiment indicator	Consists of 15 individual components of the confidence indicators computed for business sectors and consumers, standardized and weighted by:	Industry: 40% Services: 30% Consumers: 20% Construction: 5% Retail trade: 5%
Business climate indicator	Consists of five balances of opinion from the industry survey: production trends in recent months, order books, export order books, stocks, and production expectations	

Source: European Commission, 2007, pp. 15–22.

number of business conditions, including the five indicators that make up the PMI. Like the EU business indicators, the response categories offered are based on an ordinal scale. The responses reflect changes between the current and previous months.

From a sociological perspective, data from businesses are important to gain understanding of the social aspects such as internal business structures, how

Table 1.4 Institute for Supply Management, Purchasing Managers Index (March 2012)

Manufacturing at a Glance, March 2012

Index	Series Index Mar	Series Index Feb	Percentage Point Change	Direction	Rate of Change	Trend[a] (months)
PMI	53.4	52.4	+1.0	Growing	Faster	32
New orders	54.5	54.9	−0.4	Growing	Slower	35
Production	58.3	55.3	+3.0	Growing	Faster	34
Employment	56.1	53.2	+2.9	Growing	Faster	30
Supplier deliveries	48.0	49.0	−1.0	Faster	Faster	2
Inventories	50.0	49.5	+0.5	Unchanged from contracting		1
Customers' inventories	44.5	46.0	−1.5	Too low	Faster	4
Prices	61.0	61.5	−0.5	Increasing	Slower	3
Backlog of orders	52.5	52.0	+0.5	Growing	Faster	3
Exports	54.0	59.5	−5.5	Growing	Slower	5
Imports	53.5	54.0	−0.5	Growing	Slower	4
Overall economy				Growing	Faster	34
Manufacturing sector				Growing	Faster	32

Source: Manufacturing Institute for Supply Management, *Report on Business*, March 2012.
[a]Number of months moving in current direction.

decisions are made in the business, and employee satisfaction. The data collected from businesses are often used, for example, to develop organizational theories and/or understand employee satisfaction. Such studies generally collect data using qualitative research methods such as observation, indepth interviews, or focus groups, although surveys can be used. One of the most famous studies of employee satisfaction was carried out by F. J. Roethlisberger and W. J. Dickson. In the late 1920s and early 1930s they studied employee working conditions in the telephone "banking wiring room" of the Western Electric Works in the Chicago suburb of Hawthorne. They attempted to study how changes in working conditions would improve employee satisfaction and productivity and discovered a linear association between working conditions and productivity. As conditions improved further, so did productivity. The sting in the tail was at the end of the study when working conditions were reduced and productivity continued to increase. This was attributed to the fact that the employees were receiving attention, which contributed to the increase in productivity (Babbie 1995).

1.5 USE OF BUSINESS DATA OUTPUTS

Business data can be used for a wide variety of purposes. A major objective is the production of official business statistics. Other objectives include policy-oriented research (e.g., into working conditions), or theory-testing academic research (e.g., to study growth conditions for small and medium-sized businesses). In this section we discuss the different uses of business data outputs.

As a country's economic structure changes, so does the importance of certain industries to the country's economic growth; this shapes the need for business statistics. In writing this chapter, one of the questions we focused on was how did we get to where we are in relation to the production of business statistics? To assist us in addressing this question, the appendix at the end of this chapter provides a timeline of the development of official business statistics in the United Kingdom. What is evident from this timeline is the increasing demand for business statistics, to enable monitoring, decisionmaking, and policymaking as the economy changed, and consequently an increased demand for business data.

More recently there has been a progressive change in the industrial makeup of countries, with many countries seeing an increase in the service industry. In 1970, at 1995 constant prices, the service industry contributed 53% to UK GDP; by 1995 this had increased to 67%. In terms of employee jobs, in 1970, the service industry accounted for around 54% of total UK employee jobs, compared with 72% in 1992 (Julius and Butler 1998). Many have referred to this structural change as "deindustrialization" (Besley 2007). With the increasing contribution of the service industry, this led to the need for developing and implementing a UK *index of services*, to measure the economic contribution from the service industry.

Business statistics are used for a number of reasons: to monitor national, regional, and local economic performance; and to monitor individual business and policy-making. Business statistics are also used in our daily lives.

1.5.1 National, Regional, and Local Economic Performance

Business statistics are required to monitor the economic performance of a nation, region, or local area (the latter are more difficult unless administrative data are available and compiled statistics are not disclosive). Compilation of these statistics requires information on how businesses are performing, including their levels of sales, levels of stocks, the amount of money they are investing in capital, and the number of people they employ. The produced statistics provide essential information on levels and change in economic performance that are used to both monitor and inform economic decisions and policymaking. For example, in the United Kingdom (UK) the Monetary Policy Committee (MPC) of the Bank of England meets every month to review the economic statistics produced by National Statistical Institutes, the Bank of England, Her Majesty's Treasury, other government departments, other survey organizations, and international organizations such as the International Monetary Fund (IMF) and the World Bank. Based on these statistics, the MPC make their decision on the national interest rate for the forthcoming month, and in times of recession or economic downturn, decisions on whether to apply quantitative easing. In 2011, following the financial crisis, a similar Financial Policy Committee (FPC) was also setup by the Bank of England to monitor the financial sector.

Business statistics need to keep pace with economic changes. The response to the 2008–2009 global financial crisis is a good illustration of this. Prior to the crisis the financial sector was monitored using traditional asset and liability approaches shown in the balance sheets of individual organizations and each country's national accounts. However, the asset and liability statistics did not always fully include the financial sector's increased use of new financial instruments, for example, derivatives and subprime mortgages. For the majority of countries there was also little information on flows of funds between sectors and the effects of increased globalization. So from the perspective of their balance sheets, the financial sector appeared healthy, but this was not the full picture. When information on new financial instruments etc. were included in the balance sheets, many organizations were actually in dire financial straits and unable to meet their financial obligations. In the UK, following the global financial crisis there has been a statistical drive (macroprudentials) to ensure that financial statistics keep pace with financial innovation in the collection and presentation of statistics (Walker 2011); this is also mirrored at the international level by the IMF's G20 data gaps initiative (Heath 2011). Improved financial statistics should in turn, improve monitoring with the aim of providing early indication of any issues in the financial stability of the financial sector, in an attempt to prevent another financial crisis.

1.5.2 Individual Business Monitoring and Policymaking

For individual businesses there is a need to complement their internal performance indicators with statistics that provide insight into how their business is doing relative to competitors. Business managers may use statistics to assess the viability, stability, and profitability of their business in comparison to others. Feedback from surveys may alter production or service goals, and statistics that forecast future trends may be used to identify new business opportunities. Some types of businesses are more reliant on business statistics than others; for example, investment companies, such as securities dealers and hedge funds, use business statistics as part of their assessment of where to make investments and where to withdraw from investing.

1.5.3 Everyday Decisions

Although we might not realize it, data from businesses are essential to our everyday decisions. Virtually every day each of us interacts with businesses such as when buying groceries, withdrawing money from a bank account, or filling a car with gas or petrol. Every year we interact with a number of different types of businesses depending on our work and home circumstances, for example, when buying or selling a house, applying for a mortgage or loan, or going for job interviews. What we might not realize is the extent to which our everyday decisions are based directly or indirectly on data available to us from businesses. For example, macrolevel economic indicators such as GDP and CPI may influence, our decisions on how much to spend and save, whether to purchase a new car, and/or whether to change jobs especially if the decision involves moving to a job in another industrial sector. On a day-to-day basis, we might not even realize that these economic indicators are the sources of TV, radio, newspaper, and news website headlines and stories (see Figure 1.4).

Savings Go to Pay Bills – Families Squeezed by Rising Prices
(*Daily Mirror*, March 21, 2012).

When the Going Gets Tough, Britain Goes Shopping: Retail Sales Surprise
(*The Guardian*, February 18, 2012).

Investment Collapse Hits UK Recovery
(*The Daily Telegraph*, February 25, 2012).

Figure 1.4 Economic indicator newspaper headlines.

1.6 OBTAINING BUSINESS DATA

So far, we have introduced the statistical production process (using the GSBPM), surveys, and the survey process, and we have discussed the types and uses of business data outputs. Now, we return to the production process, and look at different methods for obtaining business data. Using the GSBPM, it is evident from the "specify needs" subprocesses that once data needs are confirmed, one must decide how the data will be obtained to meet the need for information. The following section provides an overview of how data can be collected from businesses and questions that need to be considered to determine whether a survey is needed.

1.6.1 Business Data Collection Methods

There are five possible methods for obtaining business data: (1) surveys, (2) administrative data, (3) electronic data interchange (although this can be used as a collection method within a survey), (4) published business information (e.g., published company accounts), and (5) qualitative methods (e.g., indepth interviews, focus groups, observation). When considering what method to use to obtain the required business data, it is imperative to determine why the data are required, that is, which issues the data will address and whether the data will be generalizable to the total population, as this will help to determine the most suitable method of data collection (see Table 1.5).

Administrative data are data that have been collected for another purpose. For example, each country has a taxation system that collects tax data from individuals and businesses; this data can be, and often is used, instead of collecting information on people's income or profits from businesses. Electronic data interchange is used to electronically retrieve data directly from a business's record system and transfer these to the survey organization. Business records may differ in terms of the concepts that they use and the time periods they relate to, compared to the concepts and time periods required by the survey organization. There are two possible approaches for dealing with this issue; either the original data are transferred to the survey

Table 1.5 Examples of Questions and Associated Possible Data Collection Methods

Question	Possible Associated Data Collection Methods:
1. What is the economic performance of individual businesses, businesses in a specific industry, and/or all businesses?	Quantitative data from surveys, electronic data interchange, administrative data, and/or published business data
2. What are the perceptions of businesses in relation to the future economic prospects?	Quantitative data from surveys; or qualitative data from (e.g.,) indepth interviews or focus groups
3. What are the business policies and practices in relation to flexible working hours?	Quantitative data from surveys; or qualitative data from (e.g.,) indepth interviews, focus groups, or documentary analysis
4. How satisfied are employees with their working conditions?	Quantitative data from surveys; or qualitative data from (e.g.,) indepth interviews, focus groups, or observation

organization who then matches the data onto their conceptual definitions, or the data are matched and checked to the required definitions of variables by the business. Either approach requires careful preparation and can be time-consuming (Keller 1996; Roos 2010). Data provided from the businesses themselves are disseminated, for example, via annual reports, stock reports, and the release of profit figures. Qualitative data are obtained using methods such as indepth interviews, focus groups, observation, or documentary analysis.

As the focus of this book is on business surveys, we will not

- Deal specifically with the use of administrative and register data; for those interested in these methods, Wallgren and Wallgren (2007) and a special issue of *Statistica Neerlandica* (Bakker and Van Rooijen 2012) are useful texts. However, although this book focuses on business surveys, it must be noted that administrative data can play a role at different stages of the survey process, such as by, creating and maintaining the statistical business register, quality assessment, and estimation. Nowadays, many statistical offices make use of these data next to survey data, to produce statistics (Lane 2010).
- Discuss electronic data interchange (EDI) [Swatman and Swatman (1991) provide a useful overview of EDI; Clayton et al. (2000) provide examples of the use of EDI]. Electronic data interchange can, however, be incorporated in surveys as a data collection technique using Internet technologies [see, e.g., Roos (2010) and Haraldsen et al. (2011)].
- Cover qualitative research [see, e.g., Boeije (2010)], except for the use of qualitative methods as part of the development of business survey designs. Qualitative methods are essential in the development of business survey data collection instruments and survey communication and play an important part in Chapter 7, which discusses development and testing.

Table 1.6 Questions to Ask When Considering the Use of Existing Data

Does there appear to be a possible existing source for the data you require?
If yes:
- Does the timing of its availability match with your timing requirements?
- Does the existing source(s) use the same definitions and reporting periods?
- What was the objective of collecting the existing data source—links to possible errors (e.g., possible underreporting if from tax data or overreporting if from profit data)?
- How comprehensive is the existing source(s) in terms of businesses that it covers?
- How "clean" are the data? How were they processed? Were consistency checks carried out?
- Is it likely that the existing source(s) will continue?
- Are you able/allowed to access the existing source(s)?
- Are there security issues for transferring the data? If so, can these be overcome?

When business data are needed, a general guideline is to explore the availability of existing data in either registers or business publications before considering whether to collect "new" data (i.e., from a survey or using qualitative methods). Questions to ask, when considering the feasibility of using existing data, are listed in Table 1.6. It is evident that the mere availability of existing data does not necessarily mean that it will meet your data needs. For example, the definitions, reporting period, and/or when the data are available, may not meet your statistical needs. Data checking, cleaning, and sometimes coding may also be needed before the data can be used. An example of this is Statistics Canada's use of tax data, which they receive from the Canada Revenue Agency. The records in the files do not contain industrial classification codes, so on receipt of the files, Statistics Canada codes the data. When the coding is complete, a separate file providing the industrial codes is sent back to the Canada Revenue Agency with a stipulation that the file is to be used for "statistical purposes only". In conjunction with this, Statistics Canada have developed an algorithm to determine when a business should be removed from the business register as the tax department leaves businesses on their files longer than required for business register purposes, that is, no activity for a set period of time (Sear 2011).

The questions in Table 1.6 are not the only questions that you need to ask yourself. There are also benefits and losses from using existing data sources. The benefits include reduction of response burden, reduced data collections costs, and generally electronic availability of data. In contrast, the losses include limited control of the data, processing errors, and coverage error (Lavallée 2002). So, when considering using existing data, the questions posed in Table 1.6 should also be considered in relation to the benefits and losses associated with using the data. We refer to Bakker and Van Rooijen (2012) who discuss the methodological challenges of register data statistics.

In many countries such as the United States and United Kingdom there are also legal barriers to accessing and using administrative data. In these countries there is often greater reliance on data collection using surveys. The UK 2007 Statistics and

Registration Service Act has provisions for secondary legislation that allows for the sharing of data between government departments. However, using these provisions can be laborious and time-consuming as they require the agreement of all parties and the passage of legislation in either parliament or one of the devolved administrations. Hence there is greater reliance on acquiring data using surveys, and countries often have legislation that allows them to collect business data using surveys for the production of official statistics. For example, in the UK business data are collected under the 1947 Statistics of Trade Act.

On the other hand, in many countries, such as Finland, Sweden, Norway, and Denmark (United Nations Economic Commission for Europe, 2007) and the Netherlands (Snijkers et al. 2011a), NSIs have a strict data collection policy. These policies state that existing data should be considered first both for business and social statistics. If existing data cannot be used, for any of the reasons mentioned above, a survey can be considered.

1.7 BUSINESS SURVEYS

Following the decision to use a business survey to obtain the required data, the survey needs to be planned, designed, and conducted. This is more easily said than done. Although business surveys follow the same high-level survey process as in any other type of survey (see Table 1.1), they have some additional constraints and considerations (e.g., heterogeneity of businesses, the business context, the labor-intensive response process), which in combination with general survey constraints and considerations (e.g., financial resources, time, survey errors, response burden) result in some unique survey design and conduct features. This section provides an overview of the types of business surveys and some unique features that will be discussed later in the book.

1.7.1 Types of Business Surveys

Survey organizations conduct different types of business surveys to collect different types of data from either the business unit (e.g., production, financial, business perceptions, or policies) or employees (e.g., earnings, pension contributions, working conditions). The latter is basically a social survey within a business survey. The data collected in business surveys can include: (1) economic data (e.g., flows, balance sheets, products, or employment), (2) business characteristics (e.g., contact information, organizational structure, ownership), (3) business perceptions, (4) data on business policies, or (5) data on business practices (see Chapter 8 for further information on questions commonly found in business surveys). The type of business survey (i.e, what it collects, from whom—i.e, type and/or size of business, and using what methods, e.g., self-administered or interviewer-administered) will ultimately be determined by the data requirements, how the survey data will be used, and survey constraints and considerations, such as costs and business context.

Business surveys can be categorized according to a number of characteristics. The most obvious are periodicity, industry, business size, and mode. Generally the characteristic of the survey will be determined by

1. The information needs (e.g., one-off information requirement or continuous requirement), one specific industry or a number of industries, and a specific size of business or all sizes of businesses
2. The available budget and time

Furthermore, not just one design is used [the one-size-fits-all approach (Snijkers and Luppes 2000)]. Instead, business surveys are often tailored to the business context (Dillman et al., 2009). There may be different designs (or questionnaires) for different industries and/or sizes of business, reflecting the "heterogeneity" of the business world (Rivière 2002).

1.7.2 The Business Survey–Output Production Process

Figure 1.5 provides an overview of the business survey and output production process. At the production level, the production of statistics from business surveys follows the phases in the GSBPM (left side of Figure 1.5), and the same survey process as any other type of survey. Moving from left to right in the figure, what distinguishes the business survey from other types of surveys are some of the constraints and considerations (especially in relation to the characteristics of the business world, sample frame, sample, mode, questions and questionnaires, respondent communication, and running the collection), and obviously the unique features (note that some of the unique features are more commonly used, rather than unique features of business surveys).

1.7.2.1 Business Survey Constraints, Considerations, and Tradeoffs

Designing and conducting a business survey entails the same constraints, considerations, and tradeoff decisions as in any other type of survey, for example, financial resources, relevant expertise, time, the production environment, survey errors, response burden and trust (see Section 1.3.2). In addition to these, the production of statistics from business surveys includes specific constraints and considerations such as the heterogeneity of businesses, the business context, and the labor-intensive response process (see Figure 1.5: constraints and considerations).

Heterogeneity of Businesses The heterogeneity of businesses is characterized by the type of activities they undertake and their size (defined either by turnover or number of employees) (Rivière 2002). In addition, the business world is characterized by enterprises [a business under autonomous and single-person control, usually producing a single set of accounts], local units [a single site (geographic location) where a business operates], and enterprise groups [a group of enterprises under common ownership]. These constraints and considerations need to be considered in relation to aspects such as the register, sample frame, sample design, and survey communication. Chapters 3 and 5 provide further information on the heterogeneity of businesses; Chapter 11, on business classifications.

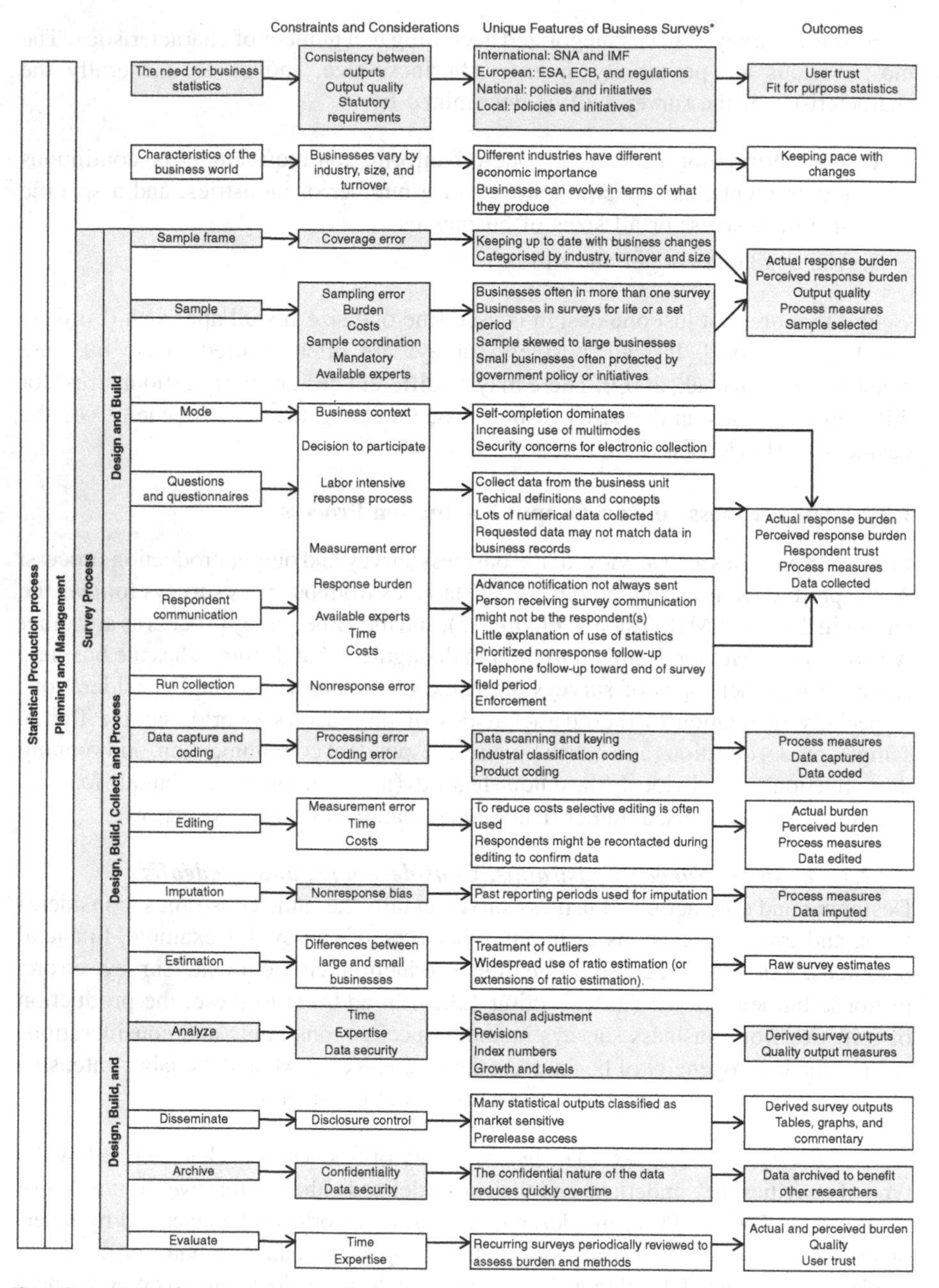

Figure 1.5 An overview of the business survey and output production process.

The Business Context–Response Process Collecting data from businesses requires the establishment of a data collection process within the business surveyed: the response process. Two stages in this process can be distinguished: the decision to participate and the performance of the response tasks. To collect good survey data with minimum production and compliance costs (i.e., response burden), it is vital that the survey design be tailored to this process. For example, if businesses use definitions of variables other than those used in the questionnaires, this will affect the quality of the data, and will also increase response burden in mapping the definitions; if the questionnaire is sent at a time when the data are not yet available, this may result in additional nonresponse follow-ups, and increased survey costs and response burden. Chapter 2 discusses the business context in detail from a multidisciplinary perspective.

1.7.2.2 Unique Features of Business Surveys

Figure 1.5 includes some of the unique features of business surveys that are derived primarily from the need to deal with general survey and business survey constraints and considerations. The outcomes of each survey and production process are then highlighted in the right column of the figure.

From the discussion thus far and further detailed discussion in this book, the distinguishing features of business surveys appear to be as follows [see, e.g., Snijkers and Bavdaz (2011)]:

- Registers and sample frames have to be kept up-to-date as the business world can change rapidly (e.g., business births, deaths, mergers, takeovers, and business size; see Chapters 3 and 5).
- Samples are stratified according to business size, as large businesses contribute more to economic output (see Chapter 5).
- Recurring surveys with overlapping sample designs are common (see Chapter 5), resulting in recurring contacts (see Chapter 9).
- Surveys are often mandatory (e.g., NSI business surveys), which can make response quality a more prominent quality issue than response rates (see Chapter 3) and lead to enforcement for noncompliance (see Chapter 9).
- Businesses can be classified to more than one industry depending on whether a sales or value-added approach is undertaken (Chapter 5, Box 5.1); and how the coding is carried out (e.g., respondents, automatic coding, or expert coders; see Chapter 11). The quality implications of this aspect are discussed in Chapter 3.
- Response burden is a political and methodological concern (see Chapter 6).
- Business surveys collect data about business units and not about the respondent. The important consequence of this is that while surveys of individuals focus very much on the psychological information processing processes, business surveys also need to factor in social processes, as well as how information is recorded, stored, and retrieved in a business (see Chapter 2).
- Survey communication may not be seen by the respondent(s), as it is often sent to a gatekeeper in the business (see Chapters 2 and 9).

- Self-administrative designs (e.g., paper, web, telephone data entry) are more common than surveys conducted using interviewers (see Chapter 8).
- Data often have to be retrieved from business records or work colleagues (see Chapter 2).
- There may be more than one respondent to a survey, and one respondent may report for more than one survey (see Chapters 2 and 8).
- In NSI business surveys, most questions concern production activities and financial results (see Chapter 8).
- Responses to questions often require calculations (see Chapter 8).
- To reduce survey organization costs, selective editing approaches are often used, so data errors are prioritized according to their impact on the final output (see Chapter 11).
- During the data editing phase respondents are often recontacted to check the submitted data (see Chapter 11).
- For recurring surveys, data for past reporting periods for either individual businesses or similar businesses can be used in imputation (see Chapter 11, historical imputation).

In relation to the statistical production process, key points to remember are that in NSIs, the need for business statistics is generally dictated by the requirements of national accounts, structural business statistics, and short-term statistics, which, in turn, are determined largely by keeping pace with economic changes to assist decisionmaking, monitoring, and policymaking needs. For NSIs a substantial amount of business statistics will be determined by regulations, including European regulations such as the European system of accounts (ESA), the structural business statistics (SBS) regulation, and the short-term statistics (STS) regulation.

1.7.3 Perspectives of the Business Survey Process

Having identified the processes involved in carrying out a business survey, constraints and considerations, and the unique features, we will now discuss the business survey process from the perspective of the survey organization, respondents, and users.

1.7.3.1 Survey Organization Perspective

From the survey organizations perspective the business survey process, just like any other type of survey, needs to be cost-efficient, minimize response burden, be optimized to produce statistical outputs that meet users' needs, and maintain or build trust in the statistics produced. Developing and implementing an appropriate survey design is therefore challenging. Here, *design* is defined as "the way in which something has been planned and made, including what it looks like and how well it works" (Oakland 2004, p. 205). Before implementing the survey, it is important that the survey organization design, build, and test the survey process. This strategy should include

1. Agreeing and documenting data requirements with users
2. Considering the costs and benefits of different survey designs (e.g., sample size, data collection modes, numbers of questions and reminders, method of processing data)
3. Agreeing with users on the most cost-efficient design that will produce fit-for-purpose statistical outputs; acknowledging the need for balance between costs and quality
4. Involving the relevant people (e.g., system builders, survey methodologists, potential data suppliers) in developing the components of the survey process
5. Testing the different components of the survey process (e.g., sample frame for coverage, sample design for minimizing sampling error, questions that can be understood and responded to, data capture and coding to minimize error, editing to identify errors but minimize costs)
6. Documenting the survey methodology and processes
7. Maintaining user engagement throughout the development–testing process

All these points should always be considered from both user and respondent perspectives and undertaken within the frameworks of quality and project management; see Chapters 3 and 4 for more detail.

The ultimate goal of the survey organization is to compile survey outputs that meet user needs. However, tradeoff decisions often have to be made by survey organizations when developing and running surveys. For example, from 2003 to 2005 ONS redeveloped the *new earning* survey (now known as the *annual survey of hours and earnings*). Drivers for the redevelopment included new data requirements and known data problems. The redevelopment and testing resulted in increasing the proposed length of the questionnaire from two to six sides of A4 paper. Following a field test ONS had a tradeoff decision to make between increased costs of the questionnaire (paper, printing, postage, and scanning) versus reduced actual and perceived response burden; even though the questionnaire had increased in length and more questions asked, the improved questionnaire and visual design actually resulted in a reduction in respondent burden. The final tradeoff decision was to limit survey costs by reducing the length of the new questionnaire to four sides of A4 paper (Jones et al. 2008a).

From the survey organization perspective, however, respondents are key to any survey process, as they provide the most valuable contribution—their data. In business surveys an important part of the production process is outside the survey organization: the data collection process that goes on within businesses (i.e., the response process).

1.7.3.2 Respondent Perspective

Most respondents would prefer no surveys, but unless existing data are available for all data requirements, then a world of "no" surveys is impossible. Respondents therefore attempt to tolerate the inconvenient requests from survey organizations. From their perspective, survey requests

- Come from various survey organizations, including NSIs, other public bodies, and private survey organizations; and from multiple surveys from the same organization. For example, for NSI surveys, large businesses can be selected for many survey samples at the same time.
- Often come at a time before the requested data are available; for example, requests for monthly retail turnover before the monthly figures have been fully compiled.
- Come at a time when the business is occupied with other activities (e.g., end-of-year financial accounts).
- Are tailored to the business but people in the business have to respond on behalf of the business. This may involve more than one person in the business if the survey includes questions on different business activities, such as human resources (number of employees) and business accounts (turnover, stocks).

It is seldom clear to business respondents why they have been selected to participate in the survey and why the data are being requested from the business. Therefore, two central themes should be considered from the respondent perspective: (1) building trust in survey methods (especially confidentiality and disclosure control), and (2) building trust and knowledge in the outputs that are produced from business surveys. These themes should be apparent in all survey communication materials: in the questionnaire, and also in letters, brochures, and other correspondence. Chapter 9 (in discussion of survey communication to encourage survey participation) in particular deals with these issues.

1.7.3.3 User Perspective

From the user perspective there is generally a need for more and more information. This is especially true when economic conditions are changing and/or policies have changed. Often users will not have indepth knowledge of the processes and methods involved in designing and conducting a survey. They know what they need and simply want to have the data as soon as possible; they may have little knowledge of the complexities and difficulties in designing and conducting a survey, or any notion of the concept of response burden. Discussing these issues and the related tradeoffs when specifying data needs with users is therefore important.

1.8 OVERVIEW OF THE BOOK

1.8.1 The Audience

This book has been written for anyone who is involved in designing and/or conducting business surveys, wishes to conduct a business survey, produces statistical or research outputs from business surveys, or uses business survey outputs. The readers may work in NSIs, universities (e.g., business studies, social sciences, economics, methodology and statistics), profit and not-for-profit survey organizations, international statistical organizations (e.g., OECD, IMF, UN, Eurostat), or central banks. Users of business statistics, such as policymakers, analysts, and researchers, may find this book of interest to learn more about the world of collecting data from businesses, and the issues involved.

In fact, this book is of interest to anyone interested in surveys, such as social survey methodologists and survey practitioners in general. The book gives new insights into survey methodology, as well as guidelines that are relevant not only for business surveys but also for social surveys (like the design of web surveys in household surveys). This book is a useful resource tool, providing guidelines and discussing and explaining methodologies in terms of their theoretical backgrounds. It gives guidance to analysts, researchers, and survey practitioners.

1.8.2 Organization of the Book

This book focuses on the business survey process set in the context of the generic statistical business process model (GSBPM). The overarching considerations and practices for designing and conducting business surveys, and producing survey results, represented in the spine (five left columns) of Figure 1.6 (planning, management, business context, quality, and response burden) should be regarded as the foundation blocks for business surveys.

The concepts of planning and management are generic to any type of data collection (e.g., surveys, administrative data sources, electronic data interchange, published information, qualitative methods), but there are specific requirements for business surveys (see Chapters 4 and 10). The business context, quality, and response burden also have unique features in relation to business surveys (see Chapters 2, 3, and 6).

The individual components of the survey process, leading to the production of survey outputs, are represented in the vertical flow of boxes from "the need for business statistics" to "evaluate" in Figure 1.6. These chapters provide guidelines on designing and conducting a business survey: sampling, data collection, and respondent communication; right through to the capturing, coding, and cleaning of data. The book concludes with a chapter on the methods and procedures for moving from survey data to statistical or research outputs. The book is structured around the components shown in Figure 1.6. It is hoped that this will provide a coherent approach to designing and conducting business surveys and producing business survey outputs.

1.8.3 Chapter Summaries

Chapter 2. One of the main underlying themes of this book is that effectively designing and conducting business surveys relies on an understanding of the business context—the nature, motivation, and behavior of businesses, including the behavior of people involved in the business, (at work, vs. in their personal lives)—because,

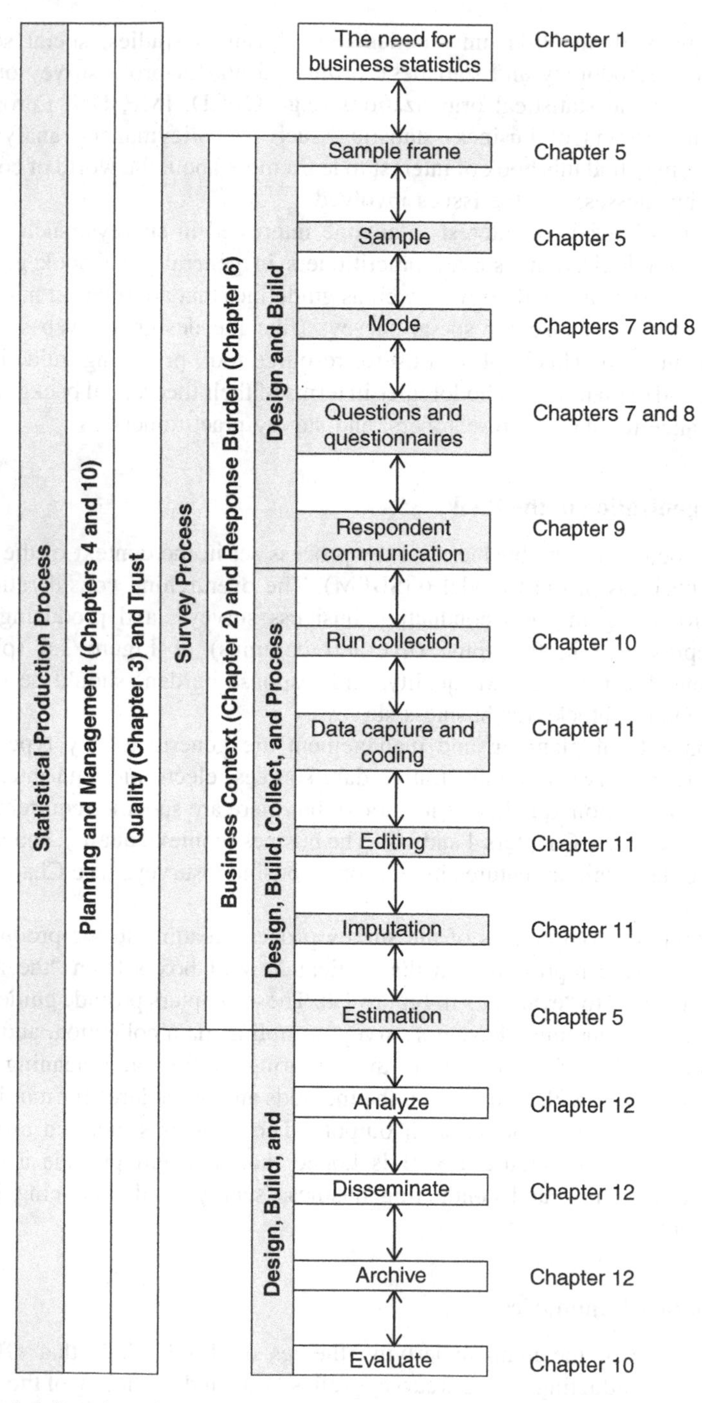

Figure 1.6 Overview of the statistical production process, survey process, and book chapters.

after all, it still takes a person to respond to a survey on behalf of the business. We also formulate a model of the business survey response process to assist survey researchers in making decisions on designing and conducting business surveys.

Chapter 3. Ultimately the objective of all kinds of surveys is to achieve maximum quality for minimum cost. Unfortunately, neither the quality nor the cost concept is straightforward. The major quality challenges and cost elements in business surveys may also differ from those in social surveys. In this chapter we discuss these issues and develop a model of business survey quality criteria. On the basis of this model, we discuss both sample-related and measurement-related quality issues in business surveys, in an attempt to identify the key challenges, that business survey designers and project managers face. These challenges are addressed in the following chapters in terms of survey planning and management, sampling, questionnaire development, and implementation of business survey designs.

Chapter 4. The initial step in the design of a survey involves planning of the survey production process and everything that is needed to collect and process the data. The main challenge is how can we increase the likelihood of achieving quality responses at low costs, knowing that an important part of the production process is outside the control of the survey organization (with compliance costs associated with that response process). These uncertainties (or risks) need to be managed. This task is even more challenging since business surveys can be regarded as complex. All components of the survey need to be ready when the fieldwork begins; the sample needs to be drawn, the questionnaire implemented, the data collection process organized, and so on, which means that many activities have to be carried out within constraints. To get that done, all activities, the people involved, material and financial resources, and scheduling need to be organized on a timetable and managed. This often necessitates tradeoff decisions. This chapter presents guidelines for dealing with these issues: project management principles applied to business surveys.

Chapter 5. It is very expensive and often unnecessary to obtain information from every business, so the usual approach is to use a sample, this chapter discusses topics related to sampling. It gives summary information on the construction and maintenance of registers and frames, and then presents guidelines on how to select the properties of a sample to make estimates with the best tradeoff of quality and cost. Once the information has been collected, then an estimation process is needed to ensure that the sampling is appropriately accounted for. This chapter also discusses considerations such as outlying observations and decisions regarding which businesses to obtain information from, and how to use this information to obtain survey estimates.

Chapter 6. Burdensome questions and information tasks should be avoided because they place survey quality in jeopardy. In business surveys, response burden is also important from an economic perspective. With respect to quality, perceived response burden is probably more important than the actual time spent on completing questionnaires. From an economic perspective, the reverse applies. Business surveys are commonly considered as a nonprofitable cost and a potential threat to the nation's productivity. Even if only a tiny proportion of total regulatory costs are due to business surveys, there is strong pressure in many countries to reduce business survey response burden. Reductions can be achieved by reducing samples or the number of variables, by sample coordination, or by improving the survey communication and measurement instruments. This chapter reviews response burden issues and measurements and suggests ways to minimize the response burden in business surveys.

Chapter 7. In this chapter we focus on the development and testing of survey questions, questionnaires, and data collection instruments. We discuss how to develop the concepts that data users wish to measure into survey questions. But survey design doesn't end there, because no survey researcher wants to spend time and resources collecting data that do not meet user needs. To ensure that survey questions capture the desired data, they need to be tested and evaluated, with respect to the business context. This chapter describes a wide variety of research methods to accomplish this, within a quality framework of reducing measurement error while also reducing or controlling respondent burden.

Chapter 8. This chapter discusses the measurement instruments used in business surveys. The main focus is on self-completion questionnaires, partly because self-administered surveys are the most common data collection method employed in business surveys. Much of the previous literature on business questionnaires has described how business survey questions should be presented. This aspect is present in our chapter as well, but we also focus closely on the content of business questions. Instead of using the common term *questionnaire design*, we prefer to discuss questionnaire communication. Questions are communicated to the respondents through the questionnaire, and as questionnaires are computerized, the communication can assume a conversational character. Another important effect of computerization is that the content of questionnaires is no longer limited to questions. The same is true for administrative tasks before and after the business survey questions are answered. This means that there is a whole set of questionnaires that need to be designed and presented together.

Chapter 9. Next to questionnaire communication, all other aspects of communication are important in business survey designs. This is what we call *business survey communication*, which includes activities mentioned in Section 9.1. The ultimate goal of these communication activities is to obtain quality responses. Traditionally survey communication has received little attention in comparison to other components of the business survey design. In books on survey methodology, survey communication is often discussed in the chapter on data collection, focusing on operational issues instead of motivating and facilitating businesses to cooperate. Chapter 9 focuses exclusively on this important component.

Chapter 10. During the fieldwork stage, the survey components are implemented and the survey launched. It is during the fieldwork that it will be clear whether the planned survey design is effective in terms of quality and cost challenges. In this stage of the production process there may remain uncertainties that need to be managed. To ensure that the targeted survey results are met, the fieldwork needs to be systematically monitored. Using real-time information from the data collection process, such as the response rate at any given moment in time. This ensures that appropriate actions can be taken early on in case the targets are not met. This chapter focuses on managing the fieldwork—monitoring and controlling the data collection process and the quality of the survey results, adopting an active fieldwork management approach—and relating the essential elements of this approach, paradata (or process data) and indicators, to survey errors and costs.

Chapter 11. During the survey field period the data will be initially captured and transferred to the survey organization for data capture. Once captured, and depending on the type of data collected, the data may need to be coded (e.g., to industrial codes). The data will then need to be cleaned using editing and imputation methods. This chapter provides an overview of capturing, coding, and cleaning survey data. It discusses the processes and methods involved, as well as potential sources of error; elements that impact on the design of other parts of the survey process (e.g., questions, questionnaire instructions, survey communication); and ways to minimize, measure, and monitor errors. As the literature on data capture and coding, and the treatment of errors is somewhat sparse, this chapter goes some way to address this issue.

Chapter 12. As this book sets the survey process in the context of the statistical production process, we felt it necessary to conclude by providing an introductory overview of the processes and methods that take you from survey data to statistical outputs. This chapter discusses different types of analysis, and analytical techniques for improving statistical interpretation, namely, seasonal adjustment, index numbers, and values and volumes. It then focuses on the prerequisites for dissemination (tables, graphs, disclosure control, statistical commentary, etc.), and finally on dissemination activities, including data archiving. This chapter merely scratches the surface of some specialized areas; key literature sources on this topic are cited throughout the chapter.

ACKNOWLEDGMENTS

In writing this chapter we would like to thank Simon Compton and Martin Brand (ONS) for their reviews, comments, and suggestions, Julie Griffiths (ONS) for her support, and the ONS media team for supplying relevant newspaper headlines.

APPENDIX[1]

(This appendix presents a timeline of the development of official business statistics in the United Kingdom.)

In writing this book, one of the questions we had was: How did we get to where we are now in relation to the production of official business statistics? To assist in answering this question, this appendix provides a general, nonexhaustive timeline of

[1] Appendix written by Jacqui Jones.

the development of official business statistics in the UK; what is evident from this timeline is the increasing demand for business statistics, to enable monitoring, decisionmaking and policymaking, and consequently the increasing need for business data.

- Back as early as the 13th and 15th centuries external trade figures for textiles were recorded. In the 18th century, government regulation of the Yorkshire wool industry and Scottish linen industry provided two output series. Into the 19th century textile factory inspectors provided more useful data. However, it was not until the beginning of the 20th century that the output from the textile industry was actually measured (Mitchell 1988, pp. 324–329).
- External trade statistical series commenced in 16th and 17th centuries; in the late 16th century customs figures were used to measure the trade deficit to guide policy.
- In England in 1665 Petty, followed by King in 1696, produced the first national income estimates. The rationale for producing these estimates was to provide statistical evidence that 'the State could raise a much larger revenue from taxes to finance its peace and wartime needs and to disprove once and for all the notion that England had been ruined by the Revolution and foreign wars (Frits 1992, p. 5). Both Petty's and King's estimates included a comprehensive concept of production and income, which remain to this day in SNA. The central element to this concept is that value added is generated from the production of goods and services; in contrast to Adam Smith's view that it could be generated only from the production of goods.
- The first comprehensive production statistics were those compiled from the shipbuilding industry, beginning in 1786.
- In the 1800s the dominant industries for Great Britain's economy were textiles, coal, iron, and agriculture. It was therefore important that these industries were measured for monitoring purposes.
- During the 1830s the first statistical societies were created across Great Britain, in cities such as Manchester, Bristol, Liverpool, Glasgow, and Leeds; the most prominent one—the Statistical Society of London—was founded in 1834 (Ward and Doggett 1991).
- In 1854, the first publication of the annual *Statistical Abstract of the United Kingdom* was published, followed in 1855 with the publication of the *Annual Statistics of Trade*, and in 1856 *Miscellaneous Statistics of the UK* (ibid).
- The official *Mineral Statistics* series, containing data on the coal and iron industries, commenced in 1854; however, the official collection of sales and electricity generation did not take place until 1920 (Mitchell 1988, pp. 236–239). From the 1850s coal mine inspectors made estimates of the number of people employed in these industries. For the ironworkers, except for the census, no data collection occurred until after World War I, when national insurance statistics commenced (ibid., pp. 274–278).

- The first Agricultural Census of Great Britain was carried out in 1865, and started the continuous acreage and livestock series. At that time agriculture was the dominant sector in the economy, thus the importance of measuring the output from this sector. At this time livestock coverage was less prevalent than crops. In the first full census of agriculture production in 1908, both crops and livestock were successfully measured. In relation to the numbers of people employed in agriculture, these data first became available in the 1811 population census, and from 1921 the annual agricultural census also included a count of farmworkers (ibid., pp. 180–185).

Into the 20th century, the British government still 'collected little information about industrial activity'. The need for better industrial information was recognized when questions on tariff policy were attracting attention (Ward and Doggett 1991, p. 114). In response to this observation, the following developments in official business statistics and business surveys took place (please note that this is not an exhaustive list):

- In 1906, the UK Census of Production Act was passed, which laid down the regulatory requirements for undertaking a census of production at suitable intervals. At the same time there was concern that Parliament was asking for increasing amounts of information that may have already been available to them, which some believed would lead to distrust of public statistics (Ward and Doggett 1991). The first census was carried out in 1908 in relation to production in 1907. It included questions on "the nature of the trade or business, particulars relating to the output, the number of days on which work was carried on, the number of persons employed, and the power used or generated" (Smith and Penneck 2009). It included the manufacturing industries, public utilities, and mining industries.
- In 1914, the UK Board of Trade started a regular monthly retail prices survey. Initially this covered only food prices, but in 1916 it was extended to cover clothing, fuel, and some other items. The food prices of 14 foodstuffs were collected each month in over 600 locations; data on clothing prices and fuel and light prices were collected using paper questionnaires sent to appropriate retailers. Other item prices were generally collected from administrative sources; for example, train, tram, and bus fares were supplied by the Ministry of Transport and transport operators (Ward and Doggett 1991).
- Official statistics of new house building became available only in 1920, partly because the industry was dispersed across the country and the focus was generally on external trade statistics—a category under which household building does not fall (Mitchell 1988, pp. 382–384).
- The quarterly index of industrial production was first published in 1928. The data were voluntarily collected from organizations such as trade associations. In 1947, the Central Statistical Office assumed production of the index, which was also published more frequently (monthly rather than quarterly) (Ward and Doggett 1991).

- In 1932, Clark published *The National Income, 1924–1931* and raised concerns about the quality of British official statistics citing classification issues as a quality issue. The creation of a system of national accounts was pushed further along by the 1930s Great Depression, John Maynard Keynes 1936 book *General Theory of Employment, Interest and Money*, and Tinbergen's 1936 construction of the first econometric model of the whole economy business cycle. Keynes challenged traditional economic theory and argued for government intervention to tackle unemployment. As the Keynesian approach was adopted in countries, national accounts were introduced to provide an analytical tool for economic activity. Also in 1936 Wassily Leontief published his paper on quantitative input–output (I/O) relations in the economic system of the United States, which documented his I/O study that acted as the catalyst for I/O tables linked to national accounts (Frits 1992).
- In 1932, the Import Duties Act was passed, which "included provisions for collecting statistics about industries affected by the duties imposed under the Act." This included "quantity and value of materials used and goods produced" (Ward and Doggett 1991, pp. 118–119).
- In 1935, Roy Glenday presented a paper to the Royal Statistical Society on the use and misuse of economic statistics, where he raised concerns about the uncoordinated approach to the production of economic statistics.
- In 1939, the League of Nations commissioned the production of guidelines to improve national accounts comparability, but this was delayed until after World War II. In December 1945, work on the guidelines resumed and was finally published in 1947 as the first international national accounts guidelines (United Nations 1947). However, the guidelines were ambitious, and on the request of the Organisation for European Economic Cooperation (OEEC), a simplified system were published in 1951, which only identified a current and capital account and three sectors (government, enterprises and households). This simplified system was seen as being more realistic for countries to implement (Frits 1992).
- In 1941, a central statistical office (CSO) was created to centralize the collection of statistics from government departments, including maintaining a comprehensive collection of all statistical returns. The work of the CSO included the preparation of national accounts, and the discussion of proposed statistical questionnaires to be issued to businesses (Ward and Doggett 1991). In 1941, the first official estimates of national income and expenditure were published.
- In 1946, the first monthly digest of statistics was published (ibid.).
- In 1947, the Statistics of Trade Act was passed. Prior to its royal assent, industry representatives welcomed the information that the act would bring, it provided the opportunity to collect more information from industry, "necessary for the appreciation of economic trends and provision of a statistical service for industry and government departments" (ibid., p. 120). But it also raised concerns about the burden of completing questionnaires, especially on small businesses. The act also made provision for conducting a census of distribution

(Ward and Doggett 1991). The first British census of distribution was carried out in 1950, and included wholesale and retail distribution, with a limited number of service trades. Businesses were required to provide figures on 'sales, purchases, stocks, employment and wages and salaries' (ibid., p. 121).

- In the 1949, 1950, and 1951 production censuses, the type of information requested was changed to summary information to provide important aggregates for the national income and expenditure accounts and the changing importance of different industries (ibid). Then in 1951 sampling methods were used instead of a census; for this survey, one in three establishments employing more than 10 persons were selected and for those employing 10 or fewer persons, 1 in 20 were selected (Ward and Doggett 1991).
- In 1952, the first *National Income and Expenditure Blue Book* was published (ibid). The time delay between the latest period listed in the blue book and the date of publication was a concern, with the blue book regarded as "too late to be useful." This led to the production of quarterly national accounts (ibid., p. 61).
- In 1952, the OEEC 1951 national accounts guidelines were enhanced to contain more accounts but remained simplified in comparison to the 1947 United Nations guidelines. 1953 saw the publication of the United Nations *A System of National Accounts and Supporting Tables* (United Nations 1953), which were very similar to the OEEC guidelines and included some types of nonmarket output. Further revisions to this were published in the 1953 report (Frits 1992).
- In 1954, a committee was appointed by the Board of Trade to provide advice on future censuses of production and distribution. The committee recommended the extension of sampling methods to reduce the burden on businesses. Sampling methods were used for the 1955–1957 production and distribution censuses, and there was also a radical overhaul of the questionnaire, reducing it to a single page, and the questions reframed to match, as far as possible, data available in businesses accounts. In 1958, a full census (based on both coverage and questions) was run, but the detailed questions were sent only to businesses employing 25 or more persons. Then established was a full census every 5 years and a more basic census in the intervening years (Ward and Doggett 1991).
- In 1956, the *National Accounts Sources and Methods* was published (ibid.). In the same year, the then chancellor prepared a detailed written statement for improving statistics. Part of his statement included a plea to industry to return their questionnaire more quickly.
- In 1958, the *Monthly Production Inquiry* began.
- In 1960, the second edition of the 1953 SNA extended the framework to include flow of funds and I/O tables; in 1964, the third edition improved consistency with the (IMF) *Balance of Payments Manual* and updated references to other international guidelines (Edwards 2007).
- In 1962, publication of *Financial Statistics* began (Ward and Doggett 1991).
- In 1968, the United Nations published a revised and more detailed *System of National Accounts* (United Nations 1968). The 1968 revision included major

extensions to national accounts, and the SNA was now seen as a guide for countries to develop their systems of basic economic statistics (Edwards 2007).

- In 1969, the first UK business statistics office (BSO) was created, with responsibility for most of the government's collection of business statistics (Ward and Doggett 1991). The BSO continued running the production census but also introduced monthly and quarterly surveys to provide more timely statistics. These statistics were supplemented with surveys about capital investment, stocks, purchases and sales (ibid., p. 123).
- In 1971, the first employment census was carried out; this was the key source for workplace-based employment information.
- In 1974, the first purchases inquiry was held in conjunction with the annual production census.
- In 1988, the Pickford review recommended changes to the way macroeconomic statistics were collected and compiled; and changes to the structure of the UK Central Statistical Office (Pickford et al. 1988).
- In 1991, the quarterly *survey into the Distribution and Services Sector* started. Industry by industry, the frequency of this survey gradually progressed from quarterly to monthly in 2001.
- In 1993, there were further revisions to SNA and in 1995, to ESA. Frits (1992) discusses the development of national accounts guidelines, prior to 1992 in further detail.

More recently in the UK, there have been efficiency initiatives in business surveys such as the merger of several surveys, to effect external savings in relation to response burden—less surveys, less burden—and internal financial savings by sending out fewer questionnaires, processing fewer data, and merging internal teams. For example, in 2000, the first UK annual business survey results were published, with data relating to 1998. This new annual survey was an amalgamation of several previous annual surveys (e.g., the annual production and construction censuses, the purchases inquiry, the annual employment survey, and several surveys covering the distribution and services sectors). In 2010, the monthly production survey and monthly survey into the distribution and services industries were merged into a single monthly business survey.

There has been a further revision to SNA (in 2008), and ESA (in 2010). For European member states, ESA revisions will be implemented in 2014.

CHAPTER 2

The Business Context and Its Implications for the Survey Response Process

Diane K. Willimack[*] and Ger Snijkers[*]

As we saw in Chapter 1, businesses are different! And because they are different, researchers who use surveys to collect data from businesses need to be aware of and consider these differences when designing and conducting business surveys. Survey techniques commonly accepted and used in surveys of households and individuals may be inappropriate for collecting information from businesses; they may require some adaptation to the business setting to be effective, or they may not work at all. This is of utmost importance to survey organizations that conduct business surveys, in order to design and carry out survey and statistical procedures that result in high-quality economic statistics while containing costs for both survey organizations and businesses. To achieve this goal, one must first understand the basic fundamentals and behavior of businesses, as well as how people behave within businesses, that is, at work, versus how they behave in their personal lives.

In this chapter, we begin by identifying a number of attributes of businesses and their behaviors, and describe their implications for survey response. We also discuss some perspectives of organizational effects on how people behave within a business, as well as how work gets done. In Section 2.2, we introduce a high-level model of business survey response processes. We synthesize a variety of perspectives provided by various response models along with knowledge and implications of organizational behaviors drawn from the administrative science literature. The result will be a

[*] Any views expressed on methodological or operational issues are those of the authors and not necessarily those of the US Census Bureau or Statistics Netherlands.

Designing and Conducting Business Surveys, First Edition.
Ger Snijkers, Gustav Haraldsen, Jacqui Jones, and Diane K. Willimack.

framework and a philosophy for the approach that we advocate throughout this book with regard to designing and conducting business surveys.

2.1 THE BUSINESS CONTEXT FROM THE PERSPECTIVE OF THE ORGANIZATIONAL SCIENCES

2.1.1 Business Goals and Behaviors and Implications for Survey Response

The literature on organizational behavior agrees that the goals of any organization are twofold: (1) to produce goods or services and (2) to maintain the viability of the organization over time (Simon 1997; Katz and Kahn 1978; Porter et al. 1975; Parsons 1956). Remaining viable requires that the benefits of producing goods and services outweigh the costs. For entrepreneurial businesses, this means making a profit; that is, revenue from providing goods and services must exceed the costs of producing them. For nonprofit organizations, it means reducing costs. Thus, ultimately business decisions are driven by weighing costs against potential or actual benefits.

- *Implications for Survey Response:* Responding to surveys takes time and human resources, and thus are a cost of doing business. However, the output does not contribute directly to production of goods and services, the provision of which contributes to revenue. In other words, as Willimack and Nichols (2010) point out, survey response represents a nonproductive cost to the business.

  Conversely, the benefits to businesses of economic statistics produced from survey data are predominantly intangible and indirect. The data are used by others in a manner that affects the environment within which businesses operate. Business survey statistics describe the economy. As we have seen in Chapter 1, official statistics are used to measure and monitor the economic health of a society, so that policy interventions may be undertaken to affect economic conditions. Business statistics are used by those who interact directly with businesses, such as lenders assessing risks of lending funds to specific types of businesses at a given time. Data users within or closely associated with the business conduct market research or forecast business conditions to facilitate business planning and development. Ultimately, the "benefits" of business survey statistics accrue to businesses through economic mechanisms. Since the benefits to businesses are indirect, they are seldom recognized, and thus are easily outweighed by the very real resource costs associated with completing the surveys that support business statistics—costs that are neither intangible nor indirect.

Achievement of both goals stated above is the driver for many attributes of businesses, including the nature of their interaction with the external environment; types and frequency of data tracked in business records; structure and contents of recordkeeping systems, functional and managerial structures; specialization and differentiation of functions; the need for and degree of coordination; centralization, decentralization, and/or hierarchies of authority; methods for achieving efficiencies and effectiveness.

Let's start with the external environment. To achieve their goals of producing goods and services and remaining viable, all businesses must interact with the outside world. Whether the nature of the interaction is to provide goods and services to consumers or to meet legal and regulatory requirements, attributes of the external environment impact the continuing viability of the business. Some businesses and industries are more protective in order to maintain competitive advantage or more insulated to reduce the impact of economic volatility on their operations. Others thrive on frequent and routine interactions with their environments, such as publicly traded firms or businesses that rely on positive public relations. Moreover, few, if any, businesses are immune to changes in economic conditions, such as recessions or growth, the so-called business cycle.

In addition, in their relationships with the external environment, businesses want to be seen as good "corporate citizens." This means being law-abiding and integrated into the community, thus contributing to the "greater good." The rationale isn't necessarily altruistic per se. Businesses are compelled by legal or regulatory requirements to interact with the outside world. By being law-abiding and community-oriented, businesses are trying to demonstrate goodwill to their (potential) customers, to whom they provide the goods or services that they produce, the sales of which earn them a profit, which maintains their viability over time.

- *Implications for Survey Response:* Businesses regard survey organizations as part of their external environment, and thus will react to survey requests in a manner consistent with their interactions with similar entities. For example, governmental national statistical institutes (NSIs) may be attributed with similar authority as other legal and regulatory bodies with which businesses must interact. On the other hand, businesses may consider participation in surveys conducted by universities to be an obligation to their business community through contribution to useful research.

  Important to survey response are the frequency and routine, or lack thereof, with which a business shares information with its external environment. When sharing information externally is a common occurrence or requirement for conducting business, the business will have established the necessary infrastructure to exchange information easily and securely. For example, companies with numerous critical governmental reporting requirements may even have an office and personnel devoted to that activity (Nichols et al. 1999).

Next consider the role of records in the operation of a business. Primary reasons for businesses to keep track of information and data are (1) for managing the activities of the business so as to achieve their organizational goals (e.g., managing costs and monitoring revenues to ensure profitmaking) and (2) meeting legal and regulatory requirements that, if not met, would threaten the viability of the business and its reputation as a "good corporate citizen," not to mention increased costs due to penalties and restrictions.

- *Implications for Survey Response:* Businesses, as a rule, do not monitor data solely to satisfy statistical reporting purposes. Statitiscal requirements for data collected by business surveys are typically related to the goal of measuring and monitoring economic conditions. Their underlying concepts give rise to technical definitions that may not align well with data tracked in business records. Thus, a common expectation that data requested by business surveys can be obtained directly from business records is often a myth. Rather, there is a spectrum of availability relative to statistical reporting requirements (Willimack and Nichols 2010; Bavdaz 2007). This has implications for data quality and respondent burden, which are discussed in more detail in Chapters 3 and 6 respectively.

  Moreover, most businesses do not seem compelled to routinely record these data, despite the mandatory nature of many surveys that support official statistics. We speculate that there are several reasons for this: (1) enforcement of legal requirements to report varies widely across countries and national statistics institutes, as NSIs prefer to rely on a cooperative approach for obtaining data from businesses; (2) available data, although not exactly those requested, may suffice to satisfy mandatory reporting requirements; (3) large companies, likely sampled with certainty, have personnel in place whose jobs are to prepare financial reports, and thus have resources for answering surveys. Alternately, small businesses with small sample selection probabilities may see survey participation as an aberration, and thus not worth the cost of setting up records to support survey requirements.

Regardless of size, businesses, as a matter of course, exercise cost control and reduction measures in order to help maintain their viability over time. Two cost control measures that are particularly pertinent to survey response are the ability to plan and the use of routines. Planning enables businesses to coordinate activities and processes in order to ensure that adequate resources are available at the time they are needed, while also minimizing the risks of unforeseen and costly disruptions. Additionally, cost control is aided by formal or informal routine work processes. In particular, businesses rely on developing and implementing routines for tasks that are repetitive. In this way, they are able to devote valuable resources to activities that require monitoring, judgment, and responsiveness to maintain productive operations.

- *Implications for Survey Response:* There are two important reasons why the ability to plan and the development of routines have implications for survey response: (1) many surveys, particularly those supporting official statistics, recur with predetermined regularity (e.g., monthly, quarterly, or annually); and (2) since businesses consider response to surveys as a nonproductive cost, it is in their best interests to develop strategies that control or reduce the costs or survey participation. It only makes sense, then, that businesses selected for recurring surveys would come to expect them, permitting them to plan and devise reporting routines.

Indeed, this is true not only for recurring surveys, as businesses (particularly large ones) are accustomed to many internal and external reporting requirements and information needs. In research conducted by Willimack et al. (2002), businesses expressed a desire to know well in advance, during their annual planning cycles, when to expect surveys and what types of data would be needed, so that they could at least schedule resources, or, at best, integrate response processes with other similar reporting activities that require extracting data from records. Moreover, Bavdaz (2010a) adds that because businesses rely on routine practices for cost control, new survey requests often are directed to the same employees or departments responsible for completing other surveys or forms.

To support their reporting routines, businesses retain documentation of their survey reports, retaining copies of completed survey questionnaires along with any supporting notes or other documentation explaining their answers to survey questions. This practice serves to maintain their response strategies over time for particular items and recurring surveys. One consequence for survey response is that any changes in the reported data may be considered to reflect real changes in business or economic activity, rather than as an artifact of data collection. Another consequence, however, is that reporting errors are maintained, impacting data quality (Sudman et al. 2000; Bavdaz, 2010a).

In addition, because of businesses' proclivity for planning and routine, reporting strategies may become institutionalized, perhaps even encoded into standardized practices. Then, when survey questionnaires change or questions are added or removed, reporting routines may be disrupted. Even changes in the numbering or sequencing of survey questions within a questionnaire may go unnoticed, and reporting errors may occur. Business respondents have noted that questionnaire changes increase burden as they adjust their response strategies. Nevertheless, to be conscientious reporters, business respondents have suggested and requested that changes to questionnaires be highlighted to draw attention and be addressed proactively (Sudman et al. 2000).

As businesses grow, their organizational structures become more complex in two ways: First, they gain operational efficiencies through specialization and differentiation of functions, in order to control or reduce costs. Since data are tracked to manage these different specialized functions, records are formed accordingly, and knowledge is distributed throughout the company, residing in units that benefit most directly from its use.

- *Implications for Survey Response:* In their cognitive response model for establishment surveys, Edwards and Cantor (1991) introduce the concept of the "most knowledgeable respondent" as the desired respondent for any particular survey topic, an individual who is closest to the record formation process and is assumed to know the contents of the records and the concepts being measured by recorded data. For omnibus surveys that collect multiple types of business data on different topics, there may be more than one such respondent. For example, precise employment and payroll data may be in the human resources department;

while revenues, expenses, and liabilities may be summarized in a finance department; taxable income and tax credits, calculated in a tax department; and product orders, inventories, and detailed expenses, monitored at the manufacturing plant level. When a single survey instrument collects data on multiple topics, multiple respondents would likely be needed to provide precise, high-quality response data. Gathering data from multiple data sources and multiple most knowledgeable respondents results in a labor-intensive response process, requiring effective communication and social interaction with coworkers, along with appropriate authority to obtain data from others.

A complexity often associated with organizational growth is the development of organizational hierarchies that disperse authority. The hierarchical nature of many organizations is a consequence of the functions (and their supporting data) distributed throughout the organization, each with its own authority chain. With the differentiation of functions and the distribution of knowledge, organizational hierarchies assume the role of coordinating activities and decisions across specialized units, to facilitate the production of goods and services. Persons with authority determine the types of information tracked by the organization at various levels or in particular units, the amount of detail, and the periodicity, for the purposes of supporting organizational goals.

- *Implications for Survey Response:* Authority structures determine who has access to which data and what can be done with these data. This has numerous implications for survey collection. First, data tracked by businesses are "owned" by the business and any uses of these data are at the discretion of the business. However, a business cannot speak for itself. Rather, persons in authority positions make decisions on behalf of the business in a manner that supports its organizational goals of producing goods and services, maintaining viability, and exercising corporate citizenship.

  Since data belong to the business, survey organizations must rely on authoritative decisionmakers to permit dedication of resources to survey response and then to subsequently allow the response data to be released back to the survey organization. Individuals with this type of authority may themselves undertake the survey response task in the role of the respondent, or they may delegate the task to someone else among their subordinates.

  The ultimate consequence, then, is that respondent selection is not under the control of the survey organization. Rather, respondent selection is at the discretion of the business, or more specifically, persons who have the authority to make decisions on behalf of the business. Since respondent selection is under the control of the business, the response task may or may not be assigned to an employee who is the most knowledgeable about the data desired by the survey organization. In addition, as we noted earlier, routines associated with business survey response and other types of reporting cause business surveys to often be directed to the same individual or department, regardless of survey content and the recipient's knowledge.

Consider a designated respondent tasked with an omnibus type of survey collecting data on multiple topics, and perhaps requiring involvement of multiple data providers. To provide an adequate response to such a survey, the designated respondent would be expected to willingly and successfully take an additional, and possibly new, role—that of a data collection agent on behalf of the survey organization. However, such agents must act within the context and constraints of their employing organization, which is not the survey organization. Moreover, none of the activities undertaken by the designated respondent are under the control of the survey organization.

Ultimately, the survey response process relies on a respondent who is not selected by, and whose activities cannot be directed by, the organization responsible for conducting the survey. This seems an untenable situation for survey organizations. However, understanding the business perspective is key. Businesses achieve their goals—to produce goods and services and to remain viable—through the actions, behaviors, beliefs, and attitudes of their employees, who are people. Businesses, like any other organizations, are inherently made up of people and, thus, social behavior is fundamental to their existence. Through social systems that support organizational culture and functions, the behavior of persons in organizations is ultimately influenced by the goals of the organization. We will now examine social behaviors in organizations and their implications for survey response.

2.1.2 Dimensions of Social Behavior in Organizations Pertinent to the Survey Response Task

While literature on organizational behavior discusses the impact of the organization on a number of social behaviors, the following seem to be particularly pertinent to the task of providing data for a survey. It is important to note that these dimensions are interdependent; their attributes intermingle and commingle with one another in social behavior:

- *Authority.* Authority may be regarded as legitimate power. It "may be defined as the power to make decisions which guide the actions of another" (Simon 1997). Authority is manifested only in relationship. Katz and Kahn (1978) define authority as "power which is vested in a particular person or position, which is recognized as so vested, which is accepted as appropriate not only by the wielder of power but by those over whom it is wielded and by the other members of the system." Simon (1997) goes on to say, "It is authority that gives an organization its formal structure." One of its major functions is to facilitate the coordination of activities divided by specializations in order to achieve organizational goals, such as production of goods or services. In business survey research literature, authority is often discussed in the context of decisions to participate in a survey and to release data to the external environment. In addition, persons with authority have the legitimate power to delegate to others the work involved in answering a survey, and they potentially have the ability to coordinate activities across organizational units where dispersed data reside.

- *Responsibility.* One might regard authority and responsibility as inseparable. Someone with authority should also be responsible. However, one can have responsibility for completing a task without being given authority over resources needed to accomplish the work. This is a common issue faced by project managers—deemed "the divorce of authority and responsibility" by the literature on project management, a popular philosophy for treating "work" in the context of "projects" accomplished by cross-functional teams (Frame 1995). Responsibility consists in acceptance of the expectations for conducting an activity (viz., work), accomplishing it, and being held accountable for its acceptable completion. Responsibility can be ongoing (in terms of a worker fulfilling activities associated with the job definition), or episodic (when one receives and accepts an assignment from an authority). Completing a survey often falls within the realm of the latter.
- *Accountability.* Accountability is a corollary of responsibility. One who is responsible for an activity must answer to others regarding the acceptability, or lack thereof, of its completion, subject to positive or negative consequences. Since accountability depends on the expressed judgment of others, it has a social dimension. Others to whom one is accountable have the ability to exercise discretion in their judgment. In organizations, responsibility is a byproduct of authority, and degrees of accountability may be determined by the relative importance of alternative outcomes and their association with organizational goals. Accountability is reflected in criteria used in evaluating job performance. For those involved in activities associated with survey response, the degree of accountability likely directs their motivation and attentiveness to the response task.
- *Autonomy.* Autonomy is the degree to which one feels personally responsible for one's own work. One's sense of empowerment, self-determination, or control impacts the degree to which one has discretion over one's activities and "ownership" of the outcomes. When survey response requires interaction among multiple parties, the respondent relies on the cooperation of others, perhaps regardless of their authority channels. To the degree that others are empowered and autonomous, this cooperation is facilitated.
- *Capacity.* A person is capable of performing expected activities to the degree that s/he possesses the ability and resources needed to conduct or complete the task. Since surveys involve the collection of information, a person's capacity to adequately perform this task requires knowledge of and access to the requested data. Capacity may be a common criterion for respondent selection within the organization. In other words, delegation of responsibility for completing a survey may be based primarily on the person's knowledge of and/or access to the requested data. Also, certain types of activities may become identified with a particular person or position as routine. While we note that the same person often carries out the response tasks in subsequent iterations of recurring surveys, such as those conducted by NSIs for official statistics, Bavdaz (2010) points out

that, because businesses rely on routinized practices, new survey requests are often directed to these employees or departments.

- *Influence.* Katz and Kahn (1978) describe influence as "a kind of psychological force," manifested in "an interpersonal transaction in which one person acts in such a way as to change the behavior of another in some intended fashion." Authority is a particularly pertinent means of influencing behavior in organizations, as is leadership. Other mechanisms of influence include suggestion, persuasion, recommendation, and training. An individual who lacks authority must draw upon other modes of influence to obtain compliance or desired behavior from others. In addition to authority, Cialdini (2001) identifies five behavioral principles by which influence works to gain compliance in social interactions:

  Reciprocation—one is more willing to comply with a request on the expectation of receiving a gift, favor, or concession in return.

  Commitment/consistency—one will comply with requests that are consistent with one's past behaviors.

  Social validation—one will be more likely to comply with requests that one's peers would also comply with.

  Liking—one will be more likely to comply with requests made by others whom they like.

  Scarcity—one will comply with requests associated with rare or scarce opportunities.

  Besides being applied in nonresponse reduction strategies (Groves and Couper 1998; Edwards et al. 1993; Snijkers et al. 2007a), these compliance principles also seem pertinent for interpersonal interactions within the survey response process. Influence is particularly important when the person assigned the response task lacks direct authority over others from whom information must be obtained. The respondent must then exercise influence to gain cooperation from others toward a goal not explicitly their own, such as survey completion.

- *Allegiance/Loyalty.* Organizational loyalty occurs when organizational goals align with personal goals, and when an individual identifies with and/or personally adopts the goals, objectives, and values of the organization of which s/he is a member. Loyalty is a form of organizational influence, such that an individual's allegiance ensures that his/her decisions on behalf of the organization are consistent with its goals and objectives. Loyalty, then, may support an individual's autonomy in his/her performance of duties for the organization. In the survey response process, the respondent's success in obtaining data from sources crossing intraorganizational boundaries may be associated with the organizational loyalty of others, that is, others' perceptions that fulfillment of the request supports organizational goals.

Table 2.1 illustrates how these social dimensions may be activated in the survey response process. This hypothetical scenario represents a composite of respondent behaviors and actions observed during business survey pretesting conducted at the US Census Bureau.

Table 2.1 Illustration of Social Dimensions in the Survey Response Process

Respondent Action or Behavior	Social Dimension
Sue P. Visor, who received a survey form from the US Bureau of Surveys, delegates the task of completing the form to Ray S. Pondent.	Authority
Sue chose Ray to complete the survey because he had been in the organization for a number of years, had worked in different areas, and thus had extensive knowledge of the types of data asked for on the survey, such as employment and payroll in the pay period that includes March 12, revenues by type, health insurance costs, investment in new equipment, and purchased services.	Capacity: knowledge
Ray now is obliged to do the work required to complete the response task.	Responsibility
When Sue assigned the work to Ray, she provided Ray with some instructions regarding her expectations for the outcome of the activity, including time, effort, and importance relative to Ray's other duties.	Accountability
Now, Ray, who worked in the human resources (HR) department, could readily pull the employment, payroll, and health insurance costs directly from his department's records.	Capacity: knowledge, access
But the other data items requested on the survey are not in systems that Ray has access to, if the data exist at all.	Capacity: knowledge, access
Ray considers simply leaving those items blank and returning the form to Sue, but realizes that he needs to make his best effort because Sue gave him the responsibility to do the work.	Autonomy
Ray worries about Sue's evaluation of his work if he does not at least try.	Accountability
Ray knows that he can get the revenue data from the accounting department. He and his colleague, Fanny N. Schaal, trade information fairly regularly for routine financial reports.	Influence: consistency, reciprocity
Fanny willingly complies, because this is a routine part of her job, and she knows that Ray will return the favor in time.	Autonomy, influence: reciprocity
Ray is pretty certain that the data on investment in new equipment can be found in the finance department.	Capacity: knowledge
But there's a new accountant there, Kap Spenture, who does not know Ray or Ray's normal duties. Ray is reluctant to get Sue involved to ask Kap's boss for Kap's help, because Ray feels that this is his project and his duty on behalf of the organization.	Autonomy, allegiance/ loyalty
So Ray decides to fax a memo to Kap on HR letterhead requesting the data about investment in new equipment. He includes a copy of the questionnaire with the equipment questions highlighted so that Kap can see the actual nature of the request.	Influence: authority
Providing this type of information is part of Kap's normal job duties.	Autonomy, responsibility, accountability

Table 2.1 (*Continued*)

Respondent Action or Behavior	Social Dimension
Kap sees that the survey is from the government and that it benefits the company to reply; otherwise the company might have to pay a fine.	Allegiance/loyalty
Finally, Ray sees that all that has been left is a question on "purchased services," which is something he has never heard of. So Ray assumes that his company does not have any of that and leaves the question blank.	Capacity: knowledge
Ray returns the completed form to Sue.	Accountability
Sue reviews Ray's work.	Authority

2.1.3 The Response Process Viewed as Work

2.1.3.1 How Work Is Accomplished

Work may be defined collectively as efforts undertaken to perform formal activities that contribute to the goals of the organization, which in our case is a business, where the goals are to produce goods and services, remain viable, and exhibit good corporate citizenship. To accomplish work requires that the various labor activities be coordinated to produce goods or services. When divisions of labor and specialization have occurred to ensure efficiency, resulting in intraorganizational boundaries, informational subsystems need to be integrated to enhance coordination and to support management.

Use of matrix management and cross-functional project teams brings together staff from different organizational subsystems (e.g., departments, divisions, offices) with different specialties. As such, intraorganizational boundaries are crossed in order to accomplish a project, task, or work—to produce goods and services. But there is a formality to these quasi-structures that is sanctioned by the organization to accomplish explicit organizational goals—factors not typically found in the task of completing a survey. In other words, the task of completing a survey is typically not treated like a project, with coordination of activities explicitly and overtly sanctioned by authorities. Rather, that approval, while understood intrinsically by the person assigned the survey response task, may not be widely evident organizationally.

2.1.3.2 Social Behavioral Dimensions of Work

A necessary social dimension for the accomplishment of work is communication. The coordination of specialized activities requires communication across subsystems. Instructions and expectations must be communicated to each worker to ensure that activities are conducted in accordance with specifications for quality, quantity, timing, and other parameters. Although the exercise of authority is the obvious facilitator, it may become an impediment to efficiency if taken to the extreme.

So, how does work get done in organizations? Katz and Kahn (1978) and Porter et al. (1975) point to patterns of social behavior so taken for granted as to be overlooked in formal discussions of coordination and organizational effectiveness—cooperation and individual self-control. Katz and Kahn list "cooperative activities with fellow members" as one of several elements of "performance beyond role requirements" needed for organizations to function effectively. Since an organization cannot plan for all contingencies, "the resources of people for . . . spontaneous cooperation . . . are thus vital to organizational survival and effectiveness." Porter et al. add that organizations rely on "[individual self-control] to achieve coordination . . . If organizations had to rely entirely on . . . a hierarchy of authority [for] establishing coordination, it is unlikely that society would witness very much concerted activity." The values and expectations that individuals bring to the organization "contribute to coordination almost as a matter of course without the individual having to pay attention specifically to the problem."

Porter et al. (1975) point out that "self-direction and self-control involve the internalization of certain standards and values." They contend that, for the organization's sake, these should be compatible with the organization's goals. Invoking theories advanced by management science pioneer McGregor (1985), accomplishing this alignment requires that "individuals must see that their own goals can be best achieved by attempting to reach organizational objectives." In McGregor's words, "people will exercise self-direction and self-control in the achievement of organization objectives *to the degree that they are committed to those objectives*" [italics McGregor's]—that is, organizational loyalty.

To summarize, cooperation in the work setting emanates from two sources: (1) the attribution of social norms of cooperative behavior and (2) the adoption of self-directive behaviors. The impact of the latter depends on the individual's identification with the organization's goals, or organizational loyalty, which was previously identified as a form of organizational influence on individuals' behaviors.

2.1.3.3 Accomplishing the Work of Survey Response

Business respondents' comments and behaviors during questionnaire design research and testing at the US Census Bureau repeatedly reinforce their viewpoint that survey response is a work activity—a perspective not typically taken in relation to the response process in surveys of households and individuals. Responding to a business survey is not a matter of personal choice or effort, but is seen as an activity taken on behalf of the organization, and thus is subject to the "labor contract" between an individual and the business. However, as discussed earlier, survey response is seen as work that fails to contribute to the production or management goals of the company (Sudman et al. 2000). As a result, it may not be regarded with the same value as other work activities.

In particular, in omnibus-type surveys, when data from multiple sources are needed, the response process must rely on dimensions of social behavior in addition to, or perhaps instead of, organizational sanctions. In such cases, survey response requires the participation of multiple workers from different subsystems, much like a cross-functional project team. However, survey response is not typically treated by the organization like a project sanctioned for the purpose of meeting organizational goals.

As a result, successful completion of the survey response process relies on—and often occurs because of—social norms associated with cooperation and organizational influences that support self-directed behavior. US Census Bureau pretesting of many different economic survey questionnaires with a wide variety of types and sizes of businesses consistently shows that persons to whom the survey response task is delegated are typically responsible for other similar reporting tasks requiring financial data and accounting information. Their requests of those with access to other data sources are accepted with little hesitation because the activity is considered to be within the area of responsibility of each participant in the interaction. Cooperation, in this case, is a normalized behavior, which is a fortunate circumstance for the survey response process.

2.2 A COMPREHENSIVE APPROACH INTEGRATING THE BUSINESS CONTEXT AND THE SURVEY RESPONSE PROCESS

We will consider the survey response process to consist primarily of two steps:

1. The decision to participate
2. Performance of response tasks

A third step—review and release of the survey response—is subsumed by the other two. These steps are not separate and distinct; they are not independent of one another. Although a positive decision to participate indicates the intention to respond, performing the response tasks is still necessary to achieve survey response. Moreover, research has repeatedly shown that respondents' expectations about the response task affect their participation decisions.

For surveys of households and individuals, these steps have been the subject of much research, and a substantial body of literature exists. See, for example, the text by Groves and Couper (1998) and the edited volume by Groves et al. (2002) for examinations of the decision to participate in a survey and nonresponse reduction strategies. Studying how household respondents perform response tasks also has a long history and a broad literature (Sudman and Bradburn 1982). A response model based on cognitive psychology set forth by Tourangeau (1984)—consisting of four cognitive steps, comprehension, retrieval, judgment, and communication—has become a generally accepted framework for guiding research on processes used by survey respondents to provide answers to survey questions.

The survey response process for surveys of businesses and organizations is somewhat more complex. It's true that business survey respondents are human beings like individual and household survey respondents and thus must contemplate whether to participate in a survey, and, if so, undertake the four cognitive steps when answering survey questions. However, the actions and behaviors of business survey respondents are impacted by the context of the organizational setting, as we have seen in Section 2.1.

Drawing upon research and response models built for surveys of businesses and organizations, we now formulate the model of the business survey response process

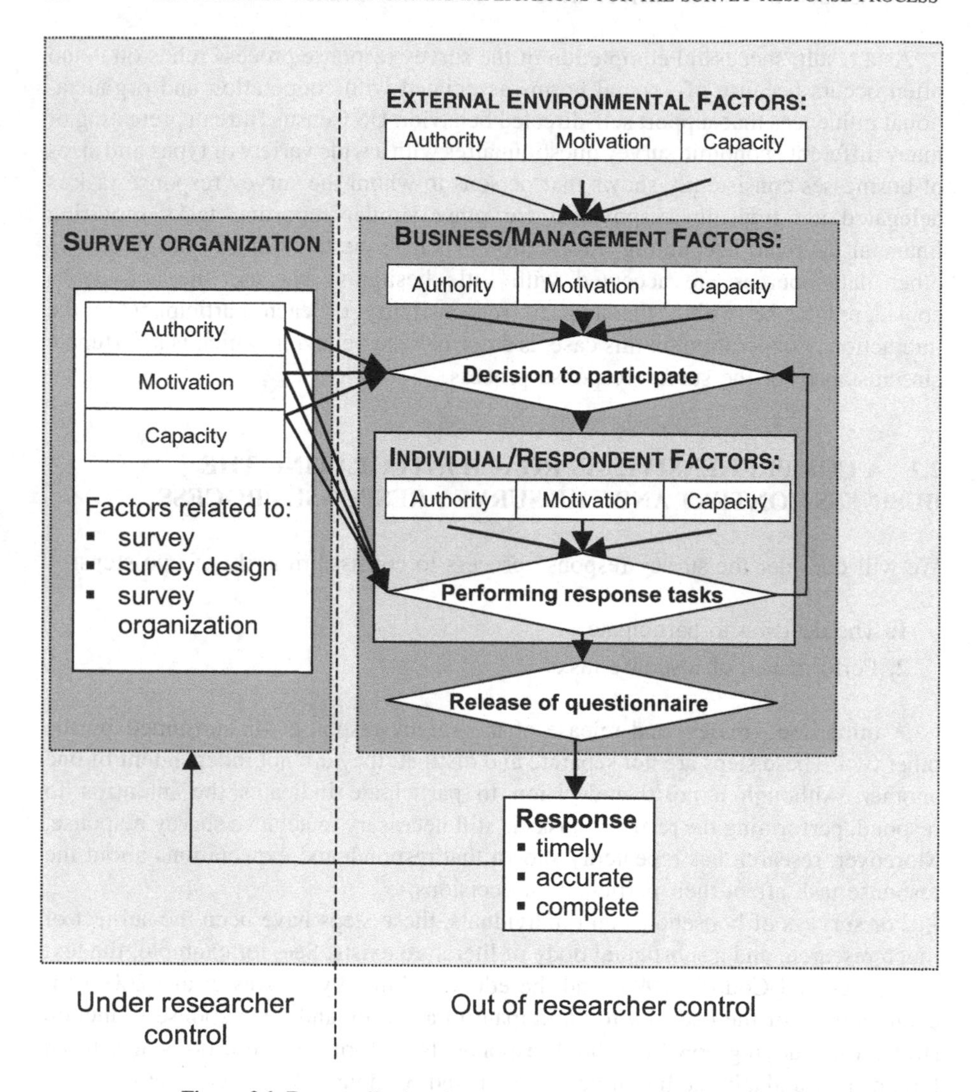

Figure 2.1 Factors affecting the response process in business surveys.

that we will use as a basis for this book. Figure 2.1 illustrates our model, providing a high-level framework that encompasses many of the activities, actions, behaviors, and perspectives described in this chapter.

Drawing upon the framework used in other response process models (Willimack et al. 2002; Snijkers 2008; Snijkers and Giesen 2010; Jones et al. 2005; Bavdaz 2010a), we first distinguish between factors that survey researchers can and cannot control. Essentially, as survey researchers, we can only control features of the design and conduct of the survey itself, the data collection procedures, processing, and so forth. Beyond our control are factors affecting the business, its approach toward surveys, the resources and processes it uses to respond to surveys, and so on. The latter has been the subject of this chapter; the former—designing and conducting business surveys—is the subject of this book.

Let us review the right side of the model in Figure 2.1—the business side—and its association with the business context within which all of this takes place, as described in Section 2.1. This high-level structure identifies three layers of factors that affect the response process—the business environment, the business itself, and the business respondent—each providing a context for the successive layers (Willimack et al. 2002; Bavdaz 2010a; Snijkers 2008; Snijkers and Giesen 2010).

The external environment contains factors outside the business that impact its operations and decisions. Notice that the survey organization resides in the business's external environment. The business layer includes attributes of the business itself, such as its size, industry, organizational structure, and management practices. The individual level encompasses personal and professional characteristics of the respondent, who is an employee acting on behalf of the business to answer survey questions.

Next, prominent in this high-level framework are factors associated with authority, motivation, and capacity. Think of these as themes that permeate each level. The question "Do I have to do this?" provides the theme for authority factors at the environmental, business, and personal levels. Likewise, the question "Do I want to do this?" sets a theme for motivational factors at each level. "Am I able to do this?" reflects the theme for capacity factors associated with each level.

We conceptualize that business decisionmaking associated with the survey response process is a function of authority, motivation, and capacity factors operating within each of the three levels: the external environment, the business, and the individual.

First, authority, motive, and capacity at the environmental level can be regarded as regulatory requirements, corporate citizenship and public goodwill, and economic conditions. These are imposed on the business from the outside, forming the obligations that the business must fulfill and framing the circumstances within which the business must operate.

Although affected by external factors, authority, motive, and capacity at the business level are determined internally. Authority is the bailiwick of decisionmakers, who determine the importance of various activities, such as survey response, along with allocating resources and prioritizing these activities. Recall that business decisionmakers are motivated by the goals of producing goods and services and remaining open and viable (making a profit). Their capacity, or ability, to undertake activities associated with survey response is affected by many factors, such as the availability of resources and data; the latter is affected by regulatory requirements and management needs, and also is distributed across functional areas and hierarchies. So business decisionmakers use their authority to weigh the tradeoffs between actions that support the business's goals relative to their capacity to carry out those actions.

At the respondent level, authority, motive, and capacity must become manifest in both cognitive and social behaviors. In Section 2.1.2 we noted, the social dimensions associated with a person's authority, motive, and capacity to complete a task, such as survey response, on behalf of the business they work for. Cognitive processes connect survey response to organizational activities that often rely on social behaviors to be accomplished.

2.2.1 The Decision to Participate

The decision to participate in a survey rests with a person representing the organization. Tomaskovic-Devey et al. (1994) conceptualized survey response propensity as a function of the authority, motive, and capacity of the designated organizational respondent, where the organizational context underpins this person's authority, capacity, and motive to respond:

- *Authority* refers to the degree to which the respondent, as a representative of the organization, is permitted to make decisions on behalf of the organization. Organizational processes endow, as well as limit, one's authority.
- *Motive* covers reasons for disclosing organizational information from both individual personal and organizational perspectives. An organization's relationship with its external environment gives rise to organizational processes that influence, if not dictate, the need for exchanging information. Likewise, organizational processes also affect the degree to which individuals identify with the organization's goals, affecting their motivation to act on behalf of the organization.
- *Capacity* encompasses the knowledge, capability, and competence of the organizational respondent with respect to the structure and sources of data pertinent to survey response. Organizational processes determine the division of labor and associated distribution of knowledge to effectively support organizational goals. As a result, the capacity of the organizational respondent impacts their ability to respond to a given survey request.

Section 2.1 demonstrated that businesses are motivated by their goals to produce goods and services and to remain open, viable, and profitable. We also showed that businesses base their decisions on cost/benefit analysis. This is true as well for decisions regarding whether to participate in a survey. The business person with the authority to make this decision weighs the cost of responding against the goals of remaining viable and profitable (Willimack et al. 2002). Costs of time and resources devoted to survey response are very tangible, as are the costs of fines and penalties that may be associated with noncompliance in the case of mandatory surveys, along with the intangible cost of a poor public image resulting from failure to be a law-abiding corporate citizen.

Both the costs and benefits of survey participation are affected by attributes of the external environment within which the business operates, internal attributes of the business itself, and attributes of the selected respondent's role within the organization—none of which can be controlled by the survey researcher—along with survey design features that are under the control of the survey organization.

2.2.1.1 External Environment Factors

Authority A business's external environment includes economic conditions, the survey-taking climate, and legal or regulatory requirements imposed on the business. Authority is imposed on businesses from the outside world in the form of legal and regulatory requirements. Not only must businesses report to taxation agencies; there are often reporting requirements associated with business impact on the earth's environment, transportation, employee safety and health, imports and exports, and foreign ownership and affiliation. Additionally, in the United States, publicly traded companies must report regularly to the US Securities and Exchange Commission. All these obligations compete with and take priority over participating in a survey, impacting the business's capacity and motivation to respond as well.

Motivation Motivational factors associated with the external environment may also affect organizational decisions about survey response. These include the survey-taking and political climates, general norms about surveys, and public goodwill. We have already explained that businesses desire a positive public image in order to engender goodwill and loyalty among current and potential customers. General norms consist of norms and attitudes about survey participation espoused by trade and business organizations, along with those of personal and professional colleagues and peers, which may influence businesses with regard to survey response.

The term *survey-taking climate* refers to general public awareness and opinions about survey organizations and official statistics. For example, in the United States, public awareness and knowledge of the Census Bureau and the role of official statistics tend to be more favorable during the population census conducted every 10 years. The political climate may also affect the survey-taking climate through public discussions among politicians about the necessity and usefulness of official statistics on one side and response burden on the other. The relative balance may affect a business's decision to participate in a survey (Snijkers 2008).

Capacity Not only does compliance with legal and regulatory requirements consume staff resources and work time; economic conditions may affect a business's capacity to respond as well. Studies have shown that during times of economic stress, whether the economy is shrinking or growing, response to business surveys has declined. Fisher et al. (2003) of the US Bureau of Labor Statistics showed that, because of poor economic conditions, businesses had to reduce staff, leaving "remaining staff with less time for daily tasks and, consequently, with less time to provide data for government surveys." Davis and Pihama (2009) not only corroborated this result but also found that economic growth led to reduced business response propensities as businesses dealt with growing markets and expanding customer bases. Willimack (2007) demonstrated similar results during periods of economic instability, when changes in labor lagged behind changes in the US GDP during both economic recessions and expansions.

2.2.1.2 Business/Management Factors

Internal attributes of the business affecting the survey participation decision include the availability of data, which, in turn, is associated with the industrial characteristics of the business, its organizational structure and complexity, and management requirements. Other internal factors include the degree to which company activities rely on information exchanged with the outside world, company policy and priorities, and the availability of resources to respond to data requests. Moreover, respondent selection is under the control of the business; and respondents' abilities to meet external data requests, including surveys, depends on their authority to release data, their access and ability to obtain the requested data, and their motivation to do so.

Authority Businesses' internal authority may be exercised through formal policies on survey participation (Willimack et al. 2002; Snijkers 2008; Snijkers et al. 2007a; Bavdaz 2010a). For example, they may have policies to comply with official government surveys, or only surveys that are mandatory, but to decline other survey requests, such as voluntary government surveys or surveys from other types of organizations. Willimack et al. (2002, p. 223) report, however, that "informal policies against reporting on voluntary surveys were primarily driven by burden and resource issues."

In addition, internal authority will be reflected in a business's priorities with regard to its actions and activities. Pertinent here are the relative priorities placed on different types and purposes for reporting (Willimack et al. 2002). We have already discussed how legal and regulatory reporting receives higher priority than statistical surveys, as noncompliance with these requirements may jeopardize businesses' viability. Moreover, internal reports for management purposes also receive priority, as this information is used for managerial decisionmaking about business operations. The provision of information to stockholders in publicly held companies is also a priority. If resources are not sufficient to fulfill reporting requests with lower priority, such as surveys, then the business may not comply with survey requests, even if the survey is mandatory.

Motivation We have seen that the goals of businesses are to produce goods and services and to remain open, viable, and profitable. As a result, the actions of businesses are guided by cost/benefit analysis, based on a rational-economic perspective (Homans 1961; Blau 1964; Weisberg 2005). Thus, motivation for a positive survey participation decision relies on business decisionmakers considering the benefits of response to outweigh its costs. Recall from Section 2.1, however, that businesses consider survey response to be a nonproductive cost and that benefits of participation tend to be indirect and often unrecognized.

From the perspective of the business, costs include the time and resources required to complete a survey questionnaire, as well as activities canceled or postponed in order to devote time and resources to survey response tasks. On the other hand, for mandatory surveys, a negative response decision may result in tangible costs in the form of fees and penalties.

Now consider the benefits side of the cost/benefit model for economic decisionmaking. Since the benefits received by businesses from survey response tend to be indirect and intangible, it is easy to see that the tangible costs of reporting outweigh the perceived benefits. Thus, in the realm of economic theory, where benefits must be at least equal to costs, the decision to not respond to a survey is obvious.

As a result, survey researchers must rely on motivators outside the economic realm to tip the balance toward a positive decision to respond. Enter social exchange theory, which argues that the exchange need not be equal in monetary value (Dillman 1978; Dillman et al. 2009). Instead, *expected benefits* refer to questions such as

"What is in it for me?" For example, the exchange concept suggests that surveys must have value and be relevant to businesses in order to interest them (HMRC 2010).

An essential part of social exchange theory is *trust* (Dillman 1978), which is reflected in questions such as

"Can this organization be trusted?"

"Will the organization keep my data secure and ensure these data are not disclosed?"

"Will I hear anything about the results of the survey?"

Businesses must trust in the survey organization; they must trust that expected rewards will accrue and that their investment in response activities will have value (Willimack et al. 2002). Trust is associated with familiarity with the survey organization. Studies have shown that survey response is more likely among those with greater knowledge of the survey organization and the uses of the collected data (Bates et al. 2009).

Likewise, past behavior is a very strong predictor of survey participation. Several studies have shown that survey response in previous survey cycles or on other past surveys is a highly significant predictor of subsequent survey response (Davis and Pihama 2009; HMRC 2010; McCarthy et al. 2006). The reason may be motivational, as the business, through its decisionmaker, demonstrates consistent behavior over time. There may also be a capacity dimension to this, as routines and documentation set up to support survey response at one point in time may reduce the cost of replying to subsequent survey requests.

Capacity At the business level, the capacity to respond is a consideration in the survey participation decision. Capacity encompasses factors associated with the business's ability to provide the data requested by the survey, such as the availability of data and resources.

It is necessary to distinguish between organizational capacity and individual personal capacity to provide a survey response. Certainly the two are linked. The personal capacity to respond is an element that contributes to performance of the response tasks, which is discussed in detail in Section 2.2.2. Note, though, that performance of response tasks is explicitly linked with the survey participation decision, as shown in Figure 2.1. Our framework links these individual factors to the participation decision through *expectations* about what survey response entails in terms of the amount and type of work involved. Expectations about the tasks required to complete the survey reflect expected resource costs, which are considered when a representative of the business decides whether to participate in the survey at all.

Organizational factors, then, underpin the capacity to respond, and expectations about the capacity to respond affect the survey participation decision. Such factors in a business include record formation, organizational structure and complexity, availability of resources, established response routines, and respondent selection.

We have already noted that businesses' records are formed around legal and regulatory requirements, along with tracking data that meet management needs for controlling costs and efficiently producing goods and services. Businesses generally do not record data solely for the purpose of answering survey questions that support statistical reporting. Capacity at the business level can be found in the availability, or lack thereof, of data that match the survey request, which, in turn, is a factor in the decision of whether to participate in the survey.

Major factors that affect a business's capacity to respond, and thus associated with the survey participation decision, are the business's organizational structure and complexity (Tomaskovic-Devey et al. 1994; Willimack et al. 2002; Bavdaz 2010a). Complexity is associated with the diversity, heterogeneity, and differentiation of business activities, as we saw in Section 2.1. It is often associated with the size of the organization, both in terms of the number of employees and the number of separate business locations or establishments. There are probably almost as many organizational structures as there are businesses, so any discussion of the effects of size on the response process should not be interpreted as absolute. However, there are some commonalities that aid our understanding of the response process.

The structure of small businesses is quite simple. It is "flat" with perhaps two or three vertical levels, consisting of one or more owners, such as a partnership arrangement, the support or administrative staff, and the employees who produce the business's goods and services (Robbins et al. 2010). The owners typically provide the management function for the business, or they may hire a manager to perform this function on their behalf. The owner(s) or the hired manager, then, is (are) the primary managerial decisionmaker(s) for the business, weighing the various costs and benefits we've been discussing in order to decide whether to answer the survey.

In contrast, the structures of large businesses may vary widely, and are often characterized by multiple production locations. Horizontal complexity occurs in the form of many different departments, each with their own specializations and authority structures, distributing various activities or functions, such that activities must be coordinated in order to produce goods and services, as explained in Section 2.1.3. Vertical complexity consists of authority hierarchies, distributing—but limiting—decisionmaking among several different levels, each answering to a higher authority. As a result, information is distributed throughout the business, and data items are defined and located to serve the specialized needs of different departments or hierarchical levels. The implication for survey-taking is that there may be different decisionmakers, depending on the topic(s) of the survey. Moreover, decisionmakers at some levels may be required to defer to those at higher levels.

So, who is the ultimate decisionmaker with regard to survey participation? Executives at the highest levels, such as chief executive officers or company presidents, are responsible for strategic decisions for their companies. Instead, the appropriate decisionmaker for survey participation should be someone who has knowledge of the requested data and also has sufficient authority to assign the response tasks and to release data (Edwards and Cantor 1991). Particularly for surveys collecting financial data, an appropriate senior manager might be a chief financial officer, although Willimack et al. (2002) found that this capacity often lies with midlevel managers, such as (assistant) controllers.

Medium-sized companies are squeezed from both sides. Their volume of economic activity often renders them subject to more reporting requirements than small businesses. Their accounting and data monitoring systems may be more complex than in small businesses, as well as more variable from business to business. As compared to small businesses, medium-sized companies may have more departments and more role differentiation. Yet, the organization may still be relatively flat, with only a few vertical levels, permitting fewer decisionmakers to exercise wider authority with respect to the survey participation decision. However, they may not have the infrastructure or a staff dedicated to internal and external reporting requirements like those found in large businesses. All in all, their capacity for survey response may be more tenuous.

In our discussion, we've revealed that a business's capacity for survey response includes having an infrastructure within the business that facilitates the exchange of information across different departments, functional units, hierarchal levels, and so forth. Bridging internal and external organizational divides are boundary spanning units, who take on the role of interfacing between organizational units with regard to administrative services and activities pertinent to the entire organization. These and other types of boundary spanning units also interact with the external environment on behalf of the business. Boundary spanners develop routines that facilitate information flows between the organization and the outside would, in both directions, into and out of the business. According to Tomaskovic-Devey et al. (1994), the purpose of boundary spanners is "to limit the impact of the environment on [the] core activities [of the business]," that is, "to shield the technical core" where goods and services are produced, thus contributing to organizational goals by enhancing efficiency and cost-effectiveness. Thus the existence of boundary spanning units, along with a level of comfort exchanging information across internal and external organizational boundaries, both factor into the decision of whether to participate in a survey.

Another factor affecting the business's capacity to respond relates to the fluidity and flux of organizational structures. For example, businesses undergoing mergers or undertaking acquisitions may have limited capacity for survey response as resources are devoted to these actions and various recordkeeping systems are not integrated, identified, or commonly understood. In contrast, the continuity offered by established and repeatable reporting routines increase the capacity to respond to a survey.

Undoubtedly, a major consideration related to the business's capacity to participate in a survey is the availability of resources to conduct survey completion tasks. Here is where we link back to an individual's capacity to respond, as decisionmakers consider the availability of appropriate personnel for answering survey questions. This is a key feature of the response process in business surveys—that it is the business, and not the survey organization, that identifies and selects the respondent.

In making this selection, research has shown that business decisionmakers consider factors such as knowledge of and access to particular data to meet the survey request. Also considered are the employee's availability and priorities associated with other work assignments (Willimack et al. 2002). When data are distributed throughout the company, additional considerations may include an employee's ability to manage intraorganizational projects and to influence others outside their direct authority.

Three time-related factors enter into the decision as well: (1) survey deadlines, (2) when the data are recorded internally and available to the business itself, and (3) expectations about the time involved in completing response tasks. The decision to participate in a survey depends on the degree to which these factors enable survey response.

Assuming that all the factors align for a positive decision by the business to participate in the survey, respondent selection and scheduling provide a segue into factors affecting and actions undertaken to perform the response task, which is the topic of the next section.

2.2.2 Performing Response Tasks

Although a simple single diamond in our visual representation in Figure 2.1, there is a lot going on in "performing response tasks." At the core of the question answering process at the personal level is the cognitive response model put forth by Tourangeau (1984), consisting of the following four steps:

1. *Comprehension*—understanding the meaning of the question
2. *Retrieval*—gathering relevant information, usually from memory
3. *Judgment*—assessing the adequacy of retrieved information relative to the meaning of the question
4. *Communication*—reporting the response to the question, such as, selecting the response category, entering the data into the data collection instrument, or editing the response for desirability.

However, in business surveys, respondents approach the survey question answering process from the perspective of work activities carried out on behalf of their companies. Recall that we conceptualize three layers of factors affecting the survey response process—external environmental factors, business/management factors, and individual/respondent factors—each providing a context for the successive layers, as shown in our model in Figure 2.1. So, when we reach the individual/respondent layer, where response tasks are performed, factors at the underlying levels impact the authority, motivation, and capacity of the employee(s) selected to complete the survey.

Willimack and Nichols (2010) and Bavdaz (2010a) incorporate external environment factors and business/management factors into their models of the question answering process—respectively the hybrid response process model and the multidimensional integral business survey response (MIBSR) model. We will use these models to guide our examination of the effect of these factors on how response tasks are performed and by whom. We shall integrate common and complementary features of both models to identify the following model of the portion of the response process involved in performing response tasks:

1. Record formation and encoding of information in memory
2. Organizing the response tasks
   a. Selection and/or identification of the respondent(s)
   b. Scheduling the work
   c. Setting priorities and motivation
3. Comprehension of the survey request
4. Retrieval of information from memory and/or records
5. Judgment of the adequacy of the retrieved information to meet the perceived intent of the question
6. Communication of the response
7. Review and release of the data to the survey organization

Note that steps 1, 2, and 7 are organizationally driven. They wrap around cognitive steps 3, 4, 5, and 6, providing the organizational context within which performance of the response tasks occur. The organizational context includes social behavioral factors revealed in Section 2.1. Related to this, we will also consider contributions to our understanding of the response process offered by Willimack's (2007) reinterpretation of principles of organizational behavior and Lorenc's (2006, 2007) application of theories of socially distributed cognition (SDC).

If this sounds complicated, it is. We have defined three contextual levels in our response process model in Figure 2.1—the external environment level, the business/management level, and the individual/respondent level. We have said that each level is a layer that impacts what happens at successive levels. Within each level we conceived that factors associated with three pervasive themes—authority, motive, and capacity—impact both the decision to participate in a survey and performance of the response tasks. At the individual respondent level, authority, motive, and capacity are paramount to performing the response tasks, as are three sources of behavioral processes that influence the respondent's actions—organizational, social, and cognitive processes.

The attributes of these three concepts—contextual levels, individual factors, and behavioral influences—are so intertwined in the steps involved in performing response tasks, that disentangling them in our description here is infeasible. Alternatively, we offer Table 2.2 as a summary of the levels, factors, and influences that are most pertinent for each step in performing the response tasks.

2.2.2.1 Organizational Context for Performing Response Tasks

Step 1: Record Formation and Encoding Since business surveys rely heavily on data in business records, the first step, *encoding*, includes *record formation*. For recorded data to be available for survey response, knowledge of those records must

Table 2.2 Attributes of Steps in Performing Response Tasks

Response Task	Contextual Levels	Individual Factors	Behavioral Influences
1. Record formation and encoding	External environment and business/ management	Capacity	Organizational and cognitive
2. Organizing response tasks	Business/management	Authority and capacity	Organizational
a. Respondent selection	Business/management and individual	Authority and capacity	Organizational and social
b. Scheduling	Business/management	Authority	Organizational
c. Priorities and motivation	Business/management and individual	Authority and motivation	Organizational
3. Comprehension	Individual	Capacity	Cognitive
4. Retrieval	Business/management and individual	Authority, motivation, and capacity	Social and cognitive
5. Judgment	Individual	Motivation and capacity	Cognitive
6. Communication	Individual	Motivation and capacity	Cognitive
7. Releasing the data	External environment, business/management and individual	Authority	Organizational

also be encoded in a person's memory. As we have seen, data recorded in business records are determined primarily by legal and regulatory requirements, management needs, and accounting standards, and not statistical purposes. As a result, data requested by survey questions may well differ from data available or known to exist in records.

Moreover, data tracked for different purposes often reside in different information systems that, although automated, may not be linked. For example, information about employees, wages, payroll, and employee benefits will typically be located in a payroll or human resources office, while financial data, such as sales receipts and expenses, used to prepare routine financial reports, such as income statements and balance sheets, will be held in a financial reporting office. Thus information and persons with knowledge about specific data are likely to be distributed across organizational units, particularly in large businesses (Groves et al. 1997; Tomaskovic-Devey et al. 1994).

Step 2: Organizing the Response Tasks The distribution of knowledge across organizational units has consequences for how response tasks are done and by whom. More broadly, since survey response is considered to be work, as we learned in Section 2.1, the associated work activities need to be organized and coordinated. *Organizing the response tasks* includes determining who will do the work and when (Bavdaz 2010a), relative to other work-related priorities and the level of attention directed by managers (Willimack and Nichols 2010). In other words, how response tasks are organized is determined by factors associated with the business and its management.

On receiving a survey request, the decisionmaker must get an idea of what needs to be done, when, and how much work may be involved. We have seen that expectations about the response tasks impact the decision whether to participate. Given a positive participation decision, these expectations also form a basis for organizing the work activities associated with survey response. To formulate these expectations, decisionmakers for the business will try to get an overview of the survey content. They may examine the survey questionnaire and materials in detail. It is more likely, however, that they will merely skim the cover letter, glance at the survey title, or, at best, do a superficial inspection of the questionnaire (Bavdaz 2010a).

After gauging the amount of work that may be involved in responding to the survey, the decisionmaker *selects the respondent(s)* and *schedules the work*. Decisions about who is to be involved in the response task and when it is scheduled are typically made in tandem. The respondent's relationship with records is paramount, in terms of knowledge of and access to them. Moreover, a particular employee or department may be responsible for filling out forms or reporting financial data on a routine basis, regardless of purpose, making them the de facto recipient of new or recurring survey requests (Bavdaz 2010a).

Once again, the size of the business affects the survey response process. In medium-sized to large businesses in particular, distributed knowledge and associated division of labor has consequences for respondent selection. In their cognitive response model for establishment surveys, Edwards and Cantor (1991) consider respondent selection implicit within the encoding step, because the preferred survey respondent is someone with knowledge of the data available in records, along with their underlying meaning. However, other organizational factors impact respondent selection, such as experience with the routine of completing forms, distributed knowledge, division of labor, organizational hierarchies, and the authority versus the capacity of a particular organizational informant to respond to a survey (Tomaskovic-Devey et al. 1994). These factors have consequences for the survey response process:

1. Response to a single survey covering multiple topics may entail gathering information from multiple people or record systems distributed throughout the business, as information is compartmentalized around specific organizational functions, particularly in large businesses. Thus, there may not be one single "most knowledgeable respondent," in the Edwards–Cantor (1991) context. Rather, there may be many "most knowledgeable respondents," each familiar with data associated to specific, but different, survey topics.
2. Institutional knowledge of multiple varied data sources typically resides in those having authority within a business, who may be removed from direct access to those sources. Instead, these authority figures delegate the response task. Although an employee's knowledge of and proximity with the requested data may help to identify possible respondents within the business, the timing of the survey request relative to the availability of not only candidate employees but also the requested data enter into respondent selection decisions. That

is, the availability of data for answering survey questions depends on when data are entered into business records and/or summarized for internal reports. Thus *scheduling the work* necessary to carry out response tasks requires that the availability of personnel and data align in time.

3. A third step in organizing response tasks involves weighing the availability of personnel and data relative to other work needs and importance. This assessment of *priorities* links organizational factors with person-level factors that affect a respondent's *motivation*, or attentiveness to the response task. Not only will priorities affect respondent selection and scheduling; the business dictates the relative importance of activities performed by its employees. The sequencing of priorities impacts the attention and resources devoted to associated activities. Willimack et al. (2002) found that respondents' assignments that receive priority are those that support business functions, while the work required to complete a survey has low priority because, as we saw in Section 2.1, it is an activity that bears a tangible cost to the business but no associated revenue generating production. In addition, an employee's personal motivation is affected by criteria determined by the business that are used to evaluate employees' job performance. Employees may also be driven by professional standards for particular activities, not to mention exhibiting pride in their own work.

Ultimately, as we pointed out in Section 2.1, it is the business, and not the survey organization, that decides who the respondent will be for a given survey. Businesses exercise varying respondent selection strategies (Nichols et al. 1999; O'Brien 2000). Consequently, the respondent's role in the business may result in variation in knowledge and actions pertinent to subsequent steps in the response process (Burns et al., 1993; Burr et al. 2001; Cohen et al. 1999; Gower 1994).

Roles in the Response Process Not only can respondent selection strategies vary; we can also identify several roles in the process of performing the response tasks, each having an impact on how these activities are carried out. These roles, as defined by Bavdaz (2007, pp. 249–255), are

- *Gatekeepers and Boundary Spanners.* Employees in these roles act as an interface between the business and the outside world or between different actors within the business, and thus their actions may serve to aid or hinder the response process (Bavdaz 2010a). Although their purpose is to protect various levels and types of business activities from distractions and disturbances, their approaches differ. Gatekeepers tend to consist of roles and positions that practice discretion in screening, perhaps even blocking, access to those carrying out business activities (Fisher et al. 2003). This includes positions like receptionists, secretaries, or personal and administrative assistants. Boundary spanners, as described in Section 2.2.1, facilitate interaction both directions, inward and outward, between the business and its external environment, and/or between different units within the business. Some common business survey respondents are themselves boundary spanners (Tuttle 2009; Willimack 2007). For example, employees in human resources or payroll departments, who answer survey questions about employment, payroll, and benefits, interact with and provide services to units throughout the business while also preparing official reports on payroll taxes submitted to taxing authorities. Likewise, employees in financial reporting departments, who answer survey questions about revenues and expenses, gather data from throughout the business in order to prepare routine internal financial reports, as well as periodic reports for external stockholders, for example.

- *Authorities.* Someone who is in an authority position in the business is enabled to make decisions on behalf of the business. In Section 2.1.2 we noted that levels of authority vary with respect to the degree of impact that decisions may have. We have already examined the traits of authority with regard to the decision of whether to participate in a survey. Persons with authority are also in a position to assign or delegate the response tasks to appropriate employees—that is, select the respondent—as well as schedule and facilitate associated work activities, set priorities, provide direction, and permit release of survey responses back to the survey organization.
- *Respondents and Data Providers.* Following Bavdaz (2007, p. 249), we shall define a *respondent* as "someone who provides data with the particular purpose of answering a survey question." Bavdaz further distinguishes a respondent as a person who has been exposed to the specific stimulus conveyed by the survey instrument, specifically the survey question and, if present, associated instructions and definitions. In contrast, a *data provider* is someone who provides data in reply to a request from another participant in the response process, who may be a respondent or someone having another role. The distinction between a respondent and a data provider lies in the degree of exposure to the actual survey question(s) and supplemental material directly associated with the survey question(s). A respondent is someone who provides information in response to such direct exposure. This need not include direct contact with the data collection instrument from the survey organization, nor must every respondent be the person who enters data directly into the survey instrument. Exposure may result from the survey question and associated material being conveyed intact from one business survey participant to another, making them both respondents. Data providers, on the other hand, react to requests from other business survey participants, presumably another respondent or response coordinator, without the benefit of this direct exposure to the survey question and associated material. Rather, the survey question is interpreted and conveyed to the data provider through some other means of office communication.
- *Response Coordinators.* The role of a *response coordinator* is to gather data from various sources, including both records and persons, located throughout the business in order to answer survey questions. Whether the other participants in the response process may be considered respondents or data providers, as defined above, depends on the mechanisms used by response coordinators to

request pertinent information from others. Response coordinators are often also respondents for some survey questions. In addition, they are typically the ones to enter answers to survey questions into the data collection instrument, and obtain any necessary approvals for releasing the data back to the survey organization. The role of response coordinator is particularly important and necessary when surveys request data on numerous different topics, such that the desired data reside in multiple locations or systems throughout the company, the consequence of divisions of labor and distributed data to support differentiated business functions. (Tuttle 2009; Keller et al. 2011). It is not unusual, then, for response coordinators to be in boundary spanning roles in the company, either formally or informally. However, research has found that it is common for response coordinators to consider it their job to answer survey questionnaires, exercising autonomy as we saw in Section 2.2.2. Thus, they are often reluctant to seek help with answering questions outside their domain, even when they know where the best data are located and whom to contact. Instead, they will try to answer the questions themselves rather than "bother" others, taking them away from their "real jobs" (Tuttle 2008, 2009); that is, as boundary spanners, they see it as their job to protect the technical core. Note that the data collection activities conducted by response coordinators, along with the successive selection of respondents and data providers, are controlled by the business, and not the survey organization, as is also reinterpretation and conveyance of the meaning of survey questions and associated materials through internal business processes to gather data to support survey response.

- *Contact Person.* A contact person is defined in relation to the survey organization. This is the person, or possibly a position title, identified by the survey organization to receive communications with regard to the survey data collection, the survey organization, or products resulting from a data collection in which the business participated. The contact person may be someone in any of the roles described above, with the exception of the data provider role. The survey organization expects the contact person to identify appropriate others in the business to whom to pass along any survey-related communication, and then to do so.

Employees involved in the survey response process may fulfill only one of these roles, or may act in the capacity of several or even all of them. This, too, is under the control of the business and typically is not known by the survey organization the first time a survey is conducted.

2.2.2.2 Organizational Context and Social Behavior

Each person involved in the response process undertakes cognitive processes and relies on social behavioral norms affected by the business context. Organizational, individual, and social behavioral factors collide within the roles of various participants in the response tasks, which move beyond the strictly cognitive realm into the social realm. Social norms direct patterns of behavior like those described in Section 2.1.2 to accomplish response tasks across multiple different participants with different roles in the response process. In addition, the social behavioral aspects of how work gets done come into play, as discussed in Section 2.1.3, where different personnel in different roles rely on communication and self-direction based on cooperation and trust among their counterparts, in order to produce adequate answers to survey questions.

Willimack (2007) draws upon organizational theories of sent and received roles (Katz and Kahn 1978; Porter et al. 1975) to guide further understanding of the social behaviors of various business participants in the actions they take toward completing business surveys. Sent and received roles involve expectations and perceptions held by each person about the other participant. These are associated with each person's position in the business hierarchy or as peers in the same or different organizational units. Through a social and communicative act, the *role sender* transmits information about the task expected of the recipient, called the *focal person*, and attempts to influence their behavior to ensure compliance. Role senders' expectations are based on their past experiences with and/or perceptions of the focal person's behavior. Focal persons receive the *sent role* through a filter of their own perceptions and personal characteristics, transforming the sent role into the *received role* from the perspective of the focal person, influencing their motivation for role performance. In addition, factors associated with the organization provide a context for this social episode involving sent and received roles.

Figure 2.2[1] illustrates an idealized interaction, called a *role episode*, between a role sender and focal person involved in survey response. Note that, as is standard practice in the organizational literature, role episodes are defined and should be interpreted from the perspective of the focal person.

For this example, assume that data are needed from multiple organizational sources to complete the survey. The role sender is a *response coordinator* (RC) and the focal person is a *data provider* (DP). The central act of the role episode involves the RC requesting data (the sent role) from a focal person, the DP, who has access to the desired data. The context for the role episode is illustrated in the oval balloons in Figure 2.2. Organizational factors bolster the role sender's request since attributes of the sender's office convey in the role sending to the DP. Attributes of the DP's personality, and interpersonal experiences with the DP, affect the role sender's expectations for and manner of the role sending behavior. Moreover, feedback loops indicate how the DP's behavior in this role affect perceptions held personally and by others. For example, fulfilling the data request may reinforce the DP's personal view as an expert, as being responsible and loyal to the organization, as well as being trustworthy, cooperative, and responsive in interactions with others.

Diagrams like the one in Figure 2.2 can be drawn for every person and to illustrate every relationship in the activities undertaken to perform survey response tasks. When linked, a path diagram will emerge demonstrating the social networks of interpersonal interactions undertaken to complete a survey response. This is the formulation of the survey response process according to principles of socially

[1] The diagram presented in Figure 2.2 is based directly on an analogous diagram found in Katz and Kahn (1978) used to illustrate their "theoretical model of factors involved in the taking of organizational roles."

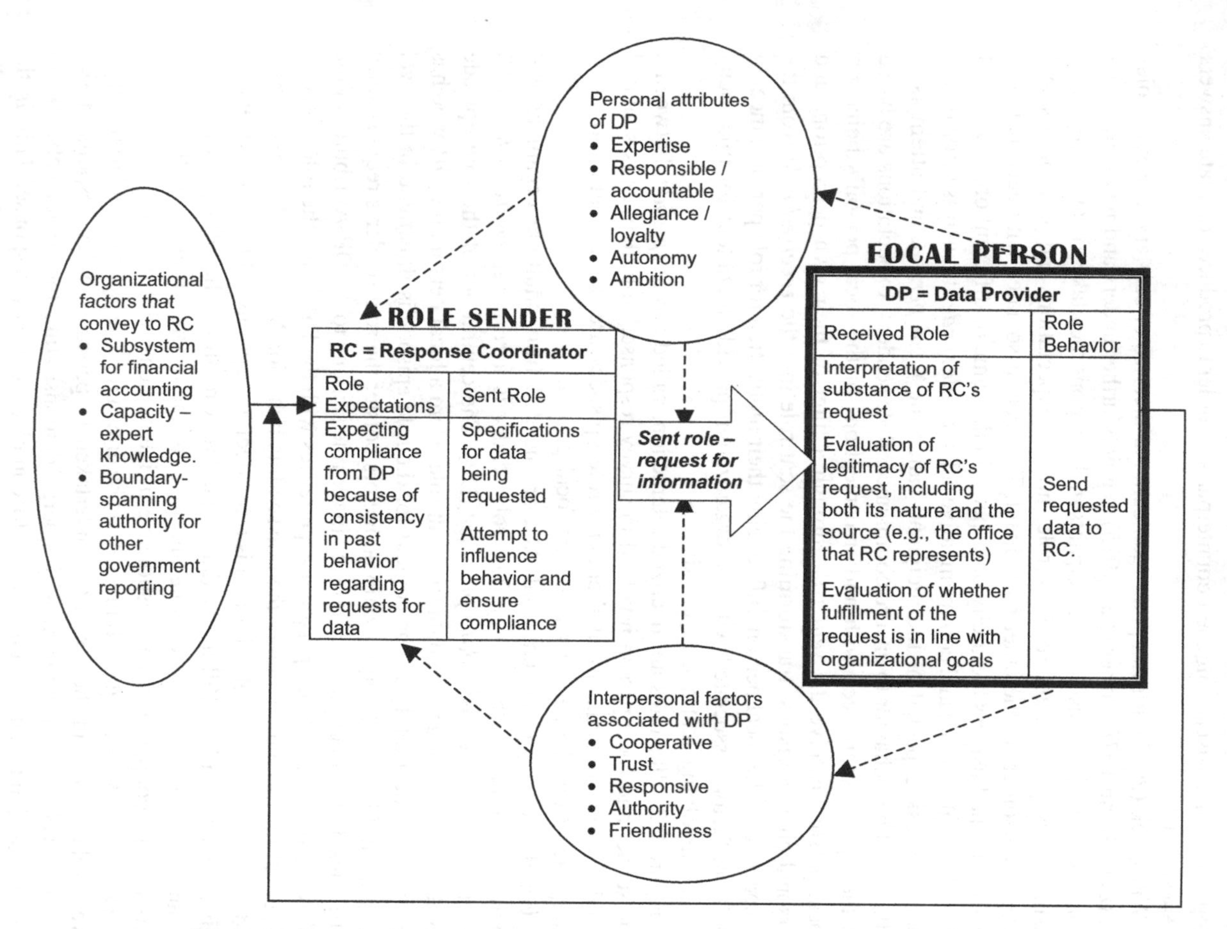

Figure 2.2 Example of role taking in the survey response process, under ideal circumstances.

distributed cognition (SDC) (Hutchins 1995) as applied by Lorenc (2006, 2007) to describe and study the response process in organizational surveys.

In Lorenc's application, SDC involves "moving the border of a cognitive system outside of an individual and letting it instead encompass a small socio-technical system." In other words, cognition is distributed between people, between people and the tools that they use, and between people over time (i.e., historically). SDC seems a particularly apt framework for considering the response process in business surveys, which, as we have seen, often requires multiple people and multiple sources of data, along with use of records or historical documentation, to reply to a survey request.

The key to SDC is in redefining the unit of analysis. It is no longer only the individual person, but the whole system, that can be regarded as "cognitive," with the social behaviors associated with accomplishing work connecting the components of the system. A SDC network begins with the person(s), tools, and systems of records involved in the creation of a particular piece of data, where data definitions, purposes, and uses are known and implemented, and the creator is the "most knowledgeable respondent" according to the Edwards–Cantor (1991) definition we saw earlier. This is the point of perfect knowledge with respect to a specific data item.

Then, as that data item moves through systems and/or is used by persons other than its creator, its meaning is subjected to variations in the cognitive processes and social interactions associated with work. Moreover, data items may be combined or manipulated for different purposes in different systems, records, or reports. As multiple people may be involved in the response process, or the response tasks could be assigned to various people in the company, accurate survey response relies on shared meaning across different possible respondents. However, the same piece of information resides differently in the minds or records of different possible respondents and may be accessed differently. The appropriateness of a data item in answering a survey question depends on the meaning and use attributed by business participants actually performing response tasks.

Moreover, since SDC includes interactions among people, between people and records, or their reliance on past behaviors, the SDC process takes place in the "open, observable, social world," rather than "in the head" (Lorenc 2006). Combining this with the organizational behavioral context embodied in role episodes described earlier, we can envision a network or path diagram representing business survey response processes, both the ideal and the actual. This approach leads us to a network analysis approach for examining and understanding internal business processes/activities to undertake and support survey response (Tuttle and Willimack 2012).

2.2.2.3 Organizational Context and the Cognitive Response Process

The steps determined organizationally—encoding/record formation and organizing response tasks (respondent selection, scheduling, and prioritizing)—provide a context within which the four personal cognitive steps occur: *comprehension, retrieval, judgment, and communication* (Tourangeau 1984). The designated respondent must comprehend the intent of the question; retrieve relevant data from memory, knowledge, and records; judge the adequacy of the retrieved information in meeting the question's intent and make any necessary adjustment; and finally, record the answer on the survey questionnaire.

The Bavdaz (2010a) MIBSR model reveals how superficial the cognitive steps are when considered in isolation from business factors associated with the response tasks that individuals must perform in order to answer survey questions. Business survey participants "take part in organizational processes while going through their own cognitive processes" to carry out the survey response tasks. Additionally, in business surveys, the cognitive steps apply not only to respondent interactions with survey questions but also to their interactions with records and other data sources within the business.

At the level of the individual, the MIBSR model underpins the cognitive response process with "three inherently linked types of knowledge" pertinent for adequate survey response. Bavdaz (2010a) calls these "knowledge of the business reality, knowledge of record formation, and knowledge of business records." The latter two have been considered in earlier steps in the response process—respectively, encoding and record formation (step 1) and respondent selection (step 2). The first one of the three, *knowledge of the business reality*, refers to an intrinsic, yet broad, understanding by the selected business respondent of the activities undertaken by the business, the goods and services produced, and the organizational structure and administrative processes supporting the distribution of these activities, goods and services, and associated information. "Knowledge of business reality thus presupposes acquaintance with every aspect of the business: who does what, what activities the business is involved in and how they are carried out, why the business situation is as it is, how it evolved through time, and so on." (Bavdaz 2010a).

This knowledge becomes critical to undertaking the cognitive steps associated with survey response in the business context, especially since many surveys conducted by NSIs tend to be omnibus in nature, asking questions about multiple varied topics. Knowledge of the business reality is necessary for the designated respondent to correctly identify the questions that apply to the business, as well as how, where, and with whom the data reside, in order to adequately answer survey questions. This understanding would appear to be critical for carrying out the business survey response process in a manner that arrives at accurate response, since survey organizations tend to design surveys with the assumption that the survey will find its way to an appropriate respondent, or at least to someone in the business with this full knowledge of the business reality. Such an expectation seems incompatible with the fact that respondent selection is under the control of the business and not the survey organization.

Step 3: Comprehension of the Survey Request Comprehension encompasses one's understanding of the presentation of the survey request and the interpretation of the words, sentence structure, and images presented as a stimulus. Business survey participants comprehend the survey request at varying degrees, depending on their roles in the response process and their goals. A contact person or someone in an authority position may skim the survey materials and questionnaires to obtain a cursory understanding of the nature of the survey request, the topics covered, and requirements in order to determine what needs to be done and how to do it. Response coordinators and respondents would be expected to undertake a more deliberate examination to apply meaning to the survey questionnaires and questions as they associate with the activities of the business. Data providers, not seeing the specific survey stimulus itself, are already subject to one level of interpretation provided by response coordinators or respondents, and will apply meaning to these requests associated with the requestor's role or position in the company.

Business surveys collect data with specific technical definitions, and questions are phrased using technical terminology typically not common to the vernacular of one's personal daily life. Thus, business survey questions presume respondents' familiarity with technical or professional terminology associated with the concepts being measured by the survey instrument. This, however, is not always the case. As we have seen, respondent selection is under the control of the business, and selected respondents may exhibit varying degrees of knowledge and competence with respect to survey topics and the language used in survey questions.

Moreover, as in everyday vernacular, the same words often have different meanings depending on the context. For example, the intent of "employment" questions on a labor force survey is very different from that of employment questions on a business survey. When the terminology used in survey questions is familiar, respondents assume that the meaning is the same as the meaning they traditionally attribute to the language, regardless of whether it actually is the same (Shackelford and Helig 2011). When respondents are unfamiliar with the terminology, they may assign meaning from their own frame of reference, such as the "native" language of their environment.

Comprehension problems are further compounded in surveys requesting data on multiple topics, particularly for respondents in companies with differentiated functions and distributed information. Possessing knowledge of the business reality may not be sufficient in helping the respondent interpret unfamiliar economic or accounting concepts and terminology often found in survey questions, or preventing respondents from confounding these with familiar concepts or with those using the same or similar terminology.

Step 4: Retrieval of Information from Memory and/or Records Willimack and Nichols (2010) note that the retrieval step in the business survey response process has traditionally focused on data availability as the result of record formation. However, record formation is only one dimension of data availability. Others include the selected respondent's access to data sources and ability to retrieve data from these sources, either records or coworkers with firsthand knowledge of them. Thus, availability may be conceived to be a function of record formation, respondents' access to records, and ability to retrieve data from them.

Retrieval, in fact, incorporates three components:

- The cognitive act of retrieving from memory knowledge of data sources, company records, information systems, company structure, or appropriate personnel

- Access to appropriate records or knowledgeable personnel
- The physical act of retrieving data from records and/or information systems, which includes extracting information from computer and paper (hardcopy) files; consulting multiple sources, both people and records, because of distributed knowledge; and compiling information.

As a result, the function defining data availability suggests particular inferences and conclusions. For example, if data exist in records and the selected respondent knows this but does not have access, then the data are not available. If data exist in records and the selected respondent does not know this, then access is irrelevant, and the data are not available. The same is true if the respondent lacks the necessary skills to extract, obtain, and compile data from data sources. Moreover, access is a necessary but not sufficient condition for retrieval, and thus data availability.

Consequently, data availability requires knowledge of and access to records, data sources, and/or knowledgeable personnel, along with relevant retrieval skills. These three components—source knowledge, records access, and retrieval skills—constitute a respondent's *capacity* to respond. Because knowledge, access, and skill sets vary with the respondent, data availability varies with the respondent.

Retrieval relies not only on the respondent's personal knowledge of and access to business records. When survey content covers a variety of topics that may not be directly under the auspices of the survey recipient, retrieval requires the respondent to be cognizant of "the business reality." Thus, the respondent must be familiar with the breadth of activities, goods and services, and associated information beyond the respondent's own domain, in order to identify and involve other sources and business personnel in the response process. If a respondent lacks this capacity, then *motivation* must be present to stimulate efforts to locate knowledgeable data providers elsewhere in the business. Such motivation may be found in the alignment of organizational and personal goals, through loyalty and allegiance, as described in Section 2.1.2.

Response coordinators draw upon their "knowledge of the business reality" and motivation to identify, contact, and request assistance from employees elsewhere in the business with knowledge of and access to the specific data needed to answer survey questions—questions that the coordinator is unable to answer. Whether these other employees are "respondents" or "data providers" depends on the mechanisms used by the response coordinator to request information from these others, according to our definitions given earlier in this section.

When the retrieval process throws respondents into a coordinator role, the merely cognitive expands into the social realm. Response coordinators must not only utilize their knowledge of the business reality to identify appropriate respondents and/or data providers, but also draw upon the social dimensions of performing work described in Section 2.1.2, such as having the authority to make such requests of other employees, the ability to influence other employees to meet the coordinators' needs, and sufficient communication skills to articulate the requests. The social exchange between a response coordinator and a downstream respondent or data provider may be considered a role episode, like that discussed earlier in this section, in which the response coordinator is a role sender and the respondent or data provider is the focal person, that is, the recipient of the communication. It is not difficult to see how the response coordinator's position in the organization and other organizational responsibilities are integrated with the request conveyed to the focal person.

Step 5: Judgment of the Adequacy of Retrieved Information to Meet the Perceived Intent of the Question Earlier we showed that data recorded for the business's purposes may not align with the concepts and definitions for data elements requested in surveys. Instead, there are degrees of data availability. Respondents will attempt to devise answers to survey questions based on available data, despite the mismatch (Willimack and Nichols 2010). This is the nature of the judgment step in the cognitive response process. Respondents will consider the makeup of the information retrieved from records or memory relative to what they think the question is asking for, and make adjustments as they deem necessary or are able. The link between data availability and response, then, consists of judgment strategies.

Bavdaz (2010) identifies various strategies and actions taken by business respondents to use available data, including manipulation, to formulate responses. Depending on how well the available data align with the survey request and degree to which they may need to be manipulated to meet the survey request, data may be defined as accessible, generable, estimable, inconceivable, or nonexistent, as follows (Bavdaz 2010, p. 86):

(a) A datum is accessible—the required answer may be readily available.
(b) A datum is generable—the required answer is not readily available to any person; the available data represent a basis for generating the required answer through manipulation.
(c) A datum is estimable—the required answer is not readily available to any person; the available data represent an approximation of the required answer or a basis for estimating the required answer through manipulation.
(d) A datum in inconceivable—no available data lead to the required answer or its approximation; some bases for generating or estimating the required answer exist but require an unimaginable effort to produce it.
(e) A datum is nonexistent—there are no bases for estimating the required answer.

Items (b)–(e) constitute judgment strategies used by respondents to formulate answers to survey questions. The link between these judgment strategies and likely response outcomes is illustrated in Figure 2.3. The data quality implications are considered in Chapter 3. In addition, this taxonomy will be discussed further in Chapter 7 as a tool for evaluating the effectiveness of survey questionnaires.

Other judgment strategies are common, such as maintaining continuity in the response strategy across recurring surveys, regardless of any reporting errors discovered in between reporting periods (Willimack and Nichols 2010; Bavdaz 2010a). Business respondents are concerned that survey organizations might

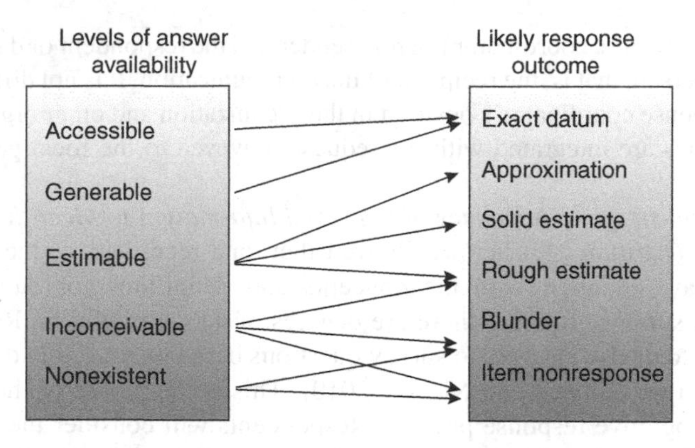

Figure 2.3 Levels of answer availability and likely response outcomes. *Source:* Bavdaz (2010a, p. 86).

misconstrue any discrepancies, while also wanting to ensure a consistent image of their companies.

Bavdaz (2010a) identifies other strategies that guide respondents' judgment. Respondents will also maintain consistent response strategies within a survey questionnaire, using the same algorithms to adjust data despite definitional differences. They may also choose to exclude one-time aberrations or temporary conditions that should modify reported information, because they see these events as not being representative of their business operations. Likewise, they may ignore activities that are marginal to the business, considering the effect on reported data to be inconsequential.

Step 6: Communication of the Response The cognitive portion of the response process is completed with respondents entering their answers to survey questions onto the questionnaire or into a data collection instrument. This entails mapping their answers onto prespecified response categories or formats. This seemingly straightforward act may require additional revision to their responses in order to meet the requirements of the data collection instrument.

By this point in the response process, respondents expect to be able to record their answers with ease. Business respondents might expect the instrument to have a familiar look and feel, similar to other forms they have to fill out, or other automated systems they use routinely. For example, research has shown that typical respondents in the business setting are quite facile with spreadsheets and associated tools (Willimack and Nichols 2010; Anderson and Morrison 2005).

Professional characteristics of business respondents may also contribute to the manner in which data are reported. Many business respondents are accountants, who value precision, balance, alignment, and consistency when recording figures. They may be uncomfortable with rounding figures to whole numbers, reporting percentages, or omitting components they believe to be critical in presenting a full and accurate picture of their companies' activities, failing, in general, to understand the consequences, or lack thereof, associated with the ultimate use of their data to generate statistical summaries (Willimack and Nichols 2010; Bavdaz 2010a).

2.2.3 Release of the Questionnaire

Step 7: Review and Release of Data to the Survey Organization The final step in the model, *reviewing and releasing the data*, returns the response process to the organizational level, since survey response is among the ways a business interacts with its external environment, where survey organizations reside (Tomaskovic-Devey et al. 1994; Willimack et al. 2002). Thus, returning a completed questionnaire back to a survey organization is, to some degree, treated like a public release of information about the company. In this context, factors related to authority, motivation, and capacity come into play again.

Authority and consequences imposed by legal and regulatory requirements that mandate response, particularly for surveys that support official statistics, may help to ensure release of the questionnaire. We have discussed the desire of businesses to appear to be law-abiding and to contribute to the wellbeing of society. Moreover, consequences in the form of monetary penalties for not reporting or for inaccurate reporting incur additional costs to the company while still imposing the requirement to respond.

Nevertheless, factors and behaviors associated with *motivation* and *capacity* have some bearing on the release of the questionnaire. Reported figures may be reviewed and verified against other externally available figures to ensure a consistent picture of the business for the outside world. Since data requested in business surveys are often proprietary in nature, concerns for confidentiality and security come into play as well. Despite legal assurances of confidentiality for reported data, publicly held businesses may delay their response until information has been released to management and shareholders (Willimack and Nichols 2010).

In practice, organizational actions associated with releasing the questionnaire may be passive. While many official survey forms request signatures of company personnel certifying the correctness of the reported data and granting official approval authorizing their release to the survey organization, research has shown this act is frequently treated as routine or inconsequential (Bavdaz 2010a; Giesen 2007). Business personnel may sign survey forms as a formality without further verification or review, motivated by an intrinsic trust that the respondent had the capacity to do the work correctly while acting in the best interests of the company. Bavdaz (2010a) found that often those assigned the response task were simply permitted to sign and release completed questionnaires, suggesting that this authority was conveyed with the assignment of the response tasks to the designated employee.

2.2.4 Role of the Survey Organization in the Business Survey Response Process: Implications for Survey Design

Let us return to our model of the business survey response process presented in Figure 2.1, and consider the left side of the model, which regards factors that are seemingly under the control of the researcher. The diagram illustrates that features associated with the survey, the survey design, and the survey organization—decisions made by the survey organization about how the survey is designed and carried out—impact a business's participation decision and the actions/activities necessary to provide a response. Other models of the business survey response process tend to treat the survey itself as both an object acted on and a stimulus for the process. Our model in Figure 2.1, as does Bavdaz (2010a), treats the survey in a manner that makes it explicit that elements of the survey design impact and influence the response process. This is illustrated in the figure by the arrows pointing from the survey organization to the business.

We begin by distinguishing among features associated with the survey, the survey design, and the survey organization. First, survey features are specified by the stakeholders, such as the objectives of the survey, its topic, the target population, the budget, applicable statutes, the timetable, and expectations about data quality. Components of the survey design consist of the sampling plan and resulting sample, the survey questions, the data collection instrument and procedures, communication and nonresponse reduction strategies, data capture and cleaning, estimation, nonresponse adjustments and imputation methods, data security and confidentiality, and dissemination. Features of the survey organization affecting survey design include procedures, policies, standards, guidelines, systems, and ethics.

This is the production environment within which the design and conduct of business surveys must be planned, which is the topic of Chapter 4. Survey quality must also be ensured while taking into consideration the business context as well. This is addressed in Chapter 3, and monitoring quality during survey production is the focus of Chapter 10.

Let us now consider how authority, motivation, and capacity implicated by features of the survey, the survey design, and the survey organization have an impact on a business's decision to participate in a survey and their performance of the response tasks.

Authority Authority exercised through legal mandate is a survey feature that has a major effect on businesses' decisions to participate in surveys. Willimack et al. (2002) report that mandatory surveys achieve higher response rates than do voluntary ones. An inquiry among respondents for the 2005 Dutch Annual Structural Business Survey showed that about 90% of the businesses reported that the major reason for participation was the legal obligation (Snijkers et al. 2007b).

Lacking mandatory authority, survey organizations may invoke a sense of authority among respondents through their reputation and legitimacy. Even voluntary surveys conducted by government organizations benefit from their authority (Willimack et al. 2002). If business respondents consider a survey and its topic as a legitimate undertaking of a particular organization, then authority is conveyed.

Authority, legal mandates, legitimacy, and reputation must be evident to business respondents. Survey design features must successfully communicate these authority factors in order to be effective. Survey communication strategies are the subject of Chapter 9.

Response tasks may also be affected by authority associated with the survey and the survey organization, by way of respondents' level of attention. Thus authority impacts motivation.

Motivation Along with or instead of authority factors providing motivation for the survey response process, motivational features of the survey, the survey design, and the survey organization come into play (Paxson et al. 1995). These motivational factors include a variety of features, such as the topic of the survey, the design of the questionnaire and the survey communication, and the corporate image of the survey organization.

The topic is a survey feature that may be relevant to business survey participants, motivating their participation and engaging them in performing response tasks. However, business respondents are unfamiliar with and often unconcerned about the published statistics and results of the survey (Willimack et al. 2002). Instead, the topic manifests itself by means of the questionnaire. Thus the questionnaire design is an important motivating factor for both the participation decision and the performance of response tasks.

Aspects of the questionnaire design, such as length, question wording, visual design, and navigation, as well as the data collection mode, have an effect on respondents' motivation to comply (Willimack et al. 2002). To aid in motivating response, the questionnaire should be consistent in its design, along with being quick and easy to understand and to work with (Snijkers et al. 2007b; HMRC 2010). Questionnaire design and mode selection are covered in Chapter 8, where we discuss guidelines for preparing respondent-friendly questionnaires that support data quality while minimizing respondent burden. In Chapter 7, we discuss methods to develop, pretest, and evaluate survey questionnaires to aid their design.

The questionnaire design is also related to actual and perceived survey response burden. In addition, as we have seen in Chapter 1, many businesses, especially larger ones, are selected for many surveys, because they contribute disproportionately to survey statistics. They are often sampled in recurring surveys, adding to their reporting burden. These factors should also be considered in survey design. Surveys should be consistent in their content and design by using consistent design guidelines, maintaining common conceptual definitions, and avoiding overlap in content across questionnaires, to the degree possible. In Chapters 5 and 6, we will see how to deal with this with regard to sample design and response burden, for example, by applying the same unit definitions, coordinating across samples in order to reduce overlap, and rotating sampled cases over time by introducing "survey holidays."

The communication strategy is another survey design feature that may be used to affect respondents' motivation to participate in the survey and also how they perform response tasks. In Chapter 9, we will see that the communication strategy is aimed at getting response by using motivation and facilitation strategies. We will demonstrate how tailoring these strategies to the business context is important to their effectiveness.

Let us also consider motivational effects of features associated with the survey organization. Here, trust is a key aspect, as we noted in Section 2.2.1 with regard to reliance on social exchange theory in supporting the business survey participation decision. One mechanism to aid in creating trust is to ensure the confidentiality of reported data, and to assure businesses of this pledge in both the survey design and the dissemination of statistics derived from reported data. While confidentiality pledges must be addressed in the communication strategy, as explained in Chapter 9, an overview of methods to ensure disclosure avoidance at dissemination is included in Chapter 12.

Capacity Numerous features of the survey and its design impact the capacity of business respondents to participate in a survey and to perform the response tasks. We have already seen that some of these features, such as the questionnaire design, also affect respondents' motivation. These features, such as detailed content, definitions and terminology, visual layout, formatting cues, length, and navigation, may aid or hinder a respondent's capacity to complete the questionnaire. The consequences of poor questionnaire design are often addressed in postcollection processing activities, such as data capture, editing, and classification procedures. These are considered in Chapter 11.

Closely associated with the effect of questionnaire design on response capacity is the mode of data collection. Offering multiple modes concurrently, for example, permits business participants to select the mode by which they respond to the survey. They may prefer a particular mode because they find it to be more convenient, efficient, or secure, or simply because it is the mode they are used to.

Business surveys have been at the forefront of developing and implementing electronic or web surveys (Anderson and Tancreto 2011; Jones 2011; Snijkers 2008), as businesses believe that their use reduces response burden and facilitates the response process (Dowling 2005; Dowling and Stettler 2007). Web reporting has been shown to be quicker than response via paper questionnaires on a recurring monthly survey (Hoekstra 2007). However, in 2011, Anderson and Tancreto noted that web surveys were not yet universally accepted and available, and, thus, alternative response options tended to be offered in mixed-mode designs. Effective development and testing of electronic data collection instruments are discussed in Chapter 7; current best practices in their design can be found in Chapter 8.

Mixed-mode designs are considered in Chapter 9, along with other components of a coordinated survey communication strategy that not only targets motivation, as we saw earlier, but also seeks to aid business participants' capacity to respond. Other important components include identifying the right contact person, which we have seen is critical to obtaining survey participation and carrying out the response tasks. Additionally, providing help by way of telephone and/or Internet help desks is a feature that enhances the public image of the survey organization, along with activities that build and maintain relationships with business respondents, as past behavior has been shown to be an important indicator of future participation decisions (Davis and Pihama 2009; McCarthy et al. 2006; HMRC 2010).

As will be demonstrated throughout this book, numerous features associated with the survey, the survey design, and the survey organization may positively or negatively impact the authority, motivation, and capacity of business survey participants in deciding whether to participate in a survey and in performing response tasks. Survey organizations can impact the costs and benefits of survey participation through selecting survey design features that support business goals and reduce response burden. Nearly every survey design decision has a potential impact on a business's decision as to whether to participate in a survey and the ability to do so, from the sample design to data collection strategies to confidentiality pledges. While this is indicated in our basic model in Figure 2.1 by way of arrows pointing from factors associated with the survey organization to critical business decisions and behaviors with regard to survey response, our main point and perspective actually reverses the direction of these arrows, pointing them from the business toward the survey organization. Throughout our book, we advocate and demonstrate how the business context must be considered by the survey organization in making design decisions, in order to successfully develop and carry out effective business surveys.

2.3 SUMMARY

The purpose of this chapter was to provide an understanding of the context within which the business survey response process happens, and to encourage the reader to consider this context when making decisions about how to design and conduct business surveys. Key points to consider and their implications include

- Businesses' goals are to produce goods and services and to remain open and viable over time. To ensure that these goals are met, people who make decisions on behalf of the business weigh the costs of an action or activity against actual or expected benefits.
    - ▸ Survey response represents a cost with no associated production.
- Businesses track data in order to meet legal and regulatory requirements and management needs.
    - ▸ Data maintained in business records may not align with survey requests.
- These data are "owned" by the business, and any uses of these data are at the discretion of the business.
    - ▸ Survey organizations must rely on businesses to permit their own resources to be devoted to survey response.
    - ▸ Respondent selection is at the discretion of the business; it is not under the control of the survey organization.
- A consequence of specialization and differentiation of functions and authority hierarchies in businesses is that different types of data are distributed throughout the company; they are created and located to meet the needs of the various units.
    - ▸ Multiple respondents and multiple data sources may be needed to fulfill survey requests for different types of data.
    - ▸ The survey response task is labor-intensive.

- Business activities are carried out by people, and thus are subject to the norms and systems of social behavior, which determine how work gets done. However, social behavior is subject to organizational constraints.
  - The survey response task is viewed as work.
  - The response process in business surveys is not only cognitive, as in surveys of individuals, it also relies on social behaviors necessary to conduct work activities.
  - The cognitive processes and social behaviors associated with business survey response take place in the organizational context of a business, where behaviors and decisions are guided by goals to produce goods and services and remain viable over time.

With these key points in mind, we developed a model, presented in Figure 2.1, of the survey response process, encompassing two primary steps:

1. The decision to participate
2. Performance of response tasks

We distinguished factors that survey researchers can control versus those that are controlled by the business. We defined three layers of factors—the external environment, the business/management, the individual/respondent—which are, along with the survey design, associated with the business survey response process. We also demonstrated how factors related to authority, motivation, and capacity associate with each of these layers, to impact the steps in the response process. These are summarized in Table 2.3.

We believe that our model is sufficiently general to be applied to businesses of different sizes, industries, and other characteristics. We expect, though, that the manner in which the model is applied will vary, because there are differences in how these characteristics affect the authority, motivation, and capacity of the business's approach to the participation decision, along with how they affect the authority, capacity, and motivation of individuals carrying out response tasks. Through our detailed descriptions, we further demonstrated how the business context permeates the steps of the response process in business surveys.

The importance of understanding the business context is that businesses exercise control over completing surveys, not survey organizations, and not those who design and conduct business surveys. Businesses will do whatever they will do. They are in control of the survey response process; we survey researchers cannot assume that they will do what we want them to do in the manner that we want. What we can do, though, is to design and conduct surveys in a manner that facilitates, or at least does not hinder, their native and usual businesses processes.

This is our philosophy and our point of departure for the remainder of this book, starting in the next chapter with defining our perspectives on survey data and process quality.

Table 2.3 Summary of Factors Associated with the Business Survey Response Process

Layers	Authority	Motivation	Capacity
External environment	• Legal and regulatory requirements	• Public goodwill • General norms • Survey-taking climate • Political climate	• Economic conditions
Business	• Policy on survey participation • Reporting priorities	• Cost-benefit analysis • Trust: familiarity with survey organization • Past behavior	• Expectations about response tasks • Record formation and data availability • Organizational structure and complexity: infrastructure • Availability of resources • Established routines • Respondent selection
Respondent	• Authority/mandate to perform tasks	• Employee attitude • Organizational and management priorities • Job performance evaluation criteria • Professional standards	• Employee's role in response process • Routine • Capacity: time, knowledge of data, competence/skill • Workload: competing activities • Data: availability and access • Timing of survey request • Survey-related factors
Survey organization and survey design	• Legal obligation • Reputation and legitimacy	• Survey topic • Confidentiality • Image of the survey organization • Actual and perceived response burden - Sample coordination - Recurring surveys • Questionnaire design • Data collection mode • Communication strategy	• Questionnaire design: - Level of detail - Visual layout • Data collection mode(s) • Communication strategies - Contacts - Coordination - Help desks

AKNOWLEDGMENTS

Portions of this chapter were drawn from an invited paper by Seymour Sudman, Diane K. Willimack, Elizabeth M. Nichols, and Thomas L. Mesenbourg, Jr., which was presented at the Second International Conference on Establishment Surveys (Sudman et al. 2000), based on a research collaboration while the late Professor Sudman was at the US Census Bureau under an Intergovernmental Personnel Agreement, 1998–1999. In addition, Sections 2.1.2 and 2.1.3, as well as the role episode model in Figure 2.2, were previously included in an invited paper by Diane K. Willimack, presented at the Third International Conference on Establishment Surveys in 2007 (Willimack 2007). We also acknowledge helpful reviews of this chapter by Elizabeth M. Nichols, Alfred D. Tuttle, and Jennifer Whitaker of the US Census Bureau.

CHAPTER 3

Quality Issues in Business Surveys

Gustav Haraldsen

Business surveys possess certain specific characteristics that differentiate them from other kinds of surveys. Hence, although the discussion in this chapter is structurally similar to any discussion of survey quality in general, we will highlight some quality issues that are unique, to, or at least more prominent in, business surveys compared with social surveys.

In any kind of production we can distinguish between *process quality*, which is the quality of our efforts, and *product quality*, which is the quality of the outcome of our efforts (Ehling and Körner 2007). One characteristic that is specific to the production of survey data compared with many other production processes, is that the product quality is the result of not one, but two, different working processes; the internal activities before, during, and after the data are collected; and the response process that takes place outside the surveyor's office during the field period (Biemer and Lyberg 2003). Moreover, even if the data quality is clearly affected by how well the survey is planned, monitored and managed by the surveyor, the quality is still decided primarily by how respondents process the questions posed to them.

The response process in business surveys differs somewhat from that in social surveys. The three most important differences are that business respondents generally do not report on behalf on themselves, but act as informants, that much of the information is collected from administrative records and that the social context has a more prominent influence on the response process than what is common in social surveys (see Chapter 2).

While most of the following chapters discuss process quality, how the response process is planned for and supervised, this chapter focuses on the product quality that we aim for in business surveys. The chapter is based on a general quality model, and some of the issues discussed are simply business survey versions of common quality issues. Other issues, however, are more specific to business surveys. As we hope the

Designing and Conducting Business Surveys, First Edition.
Ger Snijkers, Gustav Haraldsen, Jacqui Jones, and Diane K. Willimack.

reader will see, these issues are linked to the just-mentioned characteristics of the response process in business surveys. Process and product quality cannot be discussed independently of one another. While the quality issues highlighted in this chapter should be the reference point in the following chapters focusing on different survey planning and management activities, characteristics of the business survey process is the reference point for the discussion of product quality issues in this chapter.

3.1 SURVEY QUALITY FROM A USER/PRODUCER PERSPECTIVE

In its most general form, quality can be defined as "fitness for use" (Juran and Gryna Jr. 1980). The European statistical system, the system to which European national statistical agencies are committed, operates with eight general basic quality dimensions for statistical products,[1] which can be considered as specifications of fitness:

1. *Relevance*—the degree to which the data meet user needs in terms of coverage, content, and detail
2. *Accuracy*—the closeness between an estimated result and the unknown, true value. Accuracy is a central part of the surveyors' professional approach to survey quality, which will be discussed in more detail later in this chapter
3. *Timeliness*—the degree to which data produced are up-to-date
4. *Punctuality*—the time lag between the actual delivery date of the data and the target date when the data should have been delivered
5. *Accessibility*—the ease with which users are able to access the results, also reflecting the format in which the data are available and the availability of supporting information
6. *Clarity*—the quality and sufficiency of the data documentation, illustrations, and additional advice provided
7. *Comparability*—the degree to which data can be compared over time, spatial domains, and subpopulations
8. *Coherence*—the relative similarity between data derived from different sources or methods but that refer to the same phenomena

More recent quality guidance publications from Eurostat (2009b) also include four additional perspectives:

1. Tradeoffs between some of the dimensions listed above, such as tradeoffs between relevance and accuracy, timeliness, or coherence

[1] Usually they are presented as five quality dimensions, but three of them are in fact two dimensions presented as one.

2. Assessments of user needs and perceptions (e.g., measured by user satisfaction surveys.)
3. Assessments of performance, cost, and response burden
4. Documentation of confidentially, transparency, and security

These are not quality dimensions in the same sense as those listed above, but are useful quality aspects to consider when reporting on quality. Later in this chapter we will link all these quality dimensions and aspects together in a model that distinguish between quality perspectives of users and producers of business statistics, and between different quality criteria and different kinds of tradeoffs.

Lists of quality aspects similar to those listed above are also found and implemented in strategy documents and frameworks for quality reports of NSIs around the world (ONS 2011; Statistics Canada 2002; Statistics Norway 2007; Statistics South Africa 2010).

Formally, it is the producers of statistics, those who collect and analyze the data, who are responsible for a delivery that meets these requirements, while it should be the users of statistics who judge whether the quality standards are met. In practice, these two roles are intertwined. A fundamental prerequisite, both from producers' and users' perspectives, is that the user needs are specified and acknowledged. Statistics that do not meet important user needs is inadequate. Thus it is an essential part of the data collector's responsibility to investigate user needs before a new survey is planned and conducted. Moreover, a prerequisite for a fair user evaluation is that the users know the intended purpose and quality standards of the statistics. One and the same statistics will normally have many different users, most of whom have negligible or no influence on the original survey design and statistical analysis. Therefore we think that "clarity," an informative methodological documentation, is the most fundamental and important quality criterion for the users of statistics. In an evaluation of the first 100 quality audits of statistical products done by the UK Statistical Authority, they list improved engagements with users, richer commentaries, and better documentation as first priorities (Macintyre 2011).

Furthermore, we think it is important to distinguish between performing *quality evaluations* and producing *quality indicators*. For many of the criteria listed above, quality measurement is not straightforward. So, even if we say that the survey organization should be evaluated by the users of their products, the survey organization will still provide the measurements. This is particularly true for accuracy. Behind this criterion lies a conceptual model that forms a cornerstone of survey methodological quality assessments that we cannot expect statistical users to know and master. Thus, particularly for accuracy, and also for quality criteria such as comparability and comprehensiveness, the survey organization is responsible for documenting the quality. Only then the user can evaluate whether the quality is good enough.

So far we have argued that accuracy is the basic quality criterion from the survey producer's perspective, and that clarity is a fundamental quality criterion for any user evaluations.

3.1.1 Users and User Evaluations of Business Surveys

As long as we claim that surveys end when the data collection period is closed and the datafile is handed over to those who produce and disseminate statistics, economic statisticians and researchers are the primary users of business survey data. These are consequently the professionals who set the basic user evaluation standards for business surveys. The most important of these users are those who produce and update the national accounts, the GDP, and other key economic indicators. In Europe, which economic indicators that are produced and how they are defined are also to a high degree governed by Eurostat regulations (see Chapter 1). Hence, much of the business statistics presented in Chapter 1 were gathered from Eurostat publications.

The most important users of economic indicators and other statistical products based on business surveys are probably politicians, the national banks, and others who shape the economic policy. Other important users are investors, stockbrokers, business interest organizations, and economic research institutes. For these users trust to the agencies that produce the statistics is probably more important than methodological documentation. This trust is probably more dependent on the ability of statistical institutes to predict the forthcoming economic development, than on a quality assessment of the data that go into the prediction models. As we pointed out early in Chapter 1, everyday people commonly also base some of their private decisions on economic indicators and statistics. Again, confidence in the statistical institution is why they do so.

The businesses themselves also need general information about the economic climate and prospects and are generally very concerned about employment rates, production, price, and sales indices. Price statistics is used to set the prices on services and goods that the businesses are buying or producing. Salary statistics is used to negotiate how much the employees should be paid for their work. For more detailed market analysis, however, businesses often prefer to carry out their own surveys and research rather than utilize available business statistics. A data mining analysis of textbooks on business statistics and business decisionmaking carried out in 2010 indicated that economy students learn little or nothing about the role of official economic statistics in business decision processes (Biffignandi et al. 2011). The extent to which national statistical agencies acknowledge and tailor their statistical product to business needs is also questionable (Löfgren et al. 2011).

Between the producers of the national accounts and other economic indicators and the users of economic statistics, we find different mediators of statistical results. The most important are journalists covering business issues, information officers working in ministries, authors of economy textbooks, and teachers in business schools. Their quality assessment of business surveys and economic statistics will probably be influenced by both their perceptions of what information their readers look for and the quality criteria of their own profession.

From these reflections we see the contours of a three-stage dissemination model. In stage 1, macroeconomists, as the primary users of business survey results, define the initial evaluation criteria. In stages 2 and 3 the mediators and then decision-makers add their quality perspectives, and thereby perhaps change the priorities of the initial criteria or reach different conclusions regarding product quality. As we move down this dissemination chain, the evaluations will also become less specific and based more on confidence than on hard facts.

The quality assessment of macroeconomists is generally confined not to a particular survey but to how well individual survey results fit into a framework of economic statistics describing different aspects and levels of the economic life of society. Business surveys are also normally only one of several information sources in this system. According to Willeboordse (1997), the most important quality criteria in the framework of economic statistics are relevance, coherence, comparability, and timeliness.

For macroeconomists, relevance in business surveys is primarily how well the results fit into the concepts used in national accounts systems. The core variables in these systems are related to the principal transactions in the economic process, such as turnover, production value, value added, surplus, and profit or loss (Willboordse 1997, p. 47). The national accounts system also includes information about conditions that affect the economic process, such as investments, research and development, buildings and machinery, and, of course, the number and competence of employees. Compared to this, relevance in the minds of economic statistics mediators and political and economic decisionmakers will often tend to be much more specific. Therefore their assessment of the relevance of business survey results may differ considerably from assessments made by national accounts economists. Here is an example of the type of statistics that users further down on the dissemination chain prefer:

> In a Norwegian focus group about property sales, some estate agents and journalists took part. They were all interested in the prices of holiday property. However, the estate agent could not use the available statistics to suggest prices for his objects as long as the price statistics did not distinguish between costal and mountain areas in his district. The journalist also wanted a classification that better matched with the area covered by his paper. As long as business survey organizations do not listen and accommodate their statistics to such needs, the relevance of their products will be rather limited.

Coherence is important in two ways. First, results from surveys conducted at different times should form a sound and unbroken pattern of continuity. This is particularly important for price and production indexes. The main focus in this kind of surveys is generally not on the price or production level but on the relative reliability of indications of rising or falling prices and productivity. In addition to this trend perspective, business surveys should also be coherent in the sense that results from annual surveys should be consistent with tallied results from surveys carried out on a monthly or quarterly basis. Trends are also making headlines and are read with interest by politicians and investors, while coherence in the sense of consistency between surveys rarely receive the same attention.

The importance of comparability is also twofold. Results covering different aspects of the economy, such as employment figures and production data, should be comparable and tell the same story. Moreover, it should be possible to compare figures from different regions, industries, or other subpopulations. Business surveys should be able to answer questions such as which part of the country is doing best or what industries are most affected by changes in the international economic climate. As with trends, comparisons between different industries and parts of a country are considered important by all types of users. A problem that journalists, politicians, and others often encounter, however, is that standard classifications used in official statistics do not match with what is interesting from their viewpoint.

Last but not least, the value of results from business surveys is often heavily dependent on when they are published. The economy changes more rapidly than cultural changes measured in social surveys. Data that monitor the economic fluctuations are often of high interest to politicians, investors, managers, and others operating on the economic arena, but only as long as they are up to date and not considered to be yesterday's news. The public's wish for quick results on one hand and the surveyors' fear of releasing inaccurate figures on the other hand is probably one of the most common sources of tension between users and producers of business surveys.

These reflections about quality issues from the users' perspective suggest that business surveys, in particular those surveys carried out by national statistical institutes, sometimes tend to neglect users other than their principal contractor.

3.1.2 The Total Survey Error Approach

The surveyor's approach to quality is a combination of regard for the stakeholders' priorities and loyalty to professional standards. The cornerstone of any professional surveyor's evaluation of survey quality is the survey error approach. According to this approach, quality is defined as the absence of errors. The logic is similar to that of scientific hypothesis testing. Instead of arguing that we are right, we challenge the assertion that we are wrong. Survey errors are considered to distort the true picture of the population that is surveyed (Groves et al. 2004). Hence, the survey error approach is to search not for qualities, but for errors. The fewer errors that we find, the closer to a true picture we consider the results. There is no reason to believe that surveys contain more errors than other kinds of data collection. On the contrary, this pessimistic evaluation method is somewhat stricter and more systematic than evaluations that focus on the strong parts of a data collection method.

The errors included in the total survey approach are linked to the two main processes in data collections: (1) drawing a sample of units and addressing those who we want to participate in the survey and (2) developing a questionnaire or another measurement instrument and using it in the sample. In Figure 3.1 we term these two processes "representation" and "measure" and have identified eight types of errors that can occur in different steps of this double-barrel process model.

The errors listed in the figure can lead to either systematic or random differences between survey measurements and the "true value".[2] The risk of systematic or incidental errors varies, however, for different sources of error. If we omit the low-risk errors, we can define the commonly termed *mean-squared error* (MSE) as the sum of high risk systematic errors plus the sum of high-risk incidental errors. In the

[2] Strictly speaking, we are talking about the true value under the survey conditions. See survey-related effects in Figure 3.2.

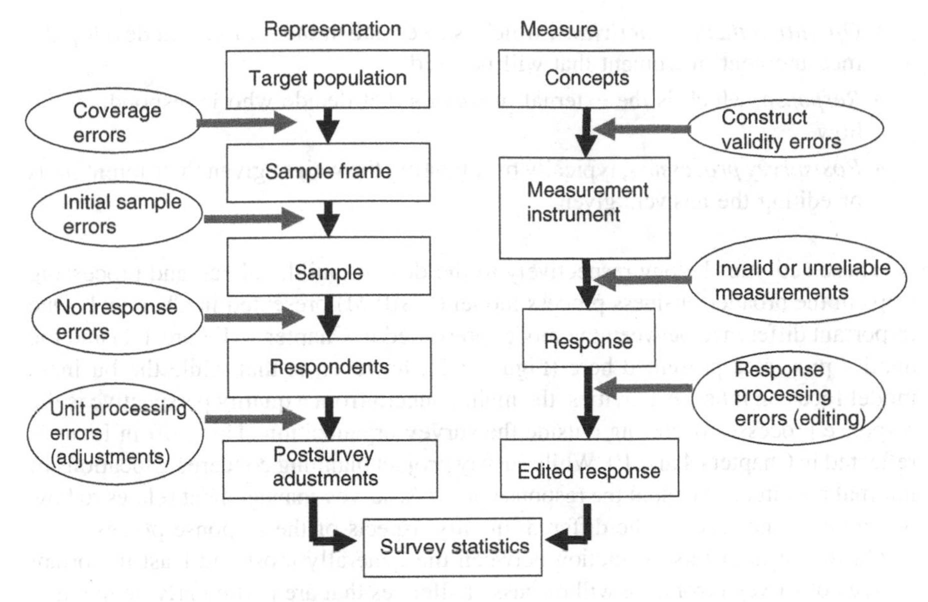

Figure 3.1 Survey steps and errors. [Based on model from Groves et al. (2004) Figure 2.5.]

MSE formula the systematic risk is termed *bias* (B) and the incidental risk, *variance*: (V) (Biemer and Lyberg 2003):

$$\begin{aligned}\text{MSE} &= B^2 + V \\ &= (B_{\text{specification}} + B_{\text{frame}} + B_{\text{nonresponse}} + B_{\text{measurement}} + B_{\text{processing}})^2 \\ &\quad + V_{\text{sampling}} + V_{\text{measurement}} + V_{\text{processing}}\end{aligned}$$

Randomly distributed errors can be estimated by using probability calculations and will, by definition, even out after several measurements. The direction and effects of systematic errors, however, are much more difficult to evaluate and will not disappear unless the errors sources are detected and repaired. Consequently, bias is usually a more serious quality problem than variance.

According to the formula, measurement and processing errors normally lead to both bias and variance, and should therefore be considered to be the most serious type of error. On the other end of the scale, sampling errors normally lead only to increased variance and are thus considered less serious. Note, however, that if the survey has a high nonresponse rate, there may be a considerable difference between the initial sample error and the sample error in the net sample.

The different steps of the sample–measurement process depicted in Figure 3.1 can also be divided into

- *Specification activities*, that is, deciding what topics that should be covered in the measurement instrument and which units are relevant in the study

- *Operationalization activities*, which is to establish the sample and develop the measurement instrument that will be used
- *Response*, which is the external processes that decide who is responding and how
- *Postsurvey processing*, typically by adjusting the weight given to different units or editing the answers given

These activities belong respectively to the design, build, collect, and processing steps in the broader business process model (GSBPM) presented in Chapter 1. One important difference between the model presented in Chapter 1 (Figure 1.1) and the quality processes presented here (Figure 3.1), however, is that while the business model focus on inhouse activities, the main concern from a quality perspective is the response processes occurring outside the survey organization. This shift in focus is reflected in Chapters 4 and 10. While survey project planning concerns allocations of internal resources to benefit the response process, survey management relates to how we monitor and react to the different quality aspects of the response process.

On the basis of this distinction between the generally most and least important sources of survey errors, we will discuss challenges that are particularly demanding in business surveys and that thus modify the initial ranking of quality issues. Before we can do that, however, we need to complement the survey error approach with two other quality aspects.

3.1.2.1 Quality Constraints

First, the degree to which the errors listed in Figure 3.1 can be reduced is limited by practical constraints. Robert Groves, in his classical book on survey errors and survey costs (Groves 1989), focused on the balance between survey errors and the cost of human and other resources. There is always a limit of how much money we can use on survey planning and implementation, and this naturally also limits the efforts we can use on minimizing errors. Thus, *cost efficiency*, defined as the relationship between quality (absence of errors) and costs (human and material), is a key evaluation criterion in survey evaluations.

Cost In a true calculation of survey costs, not only the internal cost of the survey organization but also cost to respondents to answer the survey questions should be included. In social surveys we usually recognize this cost in the sense that we are concerned about quality effects of long questionnaires. There is, however, no tradition for including the questionnaire completion time in the survey budget. In business surveys this should be different. Time is money in businesses. There is a widespread concern among politicians and employers' organizations about the cost that businesses incur with questionnaires and other reporting obligations. For this reason, most national statistical institutes are compelled to look for alternative information sources before they launch business surveys, and then to minimize the sample size and time burden of the surveys that they conduct. In this way, regard for business costs clearly is a constraint in business surveys and should be included in the survey budget and evaluations of cost efficiency. This is particularly true when business paper questionnaires are replaced with electronic questionnaires. If response and consistency checks previously performed inhouse are built into electronic versions of the questionnaires, editing costs may appearently fall without any negative effects on the response quality. However, the extra cost that business respondents incur on responding to error messages may add up to a far higher price than what was earlier paid to inhouse editors.

Time The impetus to reduce the time it takes business respondents to complete our surveys should not only be to save money. There is also a limit to how long respondents will remain motivated and produce thought-through answers. It is well known from social surveys that response quality may decrease at the end of long questionnaires (Bogen 1996), and the same is probably true for business surveys. Completion time is generally a better measure of this burden than the number of questions. This is particularly true in business surveys because they often ask for information that the respondent is supposed to collect from the firm's administrative systems. Moreover, when we treat time as an indicator of response burden, it is, strictly speaking, the perception of time rather than the measurement in minutes that really matters. We know from everyday life that when activities are interesting and absorbing, time flies rapidly. On the other hand, if they are boring and take a longer time to complete than expected, tasks feel more burdensome. Thus the perceived response burden is probably heavily dependent on the relationship between expectation and experience. We expect respondents to be more willing to accept that a questionnaire with many questions takes more time than one with only a few questions that still is time consuming to respond to. Research has also indicated that respondents prefer web (electronic) questionnaires to paper (hardcopy) forms, even when it takes a longer time to log in and complete them (Haraldsen 2004). We will return to the relationship between response burden and response quality, and how the response burden in business surveys can be minimized, in Chapter 6.

Time is an important ingredient both in survey cost and perceived response burden. In addition, timeliness during the data collection process is of vital importance for the survey result. We have already stated that for the stakeholders of business surveys, the publication date may be of utmost importance. While timeliness from the user's perspective means being newsworthy, timeliness in data collections means adhering to the timetable. Still, there is a connection between these two kinds of timeliness. The publication date is dependent on when the data collection period is over, and that date, in turn, presupposes that the data collection started on time. Delays at the start of data collection almost certainly have a domino effect on later delivery dates. Because the change from planning to implementation also means that many more people become involved in the project, a delayed start also has immediate and severe economic consequences. Hence, it is crucial that the date when the actual data collection is scheduled to start be upheld. At the same time the survey quality and the ability to deliver on time is also heavily dependent on the amount of time allowed for planning activities. When the original project plan is set up and agreed on, one should therefore be careful not to compromise too much on when the data collection should begin. One common weakness in surveys is that because of this time constraint, sufficient time is not allocated for development and testing before the field period starts. As with cost, there is a balance between the available time and the quality that we are able to achieve. And in the same way as with costs, we should evaluate the *time efficiency* of our survey design.

Technology Collecting data from respondents is a form of communication. First, we decide who we shall address, or communicate with. Next, we make contact and present our errand. If this communication succeeds, we step into a communication process governed by survey questions. All three communication activities are heavily influenced by technology. Even if the theoretical basis for probability sampling was established as long ago as the latter part of the 19th century, it was not until censuses and registers were made available for computerization that sample surveys became the dominant social science method. Personal visits and face-to-face interviewing were replaced by telephone interviewing as telephones became common property. Computerized questionnaires were first introduced to interviewers and then, with the web technology, introduced in self-completion questionnaires. This last step in the technological development of survey communication modes is about to change the main data collection instrument in business surveys from paper questionnaires to *web questionnaires*. Web questionnaires have a larger toolbox of response formats and visual aids than do paper questionnaires. Even more important, the technology can be used to imitate the dialog we know from interview surveys (Couper 2008). In this way it should be possible to combine the qualities of self-administration with some of the qualities of an interview.

Moreover, the shift from paper to information technology affects not only question–answer communication but also communication between the surveyor and the businesses surveyed. Instead of receiving a letter with a paper questionnaire enclosed, the businesses may receive a letter announcing only that the questionnaire is available at a certain web address and urge them to download it there. Later they may be reminded by email. Unlike people who are sampled in only one or a few household surveys within their lifetime, some businesses take part in several surveys every year and in many surveys that are panels. For this reason, some countries (e.g., the Nordic countries) have established *web portals* where the businesses can find lists of all the questionnaires that they are expected to fill in. In quality terms, this means that sample management quality as well as the measurement quality is affected by web technology.

We tend to accentuate how new communication technology leads to new possibilities, but implicit in this argument, we also recognize that yesterday's technologies had constraints—and there is no reason to believe that we will think otherwise about today's technology tomorrow. One obvious constraint in today's web questionnaires, for instance, is the screen size. Hence, the general point is that survey communications always is affected by the communication technology used. For an overview of the relationship between communication modes and communication technology in surveys, see Chapter 1 in Dillman et al. (2009).

Ethics A fourth type of constraint that might not be so obvious as cost, time, and technology, is ethical constraints. The surveyor's relationship with the respondents is based on trust. Research on trust has shown that knowledge of the survey organization is one of the key factors determining whether people respond (Wilmot et al. 2005). In order to decide whether the rationale for the data collection is legitimate, the respondent must know the purpose of the data collection, and what measures the surveyor has taken to prevent use of the information for other purposes. These two pieces of information therefore are the most important ingredients when a new survey is introduced in a prenotice letter or on an information sheet. If the surveyor's intentions or assurances are not accepted or trusted, this may lead to nonresponse errors or low response quality. There may be several reasons for mistrust. One may be that the information given about the purpose of the survey and data protection is insufficient and appear suspicious, or that the surveyor is not quite sincere about the surveyor's intentions. Also, if the topic and intentions are well explained, the topic simply might be controversial, or the surveyor might not be able to issue a full guarantee against undesirable access to and use of personal information. Because the information gathered in business surveys often is market-sensitive, trust, and hence ethical standards, plays an important role in business surveys.

Conclusion From these reflections we suggest that the most important constraints affecting the survey quality are costs, response burden considerations, timetable constraints, ethical considerations, and technological limitations. We also consider time to be a key aspect of several of these constraints, as it is in the user's evaluation of statistical results.

3.1.2.2 Survey-Related Effects

The term *survey-related effects* in the third corner of the total survey error triangle presented in Figure 3.2 refers to a family of quality challenges that are similar to but still different from measurement errors. The term *measurement errors* implies answers that differ from a true value. However, different answers to the same question can also be true but still differ according to the context in which they are uttered. The truth changes according to the context, so to speak. The survey itself establishes a context with specific characteristics. Thus, by running a survey, we create a situation that is different from other situations where the respondents relates to the same kind of questions as those that we present in the questionnaire, and different contexts will taint the answers given (Schwarz et al. 2008; Weisberg 2005).

We noted above that business respondents assume different roles when responding to a questionnaire and reporting to their superiors. This does not mean than one report is more correct than the other, but rather that the information taken into account and the evaluation perspective is different when one reports to management than when one reports to a statistical institute.

Different questionnaire characteristics also establish the context and indicate the perspective which the respondent is expected to take. One example that is highly relevant in business surveys is when different data collection modes are combined. Nowadays a combination of web and paper questionnaires is a common design in business surveys. What we term *instrument effects* may not mean that one data collection method yields results that are closer to an instrument-free truth than another, but rather that different communication modes affect the considerations made before the respondent comes up with an answer. As an illustration, when a question is presented on paper, it is easy for the respondent to see and relate to answers just given in previous questions. In a web questionnaire, the same question will perhaps be presented on a separate screen, but with a reminder of previous questions selected by the computer program. This difference in reference points may lead to different, but equally valid, answers. The usual way of introducing web questionnaires in business surveys is to offer it as an alternative to a paper questionnaire. In surveys that are repeated, which most business surveys are, the web pickup rate will then steadily increase from one round to the next. If the results are influenced by the data collection mode, changes in estimates may consequently either be true changes or just reflections of the growing web rate.

It is also well documented from survey research that the question order may affect the way specific questions are interpreted. Again different answers with different question orders do not necessarily mean that one order is more correct than another, just that the reflections made and consequently the answers given vary with the context in which questions are asked.

The risk of context-dependent answers is probably higher when we ask about opinions, attitudes, or evaluations than when we ask about facts. Such questions generally are rare in business surveys, but there are exceptions. Two examples discussed in Chapter 8 are business tendency surveys and surveys of working conditions. In these kinds of surveys results from different periods of time, different industries, or different countries are commonly compared. Because the surveys that are compared are carried out in different cultures, in different environments or at different times, similar evaluations may be based on quite different observations, and different results on similar observations.

These situations should not be treated as scenarios with true or false results, but they should probably not be treated in the same way, either. In the first example of the roles that respondents take, the role that we want the respondents to take, as survey participants, will usually differ from the role they normally play in the company. If this difference in roles is explained and accepted, it is probably easier to make the respondents apply different definitions and categorization principles than those they use in their daily work. Consequently, what we will try to do is to establish a common, survey-specific perspective. In the example with mode and order effects, however, we will often try to adjust for the survey effects so that they will not show up in the statistics. When this is the case, the survey effects can be estimated or controlled for by randomly assigning respondents to different modes or question orders during data collection. Randomized question order is quite easy to implement in electronic questionnaires, while randomization is more difficult to apply with different data collection modes. An alternative is to adjust for survey effects due to different communication modes using regression techniques after the data are collected.

Finally, evaluation principles change between cultures or over time. In these cases, introducing a standard evaluation perspective forcing respondents for

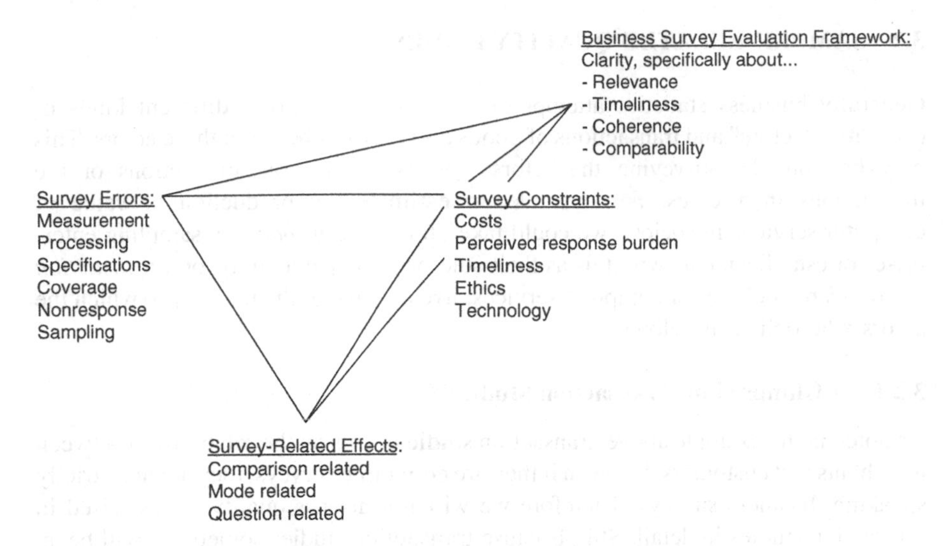

Figure 3.2 Business survey quality criteria.

instance in business tendency surveys or surveys about working conditions to use certain evaluation criteria that are different from those they normally use when they evaluate future business prospects or present working conditions seems to be wrong. Instead, we would probably recommend following up the evaluations of business prospects or working conditions by questions that map which evaluation criteria are in use.

If we combine the surveyor's total survey error perspective with the evaluation framework of business survey users, the full quality assessment model may be depicted in a model like that shown in Figure 3.2.

The model shown in Figure 3.2 is two-dimensional. The distinction between survey errors and survey-related effects and the tradeoff between these quality aspects and survey constraints form the total error approach of professional surveyors. Another, equally important dimension in the model is how users evaluate the survey results. While their main focus is on other quality aspects than the surveyor, an evaluation of the surveyor's professionalism will be a part of their evaluation. Similarly, user evaluations will be a part of the surveyor's quality assessment. In this way the professional quality evaluation and user evaluations are related to each other and are parts of a holistic quality model.

In the model we have also suggested how the different sources of error, survey effects, constraints, and user criteria should be ordered according to their importance for the business survey quality. This ranking is a first draft based on general survey methodology and some first reflections about the characteristics of business surveys. In the following section we will discuss the quality challenges that are specific to business surveys in more detail. At the end of the chapter we will then look back and decide whether this first draft needs to be modified.

3.2 SAMPLE-RELATED QUALITY ISSUES

Generally business statistics attempt to describe and analyze different kinds of economic "actors" and transactions of goods and services between these actors. This may be done by surveying the actors who perform economic actions or the transactions themselves. For example, to estimate the production or value of computer services in society, we could take an *actor approach* by sampling enterprises or establishments with this as their sole or principal activity, or a *transaction approach* by looking for computer services, irrespective of the industry to which the actors who deliver it belong.

3.2.1 A Glimpse into Transaction Studies

As noted in the example above, transaction studies focus on the relationship between merchants and customers. So, even if they are economic surveys, they are not, strictly speaking, business surveys. Therefore we will not discuss quality issues raised in transaction studies in detail. Still, because transaction studies sometimes will be an alternative to business surveys, we think it is relevant to describe a few characteristics of these studies.

One major problem with transaction studies is that we may not know where to look for the transactions in question. Moreover, we may not have a classification system or the information needed to classify the transactions that we observe. Transaction studies can be compared with traffic surveys. One can count the number of cars passing a certain point at some selected roads. It might be difficult, however, to know which roads will give a representative picture of the traffic. In the case of economic transactions, some channels, such as credit card transaction, might be possible to access, while cash or informal transaction might be inaccessible. This will obviously lead to biased results. Next, even if we know the number of passing cars, we will have only a rough idea of the proportion of travels for work and leisure unless we stop a subsample of cars and question some drivers. The equivalent challenge is found in transaction studies. An example is the Norwegian *cross-border trade survey*, in which survey credit card transactions are used to estimate how much money Norwegians spend in the shops just across the border to Sweden, while data from an omnibus survey are used to estimate the composition of goods purchased.

One of the main strengths of transaction surveys is that the transactions often are captured when, or almost when they occur. This is particularly true with modern, electronic transactions (Costa et al. 2008). While data collection in traditional business surveys takes months, transactions are registered almost at the same time as they take place and may be downloaded from transaction swifts during the following night. Another important point that we will return to later in this chapter is that it often will be transactions, and not transactional information about specific production units that are recorded in the companies' administrative registers. When this is the case, transactional data will be more readily available than the type of data we tend to ask for in business surveys.

Now we move to the sample process of business surveys.

3.2.2 Specifying Observational Units

The economic actors we normally want to base our business survey on, commonly called the statistical or analytical units, are ideally homogenous production entities that can be classified by one and only one industry code. Smaller companies may have this character, while in many other companies especially large ones, concrete units have to be split into more or less artificial units that describes different kinds of production or service, only some of which should be included in the survey. The way it is depicted in Figure 3.1, it looks like that coverage errors occur when we decide who in the target population should be included in the sample frame. This may generally be true in social surveys where the target population consists of people with certain characteristics, and where sample frame building begins with a specification of which personal characteristics should be present for inclusion in the frame. In business surveys, however, the sample frame challenge is not only to specify and next to trace the relevant target populations units but to an equal degree to delineate target population units (Choi and Ward 2004; Willeboordse 1997).

An illustrative example is a shop selling different products classified by different industry codes. The shop has a physical address, as do the enterprise that owns this, as well as other shops. Hence the enterprise and the different shop addresses can be included in a sample frame. But because the shops sell different products, there may not be any address to the part of the businesses that belongs to a specific industry code which is the production unit we want for our statistics. This analytical unit may more be an activity than a physical unit, and hence may not be possible to delineate and sample prior to the survey.

The preferred strategy in instances like these is to choose units that can be observed and that is as close to the analytical unit as possible; and then to include these units in our sample frame. How close to the statistical unit the observational units in the sample frame will be depends to a high degree on how much information we are able to collect and consider in the business register on which the sample frame is based. The most common observational units included in business registers are enterprises and establishments (commonly called *local units* in Europe). Sometimes it may also be possible to split up enterprises or establishments into smaller homogenous and identifiable production units (Choi and Ward 2004).

Enterprises are formally defined as "the smallest combination of legal units that is an organisational unit producing goods or services, which benefits from a certain degree of autonomy in decision making, especially for the allocation of its current resources" (Willeboordse 1997). The basic building block for identifying enterprises is normally legal entities. In most cases a legal unit corresponds to an enterprise, but in some cases one enterprise consists of several legal units. In these cases the criterion of *autonomy in decisionmaking* may be used to link together legal units that constitute a common decisionmaking unit. In the reverse situation, that is, when several autonomous decisionmaking units are legally linked together, this unit is termed an *enterprise group* or a *consortium*. Enterprise groups often cross national borders and are consequently difficult to include in national business registers. This, of course, also makes it difficult to produce statistics at this level. In Eurostat there is an ongoing discussion regarding the need for a supranational business register that can cover the different parts of multinational enterprises (Eurostat 2010).

An establishment can be defined as a local unit within the enterprise that can be identified geographically and that carry out at least one kind of production (Willeboordse 1997). In smaller businesses the enterprise will typically coincide with only one establishment, which will often consist of only one homogenous production unit. These are the easy cases to sample. In 2009 more than 98% of the enterprises in the Norwegian private sector had only one establishment. In The Netherlands 96% of enterprises have less than 20 employees. But even if the proportion of larger businesses is small, they account for a large proportion of economic activity, and are thus of vital importance for the economic statistics. Identifying the kind of activities taking place in larger companies, and how their production is structured, is therefore an important part of sample frame management in business surveys.

Larger businesses often have a complex mix of legal units and operational units that change continuously according to what is legally and economically most convenient. In such companies the delineation of observational units is often determined and updated by profiling teams that continuously map the organizational and operational structure of the larger businesses (Pietsch 1995). To ensure an up-to-date frame, statistical agencies that manage business registers also collect and match information from different sources to determine which units should be in and out of the frame. The information used to identify establishments is often gathered from tax or employment data that reveal salaries paid to employees working at the same premise.

Only active enterprises and establishments should be included in the sampling frame. By definition, an active enterprise or establishment will have produced goods or services during at least part of the survey's reference period. In practice this can be difficult to determine before the sample is selected and the data collection is running.

Here is an example from Norway that illustrates how long the road from the initial source to the sample frame may be:

> In September 2009, 917,000 units had an ID number registered in The Central Coordinating Register for Legal Entities. When ID numbers that obviously did not belong to any kind of business were subtracted, the first version of what may called a business register had 857,000 units. However, out of these only 52% were identified as actively producing goods or services. Consequently one ended up with a frame of 447,000 active units.

Companies are born and die every day; some are even reborn. Others merge or split up. So far we have focused on the risk of excluding units that should have been included in the survey or including units that should have been excluded. But in addition to these under- and overcoverage issues there may also be register problems with units that are incorrectly counted as separate businesses or separate businesses that are incorrectly counted as one. These are what are called multiplicity or clustering problems (Weisberg 2005). Because the basic bricks of business registers are legal units and it is quite common for managements to split up their enterprise into different legal units for administrative or tax-related reasons, multiplicity problems are probably more common in the business registers than are clustering problems. Whatever the case, descriptions of business registers should always include estimates of all these four kinds of coverage errors.

The enterprises and establishments included in business registers are classified according to their main activities. As described in Chapter 1, in Europe the Nomenclature Statistique des Activités Économique dans la Communauté Européenne (NACE) is used while the North American Industry Classification System (NAICS), developed in cooperation with Canada and Mexico, is commonly used to distinguish between industries in United States. Businesses that perform different kinds of activities can be classified by more than one industry code. Normally, however, only the primary code, referring to the main activity, is used for classification purposes. If we return to our example with computer services, this means that units that offer computer services as a secondary activity will be excluded from the samples drawn from the frame and consequently also from the survey.

The industry code is normally decided by the kind of activity that generates the most value. But sometimes, the activity that leads to the highest turnover or the activity that a majority of the employees are performing is used as an alternative way to select the industry code. However, the most labor-intensive activity does not always generate most value, nor does the activity that generates most value always generate the highest turnover. Hence, when these characteristics do not correlate, the choice of selection criteria will lead to different sample frames and at ultimately to survey results that are not comparable.

In Europe, where there are central statistical offices, you will typically find business registers that serve as a common sample frame from different business surveys. In Norway the initial source for the business frame is the Central Coordinating Register for Legal Entities, which includes employment data from the Register of Employers and Employees. In The Netherlands the samples are drawn for the businesses registered with the Dutch Chambers of Commerce and updated using, for example, data from the Tax Office. In the UK the central business survey sampling frame is the Inter-Departmental Business Register (IDBR). The IDBR is updated with both administrative data and data from the Business Register and Employment Survey.

In the United States, where the statistical production is split between different agencies, each agency must build there own frame. This leads to an extra problem with sample frames based on different sources and procedures, and consequently also of different quality. At the US Census Bureau, the source for the Business Register comes from tax data.

For those who do not have access to a business register, an alternative approach is to draw an area sample, list the businesses within the selected areas, and draw a survey sample from this list. In this case the area selection issue is an additional quality challenge. Different sampling methods are described in more detail in Chapter 5.

3.2.3 The Convenience Sample Element

Depending on how well the observational units are specified in the sample frame, a wide or narrower space is left for the businesses themselves to decide which entities fall inside or outside the business sample. The final decision is commonly based on screening questions at the beginning of the business questionnaires. These questions are used to check the business profile against register information. After doing so, management is asked only to include units and activities relevant to the survey. A screening question included in almost all business surveys is whether the business has been economically active during the reference period of the survey. There may also be questions about what kind of activities different units within the enterprise perform, combined with instructions on which units and activities the business respondents should report from. The answers to profiling questions are commonly also used to update and improve the business register.

From a methodological perspective, this introduces an element of convenience sample and a source of error in the surveys. When the businesses themselves must decide which entities belong to the sample, the surveyor loses control over the sample size. As a consequence, the survey organization is, strictly speaking, unable to calculate response rates and sample variances. Moreover, while a profiling team can ensure that the same classification and sample inclusion principles are used for all businesses, profiling questions included in the survey are prone to misunderstandings and different inclusion principles used by different businesses. The use of profiling questions are discussed further in Chapter 11.

As an illustration of how challenging screening tasks in business surveys can be, we can once more return to the example of the firm that performs computer services. We have recognized that units within the company that offer computer services as a secondary activity might be excluded from the sample frame. But in addition, the responding businesses must also recognize that units that offer internal computer services should not be included because this kind of services falls outside the definition of true transactions.

One kind of business survey that has a strong element of convenience sample is the surveys used to collect price information. In these surveys a sample of business units within an industry is usually asked to report the prices of a representative selection of the services or goods they are producing. Strictly speaking, therefore, it is the selected services and goods that are the observational units, while the businesses selected in the first step of the design are reporting units. Even if the sample of reporting units is drawn according to statistical probability principles, the sample of services or goods is drawn by intuition.

There are also reasons to believe that screening questions sometimes are used to avoid reporting. Here is a suspicious example from the Norwegian lorry transport survey:

> In this survey, which is mandatory to respond to, lorry [truck] drivers are asked to give details about their transport services during a 7-day period. On average, the response rate is calculated to be as high as 94%. However, more the $\frac{1}{4}$ of the vehicles are typically reported to be out of service during the reporting period. A follow-up study revealed that quite a lot of these should have been included in the sample and not be considered as ineligible.

The risk of convenient escape or erroneous omissions is often increased by the fact that the instructions on whom to include or exclude are often not posed as questions at all, but presented as instructions that the surveyor takes for granted that the businesses will follow. The combination of mandatory, burdensome tasks and obligation-free screening questions probably leads to an overestimation of non coverage and underestimation of nonresponse in business surveys.

3.2.4 Sampling Error Issues

Business samples are normally random samples, stratified by industry and by the number of employees or the size of the business' turnover. The number of employees is most often used when the main focus of the survey is on production, while turnover is preferred when the focus is on financial output.

Because of the economic importance of larger firms, all firms over a certain size or with revenue beyond a certain limit are usually included in the survey; that is, they are sampled with a selection probability equal to one. Hence, for these units there will be no sample variance. At the other end of the employment spectrum, the smallest companies may be omitted from the sample altogether. Information from those businesses may be discarded or estimated with the help of data from administrative registers. The sample variance will, in other words, be an issue only for medium-sized and smaller businesses included in the sample. Note that in the previous paragraph we argued for the reverse situation for coverage errors. Coverage problems apply primarily to the larger units.

Business surveys often reflect trends. What we call *short-term statistics* (STS) is therefore almost always based on business panels. The advantage of panels is that ecological fallacies that confound changes that affect individual businesses with structural changes in the composition of the businesses world, are avoided (Duncan and Kalton 1987; Trivellato 1999). On the other side, a panel sample that is not continuously revised as the business population changes will not be able to pick up economic implications of structural changes. Analysts want to evaluate how well individual businesses are doing, how the structure of the business world is changing, and how the latter changes affect the conditions and prospects for businesses to survive. Therefore a part of the panel sample must be renewed as the economic structure change. In a modern world undergoing rapid economic change, this is a quality issue that affects both the relevance and accuracy of the data collected.

As we have already stated, response burden is an important issue in business surveys. On the other hand, the temperature of modern economies needs to be continuously monitored using surveys and other data collection instruments. The number of business surveys is high relative to the business population. Therefore the sample probabilities are also high, in particular for the larger companies. Moreover, the panel character of many business surveys adds to the response burden for those that are selected. For this reason, national statistical institutes (NSIs) that conduct a number of business surveys often coordinate between samples so that businesses that have already been burdened with many surveys are excluded from the sample frame for a while or have a reduced probability of being sampled in new surveys. Different sample coordination techniques are discussed in Chapter 6. What is most important to recognize from a quality perspective, however, is that sample coordination introduces a dependence between samples that may reduce their representativity. Hence, the need for reduced response burden needs to be balanced against the need for accurate estimates.

3.2.5 Reporting Units and Informants

The reporting unit is the unit within or outside the business that actually reports the figures asked for (Willeboordse 1997). This may well differ from the units or entities that are observed. We just presented an example from price index surveys where the units sampled by the survey organization actually are reporting units. Other times enterprise managements are asked to report on behalf of different production units. Moreover, business managements quite commonly request an accountancy office outside the business itself to report on their behalf. In their communication with the businesses and depending how the questionnaire is designed, the surveyors will try to influence the decision on what the reporting unit should be and what competence the actual respondent should have. But whatever strategy is used, it is still incumbent on the business management to decide who should do the job.

There might be a conflict of interest here. The surveyors naturally want the most relevant unit and the most competent employee(s) to answer the questions. From the managers' perspective, however, filling in questionnaires might be considered a nonprofitable waste of time. If so, they might prefer to ask a lower-paid and less competent person to answer the questions. One common feature in business surveys is also that they contain questions that call for different kinds of expertise. If that is the case, the surveyor would prefer that different parts of the questionnaire be completed by different people in the company. But again, to pass the questionnaire from office to office will lead to higher costs. As a result, the questionnaire might be given to a unit or a person competent in answering most of the questions, while the response quality of less prominent topics may suffer. For example, when questionnaires are left to be completed by the business' accountancy firm, the people working there will have firsthand knowledge of the financial questions raised in the questionnaire but only secondary knowledge of how production units operate. We will return to the communication and design strategies used by business surveyors in Chapters 8 and 9.

The different business units that we have referred to in this chapter are defined by legal status, location, and kind of activity. The general dimensions in this typology are geography and production activities. In Figure 3.3 we have added local units, homogenous production units, and local homogenous production units to those already named. The distinction between some of these concepts is unclear and included here primarily for the sake of completeness. See Willeboordse (1997) for definitions of the different units. Not all of these units may be observable, and even if they are, they may not be included in the business register from which we draw our sample. Often the best we can do is to ask the business to identify observational units that are as close to our analytical unit as possible and to choose a reporting unit and actual reporters that can provide valid and reliable answers to our questions. If there is a distance between the analytical and observational units, it is a quality problem.

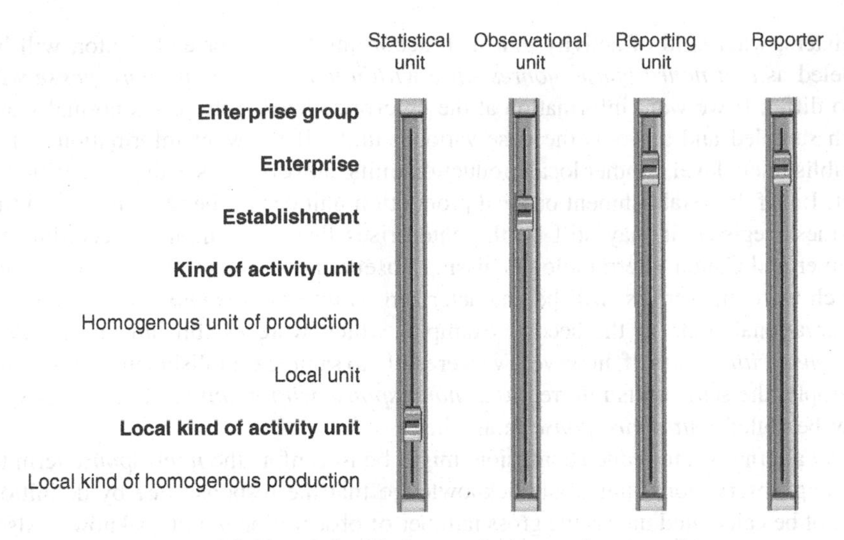

Figure 3.3 The relationship between business and survey units.

Likewise, it may also be a quality problem if the reporting unit or the reporters are on different levels or have no firsthand knowledge about the observational unit. Hence, we want all these survey units to be as close to each other as possible. If we locate each kind of units on a slide bar that ranges from enterprise group to local homogenous production units, the different positions on the slide bar can be used to identify possible unit quality problems (see Figure 3.3).

3.2.6 Response Rate Issues

Because many official business surveys are mandatory, which generally leads to a high response rate, business surveyors are often not so concerned with response rate measures. Nevertheless, we think that response rate issues are of sufficient importance to warrant discussion.

In social surveys we normally distinguish between *unit nonresponse* when the sampled person or household has not responded, and *item nonresponse* when they have responded but not answered all the survey questions. In business surveys it is not that simple, because the *observational unit*, the unit about which we want information, may differ from the *response unit*, which is the unit that provides the information. In addition, it often becomes even more complicated because business surveys may consist of more than one questionnaire that should be completed for each observational unit. We may lack either all reports from a certain reporting unit, some of the observational units within a business, or some of the questionnaires that should be completed for each observational unit. Biemer and Cantor (2007) have used the term *nonresponse within units* to refer to cases where reports from the sampled units are incomplete. Note, however, that because the sampled unit will differ between surveys according to which units are identified in the business registers, what kind of activity unit that, according to Biemer and Cantor, will be labeled as *unit nonresponse, nonresponse within units*, and *item nonresponse* will also differ. If we want information at the enterprise level, enterprises normally are both sampled and serve as the observational units. If we want information at the establishment level or other local production units, however, this is the observational unit. But if the establishment or local production unit cannot be identified from the business register, it may still be the enterprises that are sampled. According to Biemer and Cantor's terminology, missing observational units in the first example, which were enterprises, will be characterized as *unit nonresponse*; while missing observational units in the second example, which were establishments, as *nonresponse within units*. If, however, we were able to sample establishments the second example, the same units referred to as *nonresponse within units* in one design will now be called *unit nonresponse* in another.

An alternative, to reduce confusion, might be to confine the *nonresponse* term to missing observational units, and acknowledge that the response rate by definition cannot be calculated unless the gross number of observational units is known. Also, unless the reporting units are identical with the observational units, the response rate cannot be based on the number of reporting units. Moreover, if the units sampled are asked to identify the units from which we seek information, as in the example were enterprises were asked to identify relevant establishments, the number they give cannot be controlled. Hence, the stronger "the convenience sample element" is in a survey, the more uncertain the denominator in the nonresponse formula is.

We think there is a similar need for clarification of terms when more than one questionnaire has to be completed for the same observational unit. In this case it could be tempting to distinguish between *nonresponse within units* when questionnaires were missing in contrast to *item nonresponse* when questions within questionnaires were unanswered. But again this distinction depends on a certain design. If questions that were split into different questionnaires in one design were gathered in one and the same questionnaire in another survey, the scenario that was called *nonresponse within units* in the first design would be recognized as *item nonresponse* in the second design. As in the previous example, we think it is better to use the same term, *item nonresponse*, in both instances and acknowledge that the risk of a high item nonresponse rate is different when questions are split into several shorter questionnaires compared to one long questionnaire. A long questionnaire may lead to increasing item nonresponse the further into the questionnaire we come. A collection of shorter questionnaires may lead to item nonresponse on all questions posed in a missing questionnaire.

3.2.6.1 The Significance of Legal Obligations

Some comparisons between mandatory and voluntary business surveys show marked differences in response rate, while other comparisons show a surprisingly low difference. A comparison of seven US government censuses and surveys in 1998 showed a simple average response rate of 86% for mandatory surveys compared to 81% for voluntary surveys (IGEN 1998). Note, however, that these comparisons do not factor in differences in target population, the surveyor's authority, the survey topics, or survey design. Willimack et al. (2002) describe experiments that control for factors such as these and that report response rate differences between mandatory and voluntary business surveys ranging from 13% to 23%. Moreover, a comparison between surveys conducted by the US Census Bureau and the Bureau of Labor Statistics seems to indicate that institutes that normally conduct mandatory surveys (US Census Bureau) also have high response rates for voluntary surveys, while institutes that normally conduct voluntary surveys (Bureau of Labor Statistics) struggle more to make the businesses respond (Petroni et al. 2004a, 2004b). Similar results are found in a response rate comparison for the *business and consumer tendency survey*, which is a voluntary business survey conducted by different kinds of surveyors around the world (McKenzie 2005). As an example, in Norway, where the business tendency survey is one of the few voluntary business surveys conducted by Statistics Norway, the response rate was 85% in 2003, and has increased to over 90% since then (Wang 2004).

Those who seem to struggle most with nonresponse are universities, research institutes, and similar nonofficial business survey surveyors (McKenzie 2005; Paxson et al. 1995). This may be because these institutions do not have the same standing among business managers as official statistical institutes or because the more analytical topics of these surveys are met with more skepticism. Business managers probably perceive a greater need to report their production and economic results to official statistical agencies than to disclose internal procedures and relations to research institutes.

Surveyors that cannot impose mandatory business surveys apply recruiting strategies based on compliance principles known from social surveys; and, according to some of the examples cited above, quite successfully so. Mandatory business surveys may also profit from these strategies. If not noticeably on the response *rate*, good recruiting strategies may have a positive effect on the response *quality*, It is important to recognize, however, that in business surveys, respondents will be employees who are paid to do what they are told, including answering survey questions; and that often it will not be the respondent, but the business management or other gatekeepers, who decide whether a questionnaire should be answered and who should do it. Therefore, marketing efforts should probably be directed more to the management than to the respondent. Tailoring of different compliance principles to business surveys is discussed in Chapter 9.

Because of their importance for economic statistics, all larger companies are commonly selected for business surveys. In fact, as soon as the survey organization has collected data from these key participants, they are often able to publish preliminary results from the survey. As we will discuss in Chapter 4, it is therefore important to plan for a design that quickly responds to nonresponse or measurement errors in this part of the sample. If this is successfully done, preliminary results will often be almost identical to the final results after the data collection has closed. We have already noted that making business survey mandatory is an effective way to ensure a high response rate. One drawback of referring to the law, however, is that it also obliges the survey organization to ensure that the law is followed. Thus, even if a *responsive follow-up design* is used to quickly gather data from the most important businesses, it may take considerable time to finalize the data collection. In this way the mandatory character of business surveys may weaken their cost efficiency.

3.2.7 Quality Effects of Postsurvey Adjustments

Because business samples normally are stratified according to size, the results need to be weighted by the inverse of the sample probability. The most obvious quality consequence of this is that results from smaller businesses, which need to be weighed up, are less certain than results from larger businesses. Another source of adjustment errors could be that because of unit specification problems, the sample probability, and hence also the weights, are uncertain. Generally, sampling problems often are magnified by weighting techniques. This topic is discussed in more detail in Chapter 5.

3.3 MEASUREMENT-RELATED QUALITY ISSUES

Measurement errors can be divided into validity and reliability errors. Basically, with *validity errors*, there is a mismatch between the information we seek and the information given by the respondent, whereas with *reliability errors*, the answers are correctly understood but not accurately reported. In the survey cycle model presented in Figure 3.1, we distinguished between two kinds of validity. The first is *construct validity*, which is the extent to which our measurement instrument distinctively measures the phenomenon or construct that we want to gather information about (Hox et al. 2008). Next, *measurement validity*, is the extent to which our survey questions collect the intended information.

Construct validity belongs to the specification phase in survey planning. Fundamentally it is concerned with the relationship between theoretical concepts and models used in socioeconomic science and what kind of information we decide to collect in surveys. Socioeconomic properties are often not directly observable. Consequently construct validity is heavily dependent on the quality of social concepts and theories and on how well the information needed to measure the concepts embodied in a theoretical construct are communicated to the surveyor. In contrast, measurement validity belongs to the operative phase. The challenge in this phase is to translate the information needs into questions that work with the survey respondents. To succeed in this task two conditions need to be met:

1. The questions should cover all aspects of the concepts that are to be measured. This condition is reflected in the common survey requirement, which states that response options should be exhaustive and mutually exclusive (Fowler and Cosenza 2008). The same requirement applies to questions that are meant to measure different aspects of a construct.
2. The questions must be properly understood by the respondent. As the surveyor moves from the specification of information needs to questionnaire construction, the most important communication partner during survey planning consequently shifts from the stakeholders of the survey to future survey respondents.

As indicated in Figure 3.5, there is a fine line between what is perceived as valid questions and questions that collect reliable answers. Questions that collect reliable answers should give the same results as long as what is measured is stable and the respondents that are compared are equivalent. If this requirement is not met, the survey measurement will both be unreliable and invalid. Even if it is met, the answers given can still be irrelevant, and consequently give invalid answers to the questions that we want answers to. Note also that measurement validity is a part of construct validity in the sense that high measurement validity is a necessary but not sufficient condition for high construct validity.

The term *relevance* that we suggested as first priority from a user perspective coincides with the term *validity* used by survey professionals. However, what users of statistics perceive as relevant may somewhat differ from what survey professional perceive as valid information or measurements. While the professional reference point is the relationship between the measurement instruments and socioeconomic concepts and models, the user's perspective will be of a more practical character. Typically in business surveys users may not be so concerned about how well the questions measure economic constructs, but rather how well the survey results predict the forthcoming economic development. Or, in other words, they may not be so concerned about the theoretical value of the business survey results as long as the results correlate consistently with economic realities. In classical measurement theory we distinguish between construct validity, content or translational validity, and prediction validity (Carmines and Zeller 1979; Trochim 2009). These three terms may, we believe, coincide with the terms *construct validity, measurement validity*, and *user relevance* shown in Figures 3.1 and 3.2.

3.3.1 Validity Challenges during Survey Specification

The term *construct validity* is coined by psychologists who typically ask questions that are designed to measure abstract concepts such as personality traits and value orientation. Compared to this, the information needs in business statistics seem to be more straightforward and concrete. Thinking about the abstract character of the observational units in the national account, however, one is reminded on the fact that the national account is a theoretical, economic construct.

The fundamental purpose of the national account is to measure the value of the services and goods that are produced in a society. One basic validity problem related to this purpose, is that the values of services and goods are not realized before these items are consumed. Indeed, results from consumer and expenditure surveys factored into the gross national product (GNP) are calculated. But information on actual consumption is often difficult to collect because services and goods are sold through many, partly unknown channels and because consumption is not always recorded or easy for consumers to remember and report. For these reasons data from business surveys, which rather ask about deliveries or sales, are used. An underlying and questionable premise for the value estimations is then, that what is delivered or sold is also consumed. The value of statistics can be used as an example. Even if we know something about the characteristics of statistical users, there is no easy way to assess what role statistics play in the society or in the national economy. Instead, we tend to estimate its impact by measuring how much statistics that is produced (Gonzalez 1988).

Another validity challenge during the specification step is that user needs other than those defined by primary stakeholders are loosely described. Also, what might be adequate for one purpose may be flawed for another. We think that business surveys users often can be located on a scale, with socioeconomic analysts at one end and case handlers, like stockbrokers or estate agents, on the other. Analysts will normally want a rich dataset covering different aspects of the topic surveyed, while case handlers will prefer key indicators broken down into specific industries or geographic areas. The business surveys conducted are found somewhere on this scale. User needs change over time. In order to ensure the relevance of the surveys, both the user needs and the available information sources should always be investigated before new surveys are planned and carried out. Chapter 7 describes some of the techniques that we recommend for such of investigations.

3.3.2 Sources of Measurement Error in Business Surveys

Measurement errors occur during the response process. That is just as much the case in business surveys as in social surveys. The response process is commonly divided into four cognitive steps (Tourangeau et al. 2000):

1. Comprehension of the information requested
2. Retrieval of what is perceived as relevant information, from memory or documents
3. Judging and processing the retrieved information against the information requested
4. Reporting by adjusting the response to the level of detail and the measurement unit asked for in the question and by the response alternatives

As in social surveys, the establishment respondent must understand the question and decide what information is required to answer it (comprehension). Next the respondent must search for relevant information, either from personal memory or in available records. This step differs so much from the equivalent step in social surveys that we have discussed it in a separate chapter (Chapter 2). In social surveys the focus is on memory challenges. By contrast, the mental process in business surveys will typically be to determine whether the information asked for is available in the business' information system, and if so, where. If the data are available in files or documents, the respondent must access those data (retrieval), possibly by communicating the data requirement to a third party (Edwards and Cantor 1991). Hence, it is also quite common, and clearly more common than in social surveys, for more than one person to participate in the response process. Note, however, that even if business surveys, more often than social surveys, ask for information that should be recorded somewhere, they also ask for information that might be stored only in the

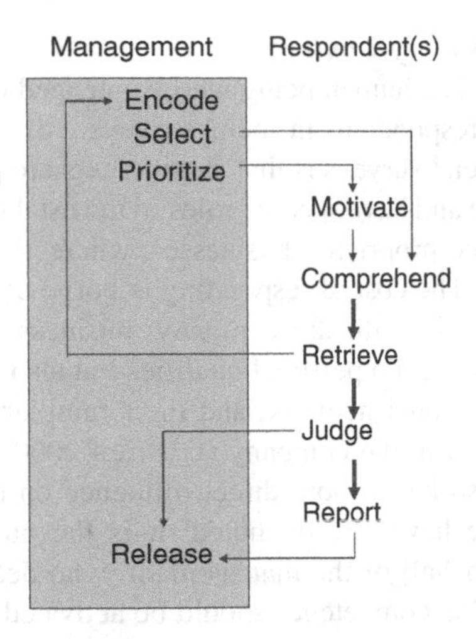

Figure 3.4 The relationship between business management and the response motivation and response process in business surveys. [Based on model from Willimack and Nichols (2010).]

respondent's or other employees' memory. Both the records and the knowledge of colleagues are parts of the business' information system (Bavdaz 2010a and b).

When relevant information is made available, the respondent must decide what to report (judgment) and compile that information. At this stage calculations or estimates may also need to be done. As we will see later in this chapter, the judgements and calculations made by business respondents will often be quite different from what is common in social surveys. Finally, the respondent must present the information in the format requested on the questionnaire (reporting) (Goldenberg et al. 1993).

The respondent's motivation sets the mood for the response process. As you will see in Figure 3.4, we have placed motivation at the top of the list of cognitive steps that will affect the response quality. If the respondents are not sufficiently motivated or meet conceptual challenges that are difficult to overcome, they tend to take shortcuts and chose what is commonly called a *satisficing strategy*. The term satisficing is a combination of *satisfies* and *suffices*; it indicates a response strategy that is sufficient to satisfy the surveyor without going through the different cognitive steps in an optimal way. This can be done by either (1) only superficially reflecting on what the questions ask for, what is relevant information, and what the answer should be, or (2) skipping cognitive steps altogether. The steps normally skipped are those two in the middle: retrieval and judgment (Krosnick 1991; Krosnick and Alwin 1987). The term *satisficing* has traditionally focused on attitude questions and order effects in social surveys, but the basic argument that difficulties might lead the respondent to look for shortcuts probably also applies to business surveys.

How the respondent addresses and masters the cognitive challenges listed above are on one hand dependent on personal and social characteristics, and, on the other hand, on the kind of questions posed and how they are presented. We start with the personal characteristics of a respondent operating within a business context.

3.3.2.1 The Business Survey Context

The business respondent is a human being who is engaged or disengaged, clever or not so clever, just as are respondents in social surveys. Still, an important difference between business and social surveys is that respondents are paid by their employers to fill in the questionnaire and have specific roles in the establishment's organization. This is even true in sole-proprietor businesses where the same person is both employer and employee. The cost of responding is borne by the company's budget. Because of this relationship with the company, business survey respondents are characterized not only by certain personal qualities but also by their positions in the organization, their professional network, and their familiarity with how the information flows and is stored in the company (O'Brien 2000).

The social context also has a more direct influence on the response process in business surveys. As we have already noted, it is the business management, or employees, who act on behalf of the management, who decide who should be the respondent, and hence what competence should be activated. The business administration also decides whether responding to questionnaires should be considered as part of the employee's normal duties and what priority the task should have. In this way the business management affects the respondent's motivation. Furthermore, administrative files that the respondent may need access to are established, updated, and controlled by administrative units. Obviously, which data are recorded and how easy it is for the respondent to access these data affect the retrieval step in business surveys. Research indicates that the respondent's perceived response burden to a high degree is the result of constraints set by the management, such as available time and access to relevant data sources (Haraldsen and Jones 2007).

If the recorded information is time consuming and difficult to access, memorizing may be an alternative, but more superficial response strategy. In panel studies, which are common in short-term business surveys, an easily accessible alternative to records may also be to slightly adjust the figures reported the last time. In fact, it is quite common in panel surveys for figures from the previous round to be included in the next round together with updating questions. In situations where recorded data are burdensome to access, this design may be particularly conducive to satisficing.

Finally, because business data may be market-sensitive, the answers given by the respondent may also be checked against the business' release policy before they are reported to the surveyor. In this way the respondent's motivation, competence, and work conditions are affected by decisions and characteristics of the employing company (Willimack and Nichols 2010). Because of this gatekeeper role, how the business surveyor addresses, informs, and communicates with the business management may have important consequences for the survey quality. In Figure 3.4 we have indicated how the relationship between business management and the motivation and response process in business surveys works. The influence of survey communication on these relationships is discussed in Chapter 9.

This is the general picture. The impact of the business context will vary between industries and, probably even more importantly, with size of the business. In larger firms the division of labor will be more formalized and the information systems, which both include human resources and available records, will be of a higher quality. In larger firms the response task will also often be given to more than one person. If so, how well the interaction between those who take part in the response process works may be an extra factor that affects the response quality (Gonzalez 1988). Those involved communicate about the questions posed in the business questionnaire and perhaps explained in separate instructions, but only some of the actors might actually have the questionnaire and the instructions on their desk. In a survey of this internal communication, Bavdaz (2010b) noted that some respondents handed over both the questionnaire and instructions to colleagues who helped them with some of the answers, while others handed over only a copy of parts of the questionnaire or simply made inquires based on their own understanding of the questions. In this way, how much of the questionnaire and appended information they had seen varied between those who provided the survey answers. This may clearly have consequences on quality. The advantage of more competent people being involved in the survey may be offset by the disadvantage of more people providing answers without actually reading the questions.

In the more numerous smaller firms, the information system may be less advanced and the different roles will more often be performed by the same person. Consequently the different concerns, such as time needed to complete questionnaires and how easy or difficult it is to provide the answers, will also be concentrated in one head. Smaller firms tend to have simpler information systems, and hence need to rely more on answers provided from memory rather than collected from records.

The combination of restricted resources and poor information systems have been observed to burden medium-sized businesses the most. While the largest businesses have more resources, the smallest businesses often have little to report and therefore do not need to spend much time on the questionnaires. They are also not so dependent on how well their activities and financial results are documented. Measurements of perceived response burden indicate that business surveys are felt most burdensome for establishments with 20–49 employees and is significantly higher for establishments of this size compared with very small businesses and those employing at least 50 people (Haraldsen and Jones 2007). Also, as establishments with more than 20 employees are more frequently surveyed than smaller businesses (see Table 6.5), response burden, and with it, also the possible quality effects of response burdens, seem to accumulate in medium-sized businesses.

In the previous paragraph on sampling issues, we recognized that business samples usually are stratified according to size groups. In other words, the samples are tailored to the size of the business. We think that the quality reflections we have just made regarding the measurement instruments used in business surveys, call for a similar tailoring of the survey communication and questionnaire design according to size. The response quality from larger companies seems to depend heavily on how the questions are communicated and handled internally among respondents and data providers. The quality challenge is therefore to determine how the surveyor can influence those individuals involved in the response process and how they work together. For medium-sized businesses, the response burden seems to be a main quality issue. Coordination between samples and simplifying the questionnaires are examples of measures that may reduce the response burden and increase the motivation to provide high-quality responses. Finally, respondents in the smallest businesses appear to be in much the same situation as participants in household surveys. They are seldom sampled for business surveys and should probably be treated as lay respondents that will answer most of the questions from memory. One point that complicates this picture, however, is that smaller firms that participate in ongoing business surveys such as employment and wage surveys, often delegate the reporting to their accountancy firm. Although most of these accountancy firms should probably be treated as medium-sized or large businesses, they differ in that their response competence is confined to accounting matters. Hence, if the surveys they receive from their customers contain questions about production or sale activities, they will seldom have firsthand knowledge of these topics and may consequently give unreliable answers or no answers at all (Bavdaz 2010a and b).

How the survey institution communicates with the business management will be a key topic in Chapter 9. Sample coordination procedures and response burden reduction measures are discussed in Chapter 6. In Chapter 8 we discuss how the questionnaire can be designed so that the response burden is minimized.

Finally in this section we also want to mention a quality problem caused by sensitivity concerns. In household surveys, anonymous microdata are commonly made accessible to social scientists for methodological inspection and alternative analysis. In contrast, access to business survey data need to be much more restricted. Because larger companies may easily be identified from the results and because data on the production and economy of the firms are sensitive, microdata are rarely made available outside the organization that has collected them. This is a quality challenge in the sense that it leads to high demands on the professional impartiality and integrity of the institutions that handle microdata from business surveys and assess their quality. Statistics Norway has its own research department where microdata from business surveys can be analyzed. The US Census Bureau makes business microdata available to their own Center for Economic Studies, a think-tank of economists. There are also data user centers at universities in different regions where academics can come and do research, However, security is tight, and the researchers must be "sworn in" to protect the data and must work onsite, at the census bureau or one of the data centers. Even with these arrangements, it may be argued that the restricted access to business survey microdata weakens the possibilities for independent quality control. Section 12.5 discusses additional issues with data archiving.

3.3.2.2 Question and Questionnaire Elements

"The medium is the message" is a famous phrase coined by Marshall McLuhan nearly 50 years ago (McLuhan 1964). In self-administered surveys the questionnaire is the medium for communication with respondents. The key element of questionnaires is, of course, what information the text tells the respondent to report. In surveys the questions consist of three pieces of information. First there are words and phrases that may be easy or difficult to comprehend. Next, survey questions communicate a request for information retrieval and processing, a task that the respondent is asked to perform. Finally, questions present a response format, which will typically be some labeled alternative or an open response box in which the

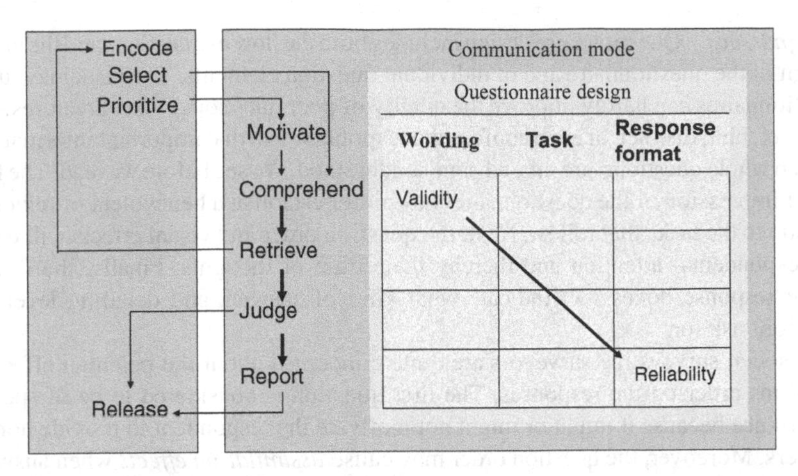

Figure 3.5 How the business context and different questionnaire elements affect the response process.

respondents are asked to report their answers. In addition to these text elements, the message is affected by how the questions and response alternatives are presented and by what mode of communication technology is used. In Figure 3.5 this scenario is depicted with the boxes labeled "questionnaire design" and "communication mode." The questionnaire constructers use these tools to address business respondents. The respondents' willingness and ability to report valid and reliable answers to our questions is affected by (1) how the questions are formulated, designed, and communicated and (2) how the response process is affected by the social context within which they operate.

The different question elements tend to be linked to different cognitive challenges. As a result, different parts of the questions also tend to be linked to different kinds of quality problems. Unfortunate question wording is a major reason for miscomprehension, and answers based on misinterpretation are, of course, not valid measurements. Retrieval and judgment problems are most often caused by complicated tasks. Depending on the specific retrieval and judgment challenges the respondents face, these problems may lead to either invalid answers or inaccurate estimates. Finally, even if the respondent has correct and precise information to report, the response format in the questionnaire may not fit with the level of detail or measuring scale used. This could lead to an answer that is less precise than it could have been with a better tailored response format.

We make use of the correlations described here when we conduct cognitive interviews with drafted questionnaires. If we are able to identify the kind of cognitive problems that questions cause, we also know what part of the question we should try to improve (Willis 2005). The reverse also applies. If we run validity or reliability tests on collected survey data, the results will indicate which cognitive steps should retraced and investigated more closely using qualitative methods. Cognitive interviews with business survey respondents will be one of the topics in Chapter 7, while validity and reliability measurements of survey results will be introduced in Chapter 10.

Design Issues Questionnaire design includes both the flow of questions and the visual design of the questionnaire and of individual question elements. The visual layout of questionnaires can hardly improve the quality of poor questions. Visual features can, however, blur, distract, or even confuse the respondents so that important information, or even whole questions, are missed or misunderstood. We see before we read. The first visual impression of the questionnaire may either establish a benevolent or reluctant attitude to the tasks that follow. Next, the question order and visual effects will catch the respondents' attention and thereby their train of thoughts. Finally, the design of the response boxes will indicate what kind of answers and detailing level the questions ask for.

In social surveys the surveyors are quite concerned about the potential effect of questions order on the responses. The first question is considered to be of special importance because it might or might not motivate the respondent to provide honest answers. Moreover, the question order may cause *assimilation effects* when answers to questions are harmonized so that they fall in line with each other or have a *contrast effect*, when respondents exaggerate the difference between questions (Sudman et al. 1996). Social surveyors are recommended to keep questions that may cause assimilation or a contrast effect as far apart from each other as possible. Because these phenomena are considered to be survey effects and not errors, order effects can also be neutralized by randomizing the order in which the questions are asked.

Similar concerns have until now been nearly absent in business surveys. Business respondents have been allowed to answer questions in the order they prefer. The reason for this is that different pieces of information may become available at different times or that different respondents should have the opportunity to answer different questions in the order that suits the establishment best. One exception of this lack of concern is when business surveys both ask for totals and for individual posts that adds up to a total. Questions like these could either be asked by starting with the total and then asking for breakdown, or progressing from the details that subsequently are added up to a total. Both methods are used [see, e.g., Figures 12.4 and 12.6 in Dillman et al. (2009)]. Going from details to a total is probably the most common order and seems sensible, as the totals that can be found in administrative records often will include items that should be excluded in business surveys.

Dillman et al. (2009) distinguish between order effects caused by cognitive processes (priming, carryover, anchoring, and subtraction) and order effects caused by normative factors (evenhandedness, consistency, and avoiding extremeness). A kind of normative pressure that could have quality implications in business surveys is when different parts of the questionnaire are completed by people of different ranks. Lower-ranking respondents will probably find it easier to report information that is consistent with what higher-ranking colleagues have reported rather than to contradict the picture given by their superiors. If the questionnaire is passed down the business line by rank, this may have an assimilation effect on the results.

One example of how visual design can affect responses that we consider particularly relevant to business surveys, is Dillman's manipulations with different response box designs and text placements in questions that ask for answers in a certain format. Dillman's tests employ different visual design of questions that ask for dates (Dillman 2007). He shows that with different designs the proportion who report dates in the format requested by the surveyor can range from 50% to almost 100%. Much of the editing done in business questionnaires addresses a similar problem, namely, that respondents fail to understand that they should not report detailed amounts, but amounts in thousands, millions, or billions. We will discuss how errors like these can be avoided by better visual design in Chapter 8.

Communication-Mode Issues Traditionally mailed paper questionnaires have been the dominating mode in business surveys (Christianson and Tortora 1995), but the possibility of presenting self-completion business questionnaires on the Internet has become increasingly popular (Haraldsen et al. 2011). The web questionnaire toolbox opens many new opportunities for quality improvements; the most fascinating one is the possibility of imitating the dialog that we know from interview surveys. In this way the dynamic character of personal communication can be combined with the flexibility of self-completion questionnaires (Couper 2008). There are reasons to believe that the introduction of web questionnaires in business surveys has positive effects on the respondent's motivation (Giesen et al. 2009; Haraldsen 2004). But the technology introduces new sources of error as well.

Computerized questionnaires are presented at a small computer screen window. One important consequence of this is that the electronic version of the questionnaire discloses less information about the number and content of the questions than a paper questionnaire. This is an example of how the data collection mode affects the questionnaire design and eventually the cognitive question–answer process and why the communication mode box is drawn as a basic framework that envelops the questionnaire design and the textual elements shown in Figure 3.5.

Note also that the data collection mode is depicted as a feature that affects not only the design and content of the questionnaire but also the respondent's motivation and cognitive processes. When paper questionnaires are moved to the Internet, the questions themselves may no longer be the main challenge for the respondents; the skills and time needed to log into the system, open the questionnaire, move from screen to screen, and finally email the answers back to the surveyors may pose additional challenges. These activities are also governed by questions of choice, response boxes, and action keys. The labels and functions of the administrative buttons found in an electronic questionnaire can thus be perceived as a kind of questionnaire within the questionnaire that requires motivation and competence in much the same way as the survey questions. In this way administrative tasks that were practical exercises on hardcopy are now computerized and governed by questionnaires with their own conceptual challenges and pitfalls. When business respondents are asked to search for information in administrative systems, this challenge may also include working in different windows.

A striking difference between the visual design of business questionnaires and questionnaires used in social surveys is that business surveys commonly are dominated by a ledger or spreadsheet matrix design with questions in the top row and items listed vertically. Tables that combine questions with units or items are generally discouraged in social surveys. Experiments that have compared household rosters with separate question sequences for each household member have shown that the roster version leads to more item nonresponse (Chestnut 2008). In establishment surveys, however, the recommendations are not conclusive. Spreadsheet design may work well for accountants and other business employees who normally use spreadsheets as their main working tool or when the spreadsheets are the information sources. For other business respondents and for questions that are more similar to questions in social surveys, a sequential design probably will work better (Dillman et al. 2009).

Computer screens are too small for spreadsheet questions may be preferable in some business surveys. Hence, if the original layout from the paper version is retained, the web survey respondents will need to move the screen window to view different parts of the matrix, and consequently may not be able to consider different parts of the matrix at the same time. Vertical and especially horizontal scrolling is generally discouraged in web questionnaires (Couper 2008). But again, these recommendations are based on research on social surveys and may not apply to the same extent in business surveys. Moreover, the usability of scrollable matrices will depend on how they are designed.

The alternative to keeping the matrix format is to replace each line in the matrix with identical question sequences. This was done when the original paper version was changed to a web version of the Norwegian lorry transport survey, One of the effects was that the sequential procedure apparently made it more difficult for some respondents to link the endpoint of one transport commission to the starting point of the next. As a consequence, more transports with no cargo were missed in the web version than in the paper matrix format (Haraldsen and Bergstrøm 2009). Whatever the best solution to these challenges is, the visual picture of the web questionnaire will be completely different from the original A3 paper format and will consequently affect the attention span, comprehension, information processing, and response in a different way.

The example above also illustrates a more general problem. In a typical data collection design in business surveys, different data collection modes are combined in the same survey. When the respondents are offered a web version of the questionnaire, those who refrain will be offered a paper version. It is also quite common for survey data to be combined with administrative data or other kinds of data from secondary sources. As new kinds of administrative data become available or the web alternative becomes more popular, the mix of modes will also change over time. If we do not correct for mode effects, changing results may not be true changes in the estimates but only reflections of changes in the composition of data collection methods.

3.3.2.3 The Content of Business Questionnaires

In a survey conducted among 21 statistical agencies in 16 countries, Christianson and Tortora (1995) distinguished between seven classes of variables in business surveys, and asked the participants to name the two most important variables and the most important measurement problems associated with these two in surveys conducted by the agency:

1. Business characteristics such as contact information, organizational structure, ownership, size, and characteristics of production premises, machinery, and

other kinds of technology. This is the kind of information asked for in most business surveys.

2. Production, measured by volume or value. This is the most important kind of questions posed to businesses in the manufacturing, service, and construction industry.
3. Financial expenditures and investments, often divided into expenditures and investments for different purposes.
4. Turnover, revenues and, volume sold are typical for retail, wholesale, and service industry surveys, but are also topics in other kinds of business surveys.
5. The number, qualifications, field of activity of employees, as well as hours worked and what the employees are paid are important information in any production survey.
6. Consumption of energy or other resources and types of waste from the production processes are highly relevant questions in surveys that focus on environmental issues.
7. The amount and value of transactions of goods or services between businesses, such as in foreign trade surveys.

The survey is useful because if offers a terminology for question types in business surveys and also lists common quality challenges recognized by the survey organizations.

In addition to the seven kinds of questions listed above, we would like to add evaluation questions posed in surveys where managers are asked to evaluate business prospects, employees are asked to evaluate working conditions, or customers are asked to evaluate products or services. Not all of these are business surveys in the sense that the respondents report on behalf of the business organization, but we will discuss them because the survey object is business activities and deliveries. Evaluation questions are also the most common kind of attitude questions in business surveys.

The problems reported with business characteristics are similar to those in household rosters and housing surveys. It might not be clear which units to include or exclude and which production facilities belong to which category. Moreover, questions regarding variables such as total floor space or total amount or type of equipment may be difficult to answer if a high number of areas or units have to be added. In other words, there may both be comprehension and judgment problems. As an example, Christianson and Tortora use farming enterprises run by landlords, tenants, and partners that include pasture, unimproved, and noncontiguous land and a variety of farm animals. The organizational arrangements in enterprises in other sectors may be just as complex. The production will often take place in different premises of different size and with different kinds of production facilities that were not previously categorized and counted.

In laborforce studies the classification task is complicated by the fact that many companies use leased employees who are formally contracted by another company and it may be unclear whether part-time or temporarily absent employees should be excluded or counted. In small, often family-driven firms, where the owner may be the only worker or work alongside other workers, she or he may also erroneously report her/himself as an employee.

Investments, expenditures, laborforce, production, and turnover figures are normally available in the business' accounting and payroll system. The basic prerequisite for good answers to survey questions about these topics is therefore how complete and accurate data in these databases are. Generally we might expect the quality of the accounting and payroll systems to vary with business size and that measurements of production volume will be of poorer quality than other figures. Moreover, the value of products that have not yet been sold may need to be estimated. Data distinguishing between different sources of energy or different kinds of waste might also need to be estimated.

Much of the accountancy and payroll data that is asked for in business surveys are also reported to the tax authorities and other governmental offices. Therefore initiatives are taken in many countries to coordinate deliveries for different purposes. One important difference between data collections for statistical and administrative purposes, however, is that while more efficient processing procedures have enabled central government authorities to allow for later deadlines, there is a consistent pressure on statistical agencies to produce more updated statistics.

Business surveys often ask for classifications or time references that are not readily available in the system. Business survey also ask establishments to specify investments and expenditures for unfamiliar purposes such as environmental improvements, integrating foreign workers, or recruiting more women into leading positions. Specific investments or expenditures for popular political purposes such as these either may not exist or be difficult to extract from the business' accountancy system.

3.3.3 Measurement Issues Summarized

In Norway in 2010, web respondents taking part in the annual structural business survey were asked to evaluate how burdensome it was to fill in the questionnaire and to identify sources of response burden. The web pickup rate was close to 90% in this survey Hence, response burden questions were posed to the majority of the sample. The main results are reported in Figure 3.6. The horizontal bars indicate which aspects of the questionnaire the respondents considered to be more or less burdensome.

Mismatch between questions or response categories and available information, need for help in searching for information available, and cumbersome calculations were reported to be the most burdensome aspects of this survey. The number of questions and layout of the questionnaire seem to be of less importance. At the bottom of the list come technical problems or complaints about the usability of the web questionnaire (Haraldsen 2010). The results should be read with some caution. Similar evaluations of questionnaires used to collect short-term statistics, nonofficial surveys, and business surveys in other countries where web questionnaires are not as well established as in Norway, may show somewhat different results. Still is seems quite clear that the mismatch between required and available information was considered to be the main problem. This result also coincides well with what Christianson and Tortora found in their survey.

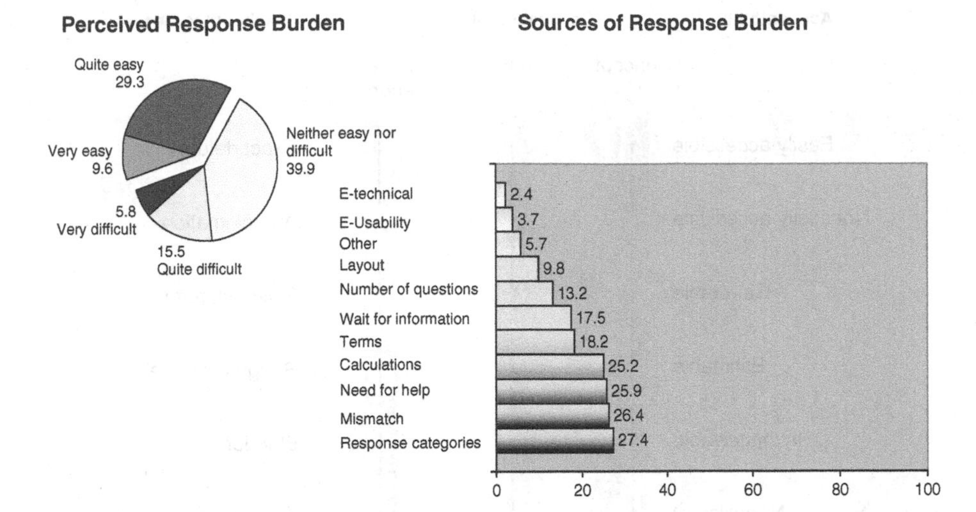

Figure 3.6 Perceived response burden and sources of response burden. Structural statistics survey 2010. Percent ($n = 16{,}572$).

Generally, content challenges seem to predominate over design and functionality issues. Next, the business respondents both give practical and substantial reasons for why they do not have the information asked for in the questionnaire. When respondents need help to collect relevant information or cannot supply the information asked for before the survey deadline, these are practical obstacles that can be solved by giving the respondents more time or help. The situation is different when the information requested in the surveys simply is unavailable and hence needs to be calculated or estimated just for the purpose of the survey. We consider these latter situations to pose the most demanding quality challenges in business surveys.

In an article about this quality issue, Bavdaz suggested a five-point typology of information availability ranging from accessible on one end to nonexistent on the other, and linked this typology with likely outcomes. The likely outcomes range from an exact datum if the information is accessible to item nonresponse if it is nonexistent. The alternatives in between are approximations, solid or rough estimates, and blunders (Bavdaz 2010a). In Figure 3.7 we have expanded Bavdaz' accessibility typology by distinguishing between three kinds of mismatch problems. The concept definitions used for business purposes may deviate from the definition used in the business survey. Then the survey may ask for information about units that are not specified in the administrative systems. This could be groups of employees or commodities or sublevels in the company that do not exist in available records. Finally, the time reference used in the business survey may not fit with the periods that are recorded in the administrative records, or the business survey may ask for information that has not yet been updated.

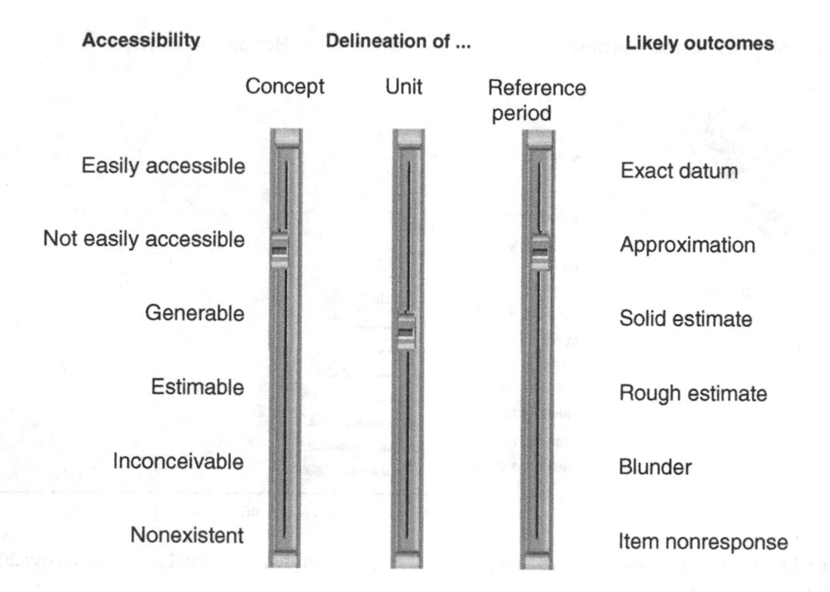

Figure 3.7 Practical and substantial information problems in business surveys. [Based on Figure 4 in Bavdaz (2010a).]

As with unit specifications discussed earlier in this chapter (see Figure 3.3), these three kinds of measurement problems can also be depicted on slide bars. The slide bars can be used to identify what kind of delineation problems survey requests pose and, in turn, what kind of response strategies which are available to the business respondent. To give an example, if the concept, the unit, or the reference period asked for in the survey is not directly accessible but is only estimable, the best the respondent can offer is a rough estimate. Or, in more general, terms, the response strategy listed on the same line as the mismatch problem will indicate the optimal way of answering the question. The options listed further up on the list are not available, and those listed further down can be characterized as different kinds of satisficing strategies.

3.4 SAMPLE AND MEASUREMENT ISSUES COMBINED

Both in the discussions on sample-related quality issues and in the previous section, about measurement-related issues, we end up with models focusing on mismatch problems. In fact, the mismatch problems between the statistical unit and the observational unit, and between the observational unit and the reporting unit and reporter in Figure 3.3, can be seen as specifications of unit delineation problems depicted in Figure 3.8. But the relationship between unit and measurement issues often seems even more interwoven in business surveys than this.

We have already stated that information relevant to the questions asked in business surveys often is stored in the company's administrative system. When this is the case, information retrieval is not about memorizing, but about identifying and utilizing available information. In a similar way as with memorizing, we think that this kind of retrieval task also needs a theoretical framework that describe its challenges. As a first step towards such a model, we lean on a model developed by Li-Chun Zhang (Zhang 2012). His model is used to discuss the various error sources in data integration when statistical agencies collect and transform data from administrative registers into registers accommodated for statistical purposes. In our mind this situation is very similar to the task given to business respondents when they are asked to accommodate information available in their business registers to our statistical requirements. The main difference between the situation Zhang describes and a business survey is that the tasks are being done outside the statistical institute and are presented in a questionnaire.

Zhang suggests a two-phase model, both with a representation and measurement cycle as in the survey cycle model that we presented in Figure 3.1. In fact, the first step is a general version of the survey cycle model that renders it applicable to both sample surveys and cases when data are collected from administrative registers. The second step describes how the administrative information collected from the first step is transformed to units and variables required for statistical purposes, in our case, requested by the survey organization. Zhang's two-phase model is shown in Figure 3.8.

By basically using the same model for retrieval from business registers as the one we used for data collection by surveys, the first phase in the model points out and name sources of errors when business respondents seek information in their administrative system. Note that the starting points of the two cycles in this first phase are termed *target set* and *target concept*. This reminds us of the fact that an exact match with the units and variable we want for statistical purposes may be unattainable. Also, the term *objects* is used instead of *units*, because the units and variables we want for statistical purposes may not be readily available, but need to be constructed during the retrieval process. Let us now first take a closer look at the representation cycle.

The accessible set of objects from relevant information sources in the enterprise corresponds to the sample frame in surveys. As we pointed out earlier in this chapter, it will vary according to the relevant information available in the business and with what information the respondent has access to. This may lead to different coverage errors; in Zhang's model these are called *frame errors*. Next, the respondents may not inspect all accessible information, but base their response to the survey questions on a sample of what is available. Hence, the term *selection error* in the model corresponds to sample errors in surveys; the major difference is, of course, that the samples that respondents draw normally will be convenience samples. Finally, there may be relevant information missing in the business documents or databases observed by the respondents that lead to errors equivalent to what we call *unit* and *item nonresponse* in surveys.

The term *construct validity* used in the survey cycle model does not seem appropriate for the kind of validity problems that may occur when a measure is extracted from an administrative source. Constructs are abstract socioeconomic phenomena that cannot be directly measured. The purpose of administrative systems, by contrast, is to serve concrete and practical tasks. The validity errors referred to in

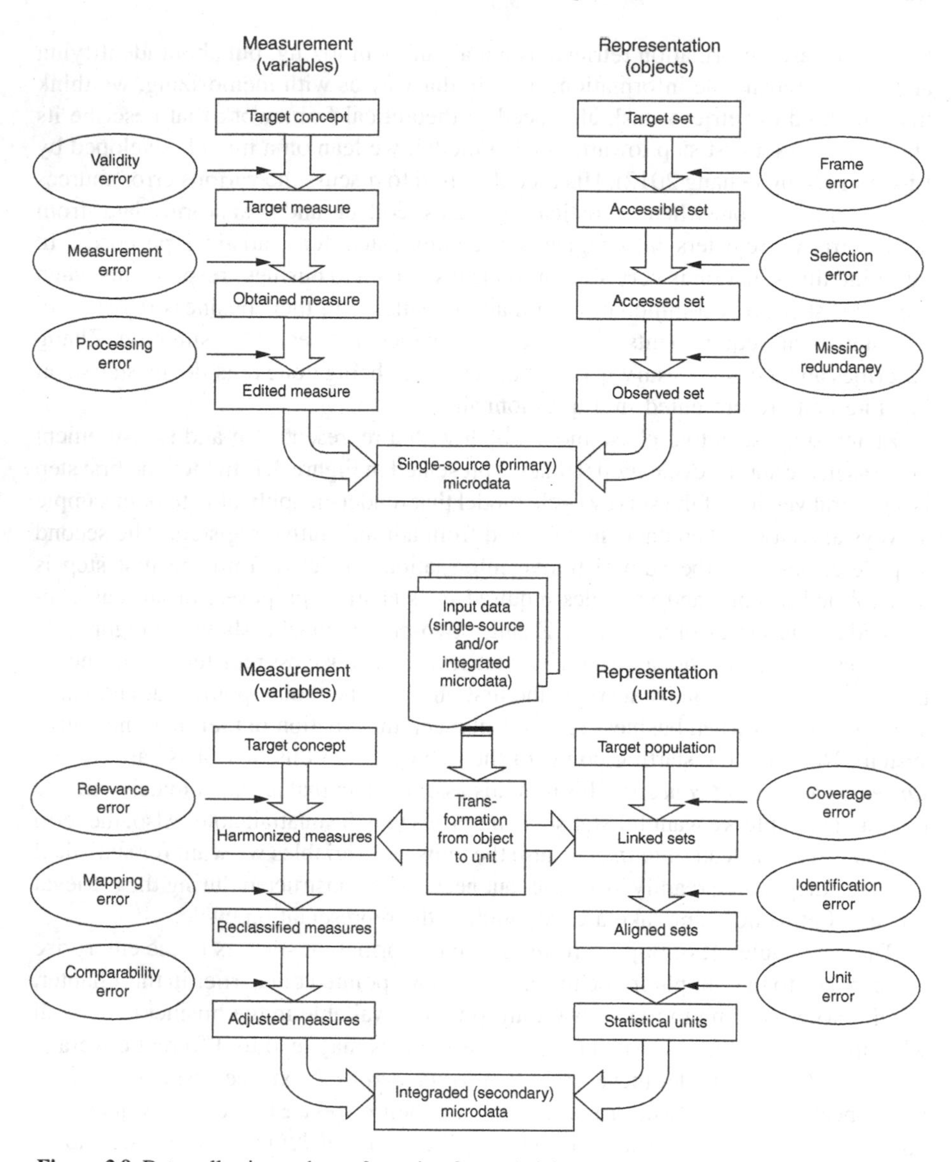

Figure 3.8 Data collection and transformation from administrative sources. [*Source*: Zhang (2012).]

Zhang's model therefore seem to be closer to what we earlier called *content validity* than to construct validity errors. The point is that measures collected from an administrative register may be invalid because they do not fully cover the information needs expressed in the business questionnaire. In the survey cycle model we used the term *content validity* to characterize validity problems caused by business survey questions, that is, linked to the survey response rather than to the relationship between concepts and the measurement instrument. This difference between the two models means that what Zhang calls *measurement errors* in the measure obtained from administrative sources will be reliability errors.

Another important difference between running surveys and collecting data from available sources is that while measurement errors in surveys are related to the measuring process, they relate to results of previous measurements when data are collected from administrative registers. This is why a term such as "obtained measure" and not "response" is used in the model.

In the second phase Zhang discuss how administrative data are transformed to statistical data in general and how information collected from different administrative sources is integrated into statistical products. Business respondents may sometimes base their answers on information collected from different sources, but normally they will probably lean on one main administrative source. Hence, integration challenges will not be as prominent as when statistical institutes produce statistics from administrative registers; or, it in other words, the balance between transformation and integration challenges will be different. This also means that we will focus more on the transformation challenges than the integration challenges in our presentation of phase two.

Objects and measurements in the administrative files may not coincide with the units and variables in business statistics. Administrative registers are commonly based on events, which means that units in administrative registers are variables in statistics and vice versa. Zhang uses an education register as an example of this. The units in this register are completed exams, and the identification of the student who has completed a specific exam will normally appear on several registrations. When we produce statistics, however, we want it the other way around. The unit in statistical tables should be students, while the number of exams they have completed should be their value on a variable. When we ask business respondents about production volumes or sale figures, we confront them with an equivalent transformation task. The units in the businesses' administrative databases are receipts that need to be sorted and summarized according to the classification we ask for before they can be reported in the questionnaire. When doing this, the units and variable change place.

The units we want reports from will also need to be constructed by linking units that meet certain criteria together into what is called "linked sets" in Zhang's model. An example might be when units producing certain goods or services should be grouped together. During such linkages two kinds of coverage errors might occur: (1) the information needed to establish the statistical units is insufficient or missing in the administrative records or (2) pairing into sets of units is done incorrectly. In both cases we will have a kind of coverage error.

Next, what Zhang calls *alignment* is a process whereby the relationships between different units in the linked data are clarified. The source of *identification errors* in the model occurs when basic units that are linked together belong to different composite units. This may lead to several identification problems. When information from different administrative sources is compared, one common identification problem is that different pieces of information contradict each other. Consequently one has to determine what information to trust before composite units can be established. Erroneous decisions when this is the case obviously lead to wrong composite units. Another identification problem we consider even more relevant in business surveys is that overlaps in basic units included in different composite units might lead to double reporting in the questionnaire. An even trickier problem is that the identification of composite units may change with the reporting unit. A typical example, described by Zhang, is when statistical units are composed of subunits with different industry codes. As we explained previously, industry codes are decided on judgments of the main activity of the unit in question. The problem is that what is considered to be the main activity of each individual subunit may differ from that considered the main activity when all the subunits are combined and judged together. As a consequence, this generally leads to different totals by industry codes, when it is aggregated over all the enterprises that have the code, versus when it is aggregated over all the subunits that have the same code.

It is worth noticing that what Zhang calls *unit errors*, which is information partially or completely missing for some of the statistical units, is introduced at the end of the representation cycle in the second phase of his model. Even if all nonresponse problems are initiated by missing information, this is a different kind of nonresponse from the nonresponse identified and named "missing redundancy" in the first phase of the model. The difference is that we are now talking about statistical units that may consist of combinations of the information collected in the first phase. Consequently, when information is missing in the first phase, a number of statistical units in the second phase may be affected. The quality problems may multiply, so to speak.

Finally, we take a look at the measurement challenges when administrative measurements are reported in statistical questionnaires. While the validity problem described in the first phase of data collection from administrative sources focused on the relationship between what the questionnaire asks for and what is available in the register, what is called a *relevance problem* in the second phase concerns the relationship between the information asked for and the statistics eventually produced. These both seem to be content validity problems; the difference is their reference point. The example of relevance issues given in Zhang's article is when employees have to be assigned to a standard categorization of occupations. Companies will use different systems and terms to identify employees in different positions. For statistical purposes, however, these have to be harmonized into a standardized set of alternatives. This can be done by presenting fixed response alternatives in the questionnaire or by applying coding standards to open questions. Often the assignment to a certain category is also decided by combining answers to different questions. Whatever procedure is used, however, the harmonizing standard employed can cover the concept eventually presented in the statistics effectively or be irrelevant. The model distinguishes between this problem and classification problems that may occur when employees are assigned to a certain category in the standard. The latter are termed *mapping problems*. Mapping problems are well known in business questionnaires, for instance, when respondents must determine the occupational categories of different kinds of consultants.

The compatibility problems referred to in the last step of the measurement cycle belong to the data editing stage in data collection, when results from different sources are compared to detect inconsistencies. The methods used to solve inconsistencies might well produce new errors. Quality issues related to editing procedures are a separate topic, which is discussed in more detail in Chapter 11.

As we see it, Zhang's two-phase model offers a conceptualization of the retrieval tasks that business respondents are faced with when they collect data for business questionnaires from administrative sources. These challenges arise in addition to the memory and judgment challenges that the respondents encounter when the requested information does not reside in administrative business registers. Furthermore it is worth noting that retrieval from administrative sources is far from straight forward, but rather may involve more error sources than memorizing.

ACKNOWLEDGMENT

The author is grateful to Anders Haglund and Anne Sundvoll, both at Statistics Norway, for reviewing the manuscript.

CHAPTER 4

Planning the Survey

Ger Snijkers, Gustav Haraldsen, and Jacqui Jones

4.1 INTRODUCTION

In Chapter 2 we described the business context within which business surveys operate and consequently need to adapt to. In Chapter 3 we discussed the quality challenges that are specific to business surveys. In this chapter we will relate these context and quality challenges to the planning of a business survey. We will consider survey planning from a project management perspective. This means that we will tailor concepts and methods taken from the project management toolbox to both general survey challenges and challenges specific to business surveys. Project management offers methods and tools that can be used to successfully organize the survey to ensure that the survey results are obtained within time and budget and according to specifications. In terms of a survey, this means procuring survey data at a preagreed level of quality at minimum cost and response burden.

Planning is the actual first step in designing and conducting a survey. In the planning of a survey a number of issues need to be considered. They are listed in Figure 4.1, which provides a framework to this chapter. The chapter is divided in two parts: Section 4.2, which briefly discusses the basics of project management in relation to the business survey characteristics, Sections 4.4–4.9, which discuss the planning of a survey in detail. We will see how to plan the business survey as a project, and how to manage the survey project to obtain the specified survey results within time and budget.

Designing and Conducting Business Surveys, First Edition.
Ger Snijkers, Gustav Haraldsen, Jacqui Jones, and Diane K. Willimack.

Basic principles:	Section	Survey planning issues:	Section
• Project management principles	4.2	**1.** Specifying the survey and planning of activities **2.** Required resources **3.** The timetable **4.** Managing, controlling, monitoring, evaluation **5.** Risk management	4.3 4.4 4.5 4.6 4.7
		• Finalizing the plan and getting started • Progress status overview	4.8 4.9

Figure 4.1 Survey planning issues.

4.2 BUSINESS SURVEYS AND PROJECT MANAGEMENT PRINCIPLES

In general, a project has the following characteristics [based on Gevers and Zijlstra (1998), Frame (2003a), and Hedeman et al. (2006)]:

- A project is a temporary organization unit (the project team) that is lead by a project manager, with a well-defined beginning and ending.
- The project team has a multidisciplinary composition mirroring the expertise needed for the project.
- The project manager is responsible for planning and managing the survey and delivering the survey results. The people in the project team are responsible for the project activities.
- The project is goal-oriented, aiming at delivering concrete, clearly prespecified results, by carrying out a number of coherent, planned activities. (In projects of a certain size this does not mean that the team members are carrying out all the activities themselves, but rather that they coordinate work done by services within their area of expertise.)
- Projects have to be realized within limited constraints (time, money, resources), while facing high levels of uncertainties to reach the goals.
- Projects should also have a clear stakeholder or client who has commissioned the project, and has agreed to accept the final results according to the preagreed specifications, and (possibly) pay for them.
- After the results have been achieved, the project is terminated.

By definition, designing and conducting a one-time cross-sectional business survey is a project. It has a clear purpose (to collect survey data), a clear beginning and end (as we have seen in Chapter 1: survey process), agreed constrained resources, and a stakeholder(s) or customer. For recurring surveys, however, the same survey is repeated every month, quarter, or year. This is not a project, but an ongoing process. However, project management principles can also be used for the planning of this process. For these surveys redesign projects may also be defined, aimed at redesigning e.g., the questionnaire.

There are four principles of project management (Gevers and Zijlstra, 1998; Frame 2003a); To those principles we add a fifth one, which involves continuous quality improvement:

1. Start with planning the survey as a project, then comes the "doing".
2. Manage the project according to the plan.
3. Control and monitor the survey project.
4. Communicate project progress.
5. Improve the quality of processes; apply the PDCA cycle (plan–do–check–act).

Compared to other projects, such as constructing a new oil platform in the Caribbean or the construction of a new metro line in Amsterdam, a project aimed at designing and running a survey may be considered to be relatively small with standard methodological procedures. Although relatively small and standardized, when studying in more detail the processes required to develop the survey components (e.g., a sample, questionnaire, communication materials), and planning the fieldwork, a survey can still be regarded as rather complex. The planning and implementation will normally be carried out by different units that need to be coordinated by different members of the project team. The fact that different data collection modes are combined adds to the complexity. Organizing the work of designing and conducting a survey becomes complex when many people are involved, and when more activities have to be carried out at the same time. As Kennedy et al. (2010, p. 575) point out: "Planning and development efforts rise with survey complexity." All the survey components need to be designed to meet the overall objectives of the survey. The different survey components also have to be ready at the same moment in time before the fieldwork starts. To meet these goals, good planning and close communication between project team members are important. The very first step in designing and running a survey is planning!

In business surveys, it is difficult to predict response rate, data quality, and response burden of the final deliverables. Many uncertainties remain because an important part of the survey production process, the response processes within businesses, is outside the control of a project manager. But also when survey components are being designed, the project may not run according to the plan and more time may be needed. These uncertainties need to be planned. Therefore it is of utmost importance that a risk assessment be included in the project plan, and that the plan include mitigating actions and active fieldwork management procedures.

In addition to complex planning and uncertainties in reaching the goals, the project manager must also deal with the environment of the project. This includes the project team, the survey organization, and the stakeholders or customers. Like any project, a business survey project is dependent on its environment; it operates within a larger organization that facilitates the survey with resources, procedures, systems, and tools. These services are ruled by standardized systems, tools, and procedures that need to be recognized irrespective of the characteristics of individual surveys. Hence, the project manager needs to balance the plan between specific needs and established standards.

Business surveys commonly have several stakeholders with different interests and expectations of the survey results. For recurring surveys, the original design may stem from work on a different project several years prior. Written, clear, preagreed specifications are often missing. These expectations need to be managed by defining the specifications and objectives early in the planning process, and communicated with stakeholders.

After the planning comes the "doing." It is the job of the project manager to make things happen, to start and manage the doing. The project will not run by itself. This is done by communicating with the project team (in e.g., project meetings). The project manager must rely on the fact that everyone is committed to the plan. A pitfall for project managers in managing a survey project is that they become involved in project execution activities, such as designing the sample or the questionnaire. A survey project manager should ideally have knowledge of the survey process, to make well-educated decisions. However, their prime focus should be on managing the project and delivering the agreed project outcomes. When project managers become involved in the project execution activities, this could jeopardize their management tasks, which may lead to failure of the project. It is the project manager's responsibility to manage, control, and monitor the project.

Once the project has started, the design and collection processes need to be managed according to the plan, and progress needs to be monitored and controlled. This holds for both the working processes and the outcome of these processes (e.g., the sample, the questionnaire, and ultimately the survey data). If the project is not running according to the plan, the processes need to be redirected in such a way that the project will get back on track. By continuous controlling and monitoring, risks that may disturb the project can be identified at an early stage, and mitigating actions can be taken and communicated.

Consider, for example, the response rate, which is a common quality indicator for the fieldwork. The required response rate should be defined at the very beginning of a survey design process. By this time the response rate will have become a quality norm. During the fieldwork the response rates should be regularly monitored. If at some point the required response rate does not seem achievable, processes that may affect it, such as plans for when and how respondents should be reminded, should be reconsidered and probably adjusted. Actions should be taken at such a point in time that it is still possible to adjust the fieldwork processes.

Within survey organizations there is a need to control and improve the quality of continuous production processes, as well as the quality of individual survey project management. The Deming cycle is aimed at getting good quality over a number of cycles, in our case getting good survey results, and improving the processes that are involved. This cycle consists of four stages: plan–do–check–act, and is also known as the *PDCA cycle* (Biemer and Lyberg, 2003); PDCA also implies: plan–run–evaluate–improve. Each time a PDCA cycle has been completed, improvements in systems and procedures should be secured to ensure that what is learned will be applied the next time. Then, the cycle starts again. It is like rolling a wheel up a hill, and ensuring that it won't roll back. The PDCA cycle involves learning and improving and requires evaluation of the project and processes, identifying issues for improvement, and taking actions.

To organize and manage a survey within this environment, and to maximize the likelihood of delivering of the results according to prespecified levels of quality and burden within time, budget, and resource constraints, a survey should be organized and run as a project. This applies to both single and recurring business surveys. The closer they can be planned and conducted according to project management principles, the better the results will be. Through application of the principles of project management, the processes of designing, building, testing, collecting, and processing the survey, as well as the quality of the survey data can be managed, controlled, and monitored, and risks minimized; and over a number of cycles these processes can be improved.

Further Reading

In this section, we have only touched on the key principles of project management. For a more detailed discussion of project management, we refer the reader to the vast body of literature on this topic (see, e.g., the Project Management Institute, `www.pmi.org`—Marketplace). Specific literature on planning and managing surveys is limited, especially for business surveys. References include studies by Willeboordse (1998), Kennedy et al. (2010), Church and Waclawski (1998, pp. 27–49, on step 1, pooling resources), Blumberg et al. (2008, pp. 55–105, Chapter 2, on the research process and proposal), Scheuren (2004), and Statistics Canada (2010, pp. 279–301, 323–385, Chapter 13, on survey planning and management, with case study; pp. 296–301, detailed planning checklist and a cost checklist).

4.3 SPECIFYING AND PLANNING THE SURVEY

Above we have discussed project management in general. Now, we will move to the actual planning of a survey. Survey planning involves the planning of *everything* that is related to the survey. The survey process consists of three stages: prefield, field, and postfield. In the prefield stage the survey is specified and planned; and the components, processes, collection, and processing designed, built, and tested. In the field stage the data collection is carried out and data processing begins; and in the postfield stage data processing is finalized, followed by the delivery of the survey results. Each stage can be divided into substages (see Figure 1.1), each with its own specific activities. A high-level overview of these activities is presented in Figure 4.2. All these activities need to be planned. In this section we will discuss the planning of each of the nine substages in detail.

Stage	Substage	Activities are focused on:	Section
Prefield stage	1. Specify the survey	- Consult with stakeholders: determine information needs - Consult with survey organization: establish business case	4.3.1
	2. Plan the survey	- Plan all stages of the survey - Write the project plan	4.3.2
	3. Design, build, and test (DBT)	All survey components that are needed for the data collection and processing are designed, built, and tested (and iteratively the survey project plan is amended): - The sample and estimation - Mode selection(s) and the questionnaire - The survey communication strategy - The data collection process - The data processing procedures	4.3.4
Field stage	4. Collect and process	- Implementation of the survey - Conducting the survey (fieldwork) - Data processing begins (capture, coding, cleaning)	4.3.3
Postfield stage	5. Process	- Data processing is finalized (coding, cleaning, imputation)	4.3.4
	6. Analyze	Analysis of the data and production of the deliverables	4.3.5
	7. Disseminate	Dissemination of the survey outputs	
	8. Archive	Archiving the survey data and outputs	
	9. Evaluate	Evaluation of the survey	4.3.6

Figure 4.2 Stages and substages in survey production process and related main activities. [Based on overviews from Czaja and Blair (2005) and Vale (2009).]

4.3.1 Specifying the Survey

The survey project starts with the appointment of a project manager. Before that, a stakeholder or customer will have commissioned the survey organization to conduct a business survey. The survey might be commissioned, for example, by a government department, Eurostat, a ministry, a university, or a business organization of a specific industry sector. To specify the survey, the project manager needs to consult with the stakeholders and the survey organization. The results of this stage are a preliminary survey project plan or a preplan, a business case for the survey organization, and a signed tender with the stakeholders. This would be input for a statement of the survey, which will be filed as the first chapter of the final project plan (see Figure 4.3).

The project manager's first job is to consult with the stakeholder or customer to agree on:

1. The survey objectives—what target and auxiliary information are needed, from what type(s) of businesses (target population) and for what purpose
2. The deliverables and product quality—what is the end deliverable (e.g., datafile, tables, or a report with a full analysis of the data) and what level of quality is required (e.g., response rate)
3. The budget, timetable, and response burden

One of the major reasons for project failure is "poorly identified customer needs and inadequately specified project requirements" (Frame, 2003a, p. 17). As Kennedy et al. (2010, p. 577) point out: "Survey objectives must be clearly specified early in the project if a survey is to be properly designed and planned." Often, however, stakeholders do not know what they really want. Also, a survey often has multiple stakeholders who may have various goals. This fact should be recognized by project managers when consulting stakeholders at the beginning of the project. It is the project manager's job to clarify the three issues listed above as much as possible.

These initial specifications mark the beginning of a dialog about the survey design, which is intensified during the survey planning phase (see next subsections). This dialog discusses primarily the following design objectives:

- The design of the sample and the sample size. At this stage it is also decided whether the survey is a one-time cross-sectional survey, a panel survey, or a recurring survey, and for what units this applies.
- The concepts of interest. The concepts that have to be measured in each wave of a panel or recurring survey are discussed.
- The mix of data collection modes that the stakeholder intends to apply. The mode(s) of data collection impacts all other aspects of the survey: sampling frames, the questionnaire design, survey communication strategy, and interactive data editing. This choice also affects the response rate and response quality, and the costs, response burden, and timetable of the survey. The choice of data collection mode(s) may also be influenced by policies and procedures of the survey organization.

Other topics to be discussed are

- The survey communication strategy and materials: how to make contact with the appropriate employees in the business and how to encourage cooperation

- Data processing, including data coding, validation, and cleaning
- External factors that affect the survey, such as mandatory status and ethical issues (e.g., not all variables might be mandatory).

After the survey design has been specified in general terms, negotiations focus more specifically on the relationship between quality issues and survey constraints (see Figure 3.2):

- The timeframe for the survey—defining period(s) for data collection (when to send out questionnaires, reminders, etc.), important milestones, and scheduling the final survey deadline
- The costs of the survey, material resources, and response burden—determining how much the stakeholder is willing to pay and what resources are needed from the stakeholder to conduct the survey
- The quality—how all this affects the quality of the deliverables, as the result of the process
- The information that is reported back to the stakeholders while the project is running, and setting milestones; include e.g., finalizing testing of the questionnaire (including a test report) or the response rates at various milestones during the fieldwork (in total and for important subgroups in the population); and selection of contact person(s) at the stakeholder's end (to whom to report).

All the information collected from discussions with the stakeholder is needed for the final contract. But before that contract can be signed, the project manager needs to put together a preliminary survey project plan from the collected information and discuss it with the survey organization to establish the business case.

The preplan, with the survey design requirements, is discussed with the survey organization to ensure that the requirements are realistic in terms of time, money, available survey organization resources, and risks. The feasibility and practicability of the survey is established. An important department to be consulted at this stage is the data collection unit. Since this department will be responsible for the actual data collection, they have to confirm that the survey can be conducted according to the specifications.

Establishing a realistic business case is important to ensure that the survey can be conducted according to the specifications. Otherwise the results will not meet the stakeholders' expectations, and the survey will incur extra costs for the survey organization. Often the plan is not realistic, and the requirements cannot be met. In these cases two situations are possible: an amber or red light. In case of an amber light, the project manager needs to renegotiate with the stakeholders on the additional issues, such as whether the timeframe is too tight or if the budget for the survey is too little. In case of a red light, the project is not considered feasible and practicable at all. In this case, there is no business case, which next has to be reported to the stakeholders. If the preplan is considered realistic, feasible, and practicable, the project manager gets a green light to start planning the survey in detail. A tender is signed with the stakeholders, awaiting a final contract.

4.3.2 Planning the Survey in Detail: The Planning Process

Once the high-level specification have been agreed on with the stakeholder(s) and the survey organization, the detailed planning needs to take place. This process is multidimensional with plans needed to design, build, and test the survey components (e.g., sample frame, questionnaire, communication material) and the survey processes (e.g., data collection process, data processing procedures). All plans need to fit into an overall plan, considering the total survey design. The result of the planning process is a survey project plan.

4.3.2.1 The Project Plan

The survey plan describes in detail the different activities during the prefield, field, and postfield phases of the survey (see Figure 4.2). It summarizes the resources needed, sets a timeline for the planned activities, and specifies how the project will be managed (see the five planning issues listed in Figure 4.1). For the plan to provide proper guidance to the survey project, the plan must be as specific, clearcut, unambiguous, and realistic as possible (Scheuren 2004).

All this leads up to a project plan with six chapters, as is shown in Figure 4.3 (Biemer and Lyberg 2003; Kennedy et al. 2010). This outline can be seen as an example of a project plan. In the figure we have also referred to chapters in this book where different aspects of a business survey are discussed in more detail.

The project plan consists of three parts: (1) statement of the survey (Chapter 1), which is the preplan (as discussed in Section 4.3.1), (2) a detailed plan for the survey components: how they should be developed and tested, and how the data collection and subsequent analysis and resulting dissemination should be carried out, as well as needed resources and the timetable (this is the main part; most of the present chapter is devoted to discussion of this part of the plan), and (3) a management—communication plan. The project manager needs to ensure that the project progress and outcomes are communicated within the project team and externally (e.g., to the survey organization and the stakeholders). In project board meetings, status reports showing the progress of the survey project should be presented and discussed. During planning of the survey and writing of the plan, negotiation and communication skills are important for the project manager, as well as obtaining commitment.

The project plan document is the product of the planning process (Frame 2003a; Hedeman et al. 2006). To succeed, it is important that this process follow a certain order (Haraldsen 1999; Kennedy et al. 2010):

1. List and number all activities in a work breakdown structure. A top–down approach should be applied, by first listing the survey components that need to

be developed, followed by a listing of the detailed activities for each of these. Start with the last activity, which is the survey evaluation in Figure 4.2, and proceed with activities that need to be completed before that activity can start. Applying this principle, move backward toward the first activity, which is the planning of the survey and the writing of the plan.

2. Determine the critical path. For all activities on the resulting list the execution time is estimated, and related activities are linked in a network diagram. This diagram gives a picture of the project structure. It will consist of activities that cannot afford delays without jeopardizing the survey deadlines and activities that have a leeway. During the prefield period a proper planning of the questionnaire design decides the length of this period (Scheuren 2004). Because of this it is essential that designing the questionnaire starts as soon as possible after the contract is signed.
3. Prepare a timetable with milestones. Next, present all activities in a *Gantt diagram*, which is a visual presentation of how the workload and division of labor is planned during the timeline of the project (Frame 2003a; Hedeman et al. 2006). The Gantt diagram also identifies deadlines for milestones. In surveys, important milestones are finalization of the survey components, including the fieldwork deadline. At this stage, budget and human and material resources are planned by linking these to activities.
4. Plan the project on a real-life calendar. This entails translating the timeline to the real-life day-to-day calendar, indicating in which weeks specific activities have to be carried out and the exact deadlines for milestones. If this is set as the final planning activity, the previous planning activities will not be dependent on when the project is realized. The plan developed in points 1–3 can be easily implemented and adjusted to different time periods.

Chapter 1: Statement of the survey (see Section 4.3.1)

This chapter provides an introduction to the survey, and includes

- Objectives of the survey
- Survey deliverables, including the quality of the survey results
- Survey specifications: the population to be sampled, the variables to be measured, and the survey design
- Project constraints:
  - The delivery date, the budget, response burden (see Chapter 6) constraints
  - Unique features of business surveys (see Chapter 1) and the response processes within businesses (see Chapter 2)
  - Total survey design and total survey error considerations (see Chapter 3)
  - The business survey production environment (see Section 4.3.2.2)

Figure 4.3 Outline of a survey project plan.

Chapter 2: The survey in detail (see Sections 4.3.2–4.3.6)

This chapter provides the work breakdown structure of the activities (and its results) that have to be carried out to meet the survey specifications and to deliver the results on time, within budget, and according to specifications. This will include

- Prefield activities—designing, building, and testing of (see Section 4.3.4):
  1. The sample and estimation → resulting in the sample (see Chapter 5)
  2. Mode selection(s) and questionnaire → resulting in a questionnaire per mode (see Chapters 7 and 8)
  3. The survey communication strategy → resulting in a communication strategy and communication items (see Chapter 9)
  4. The data collection process → resulting in data collection procedures, systems and tools, workflows, and management tools (see Chapters 9 and 10)
  5. The data processing procedures → resulting in the procedures and tools for data processing (see Chapter 11)
- Field activities (see Section 4.3.3):
  1. Implementation of the survey (see Chapter 10)
  2. Conducting the fieldwork (see Chapter 10)
  3. Starting data processing (see Chapter 11)
- Postfield activities (see Sections 4.3.5 and 4.3.6):
  1. Data analysis, dissemination, and archiving → resulting in procedures on how to analyze, disseminate, and archive the data (see Chapter 12)
  2. Survey evaluation → resulting in procedures on how to evaluate the survey (see Chapter 10)

Chapter 3: Required resources (see Section 4.4)

This chapter lists the project resources needed to produce the results according to specifications: the human and material resource and money, based on the work breakdown structure in Chapter 2 of this plan (the survey design in detail).

Chapter 4: The timetable (see Section 4.5)

This chapter presents a timetable of the survey project, from start to finish: a timeline of activities and milestones as discussed in Chapter 2 of this plan (the survey design in detail).

Chapter 5: Project management (see Section 4.6)

This chapter discusses how the project will be managed, controlled and monitored. This includes the specification of

- The organizational structure of the project, and project communication
- Process indicators for the various design, build, test, collect, and process processes, aimed at controlling time, resources, and money
- Quality indicators for the various components of the survey, aimed at controlling the quality of all components of the survey
- Project termination and evaluation

Chapter 6: Risk management (see Section 4.7)

This chapter discusses the identification of possible risks and mitigating actions.

Figure 4.3 *(Continued)*

5. Communicate the plan to those involved. The visual character of the Gantt diagram should help the project team and the services involved see where they fit in and the available time they have for the activities they are expected to perform. The Gantt diagram should also be included in the project plan document.

Planning the survey (and consequently writing the project plan) is not an easy process. It requires considerable ingenuity and creativity, and ideally knowledge of conducting surveys. It is not a linear process, but is best characterized as an iterative process. During the designing, building, and testing stage of a survey, you will inevitably end up going back and forth between the plan and the design and making amendments. This iterative approach between the survey plan and the design of the survey is less realistic when the survey field stage is reached. Here it is hardly possible to make changes without needing extra time and resources. Once the data have been collected, the design can no longer be changed (unless you start another survey, which, of course, is out of the question). Information you have forgotten about, or parts of the population not sampled will be missing! Hence, apply the first principle of project management.

When preparing the design of a business survey, the constraints, considerations, and unique features of business surveys and response processes within businesses need to be considered (as discussed in Chapters 1 and 2). The plan should also consider all sources of error that occur in business surveys (see Chapter 3), as well as response burden (see Chapter 6), in such a way that errors and burdens are minimized within the resource, budget, and time constraints. The plan needs to follow both paths of error discussed in Chapter 3: errors related to the sample and errors related to the measurement. During the planning stage, procedures or methodologies must be established to minimize these errors.

To summarize, when planning a survey in detail, three rules apply to ensure good survey results:

- Do things right from the start, instead of trying to repairing them later. Plan enough time for specifying the characteristics of a well-designed survey; good preparation is half the battle.
- Do not underestimate the time needed; plan the start of the survey project as early as possible. You will always run out of time, even if you think that the time needed is realistically planned. A rule of thumb is that the actual time you will need for designing and running a new survey is 1.5–2 times the planned time. When estimating the total time needed, add up estimates of separate activities following the work breakdown structure instead of estimating the whole in one hit.
- Plan for quality within constraints. All planning activities are aimed at producing a preagreed level of quality, at minimum costs and response burden; quality is not achieved by itself.

4.3.2.2 The Business Survey Production Environment

An important constraint we have not yet discussed is the production environment. When working in a survey organization, such as a NSI, the project manager very likely is constrained by many restrictions, since they are working in a standardized production environment. In such an environment, a single survey cannot be planned and designed separately. For instance, data collection policies apply, definitions of units and background variables may be standardized, and data collection procedures may have to be followed. Also, specific systems, and tools for designing the survey may need to be used. These policies, definitions, procedures, systems, and tools dictate to a large extend the framework for the project manager, and consequently must be factored in when planning a survey in detail. Organizational factors, when not factored in properly, are one of the major reasons for project failure (Frame 2003a).

A standardized survey production environment may consist of a large number of elements:

- *Definitions of Units.* This includes a standardized set of definitions that define business units, implemented in a standardized business register to be used as a sampling frame (see Chapter 5). This is cost-effective and facilitates the comparison of survey results.
- *Definitions of Variables and Classifications (Data and Metadata).* In every business survey a standardized set of variables and classifications should be used as background variables, such as the classification of businesses in economic activity and size. This facilitates comparison of survey results and also affects response burden and measurement error; for multisurveyed businesses coordination of variable definitions over surveys is recommended [see also Section 4.3.5 on the data documentation initiative (DDI) and statistical data and metadata exchange (SDMX)].
- *Procedures for Sampling Design.* The sampling of businesses may be guided by sample coordination procedures aimed at response burden reduction for multisurveyed businesses (see Chapters 5 and 6). This includes, rotating sample designs and survey holidays.
- *Strategies and Modes for Data Collection.* Survey organizations may have implemented general data collection policies. These policies are driven by cost and response burden reduction. In short, they may dictate that first data from administrative source should be used, and only if this is not possible or insufficient, a survey can be conducted. The modes for data collection are also dictated by including the web as the dominant mode, followed by paper, and finally interviewer-administered modes. Statistics Netherlands (Snijkers et al., 2011a), Statistics Norway (2007), and Statistics Canada (Brodeur and Ravindra 2010), for example, have implemented such a policy. The consequence of these policies is the development of mixed-mode designs. Also, tailored data collection procedures for dealing with large and multisurveyed businesses may be implemented and applied by special units, aimed at reducing nonresponse and measurement errors (as we will see in Chapter 9).
- *Procedures for Questionnaire Design.* Various standards may be implanted for the questionnaire design, ranging from standardized wordings of questions to standardized style guidelines for visual design [e.g., for the US Census Bureau (Morrison et al. 2008) and the Australian Bureau of Statistics (ABS 2006)].

Also questionnaire development and implementation systems (software) may be in place that dictates the degrees of freedom with regard to the questionnaire design. These guidelines are cost-efficient, and impact the corporate image of the organization (as various designs would have a negative impact); on the other hand, they may be very restrictive for individual surveys. (Questionnaire design is discussed in Chapters 7 and 8.)

- *Procedures for Survey Communication.* Also, the survey communication may be coordinated and standardized, ranging from standardized letter–envelope layout, the wording of letters, standardized sections in letters (e.g., on the mandatory or voluntary status of a survey), to reminding procedures. (Business survey communication is discussed in Chapter 9.)
- *Procedures for Data Collection.* Data collection procedures include a wide variety of standardized production procedures and systems that stipulate rules on data collection. This includes operations, logistics, and workflow, such as training staff, interviewing procedures, and procedures for monitoring the data collection process. (Managing and monitoring the data collection process is discussed in Chapter 10.)
- *Procedures for Data Processing.* Data processing procedures include procedures for processing the data once they have been collected, such as data capture, coding, and cleaning. Constraints mandated by the data editing procedures may, for instance, affect the questionnaire, when the editing system requires a specific set of variables. (Data capture, coding, and cleaning are discussed in Chapter 11.)
- *Individual Production Systems and Tools.* When designing a survey, the project manager has to make use of the individual systems and tools that are in place. It is the technology they have to work with. To a large extent these systems and tools dictate what they can and cannot do, similar to the process data (or paradata) that are collected to monitor the fieldwork.
- *Quality Guidelines.* In addition to these elements of the production environment, the survey organization may also have established guidelines on quality management, which may address all components of a survey and the whole process of designing and conducting a survey [see, e.g., US Census Bureau (2011a), Statistics Canada (2009), Statistics Netherlands (2008–2012), Eurostat (2011b), and ABS (2010)], and may include guidelines on dealing with respondents, customers, and the public (Groves et al. 2004). To ensure that the survey meets these quality standards, these need to be consulted and planned in as well. (Quality and risk management are discussed in Sections 4.3.6 and 4.7 of this chapter.)
- *External Conditions.* A final component of the system is external conditions that have to be factored in when designing a survey. This includes the mandatory character of a survey, survey ethics [see, e.g., Blumberg et al. (2008), ESOMAR (2008), ISI (2010), Sieber and Bernard Tolich (2012), and Singer (2008)], and confidentiality and informed consent regulations (discussed in Chapters 9 and 12) (Groves et al. 2004).

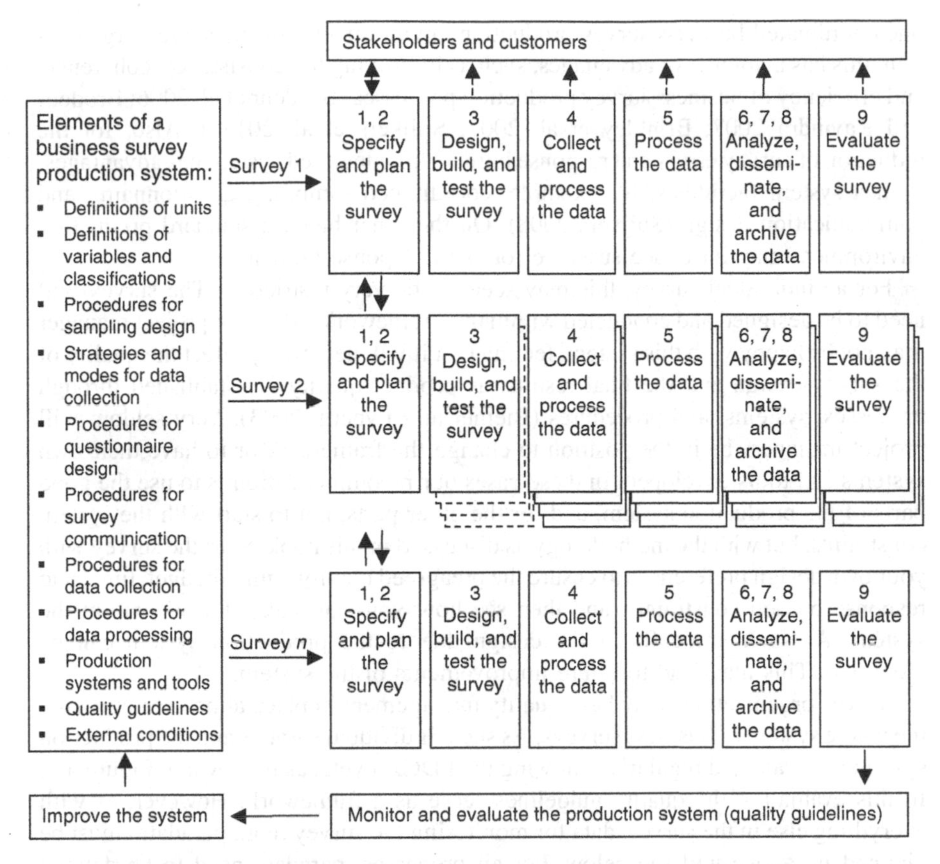

Figure 4.4 Survey planning related to the business survey production environment.

These elements have to be checked when planning the survey. It is important that they be considered upfront, not to find out in a later stage that implementation of the designed survey is impossible. The framework within which a project manager must work is shown in Figure 4.4. Within these boundaries the project manager has to plan the survey and check all relevant elements of the production system. The project manager may be responsible for a single survey, but may also be managing a number of related surveys; which may be a combination of cross-sectional and panel or recurring surveys. The latter may be planned as one survey, but conducted several times (from the one fieldwork period to the next, small changes in the design may be made). This design is indicated for the second survey in the figure. There may also be overlaps in the population and variables between surveys. This means that elements and experiences from one survey can be utilized in the next. In addition, experiences should be collected and used to improve the standard production systems, tools, and procedures (following the PDCA cycle).

The scope of such a standardized survey production environment depends on the survey organization. At best, all these procedures, systems, and tools are integrated in one coordinated business survey production environment. For the survey organization, this has a number of advantages, such as improving the consistency, coherence, and efficiency of business survey production processes (Brodeur et al. 2006; Brodeur and Ravindra 2007; Brinkley et al. 2007; Snijkers et al. 2011a). Also, for the reduction of survey errors and response burden, an integrated system has advantages; such a system facilitates, for example, coordinated sampling, questionnaire, and communication design (Snijkers 2008). On the other hand, a standard production environment may introduce survey errors and response burden.

For an individual survey, this may seem to be very restrictive. The survey will need to be designed and conducted within this framework. Also, the project manager may not be happy with this system, feeling that it will negatively affect the quality of the survey design and the final results, as product quality is established through processes, systems, and procedures (Biemer and Lyberg, 2003). Very seldom will project managers be in the position to change the framework or to have their own systems and tools developed. In these cases our recommendation is to use the good parts of the production system, and for the other parts, not to start with the system constraints, but with the methodology as discussed in this book. Start the survey with your own design preferences to ensure the preagreed quality within budget, time, and response burden constraints, and then see how you can make it work within the system. As such, tradeoffs in the design due to the production system can be identified. This may lead to future improvements of the system.

Survey organizations may have quality management in place aimed at improving the whole system of business surveys. As such, individual surveys and the production system are evaluated regularly following the PDCA cycle, as is shown in Figure 4.4. In this evaluation the quality guidelines serve as a framework. However, as with everything else in the survey, data for monitoring the survey (e.g., paradata) must be planned in, as we will see below. For all processes, paradata need to be defined upfront. Taken together, they can be used to monitor the production system as a whole (as will also be discussed in Chapter 10).

4.3.3 Planning the Fieldwork

So far we have discussed the survey specifications and the constraints that need to be considered when planning the survey (Chapter 1 of the project plan; see Figure 4.3), as well as the planning process. Regarding this process, we argued that for planning activities, we start with the last activity and then move backward. An import milestone is the final delivery date of the survey outputs; other important milestones in the planning, the startpoint and endpoint of the fieldwork, will be derived from this. The fieldwork endpoint is the final deadline for receiving data to ensure scheduling of enough time for finishing data processing and the postsurvey activities; the startpoint is marked in the plan as the day when businesses are contacted with the questionnaire. The fieldwork period should be long enough to ensure achievement of a response rate as agreed with the stakeholders.

When the fieldwork begins, all components in the design, systems, tools, and procedures need to be ready, and implemented, including data collection and data processing procedures. During the fieldwork, data processing will start as questionnaires are returned. The next section discusses the planning of the design, building, and testing of components that need to be ready before the fieldwork starts.

4.3.4 Planning the Designing, Building, and Testing of the Survey Components

Nearly all surveys will include the same survey components (see Figure 4.2). Each component will need to be planned for (which may be an iterative process) and then designed, built, and tested before the survey is conducted. Planning of the design of the individual survey components will need to be considered in tandem with the objectives of the survey, agreed survey deliverables, specified design features, project constraints, and production environment. This section provides an overview of how to plan the survey components (the work breakdown structure) in the following order:

1. Mode selections and the questionnaire
2. The sample and estimation
3. The survey communication strategy
4. The data collection process
5. The data processing procedures

When planning these components, it is important that the overall timetable be considered (see Section 4.5). We recommend starting the planning process with mode selection and the planning of the questionnaire and sample design. The mode impacts the questionnaire and initial data capture, data transfer, and data capture processing. Next, the survey communication strategy is planned. The data collection process depends on the results of these components. Finally, data processing and other postfield activities are planned. This, however, is not a linear process.

Specifying and planning the design of these components starts with the specification of the objectives, deliverables, and the constraints, like we have seen for the survey as a whole. On the basis of these specifications, the plan needs to include details on how to achieve these results, which activities should be carried out to achieve the result, how much money is needed, costs for the businesses (response burden), resources needed, who should be involved and who will do what and when, and how the survey components can be implemented in the survey organization. For each survey component the planning includes the steps shown in Figure 4.5.

4.3.4.1 Planning Mode(s) of Data Collection and Questionnaire Design

During specification of the survey, the modes to be used and the concepts of interest are set. These specifications are input for the designing, building, and testing of the questionnaire. In business surveys, a questionnaire very often is a complicated measuring instrument, with complex concepts to be measured. As we have already noted, the questionnaire design determines the critical path of the plan. Questionnaire design and communication considerations are discussed in Chapters 7 and 8.

1. Specify
   a. The objectives of a survey component as agreed to with the stakeholder
   b. The deliverables and their qualities
   c. The constraints: budget, timetable, and response burden
2. Plan the designing, building, and testing (DBT) activities for a survey component (the work breakdown structure)
   a. Study the business context, that is, the unique features of the businesses that are pertinent (as discussed in Chapter 2); these features vary according to the components (e.g., for the questionnaire, the completion process is relevant; for the communication process, the people involved in the response process and their roles are relevant).
   b. Consider the business production environment, and determine the relevant constraints (see Figure 4.4).
   c. Define the process or paradata to be collected to monitor the effectiveness of each survey component [the paradata concept is discussed in Chapter 10; see, e.g., Kreuter, Couper and Lyberg (2010)].
   d. Design, build, and test (DBT) the survey component, based on these (tradeoff) considerations, and the project requirements.
   e. Consider the consequences for the final design and tradeoffs with regard to quality (survey errors), costs, time, and response burden.
   f. Finalize the design (followed by implementation, data collection, and evaluation).
3. Required resources (e.g., people, finances, systems, software, office space)
4. The timetable: start and deadline of activities, and milestones
5. The managing, controlling, monitoring and evaluation of each design, building and testing process.
6. Identification of risks and risk mitigation.

Figure 4.5 Steps in the planning of the designing, building, and testing of survey components.

The steps involved in the planning of the designing, building, and testing of a questionnaire are shown in Figure 4.5 [see also Brancato et al. (2006), Giesen et al. (2010), and Snijkers (2002)]. The first two steps, which are important with regard to the quality of the questionnaire and need to be planned carefully, include:

1. Conceptualization of the concepts and identification of questions in the questionnaire. In this step the concepts of interest are defined in such a way that they can be measured as questions in questionnaires (Snijkers and Willimack 2011). (Conceptualization is discussed in more detail in Chapter 7.)
2. Operationalization of the questionnaire. In this step the concepts of interest are studied by checking the availability of that information within businesses (e.g., by conducting feasibility studies). Next the questionnaire is designed, built, and tested. Testing includes technical testing (to determine whether the questionnaire operates correctly), cognitive and usability testing, and piloting (see Chapter 7). Paradata are defined that need to be collected with the questionnaire to study and monitor the completion process, for both

self-administered and interviewer-administered questionnaires. These paradata include keystrokes revealing completion profiles, and the time and date as to when a questionnaire was completed, and how long it took [see, e.g., Snijkers and Morren (2010)]. Within the production environment standardized variable definition and classifications (metadata), existing questionnaires with overlapping concepts, policies on data collection modes, guidelines for questionnaire design, production systems, and tools that are used to build and implement questionnaires, as well as procedures for data capture, data coding, and data editing need to be checked.

It is difficult to predict how much time will be needed for the development of each questionnaire. This depends, for example, on the length and complexity of the questionnaire. Also, this is a cyclical process; after each test, you may need to go back and reconsider. If no feasibility studies are conducted, the issues that would have been discovered then, will be discovered during testing of the questionnaire, which is rather late in the design process. Doing a feasibility study and testing are futile if the questionnaire design is not changed afterward. Depending on the findings, the changed process may take more or less time. We have observed that designing and building of a questionnaire takes most of the time, leaving almost no time for testing, although testing is considered to be extremely important to ensure well-designed questionnaires. The American Statistical Association (Scheuren 2004, p. 44) recommends to "pretest, pretest, and then pretest some more." Also, the European Statistics Code of Practice (Eurostat 2011b, principle 8) states that questionnaires need to be "systematically tested prior to the data collection." Bradburn et al. (2004, p. 317) even advise terminating the survey "if you do not have the resources to (. . .) test," instead of spending a lot of money. We therefore recommend starting the questionnaire design as soon as possible.

When following all steps, the whole design–build–test process may take as long as 2 years. In fall 2004, Statistics Netherlands started a major redesign project to develop a web questionnaire for the Dutch Annual Structural Business Survey (Snijkers et al. 2007b). This project lasted until fall 2006, when the questionnaire was implemented, and used in the field (in spring 2007). The electronic questionnaire was developed and tested in a number of steps; initial design tested in January 2005; revised designs tested in September 2005 and January 2006. A final step was a field pilot in spring 2006, with a selected sample within the regular production process.

When planning the questionnaire design, we recommend considering each step individually to ensure a better estimate of the time needed. Estimating the time needed, as well as human and material resources and budget, can best be done in close cooperation with experts. They know what to do, and how long it will take. Experts involved in the questionnaire design include content matter specialists, questionnaire designers, fieldwork experts (interviewers, field agents, help desk staff), pretest methodologists, professional designers, and computer experts. The result of this process should be a valid questionnaire that facilitates businesses in the response process, thus minimizing response burden, nonresponse, and measurement errors.

4.3.4.2 Planning the Sample Design and Estimation

The sample provides a list of businesses that are to be surveyed (see Chapter 5 for further information on sampling design and estimation). To select a sample, five elements are needed: the target population, a definition of sample units, a register or registers, a sample frame, and a sample design (including how many businesses, with which characteristics, and the form of the estimator). Planning the sample includes the planning of all these elements. The sample planning will therefore need to include decisions on

- The population of interest (e.g., all retail businesses or all manufacturing businesses)
- The type of unit you wish to survey (e.g., statistical, collection, or analysis unit)
- Access to or creation of a register or registers that holds the required information on the population of interest and type of unit to be surveyed
- Access to or creation of a sample frame, and when this frame will be created from the register(s) to ensure that it is as up-to-date as possible
- Tradeoffs between sample size, sampling error, and costs
- How, if at all, auxiliary variables will be used in the sample design and estimation
- Whether the sample design will be stratified, probability proportional to size, a cutoff sample, or some other design
- The estimation method, such as Horvitz–Thompson, ratio estimation, or combined ratio estimation (or perhaps a more complex estimator)
- How the sample will be allocated (perhaps factoring in the population variance) and selected
- Whether "survey holidays" are applicable to certain sample units, such as small businesses

Within a coordinated production environment the project manager should check the availability of (standardized) unit definitions, registers, and frames; and for any survey coordination (processes to minimize and spread the response burden). If no standards are available, a definition of units needs to be agreed, a register(s) and a sample frame should be constructed, and activities should be planned to do so. For a one-time cross-sectional survey it may be possible to construct the frame directly, but if there is any chance that the survey might need to be repeated, a register will probably be more efficient. Large survey organizations (e.g., NSIs) have standardized business registers that are regularly updated. These registers are used to create the sample frames [as we will see in Chapter 5; see also, e.g., Ritzen (2007) and Cox et al. (1995, part A)]. The construction of a register may be time-consuming, and if there is a requirement for repeating surveys, then the register will need to be kept current.

The sample also has to be designed. One design can be used on many occasions (i.e., with many frames), or conversely one frame can support several designs. A sampling expert will be able to provide more detailed advice. The result of this whole process should be a selected sample that is representative of businesses in the population of interest, and that will result in statistics with calculated estimates of sampling errors. This process takes the least time of all survey components. The sample should, however, be drawn before the prefield data collection phase begins, which may be needed to update contact information and frame refinements (as we will see in Chapter 9).

4.3.4.3 Planning the Survey Communication Strategy Design

The survey communication strategy is aimed at obtaining accurate, complete, and timely responses. It includes all communication with businesses that is related to the survey, with exception of the questionnaire. Communication items include advance and reminder letters, telephone reminders, help desk, leaflets, and websites with frequently asked questions (FAQs). The strategy indicates communication measures and activities, and for each measure, the following: the objective, the communication mode, the message, the actors and staff involved (within both the business and the survey organization), and timing of the communication. Design of a survey communication strategy is discussed in more detail in Chapter 9.

During the specification stage, design issues may be negotiated that are input for the communication strategy, such as how respondents are contacted (e.g., by paper mail or email) and how many reminding waves are used. These aspects affect the costs, and also the response rate. These specifications must be considered when designing the communication strategy.

As with the questionnaire, the design–build–test (DBT) process determines the quality of the communication strategy. The planning steps are shown in Figure 4.5. In the planning process the response process in businesses should be studied, and the specifications established by the production system taken into account. These include prescribed communication modes, guidelines on the communication design (e.g., corporate communication guidelines, web design guidelines, or standard sections in letters, e.g., on informed consent or the statutory status of a survey), procedures and practices that are used by departments involved in the communication process (e.g., the help desk, procedures for the telephone reminder unit, or recontacting during data editing), systems used to build and implement the communication measures (e.g., the use of color), and the communication process for related surveys. To monitor the communication process, paradata need to be defined. The communication measures need to be tested individually and in coherence with all other measures to test its effectiveness in obtaining response.

The time needed to follow this process, the human and material resources, and the money needed are dependent on the complexity of the communication strategy and the number of items involved. To design communication items, communication specialists, fieldwork experts (interviewers, field agents, help desk staff), methodologists and pretest experts, and data editing staff should be consulted and planned in. In fact, all staff that are in contact with businesses in relation to the survey may be involved in the planning and the designing of the prefield, field, and postfield communication strategy.

The planning of the DBT cycle of the survey communication strategy comes very close to planning of the actual data collection, since the communication strategy also involves the timing of contacting businesses. This means that all departments involved in the fieldwork (e.g., interviewers, field agents, help desk staff) need to be consulted regarding their tasks and availability, as well as required fieldwork training and instructions.

The communication strategy may require businesses be contacted before questionnaire are sent out in the prefield period (and ends after the fieldwork has ended in the postfield stage, with businesses recontacted during data validation). This means that all materials, procedures, and staff needed should be ready even before the actual fieldwork begins. This deadline should be clearly set. Designing, building, and testing of the communication strategy as soon as possible is therefore recommended. The result should be an effective, coherent communication strategy that motivates and facilitates businesses to respond and to complete the questionnaire as thoroughly as possible, thus minimizing response burden, nonresponse, and measurement errors.

4.3.4.4 Planning the Data Collection Process

Next, the data collection itself needs to be planned, according to the steps in Figure 4.5. The data collection involves actually contacting of businesses to obtain their data, such as by sending out advance letters including a paper questionnaire or a reference to a website; the communication strategy is put into practice. The communication strategy and the (mixed-mode) survey design is input for the planning of the fieldwork. A data collection schedule is developed on the basis of these components. This is a detailed timetable of the data collection process indicating when businesses are contacted and how, and what staff is involved, for prefield, field, and postfield data collection activities. It includes dates for sending out advance and reminder letters.

Planning of the data collection process involves planning of a sequence of activities and human resources, production systems, tools, and procedures aimed at successfully conducting the fieldwork. This includes checking the feasibility of the data collection in the required period of time, that is, availability of departments involved in the in the prefield, field, and postfield stages following the data collection schedule, as well as systems and tools, and fieldwork procedures, logistics, workflows, and operations.

For data collection, a large number of production systems, tools, and procedures are needed. These include computer systems for printing and sending questionnaires; websites that are used as portals for businesses to access online questionnaires; data capture systems to collect the responses and store the data in a datafile; case management systems to monitor response, nonresponse, and communication with businesses; landline telephone systems for phone interviewing and reminding; and help desk systems to manage the incoming phone calls to ensure minimal waiting time. Procedures for handling specific situations include help desk procedures for answering phone calls and responding to standard questions such as asking for postponement, or procedures on how to introduce oneself on the phone when calling a business.

In survey organizations, these production systems, tools, and procedures are very likely already in place and do not need to be designed and developed; however, within the survey project, they need to be checked and may need to be changed according to new specifications (e.g., the collection of paradata). Thus, planning the survey incudes establishing what production systems, tools, and procedures need to be designed, redesigned, and built to conduct the survey.

For new systems and tools, technical tests need to be conducted to see if they work properly. Staff needs to be instructed and trained to work with these systems and tools, as well as trained in specific procedures. Dress rehearsals can be arranged with staff to check whether procedures work properly. New system and tools should also work within the production system and process as a whole; the whole system should work as a coherent system, without gaps, overlap, or redundancies. To ensure that all systems and tools work together as a coherent production system, a test environment can be set up to perform dry runs.

Planning of the fieldwork can be very complex, especially in case of mixed-mode designs and multiple communication modes, which includes the planning of various departments. Complicating matters even more, additional measures may be needed to ensure high sufficiently response rates. In a sequential mixed-mode design planning of a second mode is dependent on the results of the first mode. For instance, when the response for a web questionnaires is falling behind, more telephone interviewers for nonresponse follow-up may be needed than previously planned. Estimates of response rates for the sequentially used modes may be used to plan the number of contacts needed in later stages.

To carefully keep track of the fieldwork and act if necessary, an active fieldwork management approach is needed [as we will discuss in Chapter 10; see also Hunter and Carbonneau (2005) and Laflamme et al. (2008)]. This requires the collection and analysis of paradata, providing management with information on the status of the fieldwork in real time. With the use of computers and electronic questionnaires (both interviewer- and self-administered), we see an increase in paradata that can be collected. Groves and Heeringa (2006) discuss how paradata can be used to monitor and improve social survey designs during data collection. This is called a "responsive survey design". Responsive designs use paradata to guide decision making with regard to data collection features, in order to achieve higher survey data quality per unit cost. A responsive design needs careful preparation and needs to be planned in from the very start. It is defined by four planning steps (ibid., p. 440):

1. Preidentifying "a set of design features that potentially affect costs and errors of survey outputs." The design features are related to all survey components. Survey errors are associated with the sample; performance of the questionnaire; mode of data collection; communication strategy; performance of interviewers, staff, and respondents; and performance of systems and tools, as we have seen in Chapter 3. Apart from paradata related to survey errors, this also includes data on time spent on different internal operations, material resources consumption, money spent, and compliance costs.

2. Identifying "a set of indicators of the cost and error properties of those features and monitor those indicators in initial phases of data collection." A comprehensive set of indicators to monitor relevant design features needs to be defined in an early stage. (Key indicators are discussed in Chapter 10.)
3. The analysis of the paradata needs to be prepared so that the indicators can be calculated in real time during data collection. The paradata that are needed for these analyses need to be defined prior to the data collection phase, and data collection systems should be able to collect the paradata during the data collection process. In these analyses, paradata from different sources may need to be combined.
4. In subsequent phases of the data collection, features of the survey can be altered to comply with "cost–error tradeoff decision rules." Decision rules need to be decided on prior to fieldwork, and alternative designs should be designed and prepared. This means, for example, that when the response is falling behind, telephone reminders for relevant businesses will be carried out instead of sending reminder letters. This is more costly, and telephone interviewers need to be scheduled. Or, the mode of the data collection is changed from web to paper, which means that apart from a web questionnaire and a portal, a paper questionnaire should be ready.

All these measures are aimed at obtaining survey data according to the specified levels of quality, within time, budget, and level of response burden.

4.3.4.5 Planning the Data Processing Procedures

Once the responses are collected, the data can be processed. Data processing includes data capture, coding, editing, and imputation; deriving new variables; calculating weights; and finalizing the survey data files (see Chapter 11 for more details). Collected data will need to be captured from the questionnaire, possibly coded, and edited; and for missing data items or units, data may need to be imputed. When data are returned to the survey organization, the method of data capture will be dependent on the mode of data collection. For surveys using paper self-completion questionnaires, the data will need to be captured by manually keying, electronically scanning, or a combination of the two. For electronic modes of data collection the collected data will need to be collated into a datafile on a server. Depending on the type of data collected and the objective of the analysis, data coding will need to be planned. For example, will numerical data be recoded into bands? Will category data be recoded into numerical values? Will qualitative data (e.g., descriptions) be coded into numerical values? The following methods will also need to be agreed and developed:

- Methods for identifying and treating implausible data
- Methods for dealing with partial and unit nonresponse
- Estimation methods
- Derivation of variables

Planning of data processing includes planning of DBT activities for

1. Data capture procedures, based on the mode of data collection, including paradata, in such a way that the data are available when needed.
2. Coding procedures, including paradata to measure and monitor the coding process. The data coding procedures will be dependent on the availability of existing code lists, automatic coding systems, and expert coders.
3. Validation and editing checks, with methods and procedures for dealing with data that fail validation or edit checks.
4. Imputation methods for dealing with partial and unit nonresponse, including paradata to measure and monitor the imputation process. These paradata may also be disseminated with the statistical results.

The survey production specifications should be checked for the availability of departments that are involved in the data processing; for logistics, workflows, and operations (at what moment in the process will they be needed); and for the systems and tools that are used. The result of this process is a datafile that is ready for analysis.

4.3.5 Planning the Data Analysis, Dissemination, and Archiving

After the data are collected and processed, we come to the analysis, dissemination, and archiving of the data. These aspects have been negotiated with the stakeholders when the survey is specified. Now, the actual deliverables are produced according to the quality specifications, and delivered.

Again, activities to produce the deliverables need to be identified and planned. These activities may vary for each survey; a general overview is discussed in Chapter 12. Also at this stage procedures that are in place within an organization need be checked, including guidelines for the layout of reports, regulations for the archiving of datafiles, information that is stored (i.e., the data and the metadata), ensuring confidentiality and statistical disclosure control (Bethlehem 2009), and guidelines for handling customers and the public (Groves et al. 2004). Kennedy et al. (2010, p. 589) point out that universities and federal agencies generally require survey data to be "retained for a minimum number of years"; and that "archiving the final dataset and sample information must ensure confidentiality by removing identifiers." Planning of these activities again involves the planning of human and material resources, time, and budget.

Survey organizations around the world have been implementing statistical data standards such as the data documentation initiative (DDI) and statistical data and metadata exchange (SDMX) (Vardigan et al. 2008; Gregory 2011). These standards provide definitions of the data concepts required to support the entire statistical production process. The survey planners should effectively identify the range of metadata definitions required to successfully document all the processes that are carried out to produce, disseminate, archive, and evaluate statistical outputs and processes. To maximize the utility of collected and disseminated survey data and to improve the survey process, the implementation of statistical data standards could be considered (Vale 2010).

The DDI and SDMX standards are seen as complimentary to one another. In relation to the stages in the statistical process (see Figure 4.2 and Chapter 1), DDI covers the stages "specify the survey" through to "process," and then moves on to cover the "archive" and "evaluate" stages. DDI provides a way of describing microdata and processing of the data as they are integrated, aggregated, tabulated, and so on. Although the standard provides the capability to represent actual statistical data in an XML format, it is used mostly for the metadata defining the statistical data and the production process. The SDMX standard is concerned primarily with statistical data reporting and dissemination, and provides XML "containers" for the data and associated metadata. This standard is more suited to the dissemination end of the statistical process lifecycle. SDMX therefore starts at the "process" phase of the production process and touches into the "archive" phase.

4.3.6 Planning the Survey Evaluation

After the survey results have been delivered, the survey production process needs to be evaluated according to the PDCA cycle. Survey evaluation serves as improvement of the processes in all stages, including the production environment, as shown in Figure 4.4. In survey evaluation a number of questions need to be addressed; including

- Did the survey process run as planned? If not, why not? (We recommend listing these issues during all stages.)
- What issues for future improvements can be identified?
- What measures need to be taken to achieve and secure these improvements?

Survey process evaluation can be done within the project or as a postproject activity. Both evaluations need to be planned, and survey process evaluation methods need to be identified in the project plan. They include self-assessment (which is done at the end of the project), or postproject quality audits. For self-assessment evaluations, the survey design (as specified in the project plan), all kinds of paradata and indicators (see Chapter 10), the list of improvement issues, and the model on business survey quality criteria as discussed in Chapter 3 (Figure 3.2) can be used. Following this model, the survey process and its deliverables will be evaluated according to

- Criteria of the stakeholders/customers
- Survey errors and quality aspects of surveys
- Survey constraints, such as costs, and the survey production system
- Survey-related effects.

Other quality self-assessment checklists may also be used following the quality guidelines as implemented in the survey organization [see, e.g., Biemer and Lyberg (2003), Ehling and Körner (2007), and Signore et al. (2010)].

After termination of the project, quality audits may be carried out; the deliveries, the survey design, and the production process should be reviewed by independent reviewers according to quality guidelines. While self-assessments are aimed at evaluating individual processes, quality audits (as part of concern quality management) are aimed at improving the production environment. Many national statistical agencies have implemented quality guidelines and assessment procedures (e.g., auditing) as part of their quality management (Colledge and March 1997; Ehling and Körner 2007). (Quality management in relation to risk management is discussed in Section 4.7).

Further Reading

Literature on survey process and quality management and assessment includes the handbooks by Ehling and Körner (2007), which provides on overview of quality assessment methods, tools, guidelines, and frameworks, and includes many relevant references; Aitken et al. (2003), which provides an overview of indicators aimed at monitoring and improving survey processes; Biemer and Lyberg (2003), and Lyberg et al. (1997, pp. 457–600, Section D on quality assessment and control). More guidelines and standards are provided by Regan and Green (2010), Statistics Canada (2009), the US Census Bureau (2011a), and ISO (2006).

4.4 REQUIRED RESOURCES

After the survey is planned in detail, the required resources can be estimated. From the work breakdown structure, the survey project manager will plan for the human resources, material resources, and money needed for each activity. The project manager may well have to negotiate with various groups in the survey organization to obtain the experts, resources, and systems required. For example, sampling specialists, telephone interviewers, questionnaire designers, and pretest methodologists may be needed. When establishing the project team, the project manager may negotiate for the best people for each task. This is important to increase the chances of the project's success. Also, advisors and consultants can be brought in at this stage. There may also be a need for training sessions rooms, and a pretest facility for focus groups, including recording equipment. Related to the list of activities, this segment of the project plan (see Figure 4.3) presents an overview of these resources, indicating what resources are needed, when, and for how many hours. Also, the tasks of everyone involved should be described.

The work breakdown structure provides a means for estimating total costs and resources needed; these can best be estimated by adding up individual activities in the various substages. At this point in time you may find the project to be too expensive, and additional negotiations with the stakeholders may be needed. This may require changes to the plan, which may have consequences for the design and the quality.

4.5 THE TIMETABLE

Now, the project manager has an overview of all activities, human and material resources, money, and milestones. The next step is to prepare a final detailed timetable, which is Chapter 4 in our project plan (see Figure 4.3). The timetable is the core of the project plan. It lists all activities, determining the critical path, with deadlines and milestones on a real-life calendar (as we have seen in Section 4.3.2.1). Activities are scheduled in chronological order, and related to resources and money.

Although the timetable is the result of all the abovementioned planning activities, it should not be constructed at the end. The timetable should be prepared and considered as the survey components and survey stages are being planned. All components can be developed in parallel, but remember that the mode selections and the questionnaire design needs to start as early as possible, followed by the sample and the communication strategy, then the data collection process and the data processing procedures, and finally the postsurvey activities.

In this process of planning and writing, the project manager may consult stakeholders, experts, and departments again to make the plan as realistic, clearcut, concrete, and unambiguous as possible. What you do not want is to develop a fantastic survey plan only to find that there is insufficient time to carry it out. On the other hand, the timetable can be constructed only when all components have been identified. At this point you will identify activities that need to be done in parallel, and you may well find that there is not enough time for all the development and testing of survey components that you had initially planned. For example, there may be insufficient time for several iterations of cognitive testing of the questionnaire.

When the timetable of activities, people, and other resources has been developed, the survey project manager may discover that the timetable is unrealistic. People may be unavailable when they are needed, working on other projects, available for only 2 days a week, or have planned a vacation. Also, the availability of other resources and additional means need to be checked by the project manager, who might find, that the pretesting facility is already booked for another project, or that the editing staff are not available when needed. So, in negotiation with internal departments the timetable may be shifted a little, backward or forward, with the ultimate goal of running the survey and delivering data as agreed in the specification stage.

4.6 PLANNING MANAGING, CONTROLLING, MONITORING, AND EVALUATION OF THE SURVEY PROJECT

A good survey project plan needs to include details on how the project will be managed, controlled, and monitored, for a number of reasons:

- To reduce the risk of errors and to ensure that the results will be achieved in time, within budget, and according to specifications
- To monitor the project in real time, and be able to act in time, before it is too late. Here, process and quality indicators are essential.
- To assist in continuous quality improvement on project management, learn from past experiences, and avoid repeating the same mistakes.

Grouping	Overarching themes
Project	**1.** Project management **2.** Financial management **3.** Strategic management **4.** Knowledge management
People	**5.** Customer management **6.** Data provider management (businesses) **7.** Human resource management
Data, methods, and systems	**8.** Quality management **9.** Metadata management **10.** Data management **11.** Statistical framework management **12.** Statistical process management (paradata)

Figure 4.6 Overarching themes for managing a survey.

In general, we could say that the managing, controlling, and monitoring is aimed at checking whether the project is on track, to ultimately produce survey data to the required/agreed level of quality (product quality) within the project constraints. If this is not the case, measures (e.g., risk management, discussed in the next section) will be needed to get the project back on track.

The management elements for a survey are described by the general statistical business process model (GSBPM) (Vale 2009) (see Chapter 1). To effectively manage the design and execution of a survey, 12 overarching GSBPM management themes must be factored into the plan. These themes (clustered in three groups) are listed in Figure 4.6. These management themes have been discussed throughout this chapter.

The survey planning process is therefore multifaceted in that it needs to produce a survey plan to effectively and efficiently design, build, test, collect, and process the survey, which includes simultaneously factoring in the 12 overarching themes as well as the constraints, considerations, and unique features of business surveys, including details on how each stage of the survey process will be monitored, controlled, and managed.

The management chapter of the project plan (Chapter 5 in Figure 4.3) discusses how the project will be managed, controlled, monitored, and evaluated. It discusses the organizational structure of the project with roles and responsibilities of the project manager, the project team, advisors, stakeholders, a project board and its members (supervisors and stakeholders), as well as how is communicated (e.g., scheduled project meetings with the team, and when the project manager will update the project board on the progress using, e.g., status reports).

To manage, control, and monitor a project, the project manager has a number of tools:

1. The project plan, which is the anchor of the project manager
2. Project communication: meetings with project staff, and managing by walking around
3. Process indicators of the project (e.g., control charts and cumulative expenditure curves), showing the consumption of the resources and money in time compared to the plan, and progress status overviews (see Figure 4.7)
4. Quality indicators, showing the quality levels achieved as compared to the quality norms set in advance (see Section 4.7 and Chapter 10)

At regular time intervals, the progress of the project is assessed and reported by the project manager using these tools. A frequently applied approach in project management and control is management by exception; only in case the project is running off course will the project manager become involved in individual decisionmaking (Frame 2003a). Then appropriate actions need to be taken. For a more detailed discussion on project management tools, we refer the reader to the literature on project management (see Further Reading at the end of Section 4.2).

At the end of the project the project manager may evaluate the project with regard to its management. This may be done in a group meeting, but also through individual discussions, or using a questionnaire. The project plan is input for this evaluation. Following the PDCA cycle, the course of the project is evaluated, and issues for improvement and appropriate actions identified. In the management chapter of the project plan, the project evaluation is identified and planned.

The final stage in the project is the termination of the project. After the results have been delivered and both the survey process and the project are evaluated, the project comes to an end. Project termination can be done with a final meeting with the project team, the stakeholders, and everyone involved. How the project will be terminated is discussed in the management chapter of the project plan.

4.7 RISK MANAGEMENT

We have come to the end of the planning process. We now have a full picture of the survey design and associated processes, a timetable of activities (including milestones), identification of the people involved and their roles, other resources, and project management and control. There is still one thing that remains to be discussed: identification of risks, based on this picture. These risks, including mitigating measures, need to be discussed in the chapter on risk management (Chapter 6 in Figure 4.3).

Risk management is an essential part of survey planning. It needs to be considered at all stages in the planning process. When planning the survey, project managers should ask themselves at all times: What can go wrong? We can compare the project plan with a roadmap used in a journey. When preparing for the trip, you will probably also plan for events that could occur along the way (e.g., taking warm clothing during winter). This is risk management. Once you are enroute to your destination, you may be able to prepare further (e.g., buy warm boots), but sometimes you may not. In Section 4.2 we saw that in business surveys the project manager must cope with many uncertainties. These risks should be identified in the planning process as much as possible, and mitigating actions proposed. When planning the survey, workshops may be organized to discuss this. These workshops should be attended by all relevant parties in the process. These risks and actions are described in the plan. Once the project and the survey is up and running, deviations from the plan should be detected as soon as possible (by monitoring the process, as we have seen in Sections 4.3 and 4.6) so that proper actions can be taken.

The Australian Bureau of Statistics (ABS) has developed a system of quality gates "to improve the early detection of errors or flaws in the production processes" (ABS 2010). A *quality gate* is a checkpoint consisting of a set of predefined quality criteria that a survey project must meet in order to proceed from one stage of its lifecycle to the next. It can be seen as a validation point. Throughout the statistical process, a number of validation points must be passed before moving on to the next stage. "Quality gates can be used to improve the visibility of quality in the statistical processes as well as being used to measure and monitor quality in real time at various points in the end to end statistical process" (Schubert et al. 2006, p. 3).

The quality gates system consists of seven elements (ABS 2010; Gilbert 2011: compare with active fieldwork management as will be discussed in Chapter 10):

1. *Mapping*. A map of the survey process, such as the timetable of the project or the data collection schedule, is needed. Quality gates are placed on critical places onto the map.
2. *Placement and Risk Assessment*. A risk assessment is carried out to determine where to place a quality gate. This requires answering the following questions: What can go wrong? When can it go wrong? What impact can it have?
3. *Quality Measures*. These are a set of predefined indicators that provide information about potential problems at a given point in the process: How would we know if something was wrong? (This relates to controlling and monitoring of a project as discussed in Section 4.6, and the responsive design approach as discussed in Section 4.3.4.4; see also Chapter 10) (Ehling and Körner 2007; Aitken et al. 2003).
4. *Roles*. Tasks are assigned to various people or areas involved in the operation of a quality gate. Key roles are signoff by an operational staff member, a stakeholder, or a manager.
5. *Tolerance*. Tolerance levels of acceptability are predefined: What is an acceptable level of quality? They should take the form as a threshold, as is in a traffic light status report (see below).
6. *Actions*. Actions are predetermined responses to various outcomes for a quality gate.

7. *Evaluation.* This is the final component of the system of quality gates, to examine where improvements can be made in the quality gates system for future use.

Obvious places for quality gates are (Gilbert 2011):

- Where handover or integration of survey components or data between multiple areas occur
- When data transformations or changes in survey components take place
- When there are changes to processes, methods, and systems

Statistics Canada (2010) provides a list of things that can go wrong in a survey. Most issues relate to the three main causes of project failure (Frame 2003a): organizational factors, poorly specified objectives, and poorly planned projects. But still, when the survey is running, many operational problems can arise, such as questionnaires sent out too late or to the wrong addresses, or websites going down.

The predetermined actions are based on a traffic light status report. Only in case of amber or a red light are actions needed. In such cases investigations are undertaken to determine why a problem occurred. When amber occurs, the process may continue cautiously; the investigations should identify proper actions to ensure that the final results are according to specifications. In case of a red light, the process is stopped. Investigations should identify the issues at stake, and corrective actions need to be taken before proceeding to the next phase. The traffic light status report should ensure that timely actions can be taken, when the status is still amber. If the project manager cannot solve the problems, the supervisors and perhaps the stakeholders should be informed. All additional actions to guarantee quality may require more time, resources, or money.

If, more time is needed during any stages of a survey production process (see Figure 4.2), a number of actions can be taken:

1. Replan activities to speed up the process. This can be done by carrying out activities simultaneously instead of sequentially. For instances a questionnaire could be pretested by a draft of the questionnaire, or with the parts that are already finalized. Of course, this is not ideal, but considering the circumstances it is better than skipping pretesting entirely. This requires, however, that all people involved in these activities be available.
2. Get more people involved to accelerate the process. If rescheduling is not possible, more people are needed to get the work ready in time. Again, these people need to be available and have the appropriate expertise.
3. Set a new deadline at a later stage, such as prolonging the fieldwork to obtain a higher response rate. This, however, is not always possible and depends on a number of factors: (a) the stakeholders need to be consulted about a new deadline, (b) the appropriate people need to be available after the deadline, and

(c) if activities are on the critical path, this probably will result in a delay for the project as a whole.

4. Skip activities and work faster. If none of the abovementioned strategies are possible, the only option is to have the people involved work faster and reduce the number of activities. This is called *timeboxing* or *timeframing*: trying to get as many results as possible with the given resources within the period of time that is left. As a consequence, results may be produced at a lower level of quality. Project members can be asked to work overtime, such as during evenings or on weekends. Frequently during survey development questionnaire pretesting is skipped, or testing is performed only by expert reviews (Snijkers 2002). The effect of such a decision may be that during the fieldwork stage, more respondents call the help desk for assistance. So, what you are effectively doing is pushing the problems or costs downstream. If you do this, you need to identify and manage the risks. Appropriate management information needs to be collected to manage the consequences of such actions.

When more resources are needed, the needed resources can be estimated together with the project team, to see what expectations are realistic. The project staff knows best what their working agenda will be. Next, the project manager has to negotiate their availability. If it is impossible to obtain additional human resources from the survey organization, additional staff can be contracted in. This is costly, however, and contracted staff usually need training time, since they are unfamiliar with internal procedures.

If funds are being depleted, then the project was poorly planned and/or managed. The only option would be for the project manager to discuss the situation with the supervisors. If the survey organization cannot deal with the situation, this must be discussed with the stakeholders in the project board. There are two ways to solve the problem: increased the budget or reduce the quality of the survey results (e.g., response rates will be lower because telephone reminders are being skipped).

If the quality is deteriorating because of lack of time, lack of resources, or shortage of funds, the measures suggested above can be applied. If, within the planned time and resources, the quality is decreasing, a number of actions can be taken:

- Change the methods/procedures that have been established in the project plan to design, build, and test the survey components (e.g., if errors in the sample procedure have been discovered, or use other methods to test the questionnaire.)
- Change the project members. Perhaps the current project members cannot meet the quality standards as desired, because they don't have the acquired expertise. In that case the project manager has to negotiate for the release of new project members.
- Ask for more time, human resources, and/or money, so that the quality can also be improved. Money provides the opportunity to hire external experts.

	Survey components					
Survey substages	Sample	Questionnaire	Communication strategy	Data collection process	Data processing	Survey as a whole
1. Specify the survey	☐ Target population ☐ Unit definition ☐ Cross-section/recurring	☐ Data collection modes ☐ Information needs	☐ Response rates	☐ Timetable ☐ Costs ☐ Response burden	☐ Data quality	☐ Objectives ☐ Deliverables ☐ Constraints ☐ Preplan
2. Plan the survey	☐ Objectives ☐ Quality ☐ Constraints ☐ DBT ☐ Resources ☐ Timetable ☐ Risk assessment	☐ Objectives ☐ Quality ☐ Constraints ☐ DBT ☐ Resources ☐ Timetable ☐ Risk assessment	☐ Objectives ☐ Quality ☐ Constraints ☐ DBT ☐ Resources ☐ Timetable ☐ Risk assessment	☐ Objectives ☐ Quality ☐ Constraints ☐ DBT ☐ Resources ☐ Timetable ☐ Risk assessment	☐ Objectives ☐ Quality ☐ Constraints ☐ DBT ☐ Resources ☐ Timetable ☐ Risk assessment	☐ Project plan
3. Design, build and test (DBT)	☐ Sample frame ☐ Sample design ☐ Process/quality monitoring	☐ Conceptualization ☐ Operationalisation ☐ Paradata ☐ Tested ☐ Process/quality monitoring	☐ Effective and coherent communication strategy ☐ Paradata ☐ Tested ☐ Process/quality monitoring	☐ Data collection schedule ☐ Systems, tools, procedures, ☐ training/instructions ☐ Paradata ☐ Tested ☐ Process/quality monitoring	☐ Capture ☐ Coding ☐ Cleaning ☐ Imputation and weighting ☐ Paradata ☐ Tested ☐ Process/quality monitoring	☐ Pilot study/dress rehearsal

4. Collect and process	☐ Sample drawn	☐ Questionnaire implemented	☐ Communication strategy implemented	☐ Data collection process implemented ☐ Start of fieldwork ☐ Process/quality monitoring ☐ End of fieldwork	☐ Data processing procedures implemented ☐ Data processing started ☐ Process/quality monitoring	☐ Raw data file
5. Process					☐ Capture ☐ Coding ☐ Cleaning ☐ Imputation and weighting	☐ Clean data file
6. Analyze	☐ Estimation					☐ Analysis results ☐ Statistical disclosure
7. Disseminate						☐ Survey outputs delivered
8. Archive						☐ Archiving of data, metadata, paradata
9. Evaluate	☐ Evaluation of sample design	☐ Evaluation of questionnaire design	☐ Evaluation of communication strategy	☐ Evaluation of data collection process	☐ Evaluation of data processing	☐ Evaluation of survey process ☐ Evaluation of project management

Figure 4.7 Progress status overview: Overview of key milestones in substages of the survey process.

It should be noted that the assessment of quality is not the same as the assessment of the status of a questionnaire design process, for instance. It can be concluded that a questionnaire is ready, but that its quality is not what was agreed on, perhaps because no pretesting was done.

Further Reading

For a more detailed discussion on quality and risk management, we refer the reader to the following sources: ABS (2010), HM Treasury (2004) (this book establishes the concept of risk management and provides a basic introduction to its concepts, development, and implementation of risk management processes in government organizations), van Nederpelt (2012), and Frame (2003b).

4.8 FINALIZING THE PROJECT PLAN AND GETTING STARTED

The result of all this is the project plan. Before finalizing it, the plan needs to be discussed in detail with the project members, the survey organization, and the stakeholders to obtain commitment. If the plan is not approved, the planning process may start again for the parts that need to be changed. If the plan is approved, it is signed off by the survey organization, and the contract is signed by the stakeholders (the contract may be attached to the plan in an appendix). With the signing of the contract, the planning stage of the survey comes to an end, and the project is about to enter the next stage: designing, building and testing of the individual survey components. As we have seen, the plan may be amended during this stage.

The project is officially launched with a kickoff meeting, which, of course, has to be planned as well (in the management chapter). At this meeting, supervisors and stakeholders may be present to show their support and commitment to the project. Now, the project is running and everyone is expected to do their jobs according to the plan.

4.9 SUMMARY AND PROGRESS STATUS OVERVIEW

In this chapter we discussed the planning process of a survey. It is the first step in the design and execution of a survey. For all stages in the survey production process (see Figure 4.2), we identified the work breakdown structure. We also discussed the disposition of resources, money, and time. We identified management themes, and how to control and monitor a project, and finally we discussed risk management.

Now, the project is running. As we saw when we discussed project monitoring (in Sections 4.6 and 4.7), it is important to assess the status of the project at regular time intervals with regard to milestones and (sub)results. An overview of these for the substages in the process is presented in Figure 4.7 (as based on Haraldsen et al. 2010).

This overview serves as a summary of issues discussed in this chapter, but can also be used to monitor the progress of a project in the various substages of a survey production process: when a milestone is reached, it can be ticked off.

ACKNOWLEDGMENT

In writing this chapter we would like to thank Gobert Göttgens, Robbert Renssen, and Myra Wieling (from Statistics Netherlands) for their reviews, comments, and suggestions.

CHAPTER 5

Sampling and Estimation for Business Surveys

Paul Smith

In this chapter we give guidelines on many of the practical details of designing a business survey: the register and sampling frame, sample design and various aspects of estimation. The full details of the theory are well covered in other texts (Cochran 1977; Särndal et al. 1992, Lohr 2010a), and are not repeated here. We will almost exclusively consider design-based inference (including the model-assisted approach), in other words, estimation where the sampling probabilities are used explicitly. This is because the size disparity of businesses necessitates differential sampling in most circumstances, and therefore violates one of the conditions for effective model-based inference (i.e., that sampling should be ignorable). This does not mean that model-based inference could not be used in some circumstances, and interested readers can find details of this approach in Valliant et al. (2000), and of Bayesian survey inference in Ghosh and Meeden (1997).

5.1 BASIC PRINCIPLES

A piece of advice to start—a professor I know regularly accosts people with the question "What is the target of inference?"; this could advantageously be paraphrased as "What are you trying to measure?" It is well worth noting the answer to this question clearly before starting on a survey design, so that the target is always in mind; otherwise it is very easy to produce a survey whose purpose is unclear.

In the beginnings of business statistics in the early 19th century, information was generally obtained for very few variables in support of laws and tariffs, with which each business had to comply. The result was that information on businesses, where it

Designing and Conducting Business Surveys, First Edition.
Ger Snijkers, Gustav Haraldsen, Jacqui Jones, and Diane K. Willimack.

was available at all, came from a census, where every unit in the population provided data, which was then aggregated to give totals. When measurement became an important part of the process of government, it was therefore censuses that came first (e.g., the decennial US Census of Manufactures, which began in 1809).

The early 20th century saw the development of the theory of sampling, and after discussion of a number of competing systems, there was general agreement that sampling should be based on the principle of *randomization* [codified by Neyman (1934)]. Sampling involves making observations on only some of the members of a population, and using these to deduce (make inference about) characteristics of the population. The cost of collecting information is reduced, usually substantially, but this is offset by an error introduced by observing only a part of the population and needing to estimate for the remainder—this error is known as the *sampling error*. Under randomization, where the members of the population to be sampled are chosen by a random process, we are able to deduce the properties of the resulting estimates over repeated samples; that is, if we were able to repeat the sampling many times (which is not actually possible in practice), we would on average over these samples obtain results with nice statistical properties. Further, we are able to estimate the variability (that would arise from repeated samples) around these estimates, namely the sampling error. This is the design-based approach to sampling and estimation noted above.

In this chapter we are looking for ways to make the sampling efficient—to have as small a sampling error as possible for a fixed sample size, or as small a sample size as possible to achieve a fixed sampling error. So the target of a survey design is to produce estimates of the best possible quality for a suitable cost. Some compromise may be needed depending on whether quality or cost is the primary consideration. The availability of *auxiliary variables*, variables available *before* the survey is run and that contain information about the units in the population, have a major impact on the accuracy of the design and therefore how expensive a survey will be.

A design consists of both (1) the sampling information, on how many businesses and with what characteristics to include in the sample, and (2) the form of the estimator, which will be used to provide outputs that represent the population (not only the sample) of businesses. These are interdependent—changing the sample affects the output of the estimator; changing the estimator affects the properties of the sampling.

We start by describing the types of information on which a design is based (Section 5.2), which provides particularly important support for sampling in business surveys, and then go on to describe several common types of sample design processes (Section 5.3), focusing on the practical details. A box at the end of this section gives some rules of thumb for choosing a basic design. Then we describe the main estimation approaches in Section 5.4—remember that a survey design consists of both the sample design and the estimation approach, and they cannot be considered in isolation. Survey designs are generally compared by their quality measures, which means that variance estimation is the key to evaluating and choosing between designs, and a brief introduction to variance estimation is given in Section 5.6. Some of the more challenging practical issues are covered in sections on special types of design (Section 5.3.6), dealing with outliers (Section 5.5), and small-area estimation (Section 5.7).

5.2 REGISTERS AND FRAMES

One of the features with the greatest influence on sampling for business surveys (Rivière 2002) is the availability of a register or frame with quite detailed information on the population of businesses. Almost all countries have a registration system for large businesses, although the extent of coverage for smaller businesses and the particular features (including whether it is centrally or locally administered) vary widely. Sometimes this information (or a subset of it) will be available for use in business survey design, but sometimes it will be necessary to construct a frame from other sources. We cover both cases here, starting with the typical case for a national statistical institute (NSI) in Section 5.2.1, where there is full access to the frame information, and then in the remainder of Section 5.2, discussing adaptations to cope with cases where the availability of information is less.

The information on construction and use of registers is summarized, and focuses on the properties that are most important as a basis for sampling.

5.2.1 Business Registers

A *business register* is a *statistical* register, a database of information about businesses that is intended for use in survey design and operations [for an introduction to the concepts and properties of a statistical business register, see Ritzen (2008)]. It is distinct from an *administrative* register, which is used for regulatory and other purposes, but the distinction may be hardly apparent in practice. In some statistical offices the statistical and administrative databases are separate, while in others essentially the same database has multiple purposes. The statistical uses, however, need very clear rules to define what is part of the register, so it may be wisest to separate the functions to ensure clarity and control in the statistical function. For example, a business may be retained in an active state on an administrative register if it owes tax, even when it is known to have ceased trading. In this case the statistical register will want to treat the business as dead (having ceased trading). Keeping these definitions clear is more difficult on a multipurpose register.

A business is composed of different parts, and registers reflect the business' structure as far as possible, which includes a range of different types of units. We can distinguish *statistical units*, based on definitions for statistical purposes; *collection units*, which represent the business structures that we can actually approach to obtain data; and *analysis units*, which can be constructed for particular interpretation and analysis of the data. For more detail see United Nations (2007). The main types of units are

Enterprise—a business under autonomous and single control, usually producing a single set of accounts

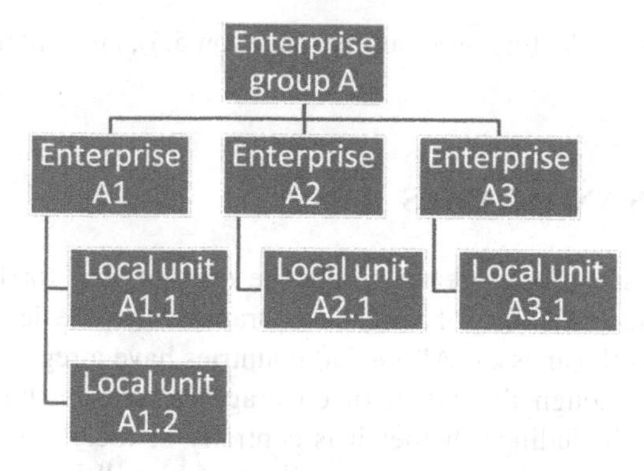

Figure 5.1 A basic structural diagram for a group of businesses under common control, with each of the three enterprises having autonomy in decisions. The first enterprise is located on two sites and therefore has two local units, while the other two have a single site each. The labeling of each level is illustrative only, but a system of reference numbers is needed.

Local unit (or *establishment*)—a single site (geographic location) where a business operates

Enterprise group—a group of enterprises under common ownership

There are many types of surveys that collect information from these three types of units. The enterprise is sometimes split to facilitate data collection from its parts, which, although not qualifying as enterprises themselves, can more easily provide the requested information; these are known as *reporting units*. In most cases, however, the reporting unit and the enterprise are synonymous. An example structure is shown in Figure 5.1 for a fairly simply organized enterprise group.

There are other types of units that are less commonly used in practice, but incorporated in many registers:

Kind of activity unit (KAU)—all the parts of an enterprise undertaking a single activity according to a classification system

Local KAU—a single activity taking place in a single geographic location (i.e., a local unit or a subdivision of it)

These units are designed to collect information on specific activities. So, if we want to measure all the retailing[1] activity covered by businesses, but *only* retailing, we would need to survey kind of activity units engaged in retailing (presupposing that these units are identifiable and that they have or can construct the information we want). This is strongly interconnected with the way that units are classified; see Section 5.2.1.2.

Many national business surveys are undertaken principally to provide information for the construction of national accounts, and the System of National Accounts

[1] *Retail* is known as *trade* in some countries—shops selling directly to the public.

(SNA 2008) (European Commission et al. 2009) also has an interpretation of units that is sometimes different from that outlined above. Therefore some adjustment may be required to keep statistics in line with the SNA concepts. According to SNA 2008, "An *enterprise* is the view of an institutional unit as a producer of goods and services" (ibid., p. 87) and an institutional unit is "an economic entity that is capable, in its own right, of owning assets, incurring liabilities and engaging in economic activities and in transactions with other entities" (ibid., p. 61). "A *local unit* is an enterprise, or a part of an enterprise, that engages in productive activity at or from one location" (ibid., p. 89). Interpretations of the term *establishment* vary, however; according to SNA 2008, "an *establishment* is an enterprise, or part of an enterprise, that is situated in a single location and in which only a single productive activity is carried out or in which the principal productive activity accounts for most of the value added" (ibid., p. 87) and "establishments are sometimes referred to as local kind-of-activity units." In Europe, however, this definition is closest to a local kind of activity unit (LKAU), which is distinct from an establishment. Finally, "a *kind-of-activity unit* is an enterprise, or part of an enterprise, that engages in only one kind of productive activity or in which the principal productive activity accounts for most of the value added" (ibid., p. 88). The SNA 2008 does not explicitly use the concept of an enterprise group.

It is possible to define specific units for particular purposes, and in some parts of the economy, for example, provision of pensions, individual pension funds may be a useful unit, but may not correspond precisely to any of the units already identified. But we will need to know how to undertake operations (sampling, updating, inclusion or exclusion from a domain[2]) on these types of units, and it will often be expedient to apply rules very similar to those for one of the main unit levels. Defining and using units in the financial sector is particularly challenging, and considerable effort has gone into the SNA (European Commission et al. 2009) to define a standard approach; the details are beyond the scope of discussion here.

In constructing a business register, it is important to ensure a good model for business structures incorporating the connections between units. This is often dependent on the information available and the processes for which it is needed; Beaucage et al. (2005) provide one example of an outline structure, for the Canadian business register. Once a model has been selected, it must be populated with information about businesses, so as to reflect as well as possible the structure of the business and how it is able to respond to surveys. In many cases businesses will be quite simple and will fit easily into the structure; however, larger and more complex businesses may not fit so easily, and their structures may change quite frequently. In these cases an activity known as *profiling* is helpful in setting up the register and keeping it current. "Profiling is a method to analyse the legal, operational and accounting structure of an enterprise group at national and world level, in order to establish the statistical units within that group, their links, and the most efficient structures for the collection of statistical data" (European Communities 2003, p. 163). It normally involves working closely with a business to

[2] Any subset of the population; see Section 5.4.2 for more detail.

understand its structure, and then using this information to create a representation of the business within the model of the business register.

Profiling has been primarily a national activity, but more recently attention has been broadened to try to improve consistency in the treatment of multinational businesses across countries, and international profiling has therefore become more important (Statistics Canada 2005), with an ESSnet project underway in Europe (Ritzen 2012).

5.2.1.1 Sources for Constructing Registers

Once the register structure has been set up, information is needed to populate the structure and its connections. This is generally derived from the administrative system by which businesses are registered. In some countries this is designed for registration purposes such as the BULSTAT system in Bulgaria, and in other countries a tax system, such as in Australia before introduction of the goods and services tax (GST). Sometimes there is more than one source; in the United Kingdom there are sources from administering value-added tax (VAT) and the employer's component of income tax [pay as you earn (PAYE)], and these are matched together to give more information and better coverage (but also additional challenges to resolve where units do not match). Whatever the main source, the register may be supplemented by other sources of information, such as on company accounts. Special sources may be needed in some sectors that are partly or entirely exempt from the administrative systems, for example, charities and the government sector.

Historically censuses have been used as a source of information [Smith and Penneck (2009) trace the development of the frame for the UK Census of Production from updating previous censuses to a register], generally not on existence of a business, which must be identified from elsewhere, but to provide other details. Although no longer widely used in more developed countries, this is still a valid, significant source. Weeks (2008) reviews the use of a range of sources of more modest coverage to construct a register, with particular application in less developed economies.

5.2.1.2 Classification

Once the units have been identified and fitted into an appropriate structure according to the units model and any profiling, they need to be classified into industries. In national accounts, "an industry consists of a group of establishments engaged in the same, or similar, kinds of activity" [SNA 2008 (European Commission et al. 2009), p. 87]. In official statistics the industry classification systems are aligned with the International Standard Industrial Classification (ISIC), although there are regional versions, such as the North American Industrial Classification System (NAICS; see `http://www.census.gov/eos/www/naics/`) and the European Union version, NACE (Eurostat 2008), which itself has country-specific implementations at the most detailed levels (see Chapter 1 for an overview of the different classifications). Classifications need to be updated periodically to take account of new activities [e.g., when compact disks (CDs) were first produced, there was no industrial classification for their production].

Some units (e.g., KAUs) are specifically defined to cover only one category in a classification, but where this does not apply, a rule is needed to classify a unit. This is generally a "wholly or mainly" approach, where the largest activity indicates the classification. However, "largest" can be assessed in different ways; for the closest connection with national accounts concepts [SNA 2008 (European Commission et al. 2009, p. 88], it should be based on the value added ("the principal activity of a producer unit is the activity whose value added exceeds that of any other activity carried out within the same unit"), but in many cases this is impossible to measure.

The information on activity used in classification varies, and may consist of

Output information by product—output is used as a proxy for value added, and the products indicate the industry to which the business is classified.

Self-classification—if the classification system is suitably available (e.g., on the web), the business may be asked to choose the appropriate classification.

Description of activity—this needs to be coded by coding software or an expert coder (for further information, see Section 11.3).

The quality of the coding directly affects the quality of the surveys that are based on it, so there are benefits in attempting to get it right. Unfortunately, the administrative processes that provide the information do not always have such classification as one of their major requirements, which are generally more centred on tax collection. In France the systems are closely linked because the accounting standards and business surveys are closely aligned, but even in countries with a strong register-based system such as Sweden there may be misclassification errors. (For an example of making a classification see Box 5.1.)

5.2.1.3 Frames from Registers

A register is dynamic, as new businesses are added and other changes occur constantly. But for a particular instance of a survey, there is a need to extract the information on those who are eligible for the survey and their characteristics, either on a specific date or to cover a particular period. This information is stored for later use in survey estimation, and constitutes the survey *frame*. It may, for example, be defined by businesses

- In a particular industrial sector
- That are in existence (i.e., removing those that are statistically dead)
- That may be undertaking the type of activity that the survey is measuring (if identifiable), such as R&D, or Internet trading

The frame is effectively a snapshot of part of the register, and should include important variables for use in sampling and estimation (see Sections 5.3 and 5.4).

5.2.1.4 Updating Registers

Registers with continuing or multiple purposes need to be updated. Often this will be accomplished by repeated use of the original sources, but the characteristics of

BOX 5.1 CLASSIFYING A SIMPLE BUSINESS

Consider a fictitious business, the Scottish Big Cheese Company, which produces three different products: cheese, scallops, and smoked salmon. It buys in milk to make cheese, and farms salmon, for which it buys fish food. The scallops are fished directly. If we could look in detail at the company's accounts for a year (including information on production and costs), we would discover

Product	Sales (£)	Input of Materials (£)	Value Added (£)
Cheese	200	150	50
Scallops	60	0	60
Smoked salmon	150	30	120

We now classify the business using three processes and the NACE rev. 2 classification:

1. *Self-Classification.* We ask the business directly what its principal activity is; given its name, the person completing the questionnaire says "cheese production," and this is duly coded as NACE 1051 (operation of dairies and cheesemaking).
2. *By Sales.* If sales by product information is available, then we can choose the product with the largest sales value—in this case cheese again. [This is only a proxy for production, which would be better, but it is rare to have information on changes in inventories (stocks) by product from which to derive production.]
3. *By Value Added.* This is the target for classification according to the SNA manual, and in this case because of the high cost of the inputs for the cheese, most of the value added comes from smoked salmon production, so the business would be classified as NACE 1020 (processing and preserving of fish, crustaceans, and molluscs) on this criterion.

So, if we run a survey of enterprises classified by process 1 or 2, all the £410 activity would be included in NACE 1051, and if we run a survey with a register classified by process 3, all £410 would appear in NACE 1020. The scallop fishing (but not scallop preserving or processing) part of the activity, however, belongs to 0311 (marine fishing).

A KAU survey would attempt to identify these activities separately.

the repeat will affect the quality of the register. Frequent updates indicate that the register stays as close as possible to its sources, which means that it is more likely to include the newest businesses, for which there is typically a lag in identification in administrative sources, and less likely to include businesses that have already ceased to trade. Including the newest businesses reduces bias from undercoverage, whereas not including dead businesses induces no bias (provided they can be identified when a survey is undertaken; see Section 5.4.2), but increases the sampling variance.

The characteristics of the administrative system affect the properties of the register and the systems used to manage it and keep it current. Registration of a new business may be a slow process, particularly if there is a tax threshold and therefore no early liability for tax. So new businesses ("births" to the register) are typically identified with a lag (Hedlin et al. 2006); for many businesses this will be only a few months (85% of new businesses have been included on the UK's business register within 4 months of their birth; see Figure 5.2), but these may be important for monthly estimates, particularly at a time when economic measurement is most difficult such as at turning points in series or during recessions. For annual estimates the impact may be less if the units are identified before sampling takes place (since annual data are collected in arrears), but there will still be an undercoverage from the most recent births. A few businesses seem to have very long registration lags (Figure 5.2), and these are potentially more serious; Hedlin et al. particularly identified businesses classified under health and social care, which are largely exempt from VAT and whose identification therefore depends only on the slower PAYE income tax registration.

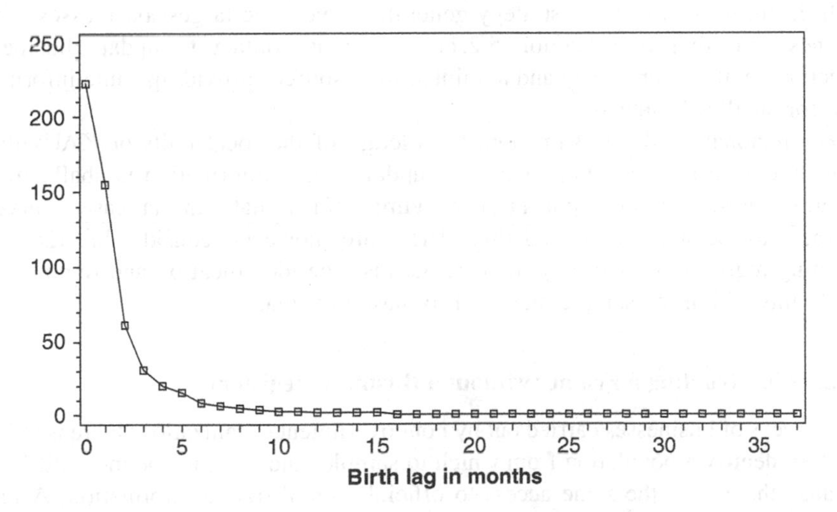

Figure 5.2 Difference between date of commencement of activity and first appearance in the business register of new births ("birth lag") for businesses added to the UK business register, based on information from January 1995 to March 1998. [Reproduced with permission from Hedlin et al. (2006).]

Similarly, businesses that stop trading may nevertheless continue to exist because there is either a lag in the deregistration process, or outstanding tax that the administrative system needs to keep track of. In this case a statistical definition of "dead" may be needed. If dead businesses are identifiable during the survey, then they can be accommodated appropriately in estimation (see Section 5.4.2); if nonresponse and death are not distinguishable, then further measures are needed to deal properly with unrecorded deaths. Faber and Gagné (2001) illustrate the use of a supplementary survey to obtain estimates of death rates.

Certain types of surveys may need special processes to identify births before they officially happen. For example, a new business will often have significant capital expenditure before it can begin to trade, and in order to measure this, the business must be contacted before it appears on the register. Special sources, such as the business pages of newspapers, can be used to identify these, but are likely to be less comprehensive than administrative processes (a simple example of trading off dimensions of quality—in this case timeliness against the completeness component of relevance).

The other challenge in updating is to keep abreast of changes in business structures. Businesses continually merge, make acquisitions of establishments or whole businesses, or sell establishments. The extent to which this affects sampling and estimation varies. Where the largest businesses gain or lose establishments, this should be reflected in the data provided without much impact on national totals. Where smaller businesses change, this can affect the accuracy of estimation, which uses these values to represent other similar businesses (see Section 5.4), and where errors in the size variables on the frame are more important. There is a tradeoff between the effort required to keep business structures current and the benefits realised from it—a suitable strategy generally covers the largest businesses with business profiling (see Section 5.2.1), a regular contact to update business structures, with other survey and administrative sources providing some information for smaller businesses.

For regional or KAU estimation, knowledge of the local units or KAUs that constitute a business are important, and updating this information is challenging because it is seldom a component of the administrative data. In that case a survey source may be needed for updating. There are, however, considerable risks in updating registers from survey sources, because the identification and removal of dead units within the sample may lead to biased estimates.

5.2.2 Constructing a Frame (without a Business Register)

For surveys of businesses carried out by nongovernment organizations there is still a need to identify a population from which to sample, and this may be more difficult because there isn't the same access to official administrative information. A few organizations construct and operate business registers from public sources, but more usually for a quite restricted population that can be tracked more easily (e.g., the largest businesses, those listed on the stockmarket, those who trade over the Internet or with some other special characteristics).

A more general problem is the need to generate a one-off sample for a particular survey purpose when there is no frame. In this case the frame can be constructed directly, using whatever data sources are available. The types of sources include

- Membership list of a trade association
- Telephone books that include businesses (e.g., yellow pages)
- Credit rating agencies
- The web

These sources also have different properties of coverage, availability, and ease of use.

5.2.3 Register and Frame Quality

There are particular quality criteria for registers and frames (general quality issues are discussed in Chapter 3), related largely to coverage. Overcoverage results from lags in identifying businesses that have ceased to trade (Section 5.2.1.4), but also from other sources, such as misclassification, which means that businesses not included in the target population are included in the register population. Overcoverage is most problematical when it cannot be identified by the survey process (e.g., because contact is not established or a questionnaire is not returned), as it then becomes indistinguishable from nonresponse. If a nonresponse adjustment is made for these businesses, this will result in overestimation.

Undercoverage is more common and more difficult to deal with in a survey. It arises from lags in identifying births (Section 5.2.1.4), and also from characteristics of the sources used in register or frame construction, which may exclude certain types of businesses (whether intentionally or inadvertently). Businesses that are not included in the sampling frame cannot be sampled, and therefore undercoverage will lead to underestimation. This will often be considered negligible (e.g., businesses too small to reach a tax threshold, while numerous, include such a small proportion of activity that their omission from a survey frame has a negligible impact on the total). Some organizations do compensate for undercoverage, however; for example, the Office for National Statistics (ONS) routinely assumes that small and medium-sized businesses that have ceased trading (an element of overcoverage) are offset within estimation by undercoverage of new businesses that have not yet been identified. The Australian Bureau of Statistics (ABS) explicitly includes a *new business provision* within its estimation, based on forecasting how many as yet unidentified businesses are likely to be present, and revising the survey estimates when the register information has caught up. One way of dealing with known undercoverage in one register is to use a dual-frame approach, where a second source provides information on units missing in the first; this is considered in Section 5.2.4.

In addition to coverage, the frame or register will contribute to the quality of estimates through the accuracy of its information—on structures, classifications, and size variables. It is not necessary for this information to be *correct* for sampling and estimation methods to work; often it need only be well correlated with the real values (see Section 5.4.6). But certainly, the better the correlation, the better the quality of the resulting estimates, and incorrect and outdated data are more likely to reduce the correlation. For some variables such as contact information, however, correct information is essential (see Section 9.6.1.1).

A final guideline is that if a business register is available, the best use of additional resources will often be to improve the quality of the register, because this has an impact on many surveys, rather than using the resources for one particular survey.

5.2.4 Dual-Frame Surveys

In some cases there is no single source that adequately covers the whole population, and in this situation one possibility is to use more than one frame, with different coverage. For example, a trade association may cover part of a population, and an optional registration system may cover another part. With information from both sources we can either (1) match the populations in advance, to give a new register with better coverage (and possibly a range of auxiliary data depending on from which source or sources the information originates), or (2) sample independently from the two sources. With independent sampling, it is important to identify as part of the information gathering whether a sampled unit could have been selected in the sample from the other source. This allows us to calculate the probability of selection appropriately to obtain unbiased estimates [see also Lohr (2007, 2010b) and Section 5.3]. In some situations the dual-frame survey can be more accurate than a single-frame one, but in general there are more challenges and more places where measurement and coverage errors can affect the estimates for dual-frames surveys. Naturally the methods are extensible to multiple frames, for an attendant increase in these quality risks.

There are few examples of strict dual-frame sampling for business surveys, but one example is the Pension Funds Inquiry in the UK in the form that ran from 1999 to 2002, which used information from the National Association of Pension Funds, a trade association list of pension funds, and an administrative register of pension schemes from the Occupational Pensions Regulatory Authority. The additional complication of this design was that pension funds consist of one or more schemes, so there was no one-to-one relationship between the units on the two frames. The design was, however, replaced because of concerns about undercoverage.

Where a frame has been constructed without the benefit of a comprehensive administrative source (Section 5.2.2), a dual-frame approach might, however, be a beneficial method for improving coverage of a population where two incomplete sources are available.

In a different variety of dual-frame survey, businesses with particular characteristics are identifiable in one source—in one straightforward example, because they have previously responded to a survey indicating that they have a certain type of activity—but the rest of the population may or may not have the characteristic. Then one frame is nested within another, and dual-frame estimation is efficient in this case

(Lohr 2007). Nevertheless, the best approach in this situation may be to use the information on the type of activity in stratification (see Section 5.3.2.1).

5.3 SAMPLE DESIGN

The way the sample is selected (the sample design) is key to making accurate estimates from business surveys. The characteristics of businesses surveys that need to be accounted for are the large size differences between businesses, and the availability of auxiliary information (from the business register) that can be used in the design to help compensate for those differences. The general idea is to select a sample, to use the sampled units to represent the unsampled units, and to adjust this "representation" using known information to better "predict" the variables to be measured in the survey for the unsampled units. The better the quality of the prediction, the better the properties of the design. So, part of sample design can be reduced to fitting a model for the outcome variables—which would be easy if we knew them, but if we did, we wouldn't need a survey! So, instead we have some proxy for the outcome—in the best situation a previous period from the same survey, but other types of information related to the survey outcome(s) will also serve. Since the predictions may not always be particularly good when using these proxy data in the design work, we will also need to factor in the *robustness* of the design, that is, how well it performs when the data we collect do not behave as expected.

5.3.1 Some Basic Ideas

For practical decisions on the type of sampling to use in a particular situation, we will refer to some of the quality measures described in more detail in Chapter 3, in particular:

1. The sampling error, for which we need *probability sampling* and a *measurable* design;
2. Bias and coverage, particularly whether the sample allows an estimate to be made for the whole of the population of interest [this issue is separate from register coverage (Section 5.2.3), as we shall assume that the estimates are made for the population as identified on the frame or register].

Probability sampling refers to the use of a random process to select the units to be included in a sample. We can then refer to the inclusion probability, the probability that a particular unit is included in the sample (or equivalently if we could draw a large number of samples, the proportion of those samples that contain a particular unit). This is fixed by the sample design, but of course in any particular sample the unit is either present or absent. It is not necessary for every unit to have the same inclusion probability, and indeed for most business surveys some of the units should be included with certainty (probability = 1) because of their importance to the final estimates. *Measurable designs* are sampling designs in which each unit has a nonzero probability of inclusion; that is, it contains no units that are always excluded from the sample. This condition is needed to ensure that estimates from the survey will be unbiased.

Randomization is the key to probability sampling; it is essential for the theory of sampling to follow, on which the designs in this chapter are based. In particular, it allows us to estimate the variability due to sampling, the most important statistic for choosing between designs. Therefore the process of selecting which units to include in a sample is important, and algorithms for selecting random samples are discussed briefly in Section 5.3.2.7.

Sampling and estimation (Section 5.4) are very closely interrelated, and a survey design must account for both. The design generally needs information about the population variance (not the sampling variance; it is the characteristics of the population that influence the design, rather than the characteristics of a particular sample). Where an estimation model uses additional information (see Sections 5.4.2–5.4.5), the population variance is derived from the estimation model. Since the population variance, in turn, influences the design, a sample design is tailored to that particular estimation method.[3] So, a survey design consists of both the sample design and the estimation method together.

Stepping back briefly before considering various aspects of sampling, we should also note that a census is a valid survey approach, effectively a "sample" in which every business has an inclusion probability of 1. The need for increasing economy in surveys militates against their use since they are very resource-intensive, but they do provide comprehensive information without sampling error (but are still prone to other errors; see Chapter 3).

We will examine three building blocks for sampling in business surveys: stratification (which we will cover in some detail since it is the most widely used), probability proportional to size (PPS) sampling, and cutoff sampling. Box 5.5 (p. 199) provides some general guidance on selection of a particular design from the options available. Section 5.3.6 covers some types of business surveys with characteristics that lead to special considerations or special applications of the methods already presented.

5.3.2 Stratified Sampling

The most commonly used sampling technique for business surveys is stratified sampling. A *stratum* (plural *strata*) is a subset of the population, and the usual approach is to *partition* the population into strata, so that each element of the population is in one of the strata. Then different sampling can be used in each stratum, as the sampling within strata is done independently. Optimal conditions for stratification are

1. When elements in the population have widely varying sizes and there is a good source of size information with which to construct the strata

[3] Unless a model-based approach is used, where some characteristics of the design are a necessary part of the assumptions. We do not cover this approach here.

2. When elements within a stratum are more similar to each other (in terms of the outcome variable) than to elements in other strata, that is, when the strata are homogeneous (Cochran 1977, Section 5.6) (or equivalently, when the variables on which the stratification is built are good predictors of the outcome variable as measured in the survey).

These conditions are frequently encountered in business surveys, since the largest businesses are much larger than the smallest, and therefore stratification by size to group the largest businesses together and separate them from the smaller ones, is almost ubiquitous. This works because the differences between businesses within strata are smaller, so the variance in response values within a stratum is smaller than that across the whole population. With a judicious choice of strata, we can then reduce the sampling variance for a fixed sample size (or reduce the sample size for a fixed variance). Designing a sample may largely therefore consist in identifying and choosing among auxiliary variables to find good stratifiers, and then implementing an appropriate stratification. There are several stages in developing a stratified design:

- Choose which variable(s) to use to construct strata.
- determine how many strata are needed (particularly when converting a numeric variable into strata).
- Determine the boundaries between (numerically defined) strata.
- Determine how many observations to sample in each stratum (this process is called *sample allocation*).

Sections 5.3.2.1–5.3.2.3 give the basic steps for determining the details of an efficient stratification, and then the remaining parts of Section 5.3.2 set out the finer details needed in the practical applications of stratified designs.

5.3.2.1 Stratification Variables

For use as the basis for stratification, the value of an auxiliary variable, such as the number of employees, must be available for each member of the population (perhaps from a registration system), but there is considerable freedom in how an auxiliary variable is defined. For example, if we have some prior information on R&D activity, we could define an auxiliary variable x^* that is

$$x^* = \begin{cases} x & \text{where } x \text{ is previous R\&D activity} \\ 0 & \text{where it is known that there was no previous R\&D activity} \\ \text{missing} & \text{where it is not known whether there was any previous} \\ & \quad \text{R\&D activity} \end{cases}$$

This auxiliary variable x^* could then be used in a stratification where "missing" is a valid value, and units with this value form one stratum. This type of approach is very useful for rare characteristics, and we will return to it in Section 3.2.5.3. Another possibility is to define the stratifier as a division of a numerical value into categories; this technique is often used for size variables (see Section 5.3.2.2).

Strata are not restricted to definitions using only a single auxiliary variable; they could be a combination of several variables. But using many stratification variables in combination quickly increases the number of strata. The population of businesses, however, is fixed in size, so each stratum then has only a few businesses, and quite a large number of them need to be sampled to support a reasonable estimate within the stratum. Some guidelines on selecting stratification variables are presented in Box 5.2.

It is equally possible to use numeric auxiliary variables in PPS sampling (see Section 5.3.3). One interpretation of PPS is that it is a very fine stratification with each business in a stratum with businesses having exactly the same auxiliary variable value.

5.3.2.2 Defining the Strata

There are three paradigms for making decisions about strata:

1. Minimize the sample size for a fixed coefficient of variation.
2. Minimize the coefficient of variation for a fixed sample size.
3. Minimize the coefficient of variation for a fixed cost.

We will largely ignore costs, as in many business surveys the cost of surveying one unit will be similar to that for another; costs can be built in through optimal allocation (see Section 5.3.2.3). Most theory is based on paradigm 1, where the user specifies in advance the coefficient of variation (CV) they would like to achieve. In many practical situations, however, the sample size for the desired CV is unaffordable, and/or the survey budget is fixed, and so paradigm 2 is used. Sample allocation (Section 5.3.2.3) is usually invertible, but methods of defining strata under paradigm 2 are relatively rare.

Once we have settled on a stratification variable or variables, preferably as a result of developing a model for the outcome variable or something closely related to it, as in Box 5.2, the next two components of the design to be determined are the number of strata needed, and which values will define the boundaries between them.

Selection of the number of strata is based on studying the changes in the (predicted) sampling variance with different numbers of strata. If there is only one stratum, businesses within it will vary greatly in size, and therefore the sampling variance will be large. Moving to two strata means that the variance within each is smaller, and this reduces the overall variance. As the number of strata increases, the overall variance continues to decrease, but less each time (see example in Figure 5.3), giving a J-shaped curve. The goal is to choose the number of strata corresponding with the point at which the variance decrease flattens out—a parsimonious tradeoff between minimal variance and keeping the number of strata small.

So, the choice of how many strata depends on predicting the sampling variance for a range of numbers of strata. Unfortunately for the hard-pressed sampling statistician, to obtain a prediction of the variance with h strata defined by a numeric auxiliary variable, it is necessary first to determine the optimal stratum boundaries,

BOX 5.2 SELECTING STRATIFICATION VARIABLES

To be available for all members of the population, the values of auxiliary variables need to be available in the frame. They can be numerical or categorical. If necessary, define an explicit category where a variable is missing—you have a lot of flexibility in how a variable is defined.

The basic stratification for a business survey covering a range of activities will typically be by industrial classification and size. The exact level of industrial classification is open to manipulation, but it is rarely excluded entirely because of the differences between businesses in different industries.

We are seeking a variable that predicts the survey outcome well. Since we don't know the outcome before we run the survey, we look for some data with which to construct a model, which we can then carry forward for our survey. The ideal is a previous instance of the same survey, if it is repeating, because that gives us the target variable and the auxiliary variables. Then we try to find the best-fitting model (from a regression-type model, which might include categorical variables, i.e., including ANOVA and ANCOVA models). Parsimonious models, with few predictors, are preferable, because we cannot use many variables for stratification—too many variables produces many strata, each with few elements (the so-called curse of dimensionality), which is not efficient. We might therefore want to use a measure that penalizes extra parameters heavily for choosing between models, such as Akaike's information criterion (AIC) or the Bayes information criterion (BIC).

Although we will divide up numerical variables into categories to make strata, it is reasonable to use the numerical version in modeling. On the other hand, a categorical variable such as industrial classification might be merged into broader categories to reduce the number of parameters in a model; this new variable should be tested as part of model selection to see whether a more parsimonious model results from the reduction in parameters. Once we have chosen a variable or variables as stratifiers as a result of the modeling, methods for breaking numeric variables up into classes are available (see Section 5.3.2.2).

A final thought—if no variables are good predictors in a model, it is probably worth including the one variable that has the best predictive power even if it is terrible, as there will likely be a gain in efficiency. It would be very unusual to have a business survey without a size stratifier[4].

[4] In fact, the PECO and DoSME projects in central Europe are a rare example of a business survey without size stratification (Smith and Perry 2001).

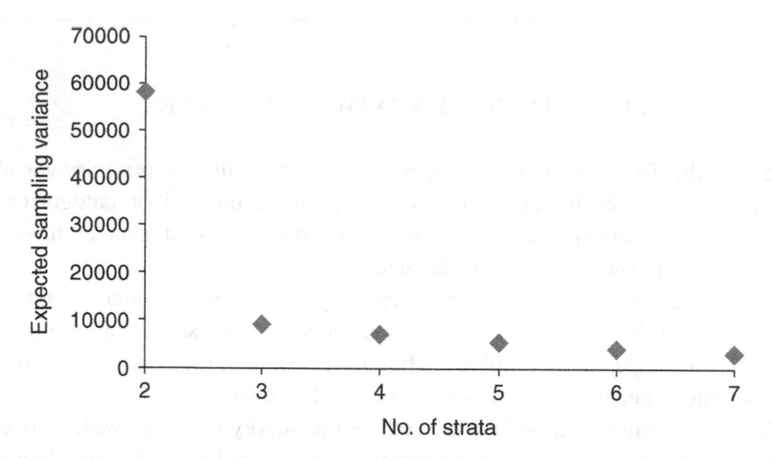

Figure 5.3 Example plot of (predicted) sampling variance versus number of strata. Most of the gain in variance from stratification is achieved with three strata.

and the boundaries therefore need to be determined across the whole range of numbers of strata, even though only one outcome will be used in practice.

The traditional approach to determining stratum boundaries has been an analytical one, based on development of a model for the stratum details, which leads to a form that is not practical to solve precisely. A solution can be obtained by making a suitable approximation, for which there are several options [see Horgan (2006) for a review]. More recently the availability of fast computing has made it more practical to use numerical approaches, with different algorithms being used to identify an optimum [for an overview, see Baillargeon and Rivest (2009, 2011)]. Comparisons have shown that the numerical approach often produces a design with smaller sample size or smaller variance, so this approach may be cost-efficient (see Section 5.3.2.6). Sections 5.3.2.2–5.3.2.5 discuss the analytical approach in detail.

Both analytical and numerical approaches are affected by nonresponse. One way to view the response process (Chapter 2) is as a sampling process that selects respondents from the full sample; this sampling process is not, however, controlled by the statistician, and therefore cannot be designed! More precisely, expected nonresponse rates can be built into a design (usually through an adjustment to the sample allocation; see Section 5.3.2.4), but the actual response rates will rarely be equal to the expected ones. Then robustness to nonresponse is a strongly desirable property—a small change in either the nonresponse rate or which units respond should not have an important impact on the quality of the design.

Where stratification is by a numerical variable with a skewed distribution in the population, there will almost certainly be a need for a completely enumerated stratum, because the largest businesses have such a significant contribution to any estimate that any sample that excludes them risks a very large sampling error.

The commonest approach to determining the division of a numerical variable into strata is the cumulative $\sqrt{f}$ method of Dalenius and Hodges (1959), which is described in Box 5.3. Using this approach on a variable with a skewed distribution

BOX 5.3 THE CUMULATIVE $\sqrt{f}$ OR DALENIUS–HODGES METHOD FOR DETERMINING STRATUM BOUNDARIES

The population is initially divided finely into groups g using the stratification variable; at this initial step more is better, since the eventual stratum boundaries will be chosen from among the boundaries of these initial groups. [For this reason, if "convenient" boundaries (e.g., rounded to certain numbers) are desired, this can easily be facilitated by choosing such rounded values for the initial groups; the impact on the accuracy of the final solution will generally be small.] In each initial group tabulate the frequency (f, number) of businesses in the population, and then take its square root ($\sqrt{f}$). These $\sqrt{f}$ values are then cumulated (see table below). If the initial groups are not of equal length, then we will have to modify the frequencies by multiplying by the ratio of the length of group g, d_g to the length of the first group d_1, $\sqrt{(d_g/d_1)f}$, and then cumulating.

Initial Groups	Frequency f	Cumulated $\sqrt{\frac{d_g}{d_1}f}$	Stratum to Which Group Belongs
1–25	108	10.4	1
>25–50	49	17.4	2
>50–75	27	22.6	3
>75–100	7	25.2	4
>100–125	5	27.5	4
>125–150	0	27.5	4
>150–175	2	28.9	4
>175–200	0	28.9	4
>200–225	0	28.9	4
>225–250	1	29.9	4

Take the total $\sqrt{f}$ [or $\sqrt{(d_g/d_1)f}$], and divide it into n equal pieces. So, in the example above, to find boundaries for four strata, if the total is t, calculate (t/4, 2t/4, 3t/4) = (7.5, 14.9, 22.4). Then find the values among the cumulated root frequencies that are closest to these calculated values, and the boundaries of the groups to which these belong are the boundaries for the strata—in this case (25, 50, 75).

In this example there are only a few groups at the lower end of the distribution, where there are the most units, and the design has little flexibility. Using the same population, we could set up the initial groups in 10s up to 50, then in 25s as before, in which case we obtain the following:

(continued)

Initial Groups	Frequency f	Cumulated $\sqrt{\frac{d_g}{d_1}f}$	Stratum to Which Group Belongs
1–10	44	6.6	1
>10–20	52	13.8	1
>20–30	23	18.6	2
>30–40	21	23.2	2
>40–50	17	27.3	3
>50–75	27	35.6	3
>75–100	7	39.7	4
>100–125	5	43.3	4
>125–150	0	43.3	4
>150–175	2	45.5	4
>175–200	0	45.5	4
>200–225	0	45.5	4
>225–250	1	47.1	4

This time the boundaries for four strata are $(t/4, 2t/4, 3t/4) = (11.8, 23.5, 35.3)$ and the boundaries of the strata are (20, 40, 75).

will result in one or sometimes more completely enumerated strata. This naive application sometimes results in a very inefficient design (Lavallée and Hidiroglou 1988). Nevertheless, it is a one-step method, is easily implemented, and does not require the designer to specify the coefficient of variation in advance (although by the same token the solution is not tailored to the desired CV, either).

Two other algorithms provide analytic alternatives to the cumulative $\sqrt{f}$ method. Hidiroglou (1986) proposes an algorithm for finding a partition into two strata in the case of Horvitz–Thompson estimation (see Section 5.4.1), one completely enumerated and one sampled, and Lavallée and Hidiroglou (1988) propose an algorithm for the optimum partition of a population into one completely enumerated and a fixed number of sampled strata using the same simple estimator and a specific allocation method (power allocation). There are some challenges with this algorithm, however; the solution is dependent on the starting values (so the algorithm does not always find a global minimum), and there can be problems of convergence, particularly for large numbers of strata (L large). The Lavallée–Hidiroglou algorithm can be adapted to other sample allocation methods (Hidiroglou and Srinath 1993), and Rivest (2002), expanding on the work of Detlefsen and Veum (1991) and Slanta and Krenzke (1994), found that its convergence was worse when using Neyman allocation. Nevertheless, in "sensible" situations with few strata and large populations, it seems to operate very effectively, and even in other situations often improves substantially on the starting values (Slanta and Krenzke 1994).

Horgan (2006) proposes a simpler method based on a geometric series, which may be useful in some situations, although it is strongly affected by the largest unit in the sample, which means that it is not well suited to business survey populations.

Using the Lavallée–Hidiroglou algorithm or Horgan's geometric series approach as a first step to separate out completely enumerated units is likely to avoid problems of (gross) inefficiency, and may provide useful initial values for iterative approaches [as in Slanta and Krenzke (1994)].

Descriptions of the cumulative $\sqrt{f}$ method, and of Hidiroglou's (1986) and Lavallée and Hidiroglou's (1988) algorithms are given in Boxes 5.3 and 5.4, respectively, but they (and some extensions) are also implemented in the *stratification* package for the R statistical software (Baillargeon and Rivest 2011), which is available free of charge from CRAN.

The various algorithms presented above have been couched in terms of an auxiliary variable denoted by x, because this is the most common situation. Ideally, we would design using the survey variable (y), but if we knew the values of the survey variable for the population, there would be no need for a survey! Instead, we assume that the survey variable(s) is (are) strongly related to the auxiliary variable. If we have prior information about the relationship between y and x, then this can be built into the Lavallée–Hidiroglou algorithm (Rivest 2002), or in any of the algorithms the stratification information can be worked out on the expected moments (Baillargeon and Rivest 2011, Section 4). But we will still have more confidence in a design if there is a strong relationship between the survey variables and auxiliary variable used in the design.

Most stratum definition in practice uses a single numeric stratifier (although this may be nested within a categorical stratifier such as an industrial classification). There may, however, be additional variables available. Two-dimensional stratification, where strata are defined according to the sizes of two auxiliary variables, was developed in New Zealand (Briggs and Duoba 2000). The idea is that businesses with large values on *either* auxiliary variable should be completely enumerated, giving rise to L-shaped strata (see Figure 5.4). Some software (Strat2D) was developed to undertake the allocation, using a search process among available boundaries. This

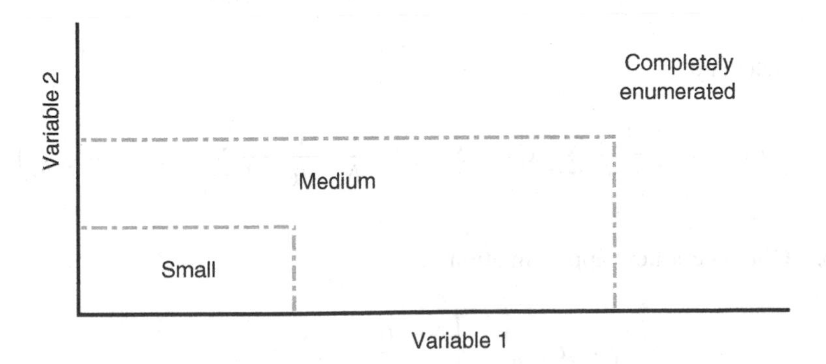

Figure 5.4 Diagrammatic representation of two-dimensional stratification with L-shaped strata as implemented in Strat2D (see text).

BOX 5.4 DESCRIPTIONS OF HIDIROGLOU'S (1986) ALGORITHM AND THE LAVALLÉE–HIDIROGLOU ALGORITHM

Hidiroglou (1986) gives a method for partitioning a population into a completely enumerated and a simple random sampling stratum with Horvitz–Thompson (HT) estimation (see Section 5.4.1). There is an exact criterion, but it is difficult to apply directly, so he also offered an approximation that can be improved iteratively (usually in only a few steps) to achieve a target coefficient of variation c (an implementation of this algorithm is available in the *stratification* package for the R statistical software (Baillargeon & Rivest 2011), which is available free of charge from CRAN):

H1. Order the auxiliary values to give $x_{(i)}$ with $x_{(1)}$ the smallest.

H2. Calculate

$$\mu_{[N]} = \frac{1}{N}\sum_{i=1}^{N} x_{(i)}, \quad S^2_{[N]} = \frac{1}{N-1}\sum_{i=1}^{N}\left(x_{(i)} - \mu_{[N]}\right)^2$$

and the population total of the auxiliary variable $X = \sum_{i=1}^{N} x_{(i)}$.

H3. Set $k=1$. Calculate a first approximation to the target value as

$$x_1^* = \mu_{[N]} + \left\{\frac{c^2X^2}{N} + S^2_{[N]}\right\}^{1/2}$$

H4. Find the number of units greater than x_1^*, call it t_1, and the number of units $n(t_1)$ that would be sampled at this first iteration

$$n(t_1) = N - \frac{(N-t_1)c^2X^2}{c^2X^2 + (N-t_1)S^2_{[N-t_1]}}$$

H5. Calculate

$$\mu_{[N-t_k]} = \frac{1}{N-t_k}\sum_{i=1}^{N-t_k} x_{(i)}, \quad S^2_{[N-t_k]} = \frac{1}{N-t_k-1}\sum_{i=1}^{N-t_k}\left(x_{(i)} - \mu_{[N-t_k]}\right)^2$$

H6. Calculate a new approximation

$$x_{k+1}^* = \mu_{[N-t_k]} + \left\{\frac{N-t_k-1}{(N-t_k)^2}c^2X^2 + S^2_{[N-t_k]}\right\}^{1/2}$$

H7. Find the number of units greater than x^*_{k+1}, call it t_{k+1}, and the number of units that would be sampled at this iteration

$$n(t_{k+1}) = N - \frac{(N - t_{k+1})c^2X^2}{c^2X^2 + (N - t_{k+1})S^2_{[N-t_{k+1}]}}$$

H8. If $0 < 1 - [n(t_k)/n(t_{k+1})] < 0.1$, then stop with the boundary x^*_{k+1},[5] otherwise increase k by 1 and return to step H5.

The Lavallée–Hidiroglou algorithm (Lavallée and Hidiroglou 1988) is more general, but does fail to converge in some situations (see main text). The number of strata L must be specified in advance, and the boundaries between the strata are designated $b_{(0)} < b_{(1)} < \cdots < b_{(L-1)} < b_{(L)}$, where $b_{(0)} = \min(x)$ and $b_{(L)} = \max(x)$.[6] The algorithm [an implementation of which is also available in the *stratification* package, Baillargeon and Rivest (2011)] is:

LH1. Order the auxiliary values to give $x_{(i)}$ with $x_{(1)}$ the smallest, and set $b_{(0)} = x_{(1)}$ and $b_{(L)} = x_{(N)}$.

LH2. Start with arbitrary boundaries satisfying $b_{(0)} < b_{(1)} < \cdots < b_{(L-1)} < b_{(L)}$.

LH3. Within each stratum formed by the boundaries from step 2 calculate

$$W_h = \frac{N_h}{N}$$

$$\bar{X}_h = \frac{1}{N_h}\sum_{j \in h} x_{(j)}$$

$$S^2_h = \frac{1}{N_h - 1}\sum_{j \in h} x^2_{(j)} - N_h\bar{X}^2_h$$

$$A = \sum_{h=1}^{L-1}(W_h\bar{X}_h)^p$$

$$B = \sum_{h=1}^{L-1}(W_hS_h)^2(W_h\bar{X}_h)^{-p}$$

$$F = Nc^2\bar{X}^2 + \sum_{h=1}^{L-1}W_hS^2_h$$

(*continued*)

[5] Hidiroglou (1986) actually uses x^*_k as the optimum as part of setting up a rule of thumb that might allow easy assessment of the boundary in some cases, but since x^*_{k+1} has already been calculated, it seems sensible to use that.

[6] Although $b_{(0)}$ and $b_{(L)}$ are not strictly needed in the design, they are needed in the algorithm to ensure that the top and bottom strata are not empty, and therefore that the correct number of strata is obtained.

LH4. To update boundaries $h = 1, \ldots, L-2$ calculate α_h, β_h, and γ_h using

$$
\begin{aligned}
K_h &= Bp(W_h\bar{X}_h)^{p-1} - Ap(W_hS_h)^2(W_h\bar{X}_h)^{-p-1} \\
T_h &= AW_h(W_h\bar{X}_h)^{-p} \\
\alpha_h &= F(T_h - T_{h+1}) \\
\beta_h &= F(K_h - K_{h+1}) + 2F(\bar{X}_{h+1}T_{h+1} - \bar{X}_hT_h) + 2AB(\bar{X}_h - \bar{X}_{h+1}) \\
\gamma_h &= FT_h\left(\bar{X}_h^2 + S_h^2\right) - FT_{h+1}\left(\bar{X}_{h+1}^2 + S_{h+1}^2\right) + AB\left(\bar{X}_{h+1}^2 - \bar{X}_h^2\right)
\end{aligned}
$$

and for boundary $L-1$, use (note that K has the same definition)

$$
\begin{aligned}
K_{L-1} &= Bp(W_{L-1}\bar{X}_{L-1})^{p-1} - Ap(W_{L-1}S_{L-1})^2(W_{L-1}\bar{X}_{L-1})^{-p-1} \\
\alpha_{L-1} &= FT_{L-1} - AB \\
\beta_{L-1} &= FK_{L-1} + 2\bar{X}_{L-1}(AB - FT_{L-1}) \\
\gamma_{L-1} &= FT_{L-1}\left(\bar{X}_{L-1}^2 + S_{L-1}^2\right) - AB\bar{X}_{L-1}^2 - F^2
\end{aligned}
$$

LH5. Calculate a new set of boundaries, $b^*_{(1)}, \ldots, b^*_{(L-1)}$ using

$$
b^*_{(h)} = \frac{-\beta_h + \sqrt{\beta_h^2 - 4\alpha_h\gamma_h}}{2\alpha_h} \quad \text{for } h = 1, \ldots, L-1.
$$

LH6. Repeat LH3–LH5 until either two sets of boundaries are identical, or they differ only by a small amount, that is

$$
\max_{h=1}^{L-1} \left| b^*_{(h)} - b_{(h)} \right| < \varepsilon
$$

for some suitable small positive value of ε.

approach is therefore closest in spirit to the numerical solutions to stratification that are discussed in more detail in Section 5.3.2.6.

There is no particular reason why strata defined by two variables need be L-shaped when using a numerical search, and the strata might be more internally homogenous if they were defined by the arcs of circles or ellipses (Figure 5.5); or they might be so well approximated by the L-shaped strata that the additional complexity is not worthwhile. This does not yet seem to have been investigated.

5.3.2.3 Sample Allocation

With paradigm (1) from the previous section, where the target is to minimize the sample size for a fixed coefficient of variation, the algorithms of Hidiroglou (1986) and Lavallée and Hidiroglou (1988) (see Box 5.4) include the allocation step, and therefore provide the stratum boundaries and the number of units to select within each stratum.

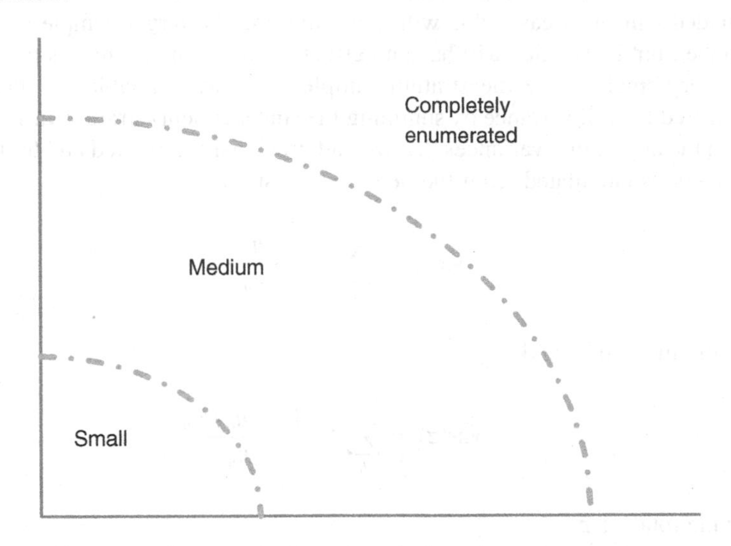

Figure 5.5 Diagrammatic illustration of elliptical two-dimensional strata.

In the situation where the sample size is predetermined (paradigm 2), however, once we have the stratum boundaries, we then need to decide how much of the sample to allocate to each stratum. There are several ways to undertake this allocation process; the most generally useful method for business surveys is Neyman allocation (Neyman 1934; Cochran 1977, pp. 98–99), which attempts to give the most accurate estimate of the overall total. To do this, it makes the allocation proportional to the product of the population standard deviation and the population size within strata. Estimates of the population variance from an existing source are therefore needed as an input. Ideally, they will come from a previous run (or a pilot) of the survey being designed. Sometimes, however, there is no previous survey, in which case a related variable might be used as a proxy—perhaps a size variable from the business register. Note that it is the relative sizes of the standard deviations that are important in the allocation, so it may be satisfactory to calculate variances of employment when turnover is the main design variable, since the proportions will be approximately similar. We will use z to represent a variable that is from a previous survey or a suitable auxiliary variable from the register.

The available element information is sorted into the new strata, and then the values of the proxy variable z are used to estimate the population variance in the strata, which for expansion estimation is

$$s_h^2 = \frac{1}{(n_h - 1)} \sum_{k \in h} (z_k - \bar{z}_h)^2$$

Note that there is no finite population correction here, as we are estimating the *population* variance. The variance used in the design must correspond to the estimation model (see Section 5.6 for further details), in which case the z_k may

$$n_h = n \frac{\quad}{\sum_h s_h N_h} \tag{5.1}$$

where n is the sample size to be allocated and N_h is the population size in stratum h. The n_h from this procedure are not, in general, integers, so they will need to be rounded—in most cases this will mean missing the target sample size by a small number, but in practice will have no effect on the estimates unless the sample sizes are very small. Once the stratum sample sizes are available, we can obtain the predicted overall variance by summing the (independent) stratum variances, derived from the population variances that we had previously estimated and finite population corrections calculated from the new sample sizes:

$$\widehat{\text{var}}(\overline{z}) = \sum_h \frac{1 - (n_h/N_h)}{n_h} s_h^2$$

for the mean of z and

$$\widehat{\text{var}}(z) = \sum_h N_h^2 \frac{1 - (n_h/N_h)}{n_h} s_h^2$$

for the total of z.

Note that because we are using an auxiliary variable in design, it is unlikely that we will achieve this variance in practice; the survey variance will additionally include the variability in the relationship between the survey variable and the auxiliary variable.

Although we have based the exposition above on Neyman allocation, other allocation approaches are occasionally useful. *Optimal* allocation (Cochran 1977, p. 96) also accounts for differences in the cost of administering the survey to different members of the population. For postal (mail) surveys, which have been commonly used for businesses, and for more modern equivalents such as web or automated phone surveys, there is little difference in the cost to the surveying organization. There may, however, be differences in interviewer-administered surveys, or if the questionnaire has different lengths for different types of business. The cost to businesses (burden or compliance cost; see Chapter 6) is also very important, and here there is much more scope for variation in costs, so it might be useful to get the best accuracy for minimum overall compliance cost, and use optimal allocation here, too.

Power allocation (Bankier 1988) is another approach, allowing allocations between Neyman allocation and equal accuracy in each stratum, and we will return to it in Section 5.3.2.5. One version of it, however, square root proportional allocation, where the sample size is proportional to the square root of the sum of an auxiliary variable $X_h = \sum_{k \in h} x_k$ (again using x as a proxy for a survey variable y), that is

$$n_h = \frac{n\sqrt{X_h}}{\sum_h \sqrt{X_h}} \tag{5.2}$$

can be very useful when the auxiliary information on which the sample is designed is rudimentary. It provides a robust distribution of the sample for multiple purposes when more detailed information is lacking, because it uses the skewed distribution of the auxiliary variable, but tempers it so that the largest businesses do not consume such a large majority of the sample size. The same approach could be taken with the population sizes N_h, but is unlikely to be useful in a business survey because of the skewed distribution of business sizes.

5.3.2.4 Extra Details of Allocation

Estimators (see Section 5.4) generally have nice properties when the sample on which they are based is reasonably large, so it may be useful to add a constraint specifying the minimum sample size in a stratum. Other conditions may also be useful in allocation. However, procedures such as Neyman allocation do not have such conditions naturally built in, so in order to use them, the allocation must be done iteratively. Once the n_h from Equation (5.1) or (5.2) have been calculated, the values of any that violate a constraint increase or decrease to meet the constraint. The sample taken up by these constrained strata is subtracted from the target sample size to give a new target, which is allocated across the remaining strata, using the same rule. The process repeats until no new constraints are breached by the remaining allocation. (Note that if there are constraints that result in samples being increased *and* decreased in the same constrained allocation, it may be necessary to include those near the constraint boundary in the next stage in case they cross the boundary in the new allocation step.)

Constraints may be used for a number of reasons; some common examples are

1. The allocated sample exceeds the population size (this is common, since there are typically few large businesses, and their population variance is very large), so we set the sample size equal to the population size; these strata are then fixed and should not be allocated in later stages.
2. A minimum sample size within a stratum. Another common condition, to ensure that the sample size is large enough for estimators based on the sample to have good properties. When the allocation exceeds the population size in some strata (as in item 1, above), since some extra sample may be available for the next step of the allocation, strata where the minimum size has been imposed may need to be reconsidered in the next step.
3. A maximum sample size is a less common condition, used when one or a small number of strata have very large values of $s_h N_h$. The idea is that we accept a small increase in the variances of the overall estimate and the estimate in the

constrained stratum for better estimates in other strata (see also *multipurpose allocation*, in Section 5.3.2.5).

4. A maximum sampling fraction may be used when a survey holiday (see Section 6.4.2) is given to ensure that the population is sampled evenly (to ensure that the maximum possible sample consistent with survey holiday period is taken on each occasion, otherwise sample sizes could vary considerably from period to period).

Sample allocation does not directly account for anticipated nonresponse, although response is unlikely to be 100% and may vary considerably according to a business's characteristics. If there is enough information, the expected response rate by stratum r_h can be estimated from past data (preferably over several runs of the survey to smooth out random variations), and then the sampling can be proportional to $N_h s_h / r_h$, which will increase the allocation in strata where nonresponse is expected to be highest. Any allocation constraints designed to give a minimum sample size for estimation would then also need to be specified in terms of n_h / r_h. This approach will make an appropriate adjustment particularly to ensure a suitable number of responses for use in estimation, which provides some control over the sampling variance, but does not necessarily impact the nonresponse bias (see Chapter 10), and may even make it worse!

We saw the need for estimates of the population variance as inputs to Neyman allocation, and some of the options for alternative variables, in the previous section. Even when there are good previous data, however, not every stratum may have enough information to calculate a variance; some may have achieved samples of 1 or 0, either by design or from nonresponse. Even where there is >1 response, the variance estimate may be quite unreliable for small numbers of responses. In this case there may be a need to "borrow" information from another source by

- Substituting a population variance estimate from another stratum that is regarded as similar (or is as similar as available).
- Using observations from several past survey occasions to obtain enough observations to make a reasonably robust estimate of the population variance.
- Summarizing population variances [e.g. by their median, or, to give more weight to recent periods, an exponentially weighted moving average], if they can be calculated from several survey occasions, but with only a few observations for each.
- Developing a model explicitly for the population variances in terms of, for example, averages of auxiliary variables within strata. If a model with a reasonable fit can be found, there is an opportunity to use information from many strata to make a prediction.

Once the strata have been defined, selection within strata (see Section 5.3.2.7) is generally by simple random sampling (where every business has the same probability of being selected); in this case the sampling probability of a unit k, namely, π_k

(which will be needed in estimation; see Section 5.4) can be simply calculated as the ratio of the sample size to the population size in the stratum h containing k, n_h/N_h. There is, however, no need to be restricted to simple random sampling as long as the estimator in due course accounts for the sample selection by using the selection probability π_k. Differences *within* strata can also be accounted for by ordering the strata in a particular way and then taking a systematic sample—a process known as *implicit stratification* (Goodman and Kish 1950).

Despite the systematic selection, the selected sample is treated as if it were a simple random sample. This has generally been regarded as a good strategy, because it avoids particularly extreme (unbalanced) samples, but Zhang (2008) has highlighted that it may generate inefficient samples more often than simple random sampling, and that simple random sampling is therefore more robust even though it initially appears more variable.

5.3.2.5 Multivariate and Multipurpose Sampling

Much of the preceding development implicitly assumes that there is a single outcome variable that is collected, or of such overriding importance that it outweighs the others and the design can focus on it exclusively. There are, indeed, surveys like this, often monthly surveys collecting information on one or a few variables to give timely information for policy monitoring, such as the Monthly Business Survey in the UK or the Retail Business Survey in Australia.

More often, however, business surveys collect information on a range of variables, of approximately equal importance. Some will be correlated with each other, but there are some variables such as capital expenditure or specific types of activity such as R&D that are not particularly well predicted by size, and are more generally not well correlated with size and activity variables. There are several strategies for designing a sample in this situation; one is to choose a small number of key variables, undertake a sample allocation separately on each, and then average the resulting allocations (a weighted average could be used if the variables have different importance). Note that if the minimum and maximum sample size constraints are used consistently, they will appear appropriately in the final design too. Skinner et al. (1994) also suggest taking separate samples, but then pooling them using methods from multiple frame estimation.

Another option is to use some combination of the target variables to produce a single variable for allocation (e.g., the first principal component). Several authors have addressed the challenge of finding a solution to allocation that in some way gives an optimum tradeoff between the different variables, based on convex or linear programming [see, e.g., Bethel (1989) and Valliant and Gentle (1997)], but these are generally more complicated to implement.

Data from surveys generally have more than one use, and while the overall total of a variable might be the primary target, there are usually many more domains (a *domain* is any subset of the population, including the population itself; for more detail, see Section 5.4.2), such as regional totals or industry totals. Neyman allocation has the accuracy of the overall total as its target, and this may mean that particular domains have rather small sample sizes, and that estimates for them are not very accurate. Sometimes we would like to trade off some of the accuracy in the total for a better spread of the sample across the domains. Adding constraints of the type described in Section 5.3.2.4 can help achieve this by guaranteeing a minimum sample size or by ensuring that one stratum does not dominate the allocation. However, there are other practical approaches; for instance, Bankier's (1988) power allocation is motivated by the need for reasonable accuracy of both national totals and regional estimates.

5.3.2.6 Numerical Methods and Software for Stratification

The availability of computing power has made it more practical to search for stratification solutions numerically, and a method was proposed by Kozak (2004). The (mathematical) space of such solutions is large and seems to be prone to local minima in some situations (Baillargeon and Rivest 2009, 2011), but the results from several studies using Kozak's algorithm indicate that such approaches can find solutions with smaller sample sizes for the same variance (or smaller variances for a given sample size) than the simpler but approximate rules given in the preceding sections. This is not *always* true, however (Baillargeon and Rivest 2011).

Numerical approaches have been used mostly to determine the boundaries between the strata, which is the part of the stratification process that has no exact solution. To determine the stratum boundaries, the number of strata and the method of allocating the sample size among the strata need to be specified. By following the same approach as described in Section 5.3.2.2, where a series of different stratum sizes is evaluated, we can choose an appropriate number of strata, but this increases the space of designs to be searched to obtain an appropriately optimal outcome. Such numerical methods allow for some compensation for nonresponse to be built into the design process, to ensure that the achieved sample is as well-designed as possible.

Various organizations have provided software for different parts of the sample design process, but they have rarely covered a full range of options. More recently, however, a package called *stratification* has been written for the R statistical software which includes a range of different options; both R and its packages are available free. See Baillargeon and Rivest (2011) for a description of its functions and some examples. It only covers Horvitz–Thompson (HT) estimation of the mean (and therefore also totals), which restricts its application for some of the estimation methods in described Sections 5.4.3–5.4.5, which utilise auxiliary information. It may be possible to design using the residuals from the estimation model as inputs, but this depends on the model (e.g., whether it is within or across strata), and in some ways the model also depends on the sample size available and therefore the allocation. If an estimation model fits well, then naively using the design from the HT estimator with this model will generally provide smaller variance than using the same design with the HT estimator. However, it is unlikely to be the optimal stratification under the more complex estimation model, and could be noticeably worse than the optimum in many cases.

5.3.2.7 Sample Selection

Once we have a design and a number of businesses to select in each stratum, we need a method for determining which businesses should be sampled, considering that the selection must be a random process in order for design-based sampling theory to work (Section 5.3.1). There are simple algorithms for selecting samples of units from a population with computer-generated random numbers [see, e.g., Särndal et al. (1992, Chapter 3) and Tillé (2006)]. One very simple approach is to allocate a (uniformly distributed) random number to each member of the population, then order the population by that random number, and choose the first n units; this will give a simple random sample with equal selection probabilities for each unit, and could be applied within each stratum of a stratified design. But in general the required probabilities of selection will vary between units, and then other algorithms will be needed. A variety of algorithms for selecting randomized samples is available in the *sampling* package for the R statistical software (Tillé and Matei 2011). There are also approaches designed to give more control over the length of time that businesses spend in the sample over multiple periods of the same survey, and for managing samples in multiple surveys. These are known as *sample coordination*, and are discussed briefly in Section 5.3.4 and in more detail in Section 6.4.2.

5.3.3 Probability Proportional to Size (PPS) Sampling

An alternative to stratification as a basic sampling strategy for business surveys is probability proportional to size (PPS) sampling. In this case the probability that a particular unit is included in the sample is proportional to some suitable numeric value (often a variable describing the size or activity of a unit, but not necessarily so). The ratio of the selection probability to the auxiliary variable can be chosen to give a fixed sample size (in which case the probability of selection is nx_k/X, where X is the population total of x_k, as long as there are no completely enumerated businesses; see text below), or to meet a predetermined quality criterion. This approach uses the auxiliary information directly to define the probability that a business is included in the sample instead of undertaking a stratification process. However, the approach to choosing an appropriate auxiliary variable from several candidate variables is the same as for stratification (see Box 5.2), although the relationship between the fitted model and the sampling is clearer since there is no pooling of different values within a group to give implicit dummy variables in the model as there is in stratification.

The design of a PPS sample is more straightforward than a stratified design, since there are no stratum characteristics to be determined. It is the sample selection that is somewhat more complicated. The usual approach is to divide the population total of the chosen auxiliary variable X by the desired sample size to give X/n. Then the population of businesses is arranged in random order $j = 1, \ldots, N$ (Hartley and Rao 1962), and the cumulative total of x, $X_{(k)} = \sum_{j=1}^{k} x_j$ is calculated with this ordering. Then a random starting value (a *seed, s*) between 0 and X/n is chosen, and the business whose cumulated x value, $X_{(k)}$ is the smallest value that exceeds this seed value is chosen as the first element in the sample. Then X/n is added to the seed, and the business whose cumulated x value, $X_{(k)}$, is the smallest value that exceeds this new value is also chosen, continuing until the end of the ordered list is reached, when n businesses should have been chosen. It is important to use the fractional parts of X/n each time to ensure that the required sample size is achieved.

This process works well if there are no businesses with $x_k > X/n$. If there are businesses that meet this condition, then they are certain to be selected, and thus should be removed to a completely enumerated stratum. Then the remaining sample can be selected as before. The new selection may generate some more businesses that are certain to be selected, so they are also moved to the completely enumerated stratum, until there are no businesses that meet this criterion, and then PPS sampling may continue as described above. This process immediately shows that PPS sampling is rarely used exclusively. Having a completely enumerated stratum is one example of combining stratification and PPS selection, but since PPS cannot accommodate nonnumeric variables, there are many other situations where an initial stratification (e.g., by industrial classification) followed by PPS sampling might be a good strategy.

The Hartley–Rao approach of using a random order is needed to ensure a good (unbiased) estimate of the sampling variance, but there may again be benefits from implicit stratification, where businesses are arranged in a particular (e.g., geographic) order to ensure that the sample is spread across this variable (see Section 5.3.2.4).

Therefore, PPS sampling has a relatively easy implementation, but then the estimation is more challenging as each unit has its own selection probability nx_k/X, particularly for estimators other than the HT estimator. In fact, although PPS sampling is widely used in survey sampling, it is quite difficult to find examples of its use in business surveys; the main applications seem to be in surveys used to collect price information (see Section 5.3.5.2). So, when would we prefer to use it? Its main benefit is in cases where the sample is rather small, in which case strata are small and some estimates are prone to bias (those that are only asymptotically unbiased); a small sample can be more satisfactorily spread across a population with PPS sampling. But then each sampled business has approximately the same contribution to the estimate, so nonresponse among any of the sample is less easily accommodated.

There are other ways to select PPS samples than the Hartley–Rao approach described above [e.g., see Ohlsson (1998), who also originally used a price index sampling context], and there are methods for coordinated sampling in PPS sampling schemes (see Section 6.4.2).

5.3.4 Cutoff Sampling

Cutoff sampling is a direct reaction to the skewness of the distribution of business size, with very many small businesses and only a few large ones, but in which the largest businesses account for a large proportion of the activity in the population (see Table 5.1). The idea is that by covering *all* the largest businesses (and none of the smaller ones), we maximize coverage of the variables of interest for a given sample size. A further variant is to include a group of the largest businesses that are all sampled, a group of the smallest in which none are sampled, and an intermediate group in which a probability sampling method is used.

There are some immediate disadvantages to this approach:

- An assumption must be made about the activity of the smallest businesses if they are to be included in an estimate.

Table 5.1 Illustration of the Skewness of the Distribution of Businesses

Employment	Number of Units (1000s)	Proportion of Total Turnover (%)
0–1	814	2.3
2–5	911	5.9
5–10	41	5.0
11–20	18	4.4
21–100	21	12.9
>100	21	69.5

Source: UK Business Register, November 2011.

- The estimate of total activity will be biased (although the bias may be small if good estimation procedures are available)
- In a continuous survey any changes in the output of the smallest businesses will go unnoticed.
- There is no sampling variance for estimates from the largest or smallest groups, so the quality of the sampling can be assessed only through the mean-squared error

These are strongly related to the basic principles discussed in Section 5.3.1—a cutoff sample is a very extreme example of a probability design in which there is no variation between samples, but it is not measurable as some selection probabilities are zero.

The properties of cutoff samples have been reviewed recently by Haziza et al. (2010). The main practical sampling challenge is how to determine the cutoff. This may be defined simplistically by the money available to run the survey(!), but if we were asked how to determine the requirement, we would want (1) to exclude only a minority of the activity and (2) information on how the estimation will work, and the assumptions on which it is based and how they hold across the range of business sizes. This information may not be available. One example where it is in electrical power generation surveys at the US Energy Information Administration (EIA) [see Knaub (2007), who also reviews applications of cutoff sampling documented on the Internet].

Where surveys collect several variables, there may be a need to design a cutoff that is efficient across the variables. This requires a tradeoff—the chosen cutoff cannot be equally efficient for all variables. Benedetti et al. (2010) propose one approach for an efficient sample in this situation, and Knaub (2011) and references cited therein give further options.

See Section 5.4.4 for comments on estimation from cutoff samples.

5.3.5 Sample Rotation and Overlap

The practical aspects of a system to manage sample rotation are described in Chapter 6, but there are choices to make that are part of designing an appropriate survey to achieve the required outputs. Many surveys will, of course, be one-off or irregular, but those conducted for regular monitoring and reporting of the economy typically have dual purposes: to estimate the level of activity, and the change in activity from period to period. The variance of a change (or difference) between two time points t_1 and t_2 is $\text{var}(t_2 - t_1) = \text{var}(t_2) + \text{var}(t_1) - 2\text{cov}(t_2, t_1)$, which means that the variance of change can be made smaller if we can increase the covariance between the estimates at the two timepoints. We can do this by including the same businesses in the sample at the two timepoints.

In annual (and less frequent) surveys the level is the dominant target, and therefore designed overlaps are typically small or none; note that the way in which business surveys tend to always have the largest units in the sample induces a covariance from this source automatically. In quarterly or monthly surveys (or even more frequent ones), the changes are often more important than the levels (at least the most recent ones!), and here extending the overlap to cover more units than the complete enumerations is helpful.

So, how should we choose how much overlap? At one extreme we could keep the same businesses in the sample on each occasion (a *panel*). But we know that business populations are dynamic, so any panel gradually loses members, as businesses cease trading, or merge. But a panel is not refreshed with new members from those businesses that start up, and therefore becomes unrepresentative of the whole population. At the other extreme we could select a new random sample on each occasion, in which case there would be no covariance. There is also a cost in data collection from changing the sampled businesses; there is typically a learning or training process where the contact in a business responsible for completing the questionnaire gains an understanding of what is required. Replacing a well-trained respondent with a fresh one can entail costs in response quality and in nonresponse follow-up.

The ideal solution is one that balances the overlap with a replenishment of the sample, to allow the sample to represent the whole population. The number of periods in the sample for a rotating design (Section 6.4.2) will also balance these. In a monthly survey, 15 months gives a monthly overlap of 93.3% and an annual overlap of 20% (before the effect of nonresponse, which may be considerable). Increasing this to say, 24 months, has a marginal effect on the monthly overlap 95.9%, but a larger one on the annual overlap which is then 50% (again ignoring nonresponse). The effect on respondent follow-up costs is, however, in the complement, so if a significant part of the cost is for "training" new respondents, the change from 6.7% of the sample being new each month to 4.1% may imply a substantial cost reduction.

Two other factors are important: (1) the effect on respondents of long periods in the sample is to alienate them from the survey, so some reasonable restriction on the number of occasions that a business is asked to respond is needed to manage the burden; and (2) the impact of nonresponse is important, because it reduces the overlap from both ends—a business that fails to respond in *either* period is not matched and therefore does not contribute to the covariance. So, 10% nonresponse each month typically leads to 15–20% nonmatches in the sample.

BOX 5.5 SELECTING A SAMPLE DESIGN

There's no substitute for experience, but you have to start somewhere in order to gain it, so here are some guidelines. They're not meant to cover every situation, but should help in making something practical. The guidelines reflect the aim of sample design, which is to make useful estimates from the survey, so some of them reflect properties of the estimators or variance estimators explained in more detail in later sections.

For general use, a stratified design gives lots of flexibility and is relatively straightforward to apply, so try it as a default design. Cutoff and PPS sampling are most useful where resources are most limited.

A completely enumerated stratum is necessary (unless the population is not skewed, which is unusual for business populations).

Make partitions (e.g., stratification) coincide with the most important domains in estimation; thus, if a key target is estimates for separate industries, including these industries as strata will be more efficient (have smaller variance) if estimation is done within strata (see Section 5.4.3).

Use a small number of stratifiers; do not stratify in so much detail that strata have small sample sizes (ideally aim for $n = 30$ for estimation [extrapolated from Cochran (1977, p. 162)], which could mean stratum sample sizes <30 if combined estimators (Section 5.4.4) are used.

If there are choices between different numbers of strata for approximately the same sample size/variance, fewer strata will generally give a more robust design.

5.3.6 Special Sample Designs

5.3.6.1 Product Surveys and Other Detailed Breakdowns of Activity

Product surveys are challenging for a sample design because the requirements are generally to obtain reasonable estimates of a range of products, which effectively means creating a design for a large, sometimes very large, number of variables, with most businesses in the population having positive values for only a few of those variables. The design is therefore multivariate (Section 5.3.2.5), and for each particular variable the sampling variance is driven by 0s and large values, and therefore tends to be large.

The European Union's PRODCOM regulation requires product surveys in the manufacturing sector to ensure coverage by specifying a minimum proportion of the population activity to be covered. This is a solution where the coverage of sample observations is large so as to reduce the impact of sampling and estimation, and indeed the regulation doesn't require estimation for the unsampled parts of the population, effectively a cutoff design (see Section 5.3.4); and the recent numerical investigations (Haziza et al. 2010; Benedetti et al. 2010; Knaub 2011) offer a route to optimisation of this type of design.

A more complex solution was used in Japan in the 1990s where an economic census provided information on which products were produced by which businesses. This information was used to construct a sample, where sample sizes for products were defined separately, but then the sample was built up by selecting businesses that produced the rarest product first, to ensure coverage, and if they also produced a more common product, then the remaining required sample size for the common product was reduced accordingly. This process continued until the full sample was selected (with some overselection for a few products necessary in the late stages). This is a nice way to maximize product coverage for a given sample size, if the product information is available to work with, although calculating the actual sampling probabilities is complicated, and approximating them by the observed sample size over population size is probably better.

If such product information is available, it may be possible to construct an algorithm to find an optimal sampling solution that is slightly more efficient than the heuristic one just described. Neither approach, however, allows for considerations of burden and sample rotation.

The UK has tried a different approach, based on multivariate allocation, to obtain an appropriate sample and make population estimates by product in its implementation of the PRODCOM survey. This has allowed some sample optimisation, unusually with a model-based estimator (Chambers et al. 1998). A comparison of this approach with another based on minimizing the average standard error, which has been used more widely in social surveys, suggested that the approach of Rahim and Currie (1993) as modified by Bell was potentially useful in these kinds of surveys [see Smith (2001) for details]. This is an area where more research would be useful before setting out any general guidance. Possibly the numerical allocation methods described in Section 5.3.2.6 would give a good solution if sufficient information on products made by businesses were available for the population.

Although this section concentrates on product surveys, any detailed breakdown of activity presents the same problems. For example, the value of exports by country of destination is regularly asked for in surveys. The approach to design is the same, although gathering any information on which to base the design may be difficult.

5.3.6.2 Price Index Surveys

Product surveys are made complex because they cover such a range of variables (products), but prices add a further layer of complexity, since here we are trying to collect information on prices of transactions by product. In practice, a price quote is generally used to represent many transactions (in a few cases there may be no transactions and the price is therefore more difficult to define; e.g., imagine obtaining a monthly price quote for a nuclear submarine).

This complexity is usually overcome by using a multistage design. For consumer price indices (CPIs), this tends to go to extremes, with six or seven stages of selection not uncommon. For example, locations (often a cutoff-type stage covering only major shopping centers, for which the local units are listed), then outlets (local units) within locations, probably stratified by outlet type; then product types within outlet, then specific products within product type.

For producer prices the sampling is less complex since transaction prices rarely vary by location (except in delivery costs). Some sample designs are derived by direct stratifications, but the survey frame seldom contains enough information on which products are produced by which business for this to result in an efficient design. The UK PPI uses a two-stage sample with the PRODCOM product survey as the first phase and with a second phase sampling from the products identified among the sample members. There is a lag between obtaining the sample information and drawing the sample for the second phase (so the survey is two-stage as well), and this imposes some challenges when sampled products or businesses are no longer available. Nevertheless, this does provide a sound basis for optimizing the design for surveys designed to measure prices.

5.3.6.3 Surveys for Rare Characteristics ("Needle in a Haystack" Surveys)

A number of characteristics of interest in business surveys are challenging when designing samples, because they occur in only a minority of the population, and are not readily identified by, for example, the classification system. Classic examples are businesses engaged in R&D activity, businesses that provide services to organizations in other countries (overseas trade in services), and businesses that have just started to trade (births).

There are several ways to deal with these situations, although none is particularly efficient—surveying for these characteristics is just difficult, and there is no way to make good estimates without resources. For a review of some approaches with examples, see Smith and Perry (2005). These are some of the strategies:

- *Satellite Registers.* Registers connected to the main register, but with different sources that better cover the rare characteristic. For example, an administrative system on credit regulation may provide a satellite register of retailers (and other types of businesses) who provide credit services to consumers. There are different degrees of integration depending on how well the units in the satellite register match the ones in the main register. In the credit example, there is no reason why the credit part of a retailer should be a local unit or other recognizable part of a retail business, although it may be.
- *Filter Questions.* Often it will be possible to use a larger survey (or a census if one exists) to collect information on the presence of a certain type of activity, and then to add this to the frame for use in design (a succession of surveys or censuses will allow the information to be maintained and added to). The ideal use of this information is likely to be in stratification—businesses identified as having the activity should be in a stratum (or series of strata) together. This could be a take-all stratum, giving a panel element to the design. Businesses not identified as having the activity still need to be covered by the sample, however, particularly if the filter question was posed to a sample—otherwise the part of the activity that was not identified is not covered, and the estimates will be biased.

- *Two-Phase Sampling.* This is quite similar to the filter question approach, but without the need to maintain a register identifier. An initial sample of business (constructed with an appropriate design) is approached, and businesses are asked whether they have the desired characteristic. A second-phase sample is selected from those that do, and sent a more detailed questionnaire with appropriate questions. This will generally be a less efficient use of resources than adding a filter question to an existing survey, although some costs are saved by having a simple initial approach and a detailed follow-up for businesses with the desired characteristic. For one-off surveys of particular unusual characteristics (e.g., to evaluate the effect of a particular policy), it may be the only available solution.

5.4 ESTIMATION

The counterpart to sampling is estimation; once we have designed a survey and selected an appropriate sample and designed and built the other survey components (e.g., questionnaire, survey communications), then the field processes described in the other chapters take place, and eventually, following data capture, coding, and cleaning (Chapter 11), produce a set of responses that constitute the data that we can use to estimate our target statistics (recall from Section 5.1 that it is wise to know what the target is before the survey is designed in order to facilitate this step).

Sample design is the first step in obtaining efficient estimates, that is, with variances as small as possible for a given cost, and with other properties such as robustness to errors in design information. In estimation we are working with the sample information that we have collected, and trying to use this in the most efficient way possible. This distinction is slightly artificial, because a survey design is composed of both the sample design and the estimation approach, and it is only sensible to choose both pieces to complement each other. The differences in costs and efficiencies between different sampling designs can be substantial, but in estimation (once the design is fixed) the cost differentials are generally smaller, although there is a cost to complexity over simplicity; we will broadly evaluate estimators according to their *variance* and *bias* (if any). However, we do not attempt to derive the variances [refer to a standard sampling text such as Cochran (1977), Särndal et al. (1992), or Lohr (2010a) for these details], and we will only comment on relative sizes of variances and biases in a general way. Variance estimation is covered in more detail in Section 5.6, but again as a practical guide, not as a theoretical development.

Business surveys are particularly susceptible to outliers because of the large differences among sizes of businesses [see Section 5.3.4; also Riviére (2002)], and this affects all forms of estimation discussed in this section, but estimation is set out assuming that there are no outliers (coping with outliers is discussed in Section 5.5).

5.4.1 Horvitz–Thompson Estimation

The most basic step in estimation is to account for the differential sampling that we built into the sample design (see the various options in Section 5.3), which is done by multiplying each observation by a weight to form an estimate. The starting point is to use the inverse of the probability that a unit was included in the sample as its weight. So, if the probability that a unit k would be selected in the sample (measured over repeated sampling) is π_k, then the weight given to that unit in estimation when it has been selected is $w_k = 1/\pi_k$. This is variously known as expansion estimation, number raised estimation, or Horvitz–Thompson (HT) estimation (Horvitz and Thompson 1953), and estimates of the form

$$\sum_{k \in D} w_k y_k \qquad (*)$$

where k denotes the sample members, D is a domain, and y is a variable for which an estimate of the total within D is required, will be unbiased if the design is *measurable*, which means that $\pi_k > 0$ for all k. Note that cutoff designs (Section 5.3.4) are not measurable, because the smallest units have $\pi_k = 0$.

Horvitz–Thompson estimation is very straightforward because it uses only information on the selection probabilities from the design, and does not account for any other information. One way to consider HT estimation is by using a sample business to represent w_k businesses in the population—itself and $w_k - 1$ others, and this is equivalent to a very basic model where the nonsampled businesses are assumed to have a mean equal to the value of the sampled business. There is a duality between the estimation methods and models for estimation, although we will seldom delve far into the model side [see Särndal et al. (1992, Part II) for more details].

Although HT estimation is the starting point for weighting approaches, if HT estimation is used for business surveys with simple random sampling, the large variation in size of businesses means that the variances of the HT estimates are typically rather large (too large for practical use). In order to reduce these variances, the sampling design must be chosen to reduce the differences in the size of businesses used in estimation, for example, by stratification. HT estimation is used successfully in business surveys in Statistics Sweden, where the size stratification is quite fine (i.e., with many strata) to control the variation in size of businesses (Davies and Smith 1999, Vol. 3, Section 4.3)—an example of how the sample design is adapted to correspond with the estimation method in an appropriate way to obtain good properties for the estimates.

Using HT estimation with PPS sampling is different, because information on the auxiliary variables has already been used at the sampling stage; we return to this topic at the end of Section 5.4.3.

5.4.2 Some Basic Ideas

The estimator (*) uses the idea of a domain, which is any subset of the population for which we would like to make an estimate. The domain is defined by a rule that, when applied to any unit in the population, states unequivocally whether that unit belongs to the domain. To produce an estimate for the total population, we simply define a domain where *all* units are members. Then the estimate is produced by summing the product of the weights and responses for all elements within the domain, as in (*). An alternative approach to making domain estimates is to define a new variable y_{dk} that is the response (y_k) if the unit is in the domain, but 0 otherwise, and then the domain estimate is the sum of the product of the weights and the new variable across the population. But (*) can be used for any domain, noting that

- If there are no sample elements in the domain, the estimate will be zero (even though it may be known that there are elements in the domain in the population).
- If there are only a few sample elements in the domain, the estimate will generally not be very accurate (i.e., the sampling variance will be large).
- Dealing with dead units identified during data collection is a very simple type of domain estimate; such dead units have zero activity by definition, so their response is zero to all questions, and this sets up a domain estimate for live businesses from among those businesses that were sampled. For this strategy to be effective, however, dead businesses must be distinguishable from non-responding but live businesses.
- If the domain has not been used in the sample design in such a way that the design controls the sample size within the domain, then we will implicitly estimate the size of the domain as well as the target statistic. The variance will therefore have two components, one from estimating the domain size and one from estimating the total from a sample. A corollary of this is that if we do use information on the size of the domain in the sampling, we can implicitly use known information about the domain size and therefore make our estimate more accurate (this is the origin of one of the guidelines in Box 5.5). For example, if we define strata to be industries defined by a suitable classification, and then we wish to make an estimate where an industry is the domain, the selection probabilities [for simple random sampling (SRS) or PPS] have N_h in their calculation, and we therefore use the known population size for the industry in estimation, leaving the variance as only the component for estimating a total from a sample.

We can naturally produce HT estimates of the totals of variables, typically denoted by x (or sometimes z to distinguish variables from an external source) and called *auxiliary variables*, where the total is already known (from the register or some other source) and if we do, the HT estimates will, in general, not be the same as the totals we already know. An estimator which, by construction, returns the known value for a population characteristic is said to be *calibrated* on that characteristic. The HT estimator is calibrated on the stratum population sizes.

There is a very wide class of calibration estimators, and if we can incorporate the known information on auxiliary variable totals into the estimator so that it is calibrated on them, this intuitively feels better because of its consistency with known information. This method generally reduces the variance of the estimator, too, and is equivalent to a model in which the auxiliary variable and the survey variables are closely related. This is the basis of the extensions of weighting discussed in the following sections.

The variables denoted by y in (*) are simply the variables collected in or derived from the survey responses. Because (*) is linear in the weights and the responses, any linear relationships are preserved in the estimation (weighting)—so components continue to sum to totals, for example.

5.4.3 Ratio Estimation

Probably the most widespread estimation method for business surveys and other establishment surveys is ratio estimation, so called because it implicitly uses a regression model through the origin (a fitted ratio) to estimate for the nonsample units. The ratio estimator is one of the *calibration estimators*, which gives the known value of a population characteristic when we make an estimate of that characteristic using the estimator and the sample information only.

Assume a simple situation of a single stratum within which we know the size of the population N and the total X of some auxiliary variable x, say, number of employees as measured by some administrative system. We also need to know the individual values of the auxiliary variable for the sample units, x_k. Then we can find the ratio r of some survey variable y to the auxiliary variable as

$$r = \frac{\sum_{k \in s} y_k}{\sum_{k \in s} x_k}$$

This estimates how much y is produced, bought, sold, and so on per employee (x), and we can make an estimate for the total t within our stratum by multiplying the total employees X by the ratio

$$t = rX = \frac{\sum_{k \in s} y_k}{\sum_{k \in s} x_k} X$$

This can be rewritten in the same form as (*) by setting the weight

$$w_k = \frac{X}{\sum_{k \in s} x_k}$$

This has an effect similar to that for a HT weight, since the numerator is a population-sized value and the denominator is a sample-sized value. Singh et al. (2001), in developing generalized software for estimation, used a framework where the design weight $d_k = 1/\pi_k$ is retained, and any further impacts of the estimator are captured in a *model weight* g_k. So, if we rewrite w_k in this format as

$$w_k = d_k g_k = \frac{1}{\pi_k} \frac{\pi_k X}{\sum_{k \in s} x_k} \tag{5.3}$$

and note that under SRS within a stratum $d_k = N/n$ we can immediately see that

$$g_k = \frac{X}{(N/n)\sum_{k\in s} x_k} = \frac{X/N}{\sum_{k\in s} x_k/n}$$

is >1 if the sample average of x is smaller than the population average, and <1 if the sample average is greater. This means that the ratio estimator compensates for the size of the sampled units in making an estimate, and this effect reduces the variance of the estimator, often quite substantially. Cochran (1977, pp. 157–158) shows that ratio estimation has a smaller variance than HT estimation when

$$\rho > \frac{1}{2}\frac{\text{cv}(x)}{\text{cv}(y)}$$

that is, when the correlation coefficient ρ between x and y exceeds half the ratio of the coefficient of variation of x to the coefficient of variation of y. In practice, where there is a good frame with some information on the size of businesses from an administrative system, ratio estimation will often be substantially better than HT estimation.

The ratio estimator is calibrated on the total X. If we continue with x and X denoting employment values from a business register, and we perform the weight calculation in Equation (5.3) in each stratum independently (*separate* ratio estimation), then in each stratum the estimated total register employment will be equal to the known total—the estimator is calibrated on register employment. This is equivalent to fitting a ratio model (regression without an intercept) $y_k = \beta x_k + e_k$ with $e_k \sim N(0, \sigma^2 x_k)$ separately in each stratum, and using it to predict the y_k for the nonsampled observations. Note that the ratio estimator is *not* calibrated on the *number* of businesses—the estimate $\hat{N}_h$ obtained with these weights is not in general equal to N_h.

The ratio estimator is (like all of a much wider class of calibration estimators) *asymptotically unbiased* (i.e., unbiased in large samples) [Deville and Särndal (1992)], but it has a bias of order $1/n$, which can be important in small samples. Inferring from Cochran (1977, p. 162), we can generate a rule of thumb that $n \geq 30$ is sufficient to render the bias unimportant, and this can be used to guide the choice of a minimum sample size in a stratum (Section 5.3.2.4). There are, however, ways to define ratio estimators to work satisfactorily across strata when the stratum sample sizes are smaller than this (see Section 5.4.4).

In multipurpose surveys, which collect information on a number of variables, ratio estimation with a particular auxiliary variable may look good for some variables (which are strongly correlated with that auxiliary variable) and not so good for others, or different auxiliary variables may be well correlated with different survey variables. Herein lies a trap, however, for changing the auxiliary variable or estimator destroys the consistency property obtained by making estimates with (*). So it is preferable to retain the same weights for all variables (noting that $w_k = d_k g_k$ is independent of y). This is particularly useful if we are trying to estimate productivity from turnover and employment information, for example.

The PPS sampling method uses the information on x and X in the calculation of the selection probabilities, and therefore the HT estimator is calibrated on X (and not N); it is therefore the sampling analog of the ratio estimator. In other words, PPS sampling + HT estimator uses the same information as simple random sampling + ratio estimator; Knottnerus (2011) suggests that in many situations the PPS sampling strategy + HT estimator is actually a more efficient choice.

5.4.4 Combined Ratio Estimation

Sometimes stratification produces rather small sample sizes within strata, particularly where a fixed budget sample is used to cover many domains with a minimum sample size requirement for each domain. In this case there may not be 30 observations in each stratum. But we can combine the information in several strata as long as we can assume that the *same* ratio model holds in all the strata, and as long as we use the information on the selection probabilities. We can again use the formulation on the left side of Equation (5.3), but if we express the form of g_k in this case under simple random sampling, we obtain

$$g_k = \frac{\sum_h X_h}{\sum_h (N_h/n_h) \sum_{k \in s_h} x_k}$$

This shows that combined ratio estimation operates in the same way as the separate ratio estimator, but now the denominator contains an estimate across the strata, and because this covers several strata, the survey weights may vary between the different strata and must be incorporated. In this way we can ensure many of the benefits of ratio estimation even when the sample sizes are rather small. Note that the combined ratio estimator is calibrated for X only in the *group* of strata; estimates of the total of x in individual strata $\hat{X}_h$ are not, in general, equal to X_h.

Estimates with the general form $\sum_k w_k y_k$ normally work only for measurable designs (where all the members of the population have a chance $\pi_k > 0$ of being selected in the sample). For the cutoff sampling designs described in Section 5.3.4, we need to choose a model to use to estimate for the part of the sample that is never selected. This will be an assumption; the model can't be fitted, because we have no data for this part of the design. An obvious assumption, however, is that the ratio model that we are using in the weighting should also apply in the cutoff part of the sample, and we can "trick" the weighting in combined estimation to do this by adding the auxiliary variable total for the cutoff part of the sample $X_{\text{co}} = \sum_{\text{co}} x_k$ to the numerator:

$$g_k = \frac{X + X_{\text{co}}}{(N/n) \sum_{k \in s} x_k}$$

to use one stratum (usually the one closest to the cutoff in size) to fit the model or

$$g_k = \frac{X_{\text{co}} + \sum_h X_h}{\sum_h (N_h/n_h) \sum_{k \in s_h} x_k}$$

to use combined strata. This increases the weights to account for the population elements that can never be sampled (effectively combined ratio estimation across the cutoff and one or more sampled strata, although the cutoff has no observations and therefore contributes no information to the fit of the ratio model). Note that variances calculated with these modified weights will not be valid, however.

Other approaches to estimate the total for the cutoff units are possible, using any model that it seems sensible to assume, but they will seldom be amenable to the trick of incorporating them in the weights; the estimate for the cutoff is calculated separately and added on as required. In some cases the cutoff may be small enough to be ignored, and Baillargeon and Rivest (2009, 2011) demonstrate that this can still lead to smaller-mean squared errors for estimates with a given sample size despite the bias of this approach.

5.4.5 More Complex Estimators

There is a wide range of extensions to ratio estimation. A fairly obvious extension is to *regression* estimators, which allow for an intercept in the model (and that are therefore calibrated on N as well as X), and for more than one auxiliary variable to be included—$y_k = \beta_0 + \beta_1 x_{1k} + \cdots + \beta_p x_{pk} + e_k$ with $e_k \sim N(0, \sigma^2)$ with p variables (although it is best to keep p small—which is the same principle as selecting a parsimonious model in regression). It is possible to introduce restrictions on the size of weights, and other constraints can be added. These types of estimators and their properties are generally beyond the scope of this book, and readers are referred to Särndal et al. (1992) and Singh et al. (2001) for more information.

There is one specific problem for which regression estimation is useful—where there is information on two types of unit, such as reporting units and local units (see Section 5.2.1), we may want to use a model that is calibrated to totals at both reporting unit level and at local unit level, for example, to calibrate reporting unit totals to x classified by reporting unit activity and local unit totals to x classified by regions. In this case we construct additional auxiliary variables for the reporting unit that are the sum of the local unit values within each region within each reporting unit (and that may therefore be 0). These extra variables are used in a regression estimator—an application of the integrated weighting of Lemaitre and Dufour (1987). Although this is a potentially useful approach, there don't seem to be many examples of its use in business surveys.

5.4.6 Accuracy of Auxiliary Information

One feature of auxiliary information that is often overlooked is that it doesn't *need* to be accurate itself. Its main function is as a predictor of the survey variable, and if it is a good predictor (i.e., its correlation with the survey variable is significantly different from zero), then using it in estimation will reduce the variances. So, note that all the auxiliary variables could be multiplied by 5, and the quality of the prediction would be unchanged (in a linear model). By a similar argument, a register containing auxiliary variables measured at different points, for example, because updating occurs only with certain events that are spread out over a number of years, does not invalidate the use of these variables. It may slightly worsen the quality of the prediction for current variables, and this will be reflected in a higher variance. But the auxiliary variables are still useful.

The process for generating the auxiliary data must be independent of the survey, however. Specifically, survey responses should not be used to update *only* the sample units on a register; this is particularly dangerous with permanent random number (PRN) sampling (see Ohlsson 1995). A similar situation arises (although it is uncommon in business surveys) if the total of an auxiliary variable is known, but the values of the same variable are unknown for individual units. The unit values can be collected in the survey and used, but any bias (measurement error) in the way the survey collects this information relative to the way the total was derived will be carried into the estimates as a bias.

5.4.7 Optimal Uses of Auxiliary Information

We have seen that auxiliary variables can be used in sample design and as part of the estimation, in both cases to improve the quality of the final estimates. So, if we have a useful auxiliary variable, where is the best place to use it? As a general guideline it is best to use available information at the design stage, as long as introducing it does not contradict parsimonious stratification [and in the case of PPS sampling, this agrees with Knottnerus's (2011) guideline for using PPS + HT estimation rather than SRS + ratio estimation]. Possibly the same information can be reused at estimation to give some additional correction and robustness to nonresponse, but it is best to design things as well as possible rather than correcting at the estimation stage. The precise answer to this question in any particular situation is not always clear, however; Cochran (1977, p. 169) gives some points to bear in mind:

1. Some information, such as location, is more easily used in stratification than estimation.
2. Using auxiliary variables in estimation works well where the relationship between the auxiliary and survey variable(s) is simple; if the relationship is complex, stratification will be an easier way to capture it.

Cochran also discusses what happens if some important variables are related to one auxiliary variable and some are related to a different auxiliary variable. In this case using different estimators may be better than stratifying, but then there are issues of inconsistency between estimates using different auxiliary variables (see Section 5.4.3).

5.4.8 Estimating Discontinuities

Many business surveys are undertaken for economic monitoring, and are therefore repeated regularly, with much of their value derived from the use of consistent methods over long time periods. It is natural that the designs of surveys should become outdated and need occasional refreshing, for a variety of reasons (e.g., to cope with changes to classification systems, new developments in methods, availability of different data sources), and when this occurs, users tend to want a consistent series. In order to produce one, we generally want to estimate the difference and then adjust one series (normally the old one) by that amount.

Unfortunately, estimating the difference between two things is generally a much trickier job than estimating the value of something, since the sample size needs to be large to provide a good estimate of the difference, and ideally the difference will be measured over multiple periods (survey occasions) to give a better estimate of the adjustment required.

In some situations if only the estimation needs to be done twice (e.g., because the data are the same but coded to two industrial classifications), it may be possible to make two estimates (van den Brakel 2010). But where something in the collection process changes (e.g., the sampling, or the way the question is asked), then the data cannot be collected in both ways. Also, since business surveys rely on obtaining information from the largest businesses, it is seldom possible to split the sample into two pieces, one for each method, and to use only the half sample to make an estimate, as this is too different from what is found during "normal" running of the survey. More details of strategies and methods for dealing with discontinuities in surveys can be found in van den Brakel et al. (2008).

5.5 OUTLIERS

The distributions of variables associated with businesses are often skewed, because there are many small and a few large observations. Sample designs for business surveys almost always ensure that the largest businesses are included (Section 5.3), because this minimizes the sampling variance, and this is equivalent to minimizing the need to estimate (rather than observe) the larger observations. In the strata (or equivalent parts of the design) that are not completely enumerated, estimation relies on the similarity of the businesses among which sampling is undertaken (possibly conditional on some auxiliary variables if a ratio estimator or one of the more complex model-assisted estimators is used). In effect, under randomized sampling (Neyman 1934), each business represents $d_k = 1/\pi_k$ businesses in the population itself, and $d_k - 1$ similar businesses.

Outlying observations interfere with this process when they are in the tail of the distribution being estimated, because there might not be $d_k - 1$ similar businesses in the population. To take an extreme example, if the largest business in the *population* were selected in our sample with $d_k = 1/\pi_k = 5$, then we would use its value to represent five other businesses even though there wouldn't be five such big businesses in the population. Of course, we never know from a particular sample what the population values are, so we can't *know* whether we have observed an unusual value. Also, there is a certain tradeoff in that many samples will have some large and some small values, so that when an estimate is made with all of them, the overestimates and underestimates at least partially cancel out (the chances of selecting a sample with *all* the largest values is very small). The last comment to make is that even when we do include these larger observations in the sample, the estimator remains unbiased over repeated sampling—it's just that we only have one sample to work with. Model-based estimation provides estimates conditional on a particular observed sample, and the interested reader could examine this approach and the influence of outlying observations on models, in, for example Valliant et al. (2000).

There is a further issue—methods for outliers are based on an assumption that the values are correct, so that they *can* be used to represent other, nonsampled businesses; Chambers (1986) calls these *representative outliers*. But there is a strong association with data editing (see Section 11.4), because both processes aim to identify unusual response values. Data editing aims to correct those that are in error (*nonrepresentative outliers*), leaving only representative outliers to be accommodated by outlier methods. In reality, this distinction is rarely clear, and some judgment may need to be made about what to do with any particular observation.

The basic principle of outlier treatment in design-based survey estimation is that we are prepared to accept a biased estimate, and that by introducing a small bias we expect to be able to substantially reduce the sampling variance. In other words, if the largest observations have smaller weights, then the estimates from samples with the largest observations will be smaller and closer to the observations from other samples, and the sampling variance will be smaller. The target is that the mean-squared error (MSE, i.e., the square of the bias plus the sampling variance) should be at least reduced and ideally minimized. Hidiroglou and Srinath (1981) give one of the first evaluations of possible approaches.

There are two challenges in implementing this idea for dealing with unusual observations: (1) to determine whether there are any outliers, and if so, which observations are outliers, and (2) to deal with them in an appropriate way in estimation. In some methods the two steps are independent, and in other methods they are linked, but in either case an outlier treatment involves both steps. We give a brief overview of approaches here, or see Gwet and Lee (2001) for more information.

5.5.1 Inspection and "Surprise Strata"

Superficially the easiest approach to identifying outliers is by inspection—looking at the values from the responding sample and identifying those that seem to be extreme. Several remarks are in order here:

- An outlier must be evaluated with respect to the estimation model (Section 5.4), so if ratio estimation is used, then the observations farthest from the fitted ratio line will be the candidate outliers (not necessarily the observations with the largest magnitude).
- In a general estimation system involving several auxiliary variables (as in Section 5.4.5), it will therefore be easier to examine the residuals from the fitted model $y_k - \hat{\boldsymbol{\beta}}\mathbf{x}_k$ to accommodate more complex estimators.
- The actual effect will be easier to see graphically!

- This approach works best for one variable, or at most a small number (when a matrix of scatterplots could be used). For surveys collecting many pieces of information, the detection of multivariate outliers will be very difficult because of the large numbers of variables. A variable-by-variable approach may then be needed, but this upsets the relationships between variables.
- The decision over what is an outlier depends both on the magnitude and the weight (in estimation) of the observation; a completely enumerated observation cannot be an outlier.

Inspection is only superficially the easiest approach, because it actually requires a complex thought process, best developed through experience, and such experience is only the accumulated understanding of what is an outlier, since in general there is no whole-population information to experiment with (an interesting exercise is to simulate a population with a skewed distribution, select samples from it, and examine the effect when the larger observations are included in the sample). There is a lack of objectivity in the inspection approach, although for the more extreme cases there may be little disagreement between different people in making the evaluation. But the bias–variance tradeoff is not explicit, and one person's appetite for bias may differ from that of another. The subjectivity of "outliering by inspection" is at odds with the need (particularly for a NSI) to have transparent, objective and trusted methodology.

When the outlying observations have been identified by inspection, their effect on the estimates is modified by a *poststratification* [a change to a new stratification based on the collected sample data (Hidiroglou and Srinath 1981)]. The outliers are moved to a *surprise stratum*, a stratum that contains *only* the outliers, and within which each observation has a weight of 1 in estimation. The original strata now have smaller population sizes and smaller sample sizes (both reduced by the number of outliers moved to the surprise stratum), and in these one of the standard estimation methods from Section 5.4 is used, with the design weights modified to reflect the new population and sample sizes within the poststrata. The estimation weights in the poststrata are therefore different from the ones that would have been obtained if the outliers were left where they had originally been sampled.

5.5.2 Winsorization

The term *winsorization* derives from an idea of Winsor in 1919, but developed by Searls (1966)—the principle is that there is a threshold value above which an observation is outlying, and that treatment involves moving outlying values toward (or to) the threshold; under suitable conditions, the threshold value is chosen to minimize the MSE. This is an internally consistent approach in that outliers are identified and treated by the same approach, and there is an objective criterion for the approach. Of course winsorization is not the panacea that this description implies; there are challenges too (see text below). But first let us describe it [for a more detailed description, see Chambers et al. (2001)].

The main idea is to replace observed outliers by a smaller value. Type II winsorization (which has the nice property that an observation at least represents itself) defines the winsorized variable by

$$y_k^* = \begin{cases} \frac{1}{w_k} y_k + \frac{w_k - 1}{w_k} K & \text{if } y_k > K \\ y_k & \text{otherwise} \end{cases}$$

Using this in place of the original value is equivalent to using the original value with a weight of 1 (as in the surprise stratum) but then using K as a replacement value in the estimation of the $w_k - 1$ similar, nonsampled observations. This is relatively straightforward, but the difficulty comes in determining a suitable value of K, since it depends on the properties of the tail of the distribution of values, where there are few observations. Kokic and Bell (1994) considered this for stratified designs, and developed a model that utilizes the information in all the strata to estimate K (using past runs of a similar survey) so as to minimize the MSE of the winsorized estimate. K also depends on the expected value under the estimation model, so it is different for ratio estimation than for HT estimation, for example.

There are a number of practical issues with winsorization, some of them common to other approaches for estimating outliers:

- *Level of Winsorization.* Where a survey is used to estimate a national total and also regional or industry totals, optimization of the MSE with K can be undertaken for only one level (i.e., there can either be one value of K for the national level, or one value for each region, but not both). The national K can be used to provide regional estimates that are consistent with the national estimate, but they are not the minimum MSE estimates for the regions. Alternatively the regionally optimal K values can be used and then the national total can be estimated by the sum of the regions, which gives optimal MSE estimates for the regions, but not for the national total.
- *Consistency.* Winsorization is optimized one variable at a time, but we often want to retain the relationships between variables, for example, when several components sum to a total. In this case we have to make a decision on whether to minimize the MSE for the total or for the components separately.
- *Updating.* The bias from winsorization is approximately $-K$, so keeping the same value of K from period to period means that estimates of change between periods are approximately unbiased. But there is also a need to update K periodically to account for changes. A conservative approach is probably a good guideline; don't update unless there is clearly a change in respondent behavior reflected in an evolution of estimates of K.

5.5.3 Other Approaches

Gwet and Lee (2001) review outlier-resistant procedures in business surveys and conclude that winsorization and Lee's (1995) adaptation of M estimation have the best properties, in terms of large-sample (statistical) consistency and ease of application. Methods that are designed to introduce only small amounts of bias do not have good MSE properties, particularly for more skewed populations.

5.6 VARIANCE ESTIMATION

It is important to be able to accompany estimates from any survey with information on how good the survey is, and this has become more widely recognized in recent times (Groves 1989; Davies and Smith 1999). Many aspects of quality do not have well-developed theory to support them, but the estimation of sampling variances, the amount by which estimates may vary because we have taken a sample instead of the whole population, does. In fact there are several approaches to estimating variances, but the design-based estimators that we described in Section 5.4 have associated design-based variance estimators. This section contains only an introduction to variance estimators—much more detail will be found in, for example, Cochran (1977), Särndal et al. (1992), and Lohr (2010a).

The first component of the sampling variance is an estimate of how variable the values in the population are. This is the familiar estimator of the population variance

$$s_y^2 = \frac{1}{n-1}\sum_{k\in s}(y_k - \hat{\mu})^2 \tag{5.4}$$

The term inside the summation is the sum of the squared differences of the sample observations from the mean; and since we do not know the population mean we use the mean estimated from the sample, $\hat{\mu}$. The divisor is $n-1$ to ensure the unbiasedness of the variance estimator since it is calculated from a sample.

In a business survey, where a range of different estimation methods may be used (see Section 5.4), it is important to make sure that $\hat{\mu}$ is the mean (expected value) under the estimation model. In Section 5.4.3 we saw that the ratio estimator could be considered as a regression through the origin, $\hat{y}_k = \beta x_k$ and we could use this model to calculate the mean for element k—so that the values in the sum in Equation (5.4) is of the squares of differences from the fitted ratio line. In this case the "mean" (expected value) is different for each distinct x value, so we can rewrite this as

$$s_y^2 = \frac{1}{n-1}\sum_{k\in s}(y_k - \hat{y}_k)^2 = \frac{1}{n-1}\sum_{k\in s_h} e_k^2$$

where $\hat{y}_k$ is the expected vaiue and e_k is the residual for unit k, both under the estimation model. This general formulation will work for any regression type estimation model.

For the sampling variance of the mean of y we need s_y^2/n. In business surveys, the sample often forms a relatively large proportion of the population, which means that a *finite population correction* (fpc) is needed and important. The fpc, $1-n/N$ (sometimes written as $1-f$), deals with the negative correlation between the selection of elements in without-replacement sample designs. Including the fpc and using Equation (5.4) gives the sampling variance of the mean

$$\mathrm{var}(\bar{y}) = \frac{1-n/N}{n}\frac{1}{n-1}\sum_{k\in s}(y_k - \hat{\mu})^2 = \frac{1-n/N}{n}s_y^2 = \left(\frac{1}{n} - \frac{1}{N}\right)s_y^2$$

If $n \ll N$, which happens only rarely in business surveys (but more frequently in social surveys), then $1 - n/N \approx 1$ and the fpc may be omitted.

The final adjustment we need is that frequently surveys of businesses aim to estimate the total for a variable y, t_y, rather than its mean, for which we need to include the square of the population size in the estimated sampling variance:

$$\mathrm{var}(t_y) = N^2\frac{1-n/N}{n}\frac{1}{n-1}\sum_{k\in s}(y_k - \hat{\mu})^2 = N^2\frac{1-n/N}{n}s_y^2.$$

In a stratified sample we apply the variance estimator separately in each stratum, and since the samples within strata are selected independently, the variances can simply be summed to get the variance for the sum of a variable over many strata. The usual estimator of the sampling variance of a total, t_y, in a stratified design is therefore

$$\mathrm{var}(t_y) = \sum_h N_h^2\frac{1-n/N}{n}\frac{1}{n_h-1}\sum_{k\in s_h}(y_k - \hat{y}_k)^2 = \sum_h N_h^2\frac{1-n/N}{n}s_{yh}^2.$$

Note the distinction between estimated *population variances* s_{yh}^2 (using the appropriate estimation model, and *without* the finite population correction) which feed into sample allocation (see Section 5.3.2.3), and the estimated *sampling variance* which also includes the fpc.

There are other ways to calculate variances, using the weights to form appropriate estimates of the average of e_k^2, but we do not give all the technical details here [see Särndal et al. (1992)]. Software for calculating variance is described in Box 5.6.

Numerical methods of variance calculation are becoming popular for surveys, and these include the jackknife, bootstrap, balanced repeated replication (BRR) and balanced repeated half samples (BRHS). In theory all these methods should work for business surveys, though the actual procedure can be quite complex (in comparison with social surveys) to ensure that the method is appropriate, mostly because of the differential size of the units (and sometimes the weights). Investigations of the differences between numerical and design-based methods using simulation studies where the true sampling variance is known (Canty and Davison 1999) have shown that not accounting for the calibration results in underestimation of the sampling variance.

Some variance estimates from business surveys, particularly of change from one period to another for coordinated samples (see Section 6.4.2), or for statistics derived from multiple surveys, become quite complex because of the need to estimate covariances, and in these situations numerical methods can be easier to implement.

BOX 5.6 SOFTWARE FOR VARIANCE CALCULATION

First a warning, that the standard variances produced by software packages are *not* the correct ones for surveys where the observations have differential sampling probabilities. The estimators must factor in the estimation weights and finite populations corrections, and various software packages do this. Some, such as STATA, account appropriately for the sample design (and, in fact, will handle more complex designs than needed for most business surveys), but do not accommodate the weighting adjustments used in ratio or regression estimators. Such variances may nevertheless be a reasonable approximation to the true sampling variances if no more specific software is available.

There are several software packages that will calculate appropriate variances, accounting for both the differential sampling and the estimation model. They include purpose-written SAS macros developed by statistical offices such as GES [Statistics Canada (Estevao et al. 1995)], CLAN [Statistics Sweden (Andersson and Nordberg 1994)]. There is also an SPSS-based package called g-Calib from Statistics Belgium. More recently appropriate procedures have been built into SAS (the PROC SURVEYMEANS and SURVEYREG procedures), and SPSS. A good if slightly old (2004) summary of the available software and their properties is available in one of the workpackages of the DACSEIS project from `http://www.unitrier.de/fileadmin/fb4/projekte/SurveyStatisticsNet/Dacseis_Deliverables/DACSEIS-D4-1-4-2.pdf`.

The free software R is beginning to be used more widely for surveys, and there are already a number of packages and functions available for variance estimation that account appropriately for complex designs, including both design-based and numerical methods.

Ideally variance estimation approaches should account for any application of imputation (Section 11.4.7) and outlier treatment. This leads to more complicated estimators in many situations, and sometimes the particular method for a situation needs to be developed. Where the methods used are objective (and therefore automatable), numerical approaches such as the jackknife and bootstrap may well be the most practical. Further details are beyond the scope of this book, but Shao and Steel (1999) provide an example of an approach suitable for business surveys.

5.7 SMALL AREA ESTIMATION

Small area estimation has become an important topic for surveys in general, where the pressure for ever greater detail from samples is driving efforts to make maximum use of sample information. *Small area* is a general term for any small domain where the number of sample observations is too small for the estimate (using the methods of

Section 5.4) to be reliable. Making credible small area estimates requires "borrowing strength," that is, using sample observations from outside the domain required in order to improve the accuracy, and making a suitable adjustment to the estimate to account for the fact that the same information is being used over many small areas. This is generally done through a model, often a multilevel model, which is a flexible way to include the appropriate information (Rao 2003). These models have been developed and applied largely in social surveys, where there are typically many similar observations. One assumption behind these models is that the errors are Normally distributed, and this is generally violated in business surveys, again because of the differences in the sizes and weights of the units. This means that small-area estimation has rarely been extended to business surveys. The few examples are where the business populations being estimated are more homogenous in size (e.g., retail outlets, agriculture) or where variables are related to employees rather than to businesses, such as earnings surveys, where there is less discrepancy in size.

Small area estimation depends on averaging, even if done in an appropriately detailed way, but business distributions tend to be very lumpy—a large factory in one area next to a residential area with no businesses will be very obvious to anyone with local knowledge, but small area models are very unlikely to yield accurate estimates for such different, but adjacent, areas. Some more recent developments have been made using the examples of farms (Chandra and Chambers 2009, 2011); these solutions are largely variable-dependent, which means that estimates for different variables will be inconsistent to some extent. Therefore the question of how to make credible small area estimates for business surveys remains an open one (Hidiroglou and Smith 2005).

Falorsi and Righi (2008) have attempted to build a sampling strategy that will provide a sound basis for a small area estimation in cross-classified designs. This may provide a different approach for developing small area estimates in businesses surveys, but the challenges are still considerable.

5.8 SUMMARY

In this chapter we have discussed the construction and maintenance of a register as a basis for a survey of businesses. The presence of a register with good coverage properties and some auxiliary information on the units distinguishes business surveys from most other types of surveys, as does the skewness of the population, and together these features both necessitate and allow sample designs with widely varying selection probabilities. This means that the choice of sample design will often be more crucial in a business survey than in other surveys—a poor design may have a very substantial impact.

The range of design options discussed in Section 5.3 is quite wide, and a design must be coupled with the estimation method that is to be used—the population variance used as an input to the design should be calculated using the appropriate estimation model. So, changing the estimator may have an effect on the quality of the design. But in some cases a poor design can be partly compensated for by strategic use of auxiliary information in estimation.

We have also briefly covered the challenges that are a normal part of running a business survey—how to deal with outlying observations (which are both relatively common and very influential compared with household surveys), and how to formulate sample designs that work reasonably well for challenging variables.

Finally, there is a framework for assessing the quality; most sampling theory concentrates on the efficiency of a design–estimator combination as measured by the sampling variance, but it is also important to take into account the effects of nonresponse (which introduces an extra element of randomness outside the control of the survey taker) and various measurement errors, particularly the inability to distinguish deaths and nonresponses in some cases. A design that is robust to these nonsampling errors will often be better than a highly optimized design that cannot be realized in practice.

ACKNOWLEDGMENTS

The author would like to thank Jacqui Jones, Mike Hidiroglou, Jelke Bethlehem, Jean Ritzen, and Martin Brand, all of whom reviewed drafts of this chapter.

CHAPTER 6

Understanding and Coping with Response Burden

Gustav Haraldsen, Jacqui Jones, Deirdre Giesen, and Li-Chun Zhang

We have emphasized several times that the monetary and cognitive burden of business surveys are two sides of the same coin and together may significantly impact the quality of survey results. Time is money in businesses. Because questionnaire completion in businesses is a cost that does not lead to any obvious financial return, time-consuming survey enquiries tend to create irritation and can be downgraded in terms of priorities. If businesses are frequently asked to participate in surveys, and surveys are time-consuming, with perplexing questions and burdensome answers to provide, there will be an increased risk of low response quality.

The term *actual response burden* is commonly used to describe the time taken to respond to a questionnaire with, *perceived response burden* describing respondents' perceptions of their survey experience. For example, Hedlin et al. (2005) found that respondents rarely equate burden with the frequency and length of times with which they respond to a particular survey. Instead, respondents perceive burden as associated with factors such as the mode of data collection, who is conducting the survey (i.e., which survey organization), and whether the produced statistics are useful to the business and/or society. Actual and perceived burden are intertwined, but still two different and relevant aspects of response burden. In businesses where the manager gets somebody else in the organization to fill in the questionnaire, different response burden aspects are also of varying importance to people playing different roles.

Even if the perspectives and evaluations are different, response burden is an important issue for business managers, business respondents, and survey organizations. As a consequence, it is also an important political issue. On one hand,

---

Designing and Conducting Business Surveys, First Edition.
Ger Snijkers, Gustav Haraldsen, Jacqui Jones, and Diane K. Willimack.

politicians are concerned about burdens, which take time away from profitable business activities. Often they are also under heavy pressure from businesses and business interest organizations to reduce the burdens imposed by society's reporting obligations. On the other hand, politicians need information to govern. This makes them eager to initiate new data collections and be heavily dependent on the quality of survey results.

6.1 A COST VERSUS A COST BENEFIT APPROACH

Strictly speaking, the term "response burden" is misleading because it is unipolar and thereby rules out the possibility that survey participation has positive effects. Responding also can have positive values if we focus on perceptions instead of time and money. Two questionnaires that take equal time and efforts to complete can still be perceived quite differently depending on how interesting or otherwise rewarding respondents find the completion task, or how they use the derived statistics. Also, from the survey organization's perspective, there are costs and benefits. A certain amount of expense and a certain cognitive cost are necessary in order to achieve results of a certain quality. These are both examples of cost benefit considerations that weight negative effects against positive ones.

Cost benefit analysis (CBA) requires identification of which costs and benefits to include, how to evaluate them, discounting of future benefits and costs over time to obtain a present-day value, and identification of relevant constraints (Prest and Turvey 1965). Initial cost benefit methodological work undertaken at the UK Office for National Statistics has focused on a CBA model for assessing proposed new data requirements (Orchard et al. 2009). The key components identified are shown in Figure 6.1. In this model two kinds of costs are balanced against two kinds of benefits. The "costs" are what the proposed data collection will cost the survey organization and the respondents. It is important that survey organizations include

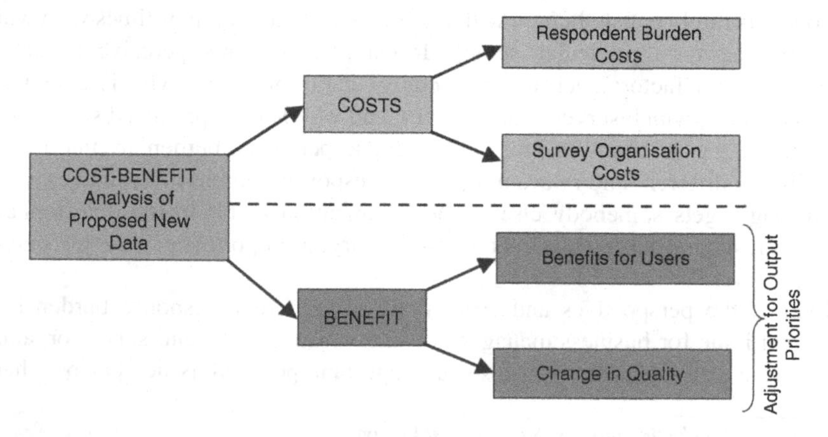

Figure 6.1 Key components of a CBA model for assessing proposed new data requirements. [From Orchard et al. (2009). © Crown copyright 2009.]

response burdens in their calculations; otherwise, they may fail to notice that internal cost reductions are outweighed by additional costs to the businesses in the sample. This is true, for example, if edit procedures are replaced by computer-initiated error checks that are activated in many electronic questionnaires. The possible benefits of data collections are improvements in the quality of statistical information and fulfillment of information needs. These benefits coincide with what we called a professional and user perspective on quality in Chapter 3. Orchard et al. also suggest how these cost components can be measured and the overall value calculated by adjusting for the benefits.

Even if politicians and businesses generally do not apply a CBA approach on survey inquires, we think that both cost and benefit considerations belong to a discussion about response burdens from a political and commercial perspective.

6.1.1 Politicians' Bird's-Eye View

As described in the beginning of this chapter, politicians often initiate data collections by their requests for information. Hence they are one important group of users, referred to in Figure 6.1, who benefit from surveys. What could sometimes be questioned, however, are politicians' ability or willingness to balance criticism toward data collectors against the anticipated benefits or their own information requests. In political rhetoric, the costs and benefits often sound disconnected.

Traditionally, the political approach to response burden is characterized by a macroeconomic viewpoint. Politicians are not so concerned about the burden of individual surveys or other reporting obligations or the burden for specific businesses. They are concerned about the total cost of official reporting obligations imposed on businesses and on the potential impact of the cost on the nation's productivity, calculated as follows:

$$\frac{\text{Real GDP}}{\text{Total hours worked by those employed}}$$

Hours spent by employees responding to business surveys will reduce the denominator and result in lost productivity (Seens 2010). Note that again this perspective does not factor in the benefits of what is produced with the reported information. It is obvious that many publicly available official statistics such as price index figures or demographic forecasts help businesses work more efficiently, for example, when negotiating contracts or exploring new markets.

Figures annually published by the Norwegian Brønnøysund Register Centre are an example of estimates calculated for such a macroeconomic purpose. In 2011 Norwegian businesses spent a total of 4760 person-years[1] on mandatory reports to official authorities. Note, however, that only a tiny proportion of this, about 2%, was due to business surveys. This relates, of course, to the fact that business surveys are

[1] Further information is available at http://www.brreg.no/registrene/oppgave/belastningsstatistikk.html.

sample surveys and consequently do not contribute as much compared to the reporting obligations that all businesses must comply with. Other examples show the same picture. Reducing business response burden has been an important target in European Union (EU) policy and national policies [e.g., UK Hampton (1995); US Paperwork Reduction Act (1980)]. In 2007 the EU Action Programme for Reducing Administrative Burdens started with the aim of cutting 25% of administrative burdens arising from EU legislation by 2012. Even if surveys for business statistics are only estimated to count for as little as 0.5% of the total costs measured, reducing the burden of statistical business surveys is one of 13 priority areas highlighted in the commision's action plan (European Commission 2008).

It may seem odd that such a high priority is given to cuts that have minor impacts on the total. The main reason may be the pressure politicians feel from the business world. In a meeting held in Brussels in 2009 The EU High Level Group of Independent Stakeholders explained that "statistics is an area of great concern for businesses, mainly due to the high perceived burden, the so-called irritation burden" (HLG 2009). More recently, politicians have moved from focusing on "just" reducing burden toward improving the regulatory environment in a broader sense, which also includes the acceptance of regulations (e.g, the European Better Regulation Agenda). A more recent OECD (2012) report on perception surveys illustrates that governments are now also interested in perceptions of regulations. This leads us to the business perspective.

6.1.2 Commercial Worries

The fact that sampling reduces the total burden imposed on businesses and industries in general does not help business managers and respondents that are sampled. On the contrary, the selected businesses may view sampling as a kind of injustice. This feeling may be intensified by the length and complexity of the questionnaire and the fact that one is enrolled in a survey panel. A traditional paper questionnaire, which some business respondents find old-fashioned compared to online information dissemination, may also add to the irritation. To make the sample burden of business surveys more fair and acceptable, methods such as "survey holidays" and sample coordination are used. In UK for instance, the Osmotherly guarantee states that businesses with less than 10 employees will remain in a monthly survey sample for only 15 months; a quarterly survey, for five quarters; and an annual survey, for 2 years (Osmotherly et al. 1996). Sample coordination means that samples are not drawn independently of each other, but in a way that controls joint survey participation. In this way the maximum number of times that a specific enterprise can be sampled is reduced. We will discuss some aspects of sample coordination in more detail later in this chapter. Irrespectively of what method is used, however, we think that the most important feature of such initiatives is how well they are marketed. Moreover, because of their economic importance, survey holidays and sample coordination have minor consequences for larger enterprises. They are normally selected in most business surveys anyway. Managers from larger businesses are also often the most influential and heeded to by politicians. As long as sample holidays and sample coordination do not curb their irritation, the pressure on politicians to reduce survey burdens will probably remain.

Business surveys are unevenly distributed during the year (as well as among the business population. The deadlines for business surveys tend to be close to other reporting deadlines, and the timing of annual surveys tends to be simultaneous with annual reporting obligations (e.g., year-end accounts). This accumulation of reporting tasks in certain periods of the year may be particularly problematic for smaller businesses where every employee participates in the production of goods or services.

There is a long distance between industrial policy based on business statistics and the initial data reports made by the businesses. If respondents feel that they have to report figures that do not match their business records and perhaps even that their answers are inaccurate, it may be even more difficult to see the link and relevance of business statistics. Statistics that have immediate value to businesses are the consumer price index (CPI), other price indices (see Chapter 12), and employment figures. In general, business respondents do not consider the results of their survey efforts as particularly valuable. In a response burden survey conducted by Statistics Norway in 2006, only 5% believed that the statistics produced from the figures they reported would be useful for their business. Moreover, only 28% regarded the statistics as useful to society (Hedlin et al. 2005). In particular, small businesses do not understand why their data are relevant for statistics. Many business respondents also complain that the information inquiries overlap with other government data request (Giesen and Raymond-Blaess 2011).

These results seem to support the discussion in Chapter 3 regarding the user perspective of NSIs, which conduct official business statistics, namely, that they concentrate heavily on their principal contractor: national accounts. There seem to be at least two needs that could be better addressed and promoted by business survey organizations: (1) the need for benchmark data that would enable the businesses to compare their productivity, wages, and prices and their customer composition with other businesses within the industry to which they belong—This is a statistical product that perhaps is particularly relevant for medium-sized and smaller businesses, which may not have their own market research units; and (2) the need for data on demographic trends, living conditions, time use, and purchasing power tailored to market considerations in different industries (Löfgren et al. 2011).

6.1.3 Perceived Response Burden

From the quality perspective of statistical organizations, cognitive response burdens probably have a more direct effect on the response quality than the time it takes to collect and complete the questionnaire. It takes time to solve demanding tasks, but it is not the time in itself, but the amount of challenges and effort felt by the respondents, which may have implications on response quality. This psychological element was first introduced by Norman Bradburn (1978), who suggested that required efforts and psychological stress should be included in the response burden concept. Later Fisher and Kydoniefs (2001) developed a model where

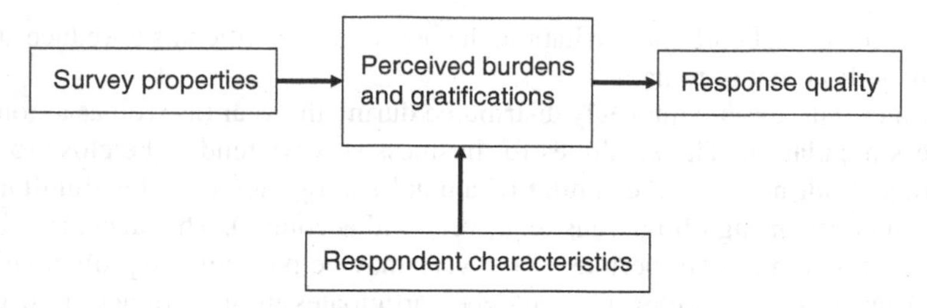

Figure 6.2 Response burdens and gratifications defined as the result of the interaction between survey properties and respondent characteristics. [From Haraldsen (2004).]

they distinguished between "respondent burden," "design burden," and "interaction burden." *Respondent burden* was defined as the personality traits and behavioral or attitudinal attributes of the respondents that affect their ability to complete the questionnaire quite independently of how the survey is designed. *Design burdens* referred to burdens linked to the mode of data collection and the content and presentation of the questionnaire. Finally, *interaction burdens* were described as the product of what happens when respondents with certain prerequisites are confronted with a survey that has certain properties. Hence, in the Fisher–Kydoniefs model, the perceived burden is influenced by the respondent's ability to answer, the design of the survey, and a combination of these elements (Fisher and Kydoniefs 2001). Haraldsen (2004) claims that respondent burden and design burden should be considered as causes of response burden rather than parts of the concept, and that what is perceived by the respondent is a feeling created by the interaction between the survey properties and the respondents' qualifications to respond. He also points out that the result of this interaction may not only be more or less burdensome, but could also have gratifying aspects (ibid.). These reflections lead up to the basic response burden model shown in Figure 6.2.

In this model the Fisher–Kydoniefs terms *design burden*, *respondent burden*, and *interaction burden* are replaced by *survey properties*, *respondent characteristics*, and *perceived burdens and gratifications*, respectively. In the extended process model introduced later in this chapter (Figure 6.4), different aspects of the respondent's business context are also discussed.

The survey properties can be divided into three main elements: the sample properties, the measurement instrument, and the survey communication procedure. Different samples, such as for different industries, will contain respondents with different qualifications to answer the survey questions. Moreover, the size and composition of the net sample will certainly affect the sample error and sample bias. Within the framework of a given sample, the tools that the surveyor has to influence the respondents' perceptions of burdens and gratifications are the measurement instrument, which normally is a questionnaire, and the survey communication procedures. In Chapters 8–10 we will discuss how the questionnaires and communication procedure can be tailored so that the response burden is minimized. Here we will only mention a fundamental difference between the information tools at hand in a questionnaire and in survey communication.

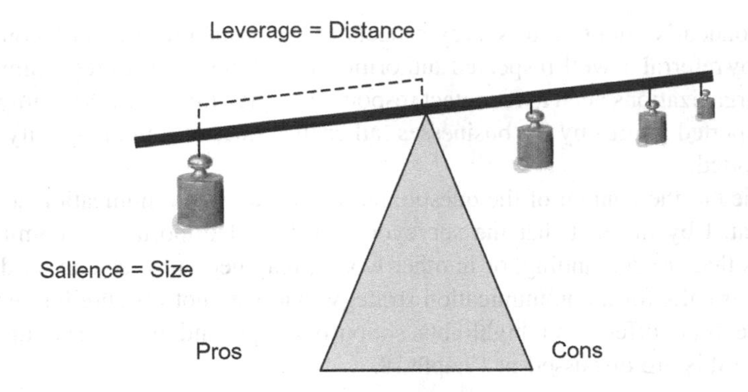

Figure 6.3 Response burden seesaw and leverage-salient theory combined.

The balance between burdens and gratifications is commonly depicted as a seesaw [see, e.g., the front cover of Dale et al. (2007)]. This illustration refer to social exchange theory, which underpins our understanding of perceived response burden (Dillman et al. 2009). In Figure 6.3 we introduce a model combining social exchange theory with leverage salience theory, which concerns how attention is drawn to important issues (Groves et al. 2000).

Social exchange theory interprets human interaction as a result of a basic feeling of trust or mistrust plus an evaluation of the burdens and benefits involved in the current relationship (Blau 1964; Dillman et al. 2009; Homans 1961). Hence, awareness of perceived response burden in business surveys means accounting for the trust or mistrust business respondents have toward business surveys in general and tasks perceived as burdensome and gratifying in specific business questionnaires. We will return to the trust issue later. What concerns us here is how the survey is perceived.

The leverage salience theory can be interpreted as extension of the seesaw model. Initially this theory focuses on survey participation. The respondent's decision to participate or not is described as less deliberate and rational than in social exchange theory. The main point, however, is that respondents' perceptions of how easy or difficult the survey completion task is, and their motivation to participate, is a result of how they feel about the aspects of the survey that catch their attention. From this observation it follows that in well-planned surveys what catches the respondents' attention should not be incidental but deliberately controlled by the surveyor. The leverage and salience of different survey aspects or participation arguments can be depicted by hooking weights of different size to the response burden seesaw model. The distance from the seesaw fulcrum to a given weight represents how important that survey or questionnaire aspect is to the respondent, while the size of the weight represents how salient this aspect is made (Groves et al. 2000).

Applied to survey recruitment, which is a main topic in Chapter 9, the model points at the challenge to identify gratifying survey aspects that matter to the respondent, and next to direct the respondent's attention to these aspects. To direct the respondent's attention to survey aspects that can inspire trust and confidence, such as by referral to well-respected authorities, is part of such a strategy. Similarly, if survey organizations need to recontact respondents, they might stress the importance of the reported figures by the businesses rather than questioning the quality of what was reported.

Applied to the content of the questionnaire, a similar communication strategy is complicated by the fact that the surveyor may find it important to communicate inquiries that are demanding, or in other words, may need to focus on burdensome tasks. This calls for a communication strategy that does not obscure burden, but at the same time offers and highlights supportive tips and tools. The means for achieving this are discussed in Chapter 8.

6.2 MEASURING AND MONITORING BUSINESS RESPONSE BURDEN

So far we have argued that both actual and perceived response burdens are important aspects of the survey burden. Which is the most important depends on the perspective taken. In Table 6.1 we have summarized how we have linked the two response burden aspects to the different perspectives held by politicians, businesses, and survey organizations. From here we will move to measurement issues.

Table 6.1 Perspectives, Reasons for Measuring, and Type of Burden

Perspectives	Reason for Measuring	Type of Burden
Political	To measure and monitor the total burden costs placed on all businesses for a specified period (e.g., annually)	Actual
Business	To measure and monitor the total burden placed on individual businesses for a specified period (e.g., annually)	Actual and perceived
Business	To measure and monitor the spread of burden across a specified period (e.g., annually)—can be undertaken at a total or individual business level	Actual
Survey organization	To identify aspects of the survey design that are burdensome to respondents and could be improved	Perceived
Survey organization	To identify aspects of the survey design that potentially could affect the quality of survey responses	Perceived and actual

Response burden estimates can be collected directly from business survey participants or estimated indirectly, as a byproduct of recordkeeping studies, pretests, or other survey development methods (see Chapter 7). Actual response burden can also be estimated indirectly by experts in the survey organization.

In 2011, a survey conducted with NSIs in 35 European countries, the United States, Canada, New Zealand, and Australia showed that 34 of the 41 institutes had measured actual business survey response burden one or several times in the period from 2006 to 2010 and

- that 20 of these had made annual estimates
- that 29 of the 34 that had measured actual burden had done so directly by asking business respondent
- that 25 had made indirect estimates based on expert evaluations, while 11 had assessed actual response burden using qualitative methods
- that 12 out of the 41 had measured perceived response burden

Since the total number of institutes that had applied different estimation methods exceeded 34, there is obviously a mix of method in many of the survey organizations (Giesen 2011). These results illustrate the following basic measurement issues:

1. Should response burden be measured?
2. If so, what kind of response burden should be measured?
3. How often should measurements be carried out?
4. Should response burden questions be asked to survey participants or estimated indirectly?
5. If both direct and indirect methods are used, what should be the selection criteria? If direct measurements are chosen, then, as in any survey measurement there will be a requirement for:
   - Clear data requirements, including specification of the relevant attributes of the concept that needs to be measured (Snijkers and Willimack 2011)
   - Methods for the measurement (sample, questions, mode, etc.) and analysis.

As in any other type of measurement, there will also be associated sampling and measurement quality issues in relation to the measurement process and outputs (e.g., accuracy and reliability).

In addition to the decisions already listed, Willeboordse (1997) identifies three pairs of concepts that need to be considered when response burden is to be surveyed:

6. Should the measures comprise the net or the gross burden, meaning should possible benefits be taken into account?
7. Should the imposed or accepted burden be measured, meaning should total response burden be estimated from the gross sample contacted or from the net sample that responded? Strictly speaking, it cannot be assumed that just

because businesses did not respond they did not spend time on the survey, such as reading the survey request and deciding not to respond, or handling nonresponse follow-up letters or telephone calls.

8. Should the measurement be maximalistic or minimalistic, meaning should tasks related to the questionnaire completion be included? Related tasks could be record formation specifically done for the reporting obligation in question, collecting and compiling requested information, and administrative tasks such as internally distributing the questionnaire. It could also be the time spent preparing (e.g., reading instructions) or on recontacts after the questionnaire is returned to the survey organization. For some types of business surveys there can be a high recontact rate during editing; for example, for more complicated financial surveys, the recontact rate can be as high as 90%. Even if the number of recontacts burdens the survey organization more than the individual businesses, it should not be ignored.

The abovementioned survey among NSIs showed that reading instructions and administrating the survey are often included in the measurement of actual response burden, while record formation and time spent on recontacts seldom are (Giesen 2011).

Another important principle which may affect the communication mode and questions asked about response burden is that:

9. The response burden questions should not incur more burden than the actual survey for which it is measuring response burden. For business surveys with just a few questions, one should consider conducting separate response burden surveys, for example, by computer-assisted telephone interviews, instead of attaching response burden questions to the actual surveys.

Given the different perspectives and with all these necessary decisions, it is not surprising that the research group that carried out the 2011 survey concluded that the heterogeneity in measurement practices was too large for international comparisons. This issue has also been highlighted by Rainer (2004), with a call to move toward international standardization to improve the quality of response burden measurement and enhance comparability (Hermann and Junker 2008).

For monitoring purposes Willeboordse recommends that actual gross burden, including the burden of administrative tasks (maximalistic), should be measured among those who respond to (accept) business surveys (Willeboordse 1997). In their 2009 *Handbook for Quality Reports*, Eurostat recommends and specifies a maximalistic approach that includes time spent assembling information prior to completing a questionnaire and the time spent by any subsequent contacts after receipt of the questionnaire (Eurostat 2011a). We would suggest that a few simple questions about perceived response burden be added to the list of monitoring questions.

Estimates of the total response burden are usually updated annually, but this does not necessarily mean that annual measurements are needed. As long as the sample, questions, or mode of data collection do not undergo significant changes, estimates from the previous years or adjustments made by experts should be sufficient. This can be formulated as a principle of periodicity stating that

10. Response burden information provided by survey respondents should be collected before and after major changes are made to the sample, questions, or data collection mode of business surveys.

6.2.1 Measuring and Calculating Actual Response Burden

Essentially, the measurement of actual burden attempts to measure the labor costs associated with responding to surveys, which is measured by labor time multiplied by hourly average wages. There are a number of methods for measuring the labor costs associated with responding to surveys. For example, the *European Statistical System Handbook for Quality Reports* (Eurostat 2009b) suggests that business survey labor costs can be calculated as

$$\text{Labor costs} = R^* T^* C$$

where R is the number of people participating in the internal data collection and completion of the questionnaire, T is the time these people spend on the survey activities, and C is the salary cost of these people.

Because salary costs are difficult to measure, individual business burden for a specific survey can simply be calculated as time taken (R^*T) rather than costs (Eurostat 2011a).

Originally Eurostat recommended the so-called standard cost model (SCM) to estimate response burden, and some European member states use this model. This model is not developed specifically to measure statistical response burden, but to measure all kinds of regulatory administrative activities that face a business. Its main focus is on the activities undertaken to provide the information for the data requirement in question. In the case of business surveys, that will typically be to identify activities such as extracting relevant data from information systems, filling in the questionnaire, and signing and sending it. One strength of the model is that it specifies activities and indicates which activities are the most burdensome. In addition to identifying activities, identifying time per delivery, the number of deliveries with a time period, and deciding tariffs are needed.

To carry out SCM investigations, detailed interviews are recommended. This is an expensive data collection method. Moreover, for business surveys, the activities needed to collect relevant information and complete the questionnaire are well known and basically the same from questionnaire to questionnaire. The main challenge is rather to get a representative view of the time burden of questionnaires of different size and complexity in businesses of different sizes and in different industries. For this purpose, the method recommended in the SCM has proved to be too expensive for implementation in large-scale and representative samples [see, e.g., Frost et al. (2010)].

An alternative recommended by Dale and Haraldsen (2007) is to either attach some additional questions at the end of business questionnaires or carry out follow-up telephone surveys with businesses that have just participated in a data collection. The response burden questions can be posed either to all participants in a business survey or to a subsample within the original sample. The main criterion for choosing between questions attached to existing surveys or a separate follow-up survey should be decided by considering the balance between subject matter questions and response burden questions (see principle 9 above).

Basically there are two questions that need to be asked, regarding the time it took to (1) collect and compile relevant information and (2) physically complete the questionnaire. The time spent on recontacts can be measured during the recontacts.

Regarding the actual wording of the response burden questions, the outset is complicated as several people may participate in both information collection and questionnaire completion. A general question asking the respondent to summarize all time spent by all involved is an example of a demanding calculation task that respondents frequently consider burdensome (see Figure 3.6). A more stepwise procedure is recommended, such as the following sequence:

1. **Was** it necessary to collect information from other persons or sources within the company before the questionnaire could be completed?
   O Yes → go to 2
   O No → go to 6
2. Approximately how much time did you spend collecting relevant information before the questionnaire could be completed?
   ☐ hours ☐ minutes
3. Did other people help you collect relevant information or completing answers in the questionnaire?
   O Yes → 4
   O No → 6
4. How many persons did you get this kind of help from?
   ☐ Number of persons
5. Approximately how much time all together do you think these persons spent helping you?
   ☐ hours ☐ minutes
6. Approximately how much time did you spend completing the questionnaire?
   ☐ hours ☐ minutes

Hedlin et al. (2005) and Peternelj and Bavdaž (2011) provide an overview and some details of the measurement of actual response burden in a number of different national statistical institutes (NSIs). Table 6.2 provides an overview of how different NSIs measure actual response burden. An interesting detail in this overview is that, apart from the UK Office for National Statistics, one of the unique features of business surveys, namely, recontacts during data validation, are not included in the measurement of actual burden.

Table 6.2 Measurement of Actual Burden in National Statistical Institutes

NSI	Measurement
Australian Bureau of Statistics	Measured as the product of the number of completed questionnaires and average completion time (directly measured from respondents); measure not provided as a monetary value; measurement and management of burden coordinated by the Statistical Clearing House
Statistics Canada	Reported annually to Parliament; burden calculated using survey frequency, sample size, average time to complete survey, and response rate; burden calculated separately for large and small businesses
United Kingdom Office for National Statistics	Burden measured as total annual number of initial contacts with businesses, multiplied by time taken to complete questionnaire (including collating the information), multiplied by an estimated data validation recontact rate, and then multiplied by an hourly pay rate for respondent (Green et al. 2010b)
Ireland (Central Statistics Office)	Measured as estimated time to complete questionnaire, multiplied by number of survey respondents and surveys; calculated time then converted to a monetary value using estimates of hourly rates
Statistics Austria	Use a *response burden barometer* that measures time taken to complete questionnaire, collate data, and determine how many people are involved and coordination time (Rainer 2004)
Germany (Federal Statistical Office)	Burden measured by collecting information on those involved in data collection, time each person spent, and origin of the data (Stäglin et al. 2005)
Statistics Netherlands	Burden measured as estimated time to complete questionnaire multiplied by number of respondents; measure is then converted into a monetary value using an hourly rate (Oomens and Timmermanns 2008)
Statistics Denmark	Annual calculations of response burden; burden measured as number of questionnaires multiplied by estimated time of completion; time estimates are based on assessments by trade organizations (Dyrberg 2003, 2006)
Statistics Finland	Burden measured as time taken for respondents to acquaint themselves with data collection procedure, read instructions, collate requested data, and complete questionnaire; number of people involved also included (Leivo 2010)
Statistics Sweden	Burden measured as estimated time taken to understand what data are being requested, and then collect and report the requested data (Notstrand and Bolin 2008)

(*continued*)

Table 6.2 (*Continued*)

NSI	Measurement
Statistics New Zealand	For each business, calculation of typical time taken for a business of that size to complete its survey obligations in a year; number of surveys in which the business is included also calculated (Merrington et al. 2009)
Statistics Norway	Burden measured as estimated time taken to collect information and complete questionnaire, reported annually to Brønnøysund Register Centre; measurement conducted among respondents to Structural Statistics questionnaires in 2006 and 2010

6.2.2 Measuring Perceived Response Burden

Taking as a point of departure that the perceived response burden or gratification that the respondent experiences is a result of the interaction between survey properties and the respondent's qualifications to respond, the obvious place to start looking for specific kinds of burden is in the progressive cognitive steps that the respondent takes to answer a question (see Figure 3.4). Next, for practical reasons the survey designer will look for links between different burdens and different aspects of the survey communication strategy or questionnaire properties. In Figure 3.5 we divided the questionnaire properties into content (wording, task, and response alternatives), layout, and functionality. This approach was taken in focus groups and interviews with business respondents in the first of two joint European Commission projects carried out by Statistics Norway, Statistics Sweden, and the UK Office for National Statistics (Hedlin et al. 2005). Based on the results from the first project, the second project resulted in the production of a handbook that provides concrete suggestions for how actual and perceived response burden can be measured and monitored (Dale and Haraldsen 2007). The core questions suggested for measuring perceived response burdens are listed in Table 6.3. When questionnaires are presented on Internet, technical problems or low usability may add to the response burden (for a discussion see Chapter 8).

In addition, the handbook suggests that parts of the respondent's motivation could be mapped by the two questions listed in Table 6.4.

In the response list of the main reasons for why the questionnaire was considered time-consuming or burdensome in Table 6.3, you will find references to different cognitive steps (e.g., unclear terms, need to collect information from different sources, complicated calculations) and questionnaire properties (e.g., messy presentation, response alternatives that did not fit). In the same way as with the response burden questions, the two questions about the usefulness of statistics indicate how successfully the survey organization has communicated the value of survey participation. Similar questions can be developed to measure other arguments used in the survey communication (for a discussion, see Chapter 9).

As the questions of perceived response burden are presented here, they may reveal problems, but do not identify which survey questions cause which problems. Still,

Table 6.3 Core Questions for Measuring Perceived Response Burden

Indicator	Question	Response Categories
Perception of time	Did you think it was quick or time-consuming to collect the information to complete the . . . questionnaire?	Very quick Quite quick Neither quick nor time consuming Quite time-consuming Very time-consuming
Perception of burden	Did you find it easy or burdensome to fill in the questionnaire?	Very easy Quite easy Neither easy nor burdensome Quite burdensome Very burdensome
Reasons for respondent finding questionnaire completion time-consuming	Why did you find it time-consuming?	Had to collect information from different sources Needed help from others to answer some of the questions Had to wait for information that was available at different times Other reasons, please specify
Conditions for burden	What conditions contributed to making the questionnaire burdensome to fill in?	Too many questions Messy presentations made the questionnaire hard to read Unclear or inadequately defined terms Questions entailing complicated or lengthy calculations Available information did not match the information asked for Difficult to decide which response alternative was the correct answer Other reasons (please specify)

Table 6.4 Questions Indicating Motivation

Indicator	Question	Response Categories
Usefulness for own business	Do you consider the statistics from this questionnaire are useful or useless for your business?	Very useful Fairly useful Neither useful nor useless Fairly useless Very useless Don't know
Usefulness for society	Do you consider the statistics from this questionnaire are useful or useless for society?	Very useful Fairly useful Neither useful nor useless Fairly useless Very useless Don't know

this may often be quite obvious. If not, there are several methods that can identify where the questionnaire needs improvement. One simple way is to add an open question about which questions caused problems to the response burden sequence. Another approach that can be used in web questionnaires is to link response burden results to paradata, such as paradata showing that certain questions took suspiciously long time to complete or were corrected several times before the questionnaire was completed. For a discussion on how paradata can be used to identify weaknesses in business questionnaires, see Haraldsen et al. (2006) and Snijkers and Morren (2010).

When the proposed methodology for monitoring and evaluating business survey response burdens was discussed at the 13th meeting of the UK National Statistics Methodology Advisory Committee (Jones et al. 2007b), the committee conclusions were that

- This work is relevant to all countries.
- A quality-driven approach to response burden is a viable objective. Actual and perceived burden must be distinguished, remembering that they are interrelated.
- Perceived response burden could be measured during the testing phase.
- A number of suggestions were made regarding the core perceived response burden question set.
- Minimize the task of measuring response burden by limiting these exercises to a reasonable size.
- A change in the questionnaire design can initially raise the response burden placed on businesses; this decreases as respondents become familiar with the new design.

Three years later the survey among NSIs showed that only 12 of the 41statistical institutes had measured perceived response burden in the 5-year period from 2006 to 2010. According to the same survey, eight institutes had measured perceived response burden every year. These results suggest that the full potential of response burden questions as a tool for quality analyses and improvements is not yet fully realized (Giesen 2011).

6.2.3 Calculations

From a political perspective, what you need to calculate is the total burden placed on businesses, commonly for all surveys that different survey organizations carry out. For survey organizations such as NSIs, which conduct a high number of business surveys each year, it will rarely be possible or necessary to measure response burden by asking respondents in all surveys all the time. Provided no major changes are made in the questionnaires, previous measurements will probably still be valid. If there are few questions that do not require much time-consuming searches for relevant information, estimates based on qualitative methods or expert evaluation may also be applicable. Whatever method is used, however, documentation is essential.

Calculations begin with estimates for individual surveys. If you have results from subsamples of surveys, the figures have to be aggregated, calculating the mean or median and then multiplying by the number of businesses in the sample or the number of businesses who responded. Adjustments due to stratification may be necessary. The estimated time or cost burden of all surveys within a certain period, normally a year, are then added together. If the same survey is run several times, estimates for one round could be multiplied by the number of times the surveys has been conducted within the measurement period, adjusted for differences in response rates. The fact that response burden normally decrease by experience should also be taken into consideration.

While the politicians are concerned with the total of these calculations, business managers will need a benchmark figure that compares the extent of their burden with that of other, comparable businesses. For this purpose it might be necessary to produce response burden figures for businesses of different sizes and industry groups.

Measures of actual burden provide a proxy for the costs to businesses; however, certain limitations should be considered when the results are evaluated. When figures of total cost are given, respondents are generally treated as homogenous groups, with no adjustments for industrial classification, size of business, or costs and benefits. One criticism of this homogenous, average approach is that it disregards any inverse relationships between burden costs and size of business. For example, Seens (2010) found that smaller businesses had a disproportionate amount of regulatory burden.

The actual response burden approach does not consider the economic perspective of opportunity cost [see, e.g., Harberger (1971) and Shaw (1992)], which is "the value of the best foregone alternative" (McAfee and Johnson 2006). Opportunity costs include monetary costs and also the value of what was forfeited. Opportunity costs are difficult to estimate and involve quantifying the cost of a particular good or activity and comparing this to the value of the next best alternative that was given up. Opportunity cost is subjective and associated with utility. Hence, in principle, participating in business surveys may actually involve a smaller opportunity cost compared to productivity in the work of the business.

One way to construct an indicator of perceived response burden based on five-point scales such as those showed in Tables 6.3 and 6.4, is to give values to the response categories, where $-1 =$ very burdensome or time consuming, $-0.5 =$ quite burdensome or time-consuming, $0 =$ neither/nor option, $+0.5 =$ quite easy or quick, and $+1 =$ very easy or quick. The resulting indicator will vary from -1 to $+1$. For an example of how trends in perceived response burden can be presented, see Figure 3.6 and Gravem et al. (2011). Results showing the level and most common source of perceived response burden should be used to tailor later survey communication strategies (see Chapter 9).

6.3 AN ANALYTICAL APPROACH TO RESPONSE BURDEN

An analytical approach to response burden focuses on aspects of the survey design that affect response burden, and thereby potentially also affect the response quality of business surveys. By feeding this kind of information back to the survey designer,

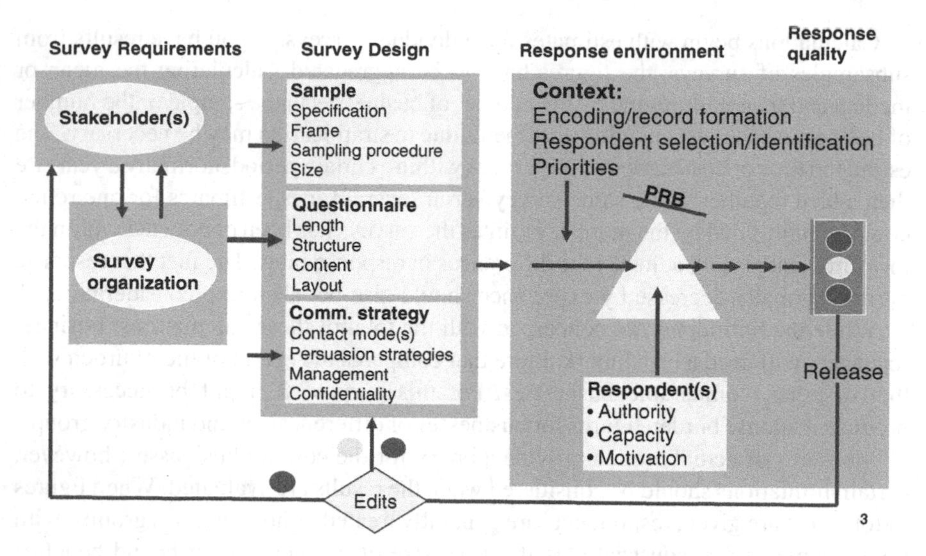

Figure 6.4 Total business survey burden model.

one can ensure that improvements to the survey communication strategy and measurement instruments more accurately address response burdens which really matters for the respondents and for the survey quality. In this way the analytical approach take both respondent and survey organization perspectives.

For this purpose we need a model that identifies different sources and effects of response burden and links these to the data collection process. This is done in Figure 6.4, which presents a general, holistic picture of the data collection process from start to end (Dale et al. 2007). In this figure we have opened the survey property and respondent characteristics from Figure 6.2 and put this basic figure into a business context, described in Chapter 2 and previously by Willimack and Nichols (2001, 2010). The seesaw character of perceived response burden from Figure 6.3 is also indicated in the model. An incorrect answer, no answer, or a correct answer is indicated by red, yellow, and green traffic lights, respectively.

The model aims to follow response burden from the conceptualization of data requirements to the receipt of data from the business, providing a holistic approach to burden, in which the respondent is only a part. In this way, burden is conceptualized as a cyclical process, which is transferred between actors in the survey process. The arrows indicate different processes. The actors involved are stakeholders, the survey organization, businesses, gatekeepers within the businesses, and respondents.

Burden is identified as originating from the stakeholders and the survey organization actors. A process of identifying and agreeing on survey requirements takes place between these two actors. The process of interaction between the stakeholders and survey organization produces a set of survey requirements, which are incorporated in the survey design by the survey organization. The sample, questionnaire design, and survey communication strategy are key elements that constitute the survey design. They are also influenced by the data collection modes, indicated by the darker frame surrounding the three key elements in Figure 6.4.

Distribution of surveys to the businesses and respondents can be divided into two phases. The first phase concerns how the business survey is distributed to the businesses (mode of data collection). The second phase is the internal distribution that takes place inside the businesses and comprises three actors: business management, gatekeepers and respondents. The characteristics and behavior of each of these actors can impact both collectively or individually on total burden.

In the response environment part of the model, respondent perceptions include actual and perceived response burdens and rewards. The evaluation made by the respondents is based on survey design properties, respondent characteristics and the context within which the respondent operates. The *context* and *respondents* characteristics originate from business survey participation work by Willimack and Nichols (2001) and Willimack et al. (2002). Respondent burden factors include respondents' knowledge of the survey organization and the particular survey, their prior exposure to the survey, timing of dispatch and return date, the number of people involved in the response process, the survey design, and the mode of data collection.

The four lines linking perceived response burden arrows with response quality refer to the four cognitive steps described by Roger Tourangeau (Tourangeau 1984; Tourangeau et al. 2000). The model attempts to illustrate that this psychological process takes part within a specific social context. The response outcome that is the result of this process then flows back to the survey organization, Where it is checked, edited, and accepted or taken back to the respondent by recontact before it is ultimately included in the statistics that goes back to the stakeholders.

The model highlights the fact that burdens are passed around from the stakeholder–survey organization interaction to respondents through an often far-from-ideal survey design. Burden may have cumulatively built up since the initial interaction process between the stakeholder(s) and survey organization. Respondents can pass this burden back to the survey organization, and ultimately the stakeholder(s) through a decrease in survey data quality and an increase in non-response rates.

In this way the survey can be described as a cyclical process that starts with the specification of information needs and ends with collected data. The basic success criterion is that the requested data match business information needs. Since the actors change and the communication is largely one-way, the risk of mismatch is high and not easily detectable. Burden is transferred between the actors, and ultimately decisions made at the beginning of the survey process come back as total burden at the end of the process.

Two important points can be drawn from the total business survey burden model:

1. The importance of recognizing that perceived response burden may originate from sources other than the survey design. If response to a particular survey is perceived as burdensome due to contextual conditions or the respondent's personal characteristics, we can try to adjust our instruments to these conditions, but they are often difficult to change.

2. The importance of recognizing that a survey design (as well as the other causal factors), consists of a number of components. What we need to identify is not only that a certain survey design affects perceived response burden but also which design components have that effect.

6.3.1 Disentangling Some Sources of Response Burden

In addition to core questions about actual and perceived response burden (see Tables 6.3 and 6.4), the handbook by Dale et al. (2007, p. 12) also offers some noncore questions that identify contextual factors of the businesses and personal characteristics of the respondents. Examples are questions about how easily the respondent could access the relevant information, how easy or difficult it was to find the time needed to complete the questionnaire, and what competence respondents felt they had about the issues raised in the questionnaire. For the full list of noncore, questions, see Dale et al. (2007, p. 24).

In an empirical study carried out among respondent to six of Statistics Norway 2006 annual business surveys, these noncore questions were used to evaluate the importance for the perceived response burden of context and personal factors against questionnaire design factors (Haraldsen and Jones 2007). Although the six questionnaires were similar, the questionnaires directed toward the construction and sea transport industries contained more questions; the lowest number of questions were on the service industry questionnaire. Under the (questionable) condition that the response burden was not affected by industry, one would therefore expect, because of the number of questions, that the construction and sea transport industry questionnaires would result in higher response burden than would the service industry questionnaire.

Questions about perceived response burden were attached to the web versions of the questionnaires. For the different industries the achieved response rate ranged form 30% (sea transport) to 45% (service industry). The answers to the core questions about perceived response burden were transformed into an indicator ranging from -1 to $+1$ as explained in the previous section. The importance of different sources of response burden was evaluated individually and combined with the help of logistic regression analysis.

It must be noted that this was an investigation into the usefulness of an analytical approach to the causes and effects of response burden. The results of the analysis should be treated with caution and cannot be generalized to the net sample, Still, we think the results are thought-provoking and illustrate the potentials of the response burden model. The overall picture from analyzing context factors, personal characteristics, and questionnaire design in association with perceived response burden were

- Calculations and mismatches between questions and available information were a common problem. The same was true when similar questions were posed in 2010. The 2010 results are shown in Figure 3.6.
- Having sufficient time available for the questionnaire was essential in terms of perceived response burden. This result underscores the importance of a survey

communication strategy which concerns the relationship with business managements and goes beyond the individual survey.

- The number of questions was probably the most important difference between questionnaires. However, poor layout and usability problems were often mentioned by those who felt that the questionnaire was very burdensome.
- Qualitative research previously carried out (Hedlin et al. 2005) concluded that response burden was highest in medium-sized businesses. The evidence was that small businesses do not have so much to report and larger businesses often have good documentation systems and people who respond to surveys as part of their job. The results from the 2006 quantitative analysis support this, with medium-sized businesses (20–49 employees) reporting the highest perceived response burden (0.06), but the differences are not large. Rather, the quality of documentation systems was probably the most important difference between industries.
- The perception of statistical uselessness accompanies perceived response burden. This correlation was particularly strong in the construction industry.

6.3.2 Suggesting a Curvilinear Correlation between Response Burden and Response Quality

The challenges business questionnaires present to business respondent can cause quality problems because they are too demanding. But if demanding tasks are not recognized, or if they are recognized but overlooked, this may be as serious a source of error as the tasks themselves. As stated when we introduced the cognitive steps of the question–answer process in Chapters 2 and 3, respondents who are not sufficiently motivated, are not sufficiently alert to what the questions ask for, or avoid challenges that are difficult to overcome, may take shortcuts in order to provide a response.

Taking a closer look at the pie graph in the upper left of Figure 3.6, one discovers that most business respondents claimed that the questions in the structural statistics survey were neither particularly easy nor difficult to answer. Moreover, almost 40% found the questions easy or very easy to answer, while just over 20% found them difficult or very difficult to answer. We suspect that this distribution to a certain extent mirrors superficial reading, perhaps combined with a satisficing strategy. A quality test on the basis of the same perceived response burden question in a smaller study of the wage and salary survey seems to substantiate this interpretation. In this test the number of violations of one item nonresponse and three consistency checks were used as a quality indicator and cross-referenced with perceived response burden measurements (Fosen et al. 2008). Figure 6.5 shows the results.

The number of test subjects was low (248) and the threshold for significant results consequently high. But even if the results should be interpreted with care, they indicate that business respondents who find the questionnaire easy to answer make just as many errors than those who found it very difficult. Also, as more respondents in this case found the questionnaire easy than difficult (17% + 49% vs. 6% + 19%), errors made by respondents who found the questions easy could be a more serious

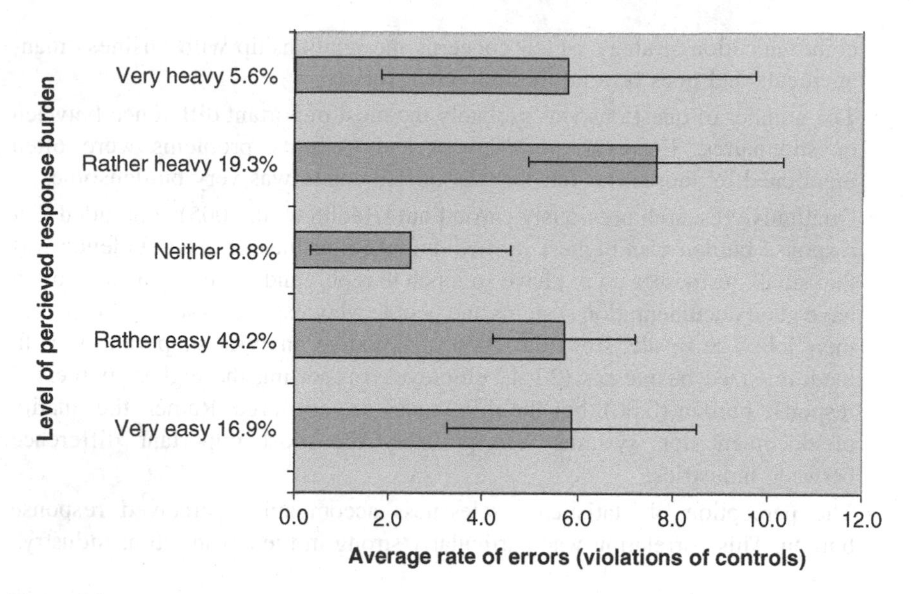

Figure 6.5 Average number of quality violations among respondents to the Norwegian wage and salary survey relative on how easy or burdensome they found the questionnaire.

quality problem than errors made by those who found it them difficult. In other words, problems with respondents who did not recognize how demanding the questions were seemed to be a bigger quality problem than problems with respondents who recognized the burdensome quality challenges posed by the questions.

Comparisons between questionnaires of different length reflect a similar conclusion. As long as the number of items is not reduced, shortening the questionnaire normally implies that the number of words and space provided for answers have to be reduced. This can be beneficial if what is taken away is superfluous information. But when the number of words and visual clues are reduced the questions often become more ambiguous and open to misinterpretation or superficial answers (Bogen 1996). An example is a comparison between short and long versions of the 2005 US Agricultural Census questionnaire (McCarthy 2007). One measure used to reduce the number of pages in the short version of this questionnaire was to pose general questions about commodities grown and produced on the farm. Instead of asking several questions about specific kinds of production, only some relevant products were listed as examples to accompany the questions. This and other textual simplifications seemed, however, to reduce response quality. This links back to the fact that survey organization methods for cost reduction, such as shortening questionnaires, may generate additional survey quality costs.

The European Statistics Code of Practice states that "The reporting burden should be proportionate to the needs of the users and should not be excessive for respondents. The statistical authorities monitor the response burden and set targets for its reduction over time" (Eurostat 2011b). It is worth while noting that this formulation underlines a fine balance between excessive response burden and burdens that must be imposed on the respondents for quality reasons. The examples just given suggest that business survey questions sometimes may be formulated or simplified in ways that invite superfluous answers, and hence that such questions should be reworded to allow respondents more time to think about what the questions ask for. Referring to the question–response process described by Tourangeau et al. (2000), this implies that it might be more important to simplify the survey tasks than to reduce the time it takes to read the questions. We will return to the fine balance between simplifications that obscure problems and simplification that makes the quality challenges easier to tackle several times in this book.

6.4 MEASURES TO REDUCE RESPONSE BURDEN

Survey organizations can reduce response burden in several ways. In the above-mentioned web survey to 41 NSIs, the participants were asked to rate 17 possible burden reduction strategies according to how common these strategies had been applied. On average, institutes had applied 12 of the 17 strategies to at least some of their business surveys. Only one institute had not implemented any of the proposed measures, and four institutes had implemented 16 of the 17 proposed strategies. These figures indicate a high activity. In contrast, the same institutions had launched hardly any quantitative research on the effects of the measures taken. Effects of the measures taken are seldom monitored, analyzed, and documented (Giesen 2011). We will return to the research and documentation issue at the end of this section. But first we will take a closer look at the measures implemented. To do this, we will split the possible burden reductions listed in the web survey into four categories: survey request reductions, sample coordination, survey communication measures, and questionnaire improvement measures.

6.4.1 Survey Reduction Measures

The actual burden approach compels survey organizations to minimize the size of their business survey samples (and, if recurring, the frequency of sample rotation), the frequency of the surveys, and the number of questions asked. Examples of this actual response burden approach can be seen, for example, in the 2007 European Commission Action Programme on reducing administrative burdens in the European Union (European Commission 2008) and in the US definition of response burden (White House 2006).

Five survey reduction measures were listed in the NSI web survey (see Figure 6.6).

Using administrative data (e.g., tax data) instead of surveys is a potentially powerful instrument for reducing response burden. However, usually special legislation is needed to grant statistical agencies access to these data. The NSI survey indicated that 9 of the 41 institutes had managed to replace most of their data collection with administrative data or other kinds of secondary sources. In some nations, like The Netherlands and the Scandinavian countries, there is strong political pressure on using alternative sources to surveys. For example, in The

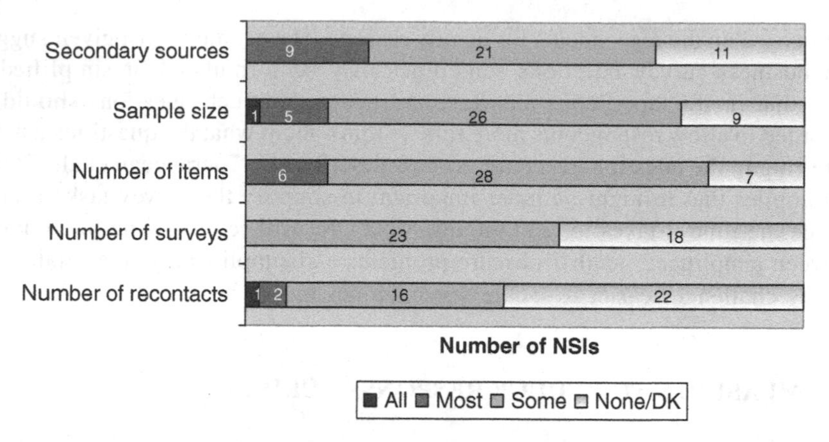

Figure 6.6 Response burden reductions by reducing survey requests. Number of NSIs that apply the listed methods for all, most, some or none of their business surveys.[2]

Netherlands the Statistics Law of 2005 gives Statistics Netherlands both the right and the obligation to use administrative data (Oomens and Timmermanns 2008). This means that survey data can be collected only if the administrative data are insufficient.

A step further than simply reusing data already collected by others is to coordinate data collection with different partners. A successful example of such a coordinated data collection is the Portuguese simplified business information system (Pereira 2011), which allows businesses to provide information to several public authorities at the same time, using one single portal.

Alternatively, to prevent double data collections, some countries keep registers at the national level to monitor and reduce burden caused by all government surveys. For example, in Australia the *Statistical Clearing House* provides a central clearance point for business surveys that are run, funded, and conducted on behalf of the Australian government with the purpose of reducing duplication, minimizing business burden, and ensuring that surveys are fit for purpose.

There are no sampling errors in registers. Moreover, because registers contain the whole population of interest, it is easier to produce statistics for small areas. On the other hand, because they are established for a purpose other than statistics, the validity of register data may be poor. Ideally the choice between registers and surveys should therefore be based on an evaluation of the need for small-area statistics against the need for valid and reliable information.

Reducing sample sizes, reducing the number of surveys or number of items belong to the same family of burden reduction measures. At first sight they all seem to include loss of information. Critical reviews may, however, disclose that a more efficient sample design could be applied or that superfluous information is collected. There are also several methods that can be used to meet quality issues raised by these methods. One example is the small-area estimation method, which utilizes information that belongs to a neighboring domain or earlier time period (Hedlin 2008).

[2] "None" also includes those who answered "I don't know."

Instead of skipping questions altogether, one should also consider whether questions can be skipped in parts of a time series or from parts of the sample. In recurring business surveys questions are sometimes replaced by estimates based on earlier forecasts (Öller 1990).

As stated before, recontacts with survey participants can be more time-consuming for survey organizations than for the individual businesses contacted. Being contacted may even have a positive effect on the attitude toward the survey organization because it tells the respondents that the organization cares about their efforts. We therefore suspect that measures taken to reduce the number of recontacts usually are internal cost-saving initiatives rather than initiatives for reducing response burden. Statistical agencies have made considerable efforts to streamline their editing process (see Chapter 11). In mandatory business surveys some have also skipped reminders, and when business surveys are computerized, edits can be outsourced to the questionnaire completion stage. With the exception of fewer reminders, these measures seem to have obvious quality gains. Their effects on the response burden, however, are more questionable.

6.4.2 Sample Coordination

Coordination of the inclusion in different samples taken from the same population can obviously affect response burden. According to the NSI survey, only 9 out of 41 institutes did not use sample coordination in any of their business surveys. One institute coordinated all their business surveys, while 12 coordinated most of their surveys and 19, some surveys. Given the dynamics of the business population as well as the large number of repeated surveys with different frequencies and sampling–rotation designs that are typical at any NSI, sample coordination in practice involves a great deal of details that are beyond the scope of this chapter. Here the discussions focuses on the prerequisites of sample coordination and its effects on response burden.

Table 6.5 presents a typical scenario of sample coordination. Assume the population to consist of all the business units (27,815) of a certain industrial

Table 6.5 Number of Units in a Hypothetical Sample and Population by Survey and Domain

	Number of Employees:							
Survey	0–4	5–9	10–19	20–49	50–99	100–249	250+	Total
1	0	373	289	177	91	23	10	963
2	0	116	241	295	181	30	10	873
3	0	275	385	118	91	15	10	1,364
4	167	466	964	590	181	30	10	2,408
5	3,115	3,726	3,854	1,179	181	30	10	12,095
6	0	112	193	177	91	30	10	613
Total participation	3,282	5,538	5,926	2,536	816	158	60	18,316
Population size	15,110	7,451	3,854	1,179	181	30	10	27,815

grouping, divided into seven domains by the number of employees. There are six different surveys at a given point in time. The domain sample size of each survey is shown for surveys 1–6 in Table 6.5, where, for example, survey 1 requires a subsample size of 373 from the domain of units with 5–9 employees, and so on. The total of all six subsamples in each domain is given as "total participation" because it is the total number of survey participations within the given domain imposed by the sampling designs. Sample coordination cannot change this scenario. What sample coordination does affect is the extent to which the different subsamples should overlap. For instance, in the domain of units with 0–4 employees, maximum overlap is achieved if all 167 units participating in survey 4 also participate in survey 5, while minimum overlap results if none of the 167 units participate in survey 5. In this way, the total response burden, which is somehow related to the total survey participations, can be distributed differently by means of sample coordination.

6.4.2.1 Common Frame and Regime

The most important prerequisite for sample coordination, as illustrated in Table 6.5, is the creation and maintenance of a *common* sampling frame, usually in close connection with the business register, which is used for all surveys that belong to the coordination system. Indeed, a coordination system will be of rather limited value for the business community, as long as the number of surveys outside the system remains notable. In other words, the responsibility of sample coordination really belongs to the entire statistical organization, and the different subject areas must cooperate in practice to make the system work. For instance, to create an overview as illustrated in Table 6.5, the survey organization needs to summarize.

- The number of surveys in the business statistical system
- The parts of the common sampling frame that are relevant for each individual survey
- All individual samples at any given timepoint.

To actually practice sample coordination, the survey organization must have information on all the individual sampling-and-rotation designs, as well as the participation history of each individual business unit in each survey. Without this information, control of future survey participation and, hence, response burden can never be as exact as possible.

Clearly formulated principles of coordination and easily understandable operational procedures are of utmost importance for gaining approval and confidence from the respective divisions in charge for migration from independent-to-coordinated sampling. The system should be "fair" to the different statistics in the sense that no one should be forced to use a frame perceived to be inferior. Different statistics may justifiably have different sampling timepoints throughout a calendar year, and everyone prefers to have a sampling frame that is as up-to-date and accurate as possible, and to keep the time lag between frame creation and data collection as short as possible. It follows that freezing of the common sampling frame, while requiring the easiest system administration, is highly undesirable for surveys conducted at timepoints relatively distant from the reference timepoint of the frozen frame. Historically speaking, this has been a major obstacle in the development of a sample coordination system from within survey organizations.

Finally, the system should be sufficiently flexible to accommodate all types of change seamlessly. On one hand, the business population is highly dynamic compared to the general lay population. Births and deaths of units, as well as changes of units in size and classification, occur all the time. On the other hand, changes of scope and sampling design are unavoidable over time in any survey. Moreover, allowance for survey-specific exceptions from the general principle of coordination, in the form of so-called exception sample units, may be necessary in practice. Clearly, there is an issue of balance here, too many exception units would effectively destroy the coordination, whereas too rigid a regime may be counterproductive to the motivation and acceptance of sample coordination across the survey organization.

6.4.2.2 Effects on Perceived Response Burden

Having discussed some of the prerequisites of sample coordination above, we now turn to the actual effects sample coordination may have on response burden. The desirable keywords are transparency, fairness, and predictability, all of which closely relate to the *perceived* response burden. In return, the business community's perception of a sensible response burden may enhance their willingness to respond as well as the quality of their response. This perception is also important from a political stand-point with regard to the various stakeholders of the survey organization.

Transparency An overview such as that presented in Table 6.5, as both a prerequisite and the result of sample coordination, provides the foundation for transparency in communication toward the business community as well as the various stakeholders of the survey organization. The respondents should be clearly informed as to *how* and *why* they are selected to participate in the surveys, as well as the efforts that have been made to limit individual response burden. The respondents should no longer be given the impression that, from time to time, they are dealing only with a part of the survey organization that happens to be running a particular survey, without reasonable coordination with other parts of the same organization. Rather, one should be able to present the respondents with a complete picture on behalf of the entire survey organization. In this way, transparency in communication will indicate that there is an effective coordination system. This can help create a more favorable perception of the response burden, even if the exact extent of fairness or predictability of response burden under the coordinated system remains debatable.

Although actual response burden cannot be measured exactly, sample coordination is particularly helpful for improving transparency and thereby the perception of the respondents, because survey participation is tangible and measurable. Actual response burden obviously varies from one survey to another; it also varies from one business unit to another, because the response effort necessarily depends on the information and even organization structure of each business unit. The matter becomes even more intricate with respect to the *joint* response effort for several surveys. For instance, the total response effort could be less for one business unit participating in two surveys on related topics than for two different units participating in one survey; or if one unit is to participate in the same survey in two successive rounds, than to participate twice with several years of "rest" in between, and so on. In contrast, the effort and result for control of survey participations can be conveyed exactly, which tangibly increases the degree of transparency.

Fairness The understanding of a fair distribution of response burden is important for setting the targets for sample coordination. It also provides an important background for communication with respondents. A simplistic interpretation of fairness is equal burden. Two general remarks are worthwhile:

1. Burden must be balanced against accuracy. Table 6.5, shows that the sampling fraction is seldom constant across the business units. Because of their importance to the accuracy of aggregated estimates, the larger units will have higher inclusion probabilities. It follows that evenness of survey participations can be sought only among *comparable* units. For instance, from Table 6.5, it is feasible, as a principle of coordination, to require evenness of survey participation in each of the seven groups of units by the number of employees. But it would be unacceptable to require the same for all the units regardless of size, due to concern for accuracy. The formal consequence is that the entire business population needs to be divided into mutually exclusive *domains of coordination*, in which evenness of survey participations is considered both desirable and acceptable. Note that this again requires agreement throughout the entire survey organization.
2. Exact evenness can be achieved only for the *expected*, not the *actual*, burden. In any sample survey, some units have to participate while the others "rest." So the actual burden can never be equal at any given timepoint. Over a sufficiently long period, the actual burden in regularly repeated surveys can in theory be made equal by sample coordination, provided every relevant aspect of the universe remains the same, which is unrealistic. Moreover, even in the hypothetical setting of an unchanging universe, the time required to "equate" actual burden will be so long as to be considered as irrelevant by the business community. So, evenness can only be exact on expectation. The problem is that expected equal burden may not necessarily be perceived as fair.

For an illustration, consider the domain of units with 0–4 employees in Table 6.5. One obvious way to even out the actual participation in survey 4 over time is to select—randomly on every occasion—among the units that have participated the fewest times so far. Moreover, suppose that there is no sample overlap over time. Then, provided the universe is fixed, it will take 15,110 repetitions of the survey before all the units have participated exactly the same number of times! This shows

Table 6.6 Expected Response Burden by (1) Independent Simple Random Sampling and (2) Simple Random Sampling with No Overlap for Domain of Units with 0–4 Employees in Table 6.5

	Participation Scenario			
Actual Burden	$(1,0), \Delta_4$	$(1,0), \Delta_5$	$(1,1), \Delta_4+\Delta_5$	Expected Burden
Probability: independent sampling	$p_4(1-p_5)$	$(1-p_4)p_5$	$p_4 p_5$	$p_4\,\Delta_4 + p_5\,\Delta_5$
Probability: no sample overlap	p_4	p_5	0	$p_4\,\Delta_4 + p_5\,\Delta_5$

clearly that long-run equality of actual burden is a rather irrelevant target in reality. Note that we have not yet taken survey 5 into account.

Let us consider then the expected burden. Assume two alternative approaches:

1. Independent simple random sampling for surveys 4 and 5 (i.e., no sample coordination)
2. Simple random sampling under the restriction that no unit is to participate in both surveys at the same time (i.e., no sample overlap due to coordination)

Let us denote by Δ_4 the actual burden of participating in survey 4, and by Δ_5, that of participation in survey 5. Suppose that the actual burden of participating in both is simply given by $\Delta_4 + \Delta_5$. Moreover, denote by (1,0) the scenario where a unit participates in survey 4 but not 5; by (0,1), if it participates in survey 5 but not 4; and by (1,1), if it participates in both.[3] The respective probabilities of each participation scenario under approaches (1) and (2), and the corresponding expected burden are given in Table 6.6, where p_4 $(=167/15{,}110)$ and p_5 $(=3115/15{,}110)$ are the respective marginal inclusion probabilities of surveys 4 and 5.

Note that the expected response burden is exactly the same by either approach. This is, indeed, a general observation. Thus, without any coordination, independent sampling across the different surveys can yield equal expected response burden, simply because in principle the approach does not "mistreat" or "disfavor" any particular units in the frame. Still, it is easily conceivable that a unit selected for both surveys at the same time may not *perceive* the approach as fair, as long as about 80% (i.e., $1 - p_4 - p_5$) of the units do not need to participate in any of the surveys. Note that while about 34 units (i.e., $15{,}110\, p_4\, p_4$) may be expected to participate in both surveys under approach (1), there will be no such units under approach (2). In other words, while the actual burden is $\Delta_4 + \Delta_5$ for about 34 units under approach (1), the maximum actual burden is only $\max(\Delta_4, \Delta_5)$ under approach (2). It follows that sample coordination can improve the evenness of response burden, by reducing the *variation* in the actual survey participations at a given timepoint, even though it may not yield a

[3] Non participation in either survey implies null response burden and may thus be omitted from the calculation.

more even distribution of the expected burden. While this does not really reduce the actual burden in any averaging sense, it can help improve the perception of burden.

Predictability Rotation of samples is necessary to even out actual response burden. This explains why predictability of survey participation over time is an important factor to the perception of response burden, although the communication of predictability is typically asymmetric. On one hand, as it seems to provide little comfort for a business unit to be informed of future survey participation, regardless of how distant or regular it may be, this kind of predictability is rarely revealed for the prospective respondents. On the other hand, where feasible, the promise of *quarantine* (or "survey holiday" period) for a business unit after it has "served its time" in a survey is usually received positively, such that the introduction of quarantine is an important measure of predictability that is expected to have a positive effect on the perceived response burden.

Consider the scenario in Table 6.5. Assume a yearly rotation at the rate of 2 years in all the surveys. In the domain of units with 0–4 employees the total number of units that need to be rotated into the sample is $3282/2 = 1641$ in each year under approach (2) with no sample overlap. Then, there are $15{,}110 - 3282 = 11{,}828$ units available in the population to be rotated into the sample. This means that it is, in theory, possible to set a *general* quarantine of 7 years, since $11{,}828/1641 > 7$, so that a unit will have a 7-year respite from surveys after it has participated in one of the two surveys. Allowing for changes in the population over time, the survey organization might, in practice, settle for a more conservative guarantee, say, of 5 years.

Approach (2) is also feasible in the domain of units with five to nine employees, but a general quarantine is impossible across all the surveys, because there are fewer out-of-sample units ($7451 - 5538 = 1913$) than the number of units that need to be rotated into the samples ($5538/2 = 2769$) on each occasion. But it is possible to introduce a *limited*, or *conditional* quarantine of, say, 5 years for all except survey 5, since $[7451 - (5538 - 3726)]/[(5538 - 3726)/2] > 6$. Indeed, it is possible to introduce another conditional quarantine of two years for survey 5 on its own, now that its sampling fraction is $\frac{1}{2}$. Even such a limited quarantine may improve the perception of burden, as the respondent is promised to a respite for the same length of time that it participated in the survey, specifically, 2 years in each case.

6.4.3 Survey Communication Measures

A study by the US National Agricultural Statistics Service (NASS) showed that attitudes toward the survey sponsor differ significantly between respondents and nonrespondents (McCarthy et al. 1999). Results like this support the notion that a general feeling of trust or mistrust affects the motivation and perceived response burden in surveys. Respondent attitudes toward survey organizations are certainly influenced in different ways, but a central arena for image building is the contact that surveyors have with their respondents during data collections. In the NSI web survey on response burden measurements and measures, the participating NSIs where asked about six kinds of survey communication contact. In Figure 6.7 we have listed these

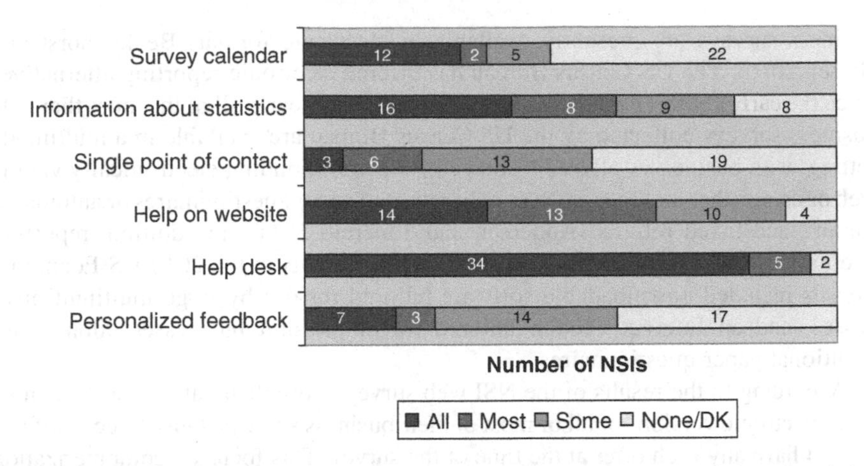

Figure 6.7 Survey communication measures for reducing response burden. Number of statistical institutes that apply the listed methods for all, most, some, or none of their business surveys.[4]

on a temporal basis, according to contacts made before data collection, during the first data collection period, or after questionnaires or no response was received. You will find the same distinction in Chapter 9, where different survey communication strategies are discussed in more detail.

Calendars showing upcoming surveys in which the businesses are expected to participate, a single point of contact in the institute, and personalized feedback were the least common forms of survey communication offered. The most common form of contact offered was a help desk for those who needed assistance. These results raise several important questions, one relating to the amount attention paid to preparing businesses for surveys in which they are selected to participate, and following them up when the data collection is finished. Other questions are related to the level of personalized and active survey communication. Survey calendars, a dedicated contact person, and tailored feedback are examples of more personal communication forms than general information given in introduction letters or on websites. Admittedly, help desks offer oral communication, but only for those who actively seek help. We also question the degree to which feedback collected by the help desk is fed back and listened to by the survey organization. What we are looking for is perhaps more active relationship building and customer service, similar to that found in the business world [see also Torres van Grinsven et al. (2011)].

6.4.4 Questionnaire Improvement Measures

More recently, developing and offering electronic survey instruments have become one of the primary initiatives taken by survey organizations to improve their business surveys. Statistics Norway introduced electronic reporting for all business surveys from 2004, and web questionnaires are now their primary mode in business surveys (Haraldsen et al. 2011). At Statistics Netherlands, more than half of the

[4] "Single point of contact" was an option for large businesses.

business surveys are presently available in electronic format (Beukenhorst and Giesen 2010). The US Census Bureau has offered electronic reporting alternatives since the early 1990s (Ambler et al. 1995). Currently, nearly all of the more than 100 business surveys collected by the US Census Bureau are available in a multimode setting, with businesses offered a choice between responding electronically via the web or using other methods, such as mail return of paper questionnaires or automated capture and faxed returns (Anderson and Tancreto 2011). In addition, reporting alternatives for approximately 5 million establishments in the 2012 US Economic Census included downloadable software tailored for use by large multiunit businesses, and an Internet version tailored for single unit businesses, along with traditional paper questionnaires.

According to the results of the NSI web survey, more than half of the NSIs now offer electronic versions of all or most of their business surveys. Only three institutes did not have any such offer at the time of the survey. This focus on computerization stands in contrast with an absence of research, which clearly back ups the notion that electronic questionnaires lead to lower response burden and higher response quality (Löfgren 2010). When given the choice, the majority of business still seem to prefer a paper version of the questionnaire before a web version, and when only web is offered, the response rate can suffer [see, e.g., Bremner (2011)].

One transition from paper to web that has apparently been fairly successful is redesign of the Dutch structural business survey (SBS). In this project a quicker response, a reduction in the time it took to complete the questionnaire by 35 minutes, and a slight reduction in the number of respondents who found the survey time-consuming were observed (Giesen et al. 2009a). We think that this project also contains one of the keys to understanding why so many other web projects have failed. Redesign of the SBS took several years and included several rounds of testing and adjusting the questionnaire (Snijkers et al. 2007b). We suspect that many of the survey organizations that introduce web questionnaires forget that it is not the technology in itself, but how it is utilized that decides the result. For this reason, too little effort is put into actually redesigning the original paper questionnaire. Technology optimism probably also partly explains why the research on effects of computerizing questionnaires has been so sparse. These reflections lead us to the next improvement measure: testing and evaluation activities.

Approximately half of the statistical institutes (19) that participated in the NSI response burden survey reported that they tested all or most of their business survey questionnaires (see Figure 6.8). The others had tested some questionnaires or none. This latter half of the statistical institutes should perhaps remind themselves on Sudman and Bradburn's (1982) saying that "If you don't have the resources to pre-test your questionnaire, don't do the survey." In its third indicator of appropriate methodology, the European Statistics Code of Practice also state that "In the case of statistical surveys, questionnaires are systematically tested prior to the data collection" (ESS 2011).

Development and testing methods tailored for business surveys are the topic of Chapter 7. In Chapter 8 we will discuss how details in paper and web questionnaires can reduce response burden in such a way that also enhances response quality.

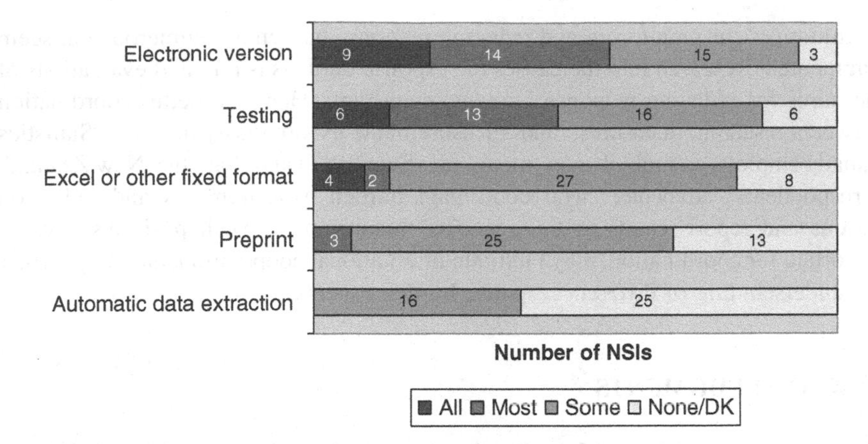

Figure 6.8 Questionnaire improvement measures for reducing response burden. Number of NSIs that apply the listed methods for all, most, some, or none of their business surveys.

6.4.5 Tailored Evaluations of Response Burden Reduction Measures

In this section we have described various methods that have been used to reduce response burden. Generally, however, evaluations of the effects of these measures are rare. We would like more evaluations and in addition, different kinds of evaluations tailored to different kinds of measures.

We have distinguished between four kinds of measures; sample or variable reductions, sample coordination, enhanced survey communication, and questionnaire improvements. When samples or the number of questions are reduced, it is rather obvious that actual response burden will also reduce. Sample reduction will affect the total response burden, while reducing the number of questions will also affect the response burden of sampled businesses. But these reductions may have quality tradeoffs, which need to be considered. Sample and variable reductions consequently call for quality analysis. This is different for sample coordination systems. As illustrated, sample coordination does not affect the expected response burden, but is rather intended to control for excessive burdens and to allocate burdens in a fair way. This objective calls for studies of how response burdens are distributed; preferably carried out in comparable groups of businesses, and comparing questionnaires of different degrees of difficulty. The ultimate goal of an enhanced communication strategy is, of course, that respondents should be more motivated to answer our survey questions in a serious way. But the path leading to this goal has several checkpoints on the way. The first is if the attitudes toward surveyors and surveys are improved. This is perhaps best studied by linking different communication strategies to user satisfaction studies. Finally, studies of perceived response burden are probably most relevant in projects that attempt to improve the business questionnaires. The relationship between questionnaire elements, perceived response burden, and response quality is complicated and calls for research based on detailed analytical models.

Initiatives for monitoring and reducing response burden are numerous but seem fragmented. Research into the causes of response burdens is rare and evaluations of measures for reducing response burdens, even rarer. Hence, a better coordination between research, measures, and measurements is obviously needed. Statistics Canada employs an ombudsman (mediator) (Sear 2011) and Statistics New Zealand, a respondents' advocate,[5] who coordinate burden measurements and reduction actions and report results to the agencies' management. Such positions may be important for coordination, may facilitate international cooperation, and may further the understanding of different response burden aspects.

ACKNOWLEDGMENTS

The authors are grateful to Mojca Bavdaz, Petra Mohorič Peternelj, Virginie Blaess-Raymond, Dan Hedlin, Tora Löfgren, Dag Gravem, and Yngve Bergström for their contributions to the BLUE-ETS work package *NSIs Practices Concerning Business Burden and Motivation* and to Trine Dale, Johan Erikson, Johan Fosen, and Øyvin Kleven for their contributions to the *Handbook for Monitoring and Evaluating Business Survey Response Burdens.*

[5] See http://www.stats.govt.nz/surveys_and_methods/completing-a-survey.aspx.

CHAPTER 7

Methods for the Development, Testing, and Evaluation of Data Collection Instruments

*Diane K. Willimack**

In this chapter, we move from sample design and coordination to methodological activities that support business survey data collection, starting with the development and testing of survey questions, questionnaires, and data collection instruments. There is an extensive literature on methods and best practices in the design, development, pretesting, and evaluation of questionnaires and data collection instruments for various modes, and much of this is applicable in business surveys as well as surveys of households and individuals. Literature on question wording and order, visual design, and mode differences, to a large extent, form a foundation for business surveys as well. However, to ensure successful collection and quality in business surveys, the business context may require that some instrument design, development, testing, and evaluation methodologies be modified or adapted, and new methods created, to effectively meet survey objectives.

Guidelines and best practices that have emerged in the design of business survey data collection instruments are the subject of the next chapter. But first, in this chapter, we will describe methods for assisting business survey designers with development, testing, and evaluation of data collection instruments to improve their effectiveness in obtaining quality response data from business survey respondents, while also controlling or reducing burden. Since many of these methods are drawn from surveys of households and individuals, we may not describe in great detail the

* Any views expressed on methodological, technical, or operational issues are those of the author and not necessarily those of the US Census Bureau.

Designing and Conducting Business Surveys, First Edition.
Ger Snijkers, Gustav Haraldsen, Jacqui Jones, and Diane K. Willimack.

procedures for their use, referring the reader to relevant texts instead. Rather, we will focus on their application in the business survey context, after highlighting unique features of business surveys and the response process that call for these adaptations.

This chapter begins by setting the context for business survey data collection instrument development, testing, and evaluation. We define a framework identifying potential contributions to total survey error to be addressed during this stage. Based on the model of the business survey response process defined in Chapter 2, we use the steps involved in performing the response task as a motivator for methodologies that evaluate these sources of error. We associate features of the business setting and context with limitations and constraints that affect ways of implementing various development, testing, and evaluation methods, often impacting their effectiveness and providing the impetus for some sort of modification or adaptation.

We then move on to listing a variety of development, testing, and evaluation methods; indicating appropriate uses; and highlighting adaptations that accommodate the business survey setting. Our focus is on self-administered data collection instruments, both paper and electronic. We also provide examples of using multiple development, testing, and evaluation methods, demonstrating how they complement one another. Finally, we describe common NSI organizational structures supporting instrument development, testing, and evaluation, along with descriptions of logistics and procedures for carrying out this research.

7.1 A FRAMEWORK FOR QUESTIONNAIRE DEVELOPMENT, TESTING, AND EVALUATION

Our framework for conducting development, testing and evaluation processes is based on a total survey error approach, where the steps in developing survey measurements are associated with sources of survey error at each stage (Groves et al. 2004). Figure 3.1 illustrates this framework.

The data collected in surveys may be considered measurements of some underlying concept of interest to a researcher. Therefore, questionnaire development begins with defining concepts associated with the research goals. Concepts can be deconstructed into specific attributes that may or may not be measurable in a practical sense. Measurements are further specified and these specifications become the raw material for survey questions. The responses to survey questions become data that are summarized or analyzed to address the research goals of the particular data collection.

These steps, which demonstrate how we get from concepts to data via surveys, are illustrated in Figure 7.1, presented as a slight variation on the "measure" side of the total survey error approach to survey quality presented in Figure 3.1. Also illustrated in Figure 7.1 are the types of nonsampling error sources we consider to be associated with each step—validity, specification errors, and measurement errors. Issues of *validity* examine questions such as "What can be measured to reflect the concept?" and "How well does the measurement reflect the concept?" *Specification error* is associated with the question "How well does the question capture the desired

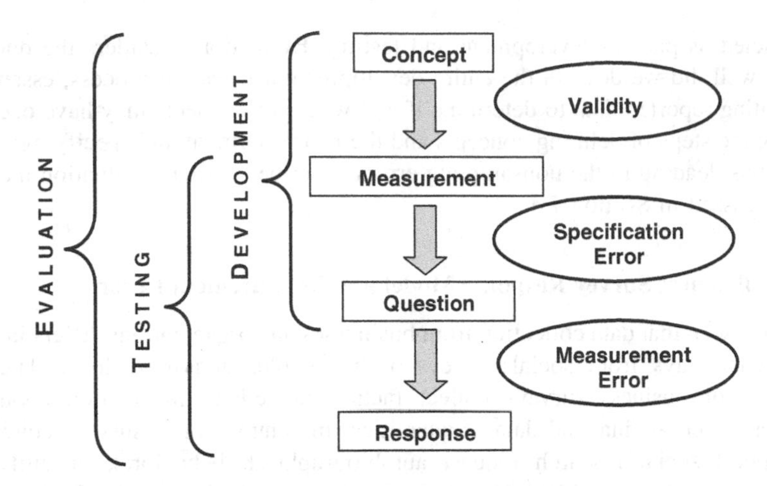

Figure 7.1 Steps and error sources in questionnaire development, testing, and evaluation.

measurement?" Finally, *measurement error* is associated with the question "How well does the response match the intent of the question?"

Finally, Figure 7.1 links the steps—starting with concepts, defining their measurement, specifying the survey questions, and obtaining responses—to the stages of questionnaire development, testing, and evaluation wherein these nonsampling errors may be encountered, investigated, and addressed.

We define question(naire) *development* to consist of the activities and processes where concepts are defined, measurable attributes are specified, and survey questions are formulated to collect these measurements. During development, our goal is to identify and define valid measurements of the underlying concepts of interest, and then to specify survey questions that are translations of the measurement definitions. Thus, during development, we are trying to ensure validity and minimize specification errors. Methods that aid this process are described in Section 7.2.

We define question(naire) *testing* to consist of research conducted prior to production data collection for the purpose of determining the degree to which responses obtained from survey questions meet the intent of the question as specified by the desired measurements. In other words, the purpose of testing is to ensure that measurements are operationalized properly in survey questions that obtain the desired data. Thus, during testing, we investigate responses to survey questions in order to identify and address problems in the questions that contribute to measurement errors in the data, and, if necessary, determine whether poorly specified measurements led to problematic survey questions. Methods to be used during questionnaire testing are covered in Section 7.3.

We define question(naire) *evaluation* as consisting of investigations conducted to assess the performance of the survey question in measuring the underlying concept as reflected in the quality of the collected response data. Evaluation is typically conducted during or following collection of the data, or at least involves data collected from actual respondents, such as evaluating field tests or experiments conducted as part of development and testing. Evaluation considers the question "How well did we do?" in the entire development and testing process, essentially evaluating reported data to determine if and where disconnects may have occurred among the steps of defining concepts and their measurement and specifying survey questions, leading to the nonsampling errors described earlier. Evaluation methods are discussed in Section 7.4.

7.1.1 Business Survey Response Model and Measurement Error

We have seen that data collection from businesses and organizations differs in some significant ways from social surveys of households and individuals. The vast majority of business surveys collect factual numerical information about the business, such as financial data or quantities. In contrast, most surveys conducted to support social research request autobiographical, behavioral, or attitudinal information about individual persons, or attributes of the household. Social survey respondents are usually answering questions about themselves, their households, or other individuals in the household, and questions can typically be answered from memory. Business surveys require a person to act as an informant on behalf of the business, often relying on data contained in business records.

As a result, the survey response process in economic surveys of businesses and organizations is somewhat more complex. The model of survey response for individuals consists of four cognitive steps—comprehension, retrieval, judgment, and communication (Tourangeau 1984). As we saw in Section 2.2.2, the model of the survey response process for businesses wraps organizational steps around the cognitive steps (Willimack and Nichols 2010; Sudman et al. 2000; Bavdaz 2010a) as follows:

1. Record formation and encoding of information in memory
2. Organizing the response tasks
   a. Selection and/or identification of the respondent(s)
   b. Scheduling the work
   c. Setting priorities and motivation
3. Comprehension of the survey request
4. Retrieval of information from memory and/or records
5. Judgment of the adequacy of the retrieved information to meet the perceived intent of the question
6. Communication of the response
7. Review and release of the data to the survey organization

Variation in the performance of these steps across different businesses and respondents contributes to measurement error. To better specify the sources of measurement error in the survey response process, let us begin with step 1, *record formation*, along with knowledge of those records *encoded* in a person's memory.

We saw in Chapter 2 that the main purpose for data recorded in business records is to satisfy management needs and regulatory requirements, and not statistical purposes. Moreover, data may be distributed across organizational units where they are needed, along with employees knowledgeable about the data (Edwards and Cantor 1991). As a result, to the extent that data requested by survey questions differ from data available or known to exist in records, there is potential for measurement error.

The business context described in Chapter 2 also discussed the actions involved in *organizing response tasks*, including *respondent selection*, *scheduling the work*, and *determining priorities and motivation*. All of these activities are under the control of the business. Thus, to the extent that businesses differ in their respondent selection and scheduling strategies, there may be variation in the quality of survey responses which contributes to measurement error. In addition, businesses set priorities for respondents' work activities, as well as criteria for their acceptability, both of which affect the attention and diligence accorded the response task. Variation in these factors also contributes to potential measurement error.

Next, recall that the four cognitive steps—*comprehension, retrieval, judgment, and communication*—occur within the context of organizationally determined factors. Thus, not only do variations in respondents' cognitive abilities contribute to the potential for measurement error; the organizational context may provide additional impact. Examples include

- *Comprehension.* Respondents may have varying degrees of familiarity with seemingly standard technical terminology, and may reinterpret questions in a manner that fits their own business environment, contributing to measurement errors.
- *Retrieval.* To the degree that respondents vary in their knowledge of or access to records, or their abilities to extract or obtain data from alternative records systems or sources, measurement error may be exacerbated.
- *Judgment.* When available data fail to match requested data, measurement error may be associated with variation across respondents, businesses, and record systems with regard to strategies used to estimate, approximate, or manipulate available data in an effort to provide a survey response.
- *Communication.* Designs for and instructions on how to enter response data onto questionnaires may be inadequate for different types of data, or unclear, permitting more than one way for reporting the same information, contributing to measurement error.

The final step in the model returns the response process to the organizational level, where *data are released* to the survey organization. As we saw in Chapter 2, businesses consider survey response to be an interaction with their external environment, and thus are concerned about consistency, confidentiality, and security. Variations in criteria for releasing data, and their impact on reported data, may contribute to measurement error.

7.1.2 Constraints and Limitations to Questionnaire Development, Testing, and Evaluation in the Business Survey Context

The complexity of the response process in business surveys, the nature of business surveys, and the context of the business setting interact in such a way as to have consequences for questionnaire design and for methods used to develop, test, and evaluate survey data collection instruments. Willimack et al. (2004) identified the following major interrelated themes with implications for how questionnaire development, testing, and evaluation methods are applied in business surveys:

- The *nature of the requested data*—measurement of technical concepts with precise definitions—has consequences for many aspects of data collection, including survey personnel involved in questionnaire design, choice of data collection mode(s), design of questionnaires, and the reporting ability of business respondents. Self-administered data collection modes dominate, and thus the survey questionnaire bears a great burden in mediating between respondent and response. Detailed instructions are provided to equalize varying degrees of respondent knowledge and motivation, and to convey needed information in a consistent manner to all respondents.
- *Survey response is labor-intensive* and represents a tangible, but nonproductive, cost to the business. Moreover, the labor-intensive nature of survey response may be exacerbated by the need for multiple respondents and data sources. As a result, completion of a long or complex questionnaire may take hours, days, or even weeks, as respondents or response coordinators spend time identifying, contacting, requesting, and waiting for data from other respondents, data providers, and data sources throughout the company.
- Extensive *respondent burden*, because of statistical requirements related to target population characteristics and data use, represents a serious constraint to the survey organization's implementation of questionnaire testing, experimentation, and evaluation methods. Moreover, involving business respondents in the testing and evaluation of questionnaires adds burden on top of the burden they already encounter in completing actual survey requests or obligations.

These factors combine with the complexity of the response process, the nature of surveys of businesses and organizations, and attributes of the business setting to challenge the application of traditional survey research methods to the pretesting of economic survey data collection instruments. First, to ensure that collected data adequately reflect technical definitions requires that researchers conduct testing with appropriate respondents knowledgeable of records specific to the topic of interest. The labor-intensiveness of the response process means that it cannot be easily observed in real time by researchers, while a paradox results in that, in order to achieve the goal of reducing response burden through improved questions, conducting questionnaire testing research adds to respondent burden.

Summarized in Table 7.1, these issues implicitly or explicitly affect many of the business survey testing methods currently in use. We now discuss these methods,

Table 7.1 Summary of Business Survey Context and Questionnaire Development, Testing, and Evaluation

Issues Associated with Business Survey Needs		Consequences for: Business	Consequences for: Questionnaire Development, Testing, and Evaluation
Technical nature of requested data	Requires precise definitions and well-specified measurements Prevalence of self-administered data collection modes to encourage respondents' use of records Since self-administered survey questionnaires bear a great burden in mediating between respondent and response, detailed instructions are provided to equalize varying degrees of respondent knowledge and motivation, and to convey needed information consistently to all respondents	Data in records are determined by business—not statistical—needs Mismatch between requested data and data in business records Knowledge and data may be distributed throughout a business, and different types of data may reside in different systems	Heavy involvement by stakeholders and subject matter specialists in instrument design and development Need for and difficulty of finding the respondent most knowledgeable of the requested data Need for recordkeeping studies associated with specific items
Labor-intensive survey response process	Surveys collect data that may or may not be easily extracted from business records Data requirements may or may not align with data available in business records Data requirements may involve calculations or adjustments	Data availability is a function of record formation, respondent access to data, and amount and type of effort required to retrieve data Survey response process may be exacerbated by the need for multiple respondents and data sources Completion of a long or complex questionnaire may take hours, days, or even weeks of elapsed time	Researchers may not be able to observe or replicate businesses' survey response processes
Response burden	Skewed target population and industry concentration require inclusion of largest businesses in multiple surveys	Heavy response burden Survey response represents a tangible, but nonproductive, cost to the business Reporting routines are developed to ease burden	Involving businesses in activities associated with questionnaire development, testing, and evaluation adds burden

highlighting adaptations and modifications that improve their effectiveness in the business setting.

7.2 DEVELOPING CONTENT, MEASUREMENTS, AND SURVEY QUESTIONS[1]

Many business surveys conducted by or for NSIs provide data that support systems of national income and product accounts and the estimation of gross domestic product (GDP). Others supply data for current economic indicators. For the most part, summary statistics based on data collected in business surveys are descriptive in nature, while survey microdata may be analyzed by economists to understand or determine the mechanisms that drive the economy (e.g., relationships among different variables). Regardless, data needs start with some kind of concept or underlying construct associated with theories or hypotheses requiring specific measurements. This is where survey questions begin.

Within a theory, underlying concepts or tenets are described in a network of relationships among attributes, which form the building blocks of concepts. An attribute or indicator may be regarded as the smallest piece of measurable information that can be identified. A specific algorithm associates one or more attributes as inputs to define a concept. More specifically, a concept C_1, may be expressed as a function of attributes ranging from 1 to n ($A_1, \ldots, A_n$), where the functional definition denotes the mapping:

$$C_1 = f(A_1, A_2, \ldots, A_n).$$

Concepts can be simple, consisting of only one attribute, and thus one-dimensional. In economic research, however, most concepts are multidimensional, indicating that a concept is composed of more than a single attribute.

While underlying concepts and their attributes may seem basic and uncomplicated, their measurement may not be at all straightforward. Consider for example, the concept of employment, one of the most fundamental constructs in describing the economy. Although seemingly straightforward, "employment" has a variety of dimensions: Is the researcher interested in a straight "head count" or the number of full-time equivalents (FTEs)? What timeframe or reference period should the respondent be considering—on a particular day, during a particular pay period, at the end of the month, or some sort of year-end average? Should part-time workers be included? What about temporary employees, leased employees, or contractors? Should employees on paid or unpaid leave be counted? These and other various attributes of "employment" need to be specified, depending on the research goals. Taking these attributes together, and following a specific algorithm, defines the concept "employment."

[1] Much of this section is extracted from a paper by Ger Snijkers and Diane K. Willimack, entitled "The missing link: From concepts to questions in economic surveys," presented at the European Establishment Survey Workshop, Neuchatel, Switzerland, September 2011.

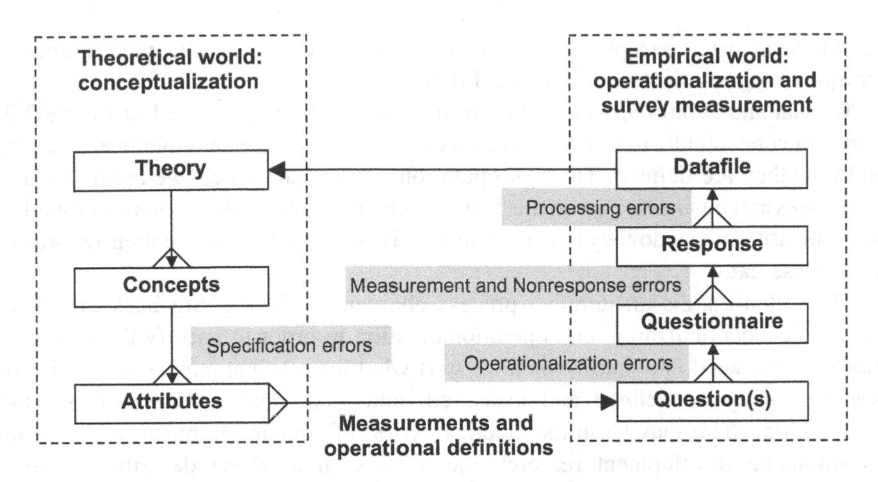

Figure 7.2 From theory to data. [*Source:* Snijkers and Willimack (2011).]

Specifying attributes is the first step toward linking concepts to the process of collecting data. In the case of a survey, questions in a questionnaire are developed to measure these attributes. Questions are the operationalization of these attributes; they are the operational definitions (Segers 1977; De Groot 1994), according to which the attributes are measured. This includes the wording of a question, response options, and instructions. A question can be based on a single attribute, resulting in a one-dimensional question; it can also be based on a combination of attributes, resulting in a multidimensional question.

The steps in the process of going from theory to data are shown in Figure 7.2. Going from concepts to attributes is what we call *conceptualization* (or *concept specification*); going from attributes to questions and a questionnaire is called *operationalization*. If the conceptualization and operationalization are done properly, we obtain valid data with respect to construct validity (Groves et al., 2004) reflecting underlying economic concepts. Poor conceptualization leads to specification errors (Biemer and Cantor 2007); likewise, poor operationalization leads to operationalization or design errors (Snijkers 2002).

To obtain valid data, construct validity is an important and necessary criterion. However, a questionnaire that is valid in the sense that it measures the underlying concepts (i.e., construct validity) may not be valid in the sense that the operationalization may lead to other kinds of nonsampling error (Snijkers 2002; Lessler and Kalsbeek 1992; Groves 1989), such as item nonresponse or measurement errors, also shown in Figure 7.2. Rounding out the error profile are processing errors that occur as responses are coded and captured to form a datafile (Groves et al. 2004).

In the end, the goal of questionnaire development is to write survey questions and design survey questionnaires that obtain valid data. Survey questions operationalize measurements associated with attributes that are decomposed from underlying concepts. To put it more simply, to get from a concept to measurement requires specification. Since our focus is on business surveys, the desired data reflect economic theory, concepts, and attributes, which, to ensure valid measurements, require clear specification of technical definitions.

We can and should approach this from both directions portrayed in Figure 7.2. Through conceptualization exercises, we determine what measurements are needed and how they are defined. Through operationalization activities, we ascertain how businesses and business respondents view a concept and how they organize around it. What measurements do they take or monitor? How do they structure their records to track these data?

This often entails an iterative process characterized by going back and forth between conceptualization and operationalization in order to specify the measurements. Often we do not know when we start what the relevant attributes may be, or how they may be defined and measured, and what nuances, alternatives, and complexities there may be in the practical world of conducting business. By using questionnaire development research methods, such as those described in Section 7.2.2, we work toward survey questions that meet, to the degree possible, their underlying intent.

But first, let us consider the roles of the various players in this process.

7.2.1 The Roles of Subject Area Specialists, Stakeholders, and Data Users

Because of stringent data requirements (e.g., the national accounts) and the technical nature of the data, stakeholders and data users play a substantial role in business questionnaire development, progressing well beyond simply providing conceptual underpinnings and survey requirements. Stakeholders may even draft survey questions themselves or collaborate closely with questionnaire design experts (Goldenberg et al. 2002a; Stettler and Willimack 2001). In addition, international statistical coordinating organizations, such as Eurostat, develop and administer regulations and set requirements for economic measurements that member NSIs must follow. As a result, business surveys rarely start with a "clean slate."

Traditionally, input is also sought from a vast array of data users, including subject area experts, academic researchers, advocacy group members, and industry experts, such as representatives from trade or professional associations. Data users not only suggest topics for survey content but also review the appropriateness of proposed measures relative to survey concepts (Andrusiak 1993; Fisher et al. 2001a; Ramirez 2002; Underwood et al. 2000). Trade association representatives may also be target respondents, and thus represent both perspectives during questionnaire development when they suggest terminology for survey questions and offer advice on the availability of requested data.

7.2.2 Methods for Content Development and Specifying Measurements

A number of methods can be used to assess data needs, investigate concepts, ascertain attributes, identify measurements, and specify questions to achieve construct validity at the design stage. Hox (1997) distinguishes between top–down, theory-driven and bottom–up, data-driven approaches. The theory-driven approach represented by the left side of Figure 7.2, starts with constructs and works toward observable variables; the data-driven approach encompasses the right side, with operationalizations and observations, and works toward theoretical concepts. We will describe how various approaches fit into this dichotomy of methods.

7.2.2.1 Top–Down Theory-Driven Approaches

Dimension/Attribute Analysis Hox (1997) calls this method *dimension/indicator analysis*, where an indicator is similar to what we've called an attribute. According to Hox, many researchers view dimension/indicator analysis as the ideal approach to bridge the gap between theory and measurement. This analysis basically follows Figure 7.2. Empirical attributes (or indicators, according to Hox) are specified for the concepts in a theory. The process of concept specification is driven by an existing theory, logical reasoning, or may also be based on results from previous research. The result is a network of concepts that are logically tied together, and collectively make up the theory. Next, appropriate empirical attributes are defined for the concepts. The process of concept specification ends when one or more attributes can be identified for all concepts. The attributes are the basis for survey questions.

Data User Needs Assessment The literature indicates the prevalent role of consultations with stakeholders, sponsors, data users, industry experts, and internal survey personnel to support questionnaire development (Goddard 1993; Freedman and Mitchell 1993; Mueller and Phillips 2000). A number of techniques involve data users, researchers, and stakeholders, in concert with survey methodologists, working to develop survey content by specifying and clarifying data needs. Several effective consultative methods described by Willimack et al. (2004) include

- *Expert/User/Advisory Groups.* These groups, consisting of representatives from the primary data users for a particular survey, along with senior survey managers, are consulted regularly during the questionnaire development process (Francoz 2002). Members prioritize new questions and changes to existing questions, and are responsible for final approval of the questionnaire. Alternatively, panels of industry experts may be convened to aid researchers in drilling down from concepts to attributes to common definitions and metrics (Mulrow et al. 2007a).
- *Workshops with Data Users.* Workshops may be conducted with data users to (1) identify how and why the data of interest are to be used, (2) identify gaps in existing data, (3) gain insight into data needs for specific issues, and (4) create a preliminary set of data priorities. For example, the US Census Bureau routinely conducts a number of data user workshops following release of quinquennial economic census data, and brings back recommendations regarding data needs. The US National Science Foundation also convened workshops with different types of data users, grouped researchers with similar interests or goals, and administered exercises to tease out data needs and priorities for the new Business R&D and Innovation Survey (Mulrow et al. 2007a).

- *Iterative Rounds of Consultations.* Data users generate a list of requested data items that are subsequently ranked by noted experts in the field. Major stakeholders then prioritize the remaining items. This is followed by contact with trade association representatives focusing on data availability, while survey staff identify survey questions that previously obtained high-quality data. The result is a questionnaire that balances needs for new data with obtainable data.
- *Large-Scale Content Review.* Review of draft questionnaires is sought by mailing sample forms to stakeholders, data users, trade associations, and key respondents. The US Census Bureau automated this activity via the Internet for its quinquennial economic census.
- *Exploratory Focus Groups with Survey Staff or Data Users.* Using focus group techniques to conduct meetings with survey personnel or data users helps to clarify concepts to be measured, output tables, and data publication goals, thus aiding in the specification of questions, definitions, and terminology (Oppeneer and Luppes 1998; Phipps et al. 1995). Focus groups with survey staff throughout the organization conducted for the purpose of content validation help to ensure that survey questions meet the scope of the underlying concepts. Focus groups with data users have also been used to assess their needs and to prioritize data requests (Andrusiak 1993; Carlson et al. 1993; Fisher et al. 2001b; Kydoniefs and Stinson 1999).

These and other similar methods aid development of early drafts of questions and questionnaires for additional examination using data-driven approaches involving respondents.

7.2.2.2 Bottom–Up Data-Driven Approaches

The following methods involve data provided by respondents, or information gleaned from their perspectives, which are typically qualitative in nature. These methods reflect varying degrees of exploration as concepts and their attributes gain specification. Each method, if used in sequence, garners increasing specificity regarding the desired data, and feeds into the next method. Alternatively, as the lines between methods blur, some may be blended together to gain efficiency, depending on survey goals and underlying concepts. Used iteratively with top–down methods, these approaches lead to the development of well-formed survey questions ripe for the next steps of pretesting described in Section 7.3.

Questionnaire Mining In *questionnaire mining* the researcher starts with existing questionnaires, from which questions are selected that seem to be relevant in the context of the study. The questions are reviewed as to how they are related to the central concept(s), namely, a metadata analysis. The review is done in close collaboration with content matter experts, since this requires subject matter knowledge, resulting in an expert appraisal of questions extracted from other surveys. Reports of pretest studies for these questions, analysis of paradata and item nonresponse, examination of the types and frequency of edit failures associated with these questions, or the data analysis itself, can be used to study the validity and measurement problems of these questions. The result is a selection of questions that may be used in subsequent data collection instruments, in the same or modified operationalizations. Since this approach starts with existing operationalizations, researchers must take care to not miss important aspects of the research domain, such as attributes that are not covered in questionnaires reviewed during mining.

Exploratory Early Stage Scoping Interviews/Focus Groups If it is at all uncertain how concepts can be measured, indepth exploratory interviews can be conducted with a small number of representatives of the target population. This may be accomplished through one-on-one, early stage scoping interviews (Stettler and Featherston 2010) or by using exploratory focus groups (Snijkers 2002; Gower 1994; Kydoniefs 1993; Palmisano 1988; Phipps et al. 1995). A primary goal is to understand the concepts from the respondent's point of view. Participants are asked how they interpret the concepts, how they define and structure attributes of the concept relative to their business activity, and what is or is not included. We learn about the terms respondents use in relationship to the concept, so that we can use these terms in the questionnaire to ensure they are interpreted as intended. The availability of the desired data in records and their periodicity are also investigated, along with how well recorded data match desired concepts. Other goals include assessing response burden and quality: Are the requested data easy to collect, or is it a burdensome process? In addition, we learn the identity or positions of various employees within businesses who should be contacted for survey participation or who have access to the desired information. Interview or focus group participants are selected in such a way that a wide variety of views is collected.

Concept Mapping A more structured way of studying the interpretation of concepts is by applying concept mapping in focus groups (Hox 1997). This process involves six steps. In the first step, the subject of the focus group is specified, and participants are selected from the target population. The second step, *statement generation*, is a brainstorming session with the participants, to generate statements that describe many relevant aspects (i.e., attributes) of the concepts under study. Each statement is printed on its own slip of paper or a card, and each participant gets a stack with statements. In step 3, the *structuring* step, the participants individually sort the statements into different stacks according to their own views. Each stack contains related statements. The stacks are combined in a group similarity matrix. This matrix is analyzed in step 4, *statement representation*, by a multidimensional scaling technique. The result is a concept map, showing clusters of the statements. These clusters may be interpreted as the attributes that are related to the concepts being studied. In the next step, *concept map interpretation*, this map is discussed by the focus group. The participants discuss possible meanings and acceptable labels for each statement cluster. The final step, *utilization*, concerns the translation of the statements into survey questions. The concept map gives guidance to structuring the questionnaire into blocks of related questions. Haraldsen (2003) applied this method to operationalize the concept of perceived response burden, aimed at developing a questionnaire to measure this concept.

Feasibility Studies Willimack et al. (2004) describe *feasibility studies*, where a small number of respondents are visited on site and asked about the information that will be requested in a survey collection: Are the data available and easy to retrieve? Survey personnel conduct meetings with business representatives who are involved in the data collection process. A prespecified topic list or agenda is followed to discuss the concepts of interest, the definitions, and the availability of the required data. Information collected during these visits helps determine whether concepts are measurable, what the questions should be, how to structure the questions and questionnaire, and to whom the questionnaire should be sent. This method does not yet require a fully developed questionnaire, but in contrast to the methods described above, it does require specification of the concepts of interest prior to application.

Accounting Expert Reviews Professional accountants aid concept clarification and definition as subject matter experts in business recordkeeping and accounting practices (Willimack and Gibson 2010). Informed by professional standards, regulatory bodies, and laws, professional accountants help to associate concepts with measurements likely to be available in businesses, through their knowledge of accounting databases and interfaces (e.g., human resources, production management) where data elements may be an input to or a product of accounting processes. During questionnaire development, accounting experts may work with survey sponsors, stakeholders, and content specialists, helping to evaluate correspondence between underlying concepts and business terms and constructs, along with the reportability and quality of the desired data. Accounting experts may also assist with defining concepts through comparisons with public information, performing analytical reviews of data elements (comparative ratios, relationships between data items, industry practices), and evaluating expected data sources and compilation methods to determine whether they are likely to produce the desired data. This method results in suggestions for alternative question wording and specification, which may, in turn, be investigated via recordkeeping studies.

Recordkeeping Studies According to Willimack et al. (2004), *recordkeeping studies* are generally associated with individual surveys to determine whether specific data exist in records, how well they match requested items, and how easily or readily they can be retrieved and by whom (Gower 1994; Kydoniefs 1993; Leach 1999; Mulrow et al. 2007b; Ware-Martin 1999; Mills and Palmer 2006). In a recordkeeping study, a small number of cases are selected from the target population and consulted about the data they have available in their business's records. Using a preliminary draft of survey questions, this method studies the availability of the information, how it is structured within the business's records, and what the underlying definitions and business purposes are for the data. Thus, we discover what data may be collectable and what they consist of, enabling us to gauge the degree to which recorded and available data align or fail to align with attributes of the central concept(s). Results are used to assess the collectability of data adequate to measure the desired concepts, and to aid in writing effective survey questions for obtaining these data. Alternatively, comprehensive formal studies of business recordkeeping practices may be conducted as occasional special projects investigating the structures and systems for keeping records, the location of particular types of data, and who has access to which data (US Census Bureau 1989).

7.3 TESTING QUESTIONS AND QUESTIONNAIRES

Testing begins when there is a draft questionnaire that survey methodologists and stakeholders agree has good potential for collecting valid data, or when viable alternative questions may be specified. Thus, it is important that a comprehensive development phase, using methods described in the previous section, has been deliberately and explicitly undertaken. Experience has shown that when pretesting begins with a questionnaire developed out of hand, backtracking frequently occurs to conduct research in order to operationalize concepts. The consequence is either lost time in the questionnaire development and testing stages, or poor quality in the collected data.

We now focus on various methods currently in the pretesting toolkit. First we will consider how knowledge of the response process may guide pretesting plans and protocols. Then, as we describe the various methods, we will highlight modifications that have been made in practice to accommodate the business environment.

7.3.1 The Response Model as a Framework for Testing

The model of the business survey response process presented in Section 7.1.1 is a useful tool for planning and conducting questionnaire testing research. The model is not meant to provide solutions for problems. Instead, like the four cognitive steps in Tourangeau's (1984) response model, the steps of the business survey response model offer a guide or a framework for investigations into activities and behaviors of respondents that may contribute to measurement error. Assessing whether proposed solutions actually work usually requires complementary evaluation methods, like those described in Section 7.4.

Several survey methodologists have described how they have used the steps in the response process to guide the topics covered in their protocols for cognitive interviews or other interactions with respondents for research purposes (Steve and Sharp 2010; Willimack 2008; Giesen 2007). These protocols are often in the form of lists of topics to discuss or observe, and the sequence in which to address them, during site visits with respondents. However, as we shall see in Section 7.3.2, conducting interviews with business respondents at their place of business often requires flexibility in administering the protocol, because, unlike a laboratory setting often used for cognitive interviews with individuals, the environment cannot be controlled by the researcher.

Protocols designed around the response model often begin with global or overview questions to learn about the organizational steps involved in record formation and availability, determining the identities of and selection criteria for the respondents, the criteria for assigning response activities, and other tasks or activities that the respondent is responsible for, and so on. Just as the model says, these questions provide a sketch of the context within which the cognitive steps and associated actions take place.

Protocols related to the cognitive steps among business respondents tend to be similar to those used in pretesting surveys of individuals and households. We will see that business survey pretesting utilizes many of the same cognitive research techniques, including think-aloud verbalizations, probing questions, retrospective debriefings, and observational methods. [See Willis (2005) for descriptions of common cognitive research methods and instructions about their use.]

Giesen (2007) reviewed several pretests of business survey questionnaires to evaluate the effectiveness of using the response model as a framework for evaluating business surveys. The reader is referred to this useful, practical paper, which describes the methods used and the goals achieved by their use. Giesen (2007) discusses in detail the typical structure of protocols used to conduct field pretest interviews with business survey respondents, including examples of probes in each segment of the interview protocol:

1. *Introduction of the visit*, to set the respondent's expectations for the purpose of and the procedures used during the meeting
2. *General questions about the respondent and the response process*, to engage the respondent while also permitting the researchers to become familiar with the business setting, the respondent's role in the company, and other background information to set the stage for more indepth questioning about specific survey questions
3. *Observation or reconstruction of the response process*, to examine cognitive, procedural, and burden issues associated with the survey questions being tested
4. *Evaluation questions*, to obtain additional detail assessing the quality of answers to survey questions, and to invite the respondent's opinion about the survey and suggestions for improvements
5. *Correcting data and answering the respondent's questions*, taking advantage of the face-to-face meeting to provide constructive feedback and instruction to the respondent to improve the quality of future reported data, while also building a long-term relationship between the statistical office and the respondent in support of recurring surveys

In addition, we recommend that pretesting studies designed around the business survey response process also consider the taxonomy proposed by Bavdaz (2010a), classifying response strategies related to the retrieval and judgment steps, and the likely outcome in terms of the degree of measurement error associated with resulting reported data. This model is repeated here in Figure 7.3, along with the following definitions (Bavdaz 2010a, p. 86):

(a) A datum is *accessible*—the required answer may be readily available.
(b) A datum is *generable*—the required answer is not readily available to any person; the available data represent a basis for generating the required answer through manipulation.

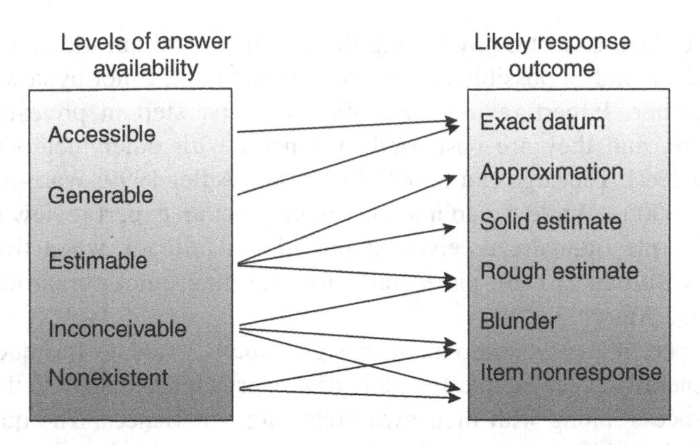

Figure 7.3 Levels of answer availability and likely response outcomes. [*Source:* Bavdaz (2010a, p. 86).]

(c) A datum is *estimable*—the required answer is not readily available to any person; the available data represent an approximation of the required answer or a basis for estimating the required answer through manipulation.

(d) A datum in *inconceivable*—no available data lead to the required answer or its approximation; some bases for generating or estimating the required answer exist but require an unimaginable effort to produce it.

(e) A datum is *nonexistent*—there are no bases for estimating the required answer.

Clearly, any outcome other than "exact datum" contains measurement error. Using the left side of Figure 7.3, the degree of data availability can be gauged for each questionnaire item during pretesting. Then, in summarizing pretest results, classifying likely response outcomes using the taxonomy on the right side of Figure 7.3 will permit empirical assessments of the severity of associated measurement errors and potential impacts on survey estimates resulting from error-prone data. Results based on this classification scheme can also aid priorities for additional research to quantify and evaluate the consequences of item-level measurement error.

7.3.2 Methods

The nature of business surveys challenges the application of traditional cognitive research methods and other generally accepted techniques to pretesting economic survey data collection instruments. We shall now describe various methods for testing questions and questionnaires, along with adaptations for use in business surveys. Much of this material is drawn from the methodology review conducted by Willimack et al. (2004).

Methodological Expert Reviews Because of the burden associated with the business survey response process, questionnaire design experts try to reduce the added burden associated with testing by conducting methodological expert reviews (also called "desk" or "paper" reviews) of the survey instrument, thus either attempting to reduce the burden of testing by having the questions and questionnaires as well-designed and sound as possible before going to the field, if not bypassing respondents altogether. Expert reviews are often the first step in pretesting a draft questionnaire, and they are best used in concert with other methods (DeMaio and Jenkins 1991; Eldridge et al. 2000; Fisher and Adler 1999; Ware-Martin et al. 2000; Jones 2003). While we do not recommend that an expert review be the only evaluation a questionnaire receives, it provides a fallback when time is short (US Census Bureau 2011b), particularly for one-time quick-turnaround surveys (Labillois and March 2000).

In an expert review, questionnaire design experts evaluate the questionnaire, applying generally accepted questionnaire design principles relative to the cognitive response process, along with their own pretesting experiences. The questionnaire expert's task is to identify potential problems that may result in reporting errors or inconsistencies, and to suggest solutions, often in a written report. Questionnaire design specialists may also collaborate with survey personnel to arrive at mutually acceptable solutions to design problems.

Cognitive Appraisals A systematic cognitive appraisal coding system developed for evaluating survey questionnaires (Lessler and Forsyth 1996) was adapted for business survey evaluation (Forsyth et al. 1999; O'Brien 2000; O'Brien et al. 2001). Expert appraisal findings may be helpful in guiding development of a protocol for cognitive interviews (Fisher et al. 2001b).

Exploratory/Feasibility Studies and Site Visits Because of the reliance of business surveys on business or institutional records, investigations into data availability play an integral role in questionnaire development. These typically take the form of exploratory or feasibility studies conducted as indepth interviews with potential data providers during company or site visits (Birch et al. 2001; Burns et al. 1993; Francoz 2002; Freedman and Rutchik 2002; Gallagher and Schwede 1997; Mueller and Phillips 2000; Mueller and Wohlford 2000; Schwede and Ott 1995). Similar to the early stage scoping approach to specifying concepts, primary goals include determining the availability of the desired data in records and their periodicity and ascertaining the degree to which recorded data match desired data. Other goals include identifying the appropriate respondent.

Sample sizes for exploratory/feasibility site visits are usually small and purposive, targeting key data providers. Survey methodologists and subject area specialists conduct meetings with multiple company employees involved in survey reporting. Usually these survey personnel follow a protocol or topic agenda and discuss definitions and the types of data that can be supplied by company reporters. Information gained during these visits helps determine whether the survey concepts are measurable, what the specific questions should be, how to organize or structure the survey questions, and to whom the questionnaire should be sent.

Since site visits are costly and telephone calls, while less costly, still tend to be laborious for staff and bothersome for respondents, self-administered approaches have been attempted (Kydoniefs 1993). An alternative used for the 2001, 2002, 2003, 2004, and 2005 Surveys of Industrial Research and Development (SIRD), jointly sponsored by the US National Science Foundation and the US Census Bureau, consisted of enclosing a supplemental self-administered questionnaire with the SIRD questionnaire in this recurring survey, and asking a small subsample of businesses to assess both the existence of data requested by proposed new questions and the quality of data sources, topics typically covered during company visits.

Respondent Focus Groups Focus group methods may be used with appropriate target populations throughout the questionnaire development, testing, and evaluation processes. We have already described their use with data users, stakeholders, or respondents during the concept development stage. Later we will describe their use during the evaluation stage to help assess the effectiveness of survey questions and questionnaires. Here we discuss their use in testing.

Detailed procedures for conducting focus groups in various settings and for different purposes can be found in a number of texts. We recommend texts by Krueger and Casey (2000) or Morgan (1997). In short, focus groups are guided discussions among a small group of people, targeted around a specific (set of) topic(s), where the data are generated by means of discussion. Focus groups are moderated by a trained facilitator, who uses a protocol, discussion guide, or outline, to lead the discussion among the group members. The role of the moderator is to encourage open discussion while managing the direction of the discussion. The discussion protocol consists of a series of questions fashioned around a *funnel approach*, where the moderator begins the discussion with a broad context for the topic of interest, and then gradually narrows the focus of the discussion through questions that become more specific as the discussion proceeds. Focus groups are usually conducted with 8 – 12 participants selected from the target population. However, makeup of the group may be homogeneous or heterogeneous around predetermined population characteristics, and may need to be adjusted relative to the research goals.

Focus groups may be conducted with respondents during the questionnaire testing phase once a draft questionnaire is available. Content evaluation focus groups examine the appropriateness and clarity of question wording, use and usefulness of instructions, general reactions to the questionnaire and its design, and compatibility of requested data with recorded data (Babyak et al. 2000; Carlson et al. 1993; Eldridge et al. 2000; Gower 1994; Snijkers and Luppes 2000). However, because businesses tend to be reluctant to reveal information to competitors, it may be difficult to recruit business respondents for focus groups (Bureau 1991; Freedman and Rutchik 2002; Gower and Nargundkar 1991). To remedy this, focus groups with business respondents may be conducted via teleconference or online chat, thereby maintaining the anonymity of participants.

Focus groups may also be conducted with respondents to test supplemental materials, such as cover letters or instructions placed in separate booklets.

Cognitive Pretesting Methods The use of formal cognitive testing for business surveys has become more prevalent since the early 1990s, particularly during the design of new surveys or when major redesigns are undertaken. Traditional cognitive methods are used in pretesting business surveys to explore the four cognitive steps in the business survey response process: comprehension, retrieval, judgment, and communication. These methods include indepth interviews, concurrent or retrospective think-aloud and/or probing techniques, retrospective debriefings, and vignettes. [See Willis (2005) for a comprehensive volume about using cognitive interviews for survey pretesting.]

Traditional cognitive probes are often augmented to explore estimation processes (judgment), request observation or descriptions of records and their contents (retrieval), or examine use of complicated structures for entering data, such as grids, often found on business survey questionnaires (communication). Because the business survey response process is more complex, cognitive interviews often take on aspects of exploratory studies as well, resulting in a cognitive hybrid that explores data availability and respondent roles along with cognitive response processes (Freedman and Rutchik 2002; Goldenberg and Stewart 1999; Sykes 1997).

Traditional cognitive methods have also been adapted to minimize additional work by respondents. Since it is often unrealistic for researchers to directly observe this labor-intensive process, probes may be phrased in a hypothetical manner to explore steps involved in retrieving data from records, such as: How *would* you go about retrieving this data? Which records *would* you consult? These probes identify data sources, discover respondents' knowledge of and access to records, recreate steps likely taken to retrieve data from records or to request information from colleagues, and suggest possible estimation strategies. (DeMaio and Jenkins 1991; Gerber and DeMaio 1999; Stettler et al. 2001).

Researchers often require flexibility and take greater latitude than may be allowed in traditional cognitive interviews that use standardized protocols, in order to react to the variety of uncontrolled and uncontrollable situations outside a cognitive laboratory setting. Researchers may apply a funnel approach like that used in focus groups, where probes become successively more directive, to help focus the business respondent's thinking.

Traditional cognitive methods, coupled with these various modifications, enable collection of more detailed information about the business survey response process (Davidsson 2002; Dippo et al. 1995; Goldenberg et al. 2002a; Laffey 2002; Rutchik and Freedman 2002). However, observation of respondent behaviors remains somewhat contrived, since the response task is rarely performed in full view of the researchers (Birch et al. 1998; Schwede and Moyer 1996). This adds to the limitations typically associated with cognitive interviews, such as the artificiality of think-aloud and probing techniques that may interrupt, rather than reveal, respondents' true cognitive processes.

Vignettes Hypothetical scenarios, known as *vignettes*, may be used during pretesting with business respondents (Morrison et al. 2004). These applications allow researchers to evaluate response errors, since correct answers associated with the vignettes are known. In much the same way as they are used in household studies, vignettes may be designed to evaluate judgment processes regarding unclear, multidimensional concepts (Stettler et al. 2001).

Another application of vignettes consists of the formulation of mock records containing simulated data, which allows researchers to observe respondents' retrieval, judgment, and communication strategies when retrieval from their own business records is too time-consuming to be completed either in advance of or during the appointed interview (Anderson et al. 2001; Goldenberg, Willimack et al. 2002b; Kydoniefs 1993; Moy and Stinson 1999; Schechter et al. 1999; Stettler et al. 2000).

Ethnographic Methods[2] The core of ethnography is the application of multiple methods to (1) achieve an indepth and holistic understanding of the subject of study, (2) corroborate and verify findings across the various methods, and (3) evaluate working hypotheses in a discursive fashion and revise them as necessary. Also central to the ethnographic approach is the location of the researcher in the subject's environment, in order to understand as fully as possible the perspective of the subjects, the resources available to them, the constraints under which they live and work, the demands and obligations imposed by other people in their social network and/or organization, and so on.

Ethnographic methods commonly used in pretesting business surveys typically consist of *indepth unstructured interviews* and, to a lesser degree, *observation*. These interviews generally take place at business respondents' worksites, which situates the researcher in the natural environment of the respondent. It allows discussion and, at times, observation of the roles played by factors external to the respondent, namely, information sources—human, electronic, or otherwise—and the advantages and/or challenges they may pose for survey response, as well as the steps involved in completing the survey.

Ethnographic research associated with organizations, such as businesses, employs two general levels of analysis, which are highly relevant to survey research:

- *Meanings/knowledge*—respondents' interpretations of survey questions; their knowledge of company attributes/activities relevant to survey; and their knowledge of the existence, structure, content, and/or location of information sources.
- *Process*—respondents' interactions with survey instruments, information systems, and formal and informal organizational networks; "knowledge in action"

The ethnographic approach incorporates multiple perspectives. In the case of business surveys, the involvement of respondents from different companies is necessary to identify common terminology that may be used to articulate survey concepts in ways that are reliable and valid across the survey population. This approach also allows researchers to discover and grasp the variability of systems of knowledge and organizational structures. This aids researchers in developing

[2] This section was written and contributed by Alfred D. Tuttle of the US Census Bureau.

instructions to help respondents achieve an accurate understanding of survey concepts that may diverge from their own ways of thinking about and accounting for their company's business activities.

Survey researchers may be able to speak with multiple personnel within each individual company, including the respondents and their supervisors and higher-level managers. In these instances, the researchers observe discussions between company personnel to better understand points of confusion in the survey instrument and hear explanations and guidance provided by more knowledgeable company personnel.

Indepth interviews facilitate discussion of respondents' interpretations of survey questions and how they answer them. Emphasis is on respondents' knowledge of structure and content of recordkeeping systems and other sources of information, individuals in the company with access to desired information not directly obtainable by the respondent, and business activities relevant to the survey.

Observation allows the researcher to collect information on

- The appearance and structure of a respondent's interface with information systems, other information sources, working papers, and other supporting documentation
- Processes involved in how respondents interact with their information systems as well as with survey instruments
- The relative proximity of colleagues and managers assisting the survey process
- Attitudes of respondents and their colleagues toward the researchers and the survey request (as an indicator of attitude toward the request and the work it entails)

The observable aspects of the respondents' context reflect respondents' internalized systems of knowledge. Therefore the observation of respondents' work environment serves to both enrich and corroborate information gained from indepth interviews.

Other ethnographic techniques useful in business survey pretesting include

- *Analysis of Artifacts.* Analysis of artifacts external to respondents' cognitive processes complements the use of in-depth interviews and can allow substantiation of respondents' self-reports. It also allows researchers to better understand the structure of data sources, company jargon, and so on. In the context of business surveys, such artifacts include:
  - Internal data products and records
  - Respondent-generated spreadsheets and other working papers used to support survey response
  - Actual survey responses, in postsurvey respondent debriefings
- *Review of Other Information Sources.* These include business news sources, company websites, publicly available financial statements, previous survey responses, working papers, company records and other internal data products, and historical documents. Such information sources allow survey researchers to

corroborate and interpret data gained from other methods, and are also useful in preparing for interviews with respondents.

- *Card Sort.* In this method, units of information are presented on individual cards and respondents are asked to sort them in various ways. This method enables researchers to understand the structure, content, and other attributes of a particular domain of knowledge. In the context of survey research, this method also allows researchers to obtain respondents' interpretation and valuation of various pieces of information, such as how to present a series of survey items in the order that makes the most sense to respondents or in crafting messages about a survey that are informative and persuasive.
- *Survey Network Analysis.* In this method, respondents are asked to build a roster of the company personnel involved in survey response, their roles in survey processes, and other attributes of interest. This method may be useful in understanding the potential for lower data quality, assessing the aggregate burden associated with a survey request, and developing typologies of companies relative to survey response for the purpose of expanding and refining our current understanding of organizational survey response.

Methods for Testing Instructions Although business surveys often rely heavily on instructions, their testing and evaluation tends to be incidental to questionnaire testing, rather than the focus. Respondents' use of instructions, where they are placed in a questionnaire and how they are formatted, may be studied in cognitive interviews or during pilot testing (Schwede and Gallagher 1996; Zukerberg and Lee 1997). One method for testing the comprehension of instructions is the paraphrasing technique from cognitive interviewing, where respondents are asked to tell the researcher how they would instruct a colleague or new employee to complete the survey question(s) (Gower 1994).

Respondent debriefings may be conducted to further reveal the degree to which instructions were used during survey response. For example, this method has been applied to evaluate redesigned questionnaires that directly incorporated instructions that previously appeared in a separate booklet (Ware-Martin et al. 2000). Tuttle et al. (2010a) added supplemental questions to a mailed self-administered questionnaire in a pilot test, asking respondents for their impressions of the redesigned version where instructions were incorporated into the form, as compared to the legacy version, where the instructions were in a separate booklet.

Pilot Studies Formal pilot studies permit evaluation of data actually reported by respondents on questionnaires, and are particularly important for new surveys or major redesigns. Descriptions of pilot studies can be found in the business survey research literature, although it should be noted that the terms *pilot* and *pretest* are often used interchangeably (Birch et al. 1999; Cox et al. 1989; DeMaio and Jenkins 1991; Fisher and Adler 1999; Fisher et al. 2001a; Goldenberg and Phillips 2000; Kydoniefs 1993; Mueller and Phillips 2000; Mueller and Wohlford 2000; Phipps 1990; Phipps et al. 1995; Schwede and Ellis 1997; Ware-Martin et al. 2000). While a *pilot* may be designed as a *dress rehearsal* to test all of the steps and systems in the data collection process (Kydoniefs 1993; Mueller and Wohlford 2000; Phipps et al. 1995), a *pretest* has been defined to be more limited in scope and questionnaire-oriented (Gower 1994).

We shall use the term *pilot* to refer to any such field test with respondents where the data collection is intended to mimic the realistic setting of the response process, without interviewers or researchers intervening in real time. Another typical characteristic of a pilot study, regardless of sample size or scope, is that the collected data are not used for official estimates, although statistical estimates may be made using the collected data in order to evaluate their quality.

Sample sizes and designs for pilot studies vary widely (Goldenberg et al. 2002a). Small-scale pilots may draw on convenience or purposive samples. Alternatively, pilot studies may be designed and conducted more formally, with sufficiently large samples to generate statistical estimates of some variables (Fisher et al. 2001a; Goldenberg et al. 2000; Kydoniefs 1993; Mueller and Wohlford 2000; Schwede and Ellis 1997). For example, the US Census Bureau conducted a pilot test of the Survey of Business Owners in 2001, using a stratified simple random sample of nearly 10,000 businesses, where the underlying target population was limited to businesses with characteristics pertinent to the redesigned questions.

Another example of a pilot test was carried out by the UK Office for National Statistics (ONS) during redevelopment of the Annual Survey of Hours and Earnings (ASHE) questionnaire. It included a sample of 480 businesses, representing 5000 employees. Because of concern that the questionnaire had increased in length, the sample included many businesses that would receive large numbers of questionnaires. (The ASHE sample is based on employees with reference to the last two digits of an employee's National Insurance Number; thus the same business can receive a number of questionnaires for individual employees). The key evaluation criteria for the pilot were response rates, compliance times, validation errors, the number of respondent queries and complaints, respondent feedback, and ONS costs (Jones et al. 2007a).

Pilot studies may be iterative or involve multiple panels targeting businesses of different sizes and industries. Kydoniefs (1993) and Mueller and Phillips (2000) provide examples of this approach, where the goals were to evaluate respondents' understanding of the questions and definitions and to test the feasibility of the data collection mode, while also investigating data sources, timing relative to survey deadlines, and the incidence of estimation. Pilot studies often take advantage of other evaluation methods, such as interviewer and/or respondent debriefings, behavior coding, or designed experiments.

7.4 EVALUATING SURVEY QUESTIONS AND QUESTIONNAIRES

7.4.1 Methods Used during Data Collection

Behavior Coding Behavior coding (also known as *interaction analysis*) focuses on observable behaviors of interviewers and respondents as they interact during the administration of a survey questionnaire. Interviews conducted in person or by telephone may be observed live, or are recorded and reviewed later by survey personnel trained to identify and classify interactional interviewer or respondent behaviors using a prespecified coding scheme. Interviewer behaviors may include descriptions of the degree to which interviewers read or fail to read a survey question as worded. Respondent behaviors include such actions as interrupting, requesting clarification, providing an adequate or inadequate answer, replying "I don't know," or refusing. Results from behavior coding can aid survey researchers in identifying problematic questions. [See Groves et al. (2004) and Willis (2005) for examples of coding schemes and how results are analyzed and interpreted.]

Behavior coding of interview interactions has been rare in surveys of businesses and organizations (Goldenberg et al. 1997), mainly because interviewer-administered surveys are rare in this target population. Interviewers for business surveys tend to be allowed greater flexibility, due to the technical nature of the requested data, so coding schemes must be modified accordingly (O'Brien 1997). In addition, such research activities are more likely to flag interviewer training issues rather than needed questionnaire modifications.

An analogous procedure for self-administered questionnaires involves the examination of completed survey forms, looking for any patterns in notes or seemingly extraneous marks relative to specific questions or sets of questions. These may indicate problematic questions or confusing visual design. This method is described further in Section 7.4.3.

Split-Sample Experiments Split-sample experiments (also known as *split panels* or *split ballots*) allow empirical evaluation of alternative questions, questionnaires, or designs. They have also been used to evaluate the positioning of instructions. In a split sample experiment, sample cases are randomly assigned to one of two or more subsample groups where some attribute of the design is being varied. These are called *treatments*. Usually one of the groups maintains current survey design/conditions; this is called the *control group*. When attributes are varied in an appropriate experimental design, then the effect or impact of the variations on the statistic of interest is evaluated using statistical comparisons. [See Kirk (2013), Keppel and Wickens (2013), or one of the numerous other texts on experimental design for details of how to construct, conduct, and analyze designed experiments.]

Split-sample experiments may be embedded into pilot studies for question evaluation (DeMaio and Jenkins 1991; Goldenberg et al. 2000; Phipps 1990; Schwede and Ellis 1997). Burt and Schappert (2002) embedded a split-sample experiment within operational data collection to empirically evaluate questions considered marginal after cognitive testing and respondent debriefings. This methodology has been used to evaluate mode selection (Dodds 2001; Manfreda et al. 2001) and to assess data quality and response burden among different data collection modes (Sweet and Ramos 1995; Vehovar and Manfreda 2000). Embedding split-sample experiments or other research methodologies into pilot studies or production data collection overtly minimizes the potential additional burden of involving business survey respondents in questionnaire design research.

7.4.2 Methods Used after Data Collection

Postcollection Empirical Evaluations Evaluation activities are routine postcollection processes typically associated with recurring business surveys. They are distinct from development and testing methodologies, which tend to blend together. Tabulating item nonresponse and imputation rates, along with examination of outliers and edit-failing records, play a significant role in quality evaluations (Birch et al. 1999; Monsour 1985; Monsour and Wolter 1989; Schwede and Ellis 1997; US Census Bureau 1987; Wolter and Monsour 1986). Results of these routine quantitative analyses should be systematically fed back into questionnaire design. For instance, high item nonresponse or imputation rates may lead to removing questions or redefining response categories.

The quality of collected data may also be evaluated through external consistency checks involving comparisons or reconciliation with data from other sources (Biemer and Fecso 1995; Erikson 2002). Professional accountants may review data reported by specific businesses, comparing them to publicly available information or evaluating them against accounting standards (Willimack and Gibson 2010). If significant discrepancies occur among the same survey questions across respondents, further investigation may be warranted to identify the source of the reporting errors relative to the survey questions (Mueller and Phillips 2000).

Respondent Debriefings and Response Analysis Surveys Respondent debriefings may be used to evaluate the quality of collected data. In *respondent debriefings*, survey respondents are recontacted and asked about the response strategies and data sources actually used to provide answers to specific questions. Thus, the degree to which the collected data meet the intent of the survey question(s) can be assessed against actual respondent behavior.

As in pretesting, there is a continuum from "informal" activities to "formal" evaluations. Formal respondent debriefings, also known as *response analysis surveys* (RASs), are conducted after data collection using structured questionnaires. The extent to which reported data meet the definitions is evaluated empirically, along with respondents' use of records and estimation strategies. Although commonly associated with pilots of new or redesigned surveys, RASs may also be used to evaluate data quality in recurring surveys. (Goldenberg 1994; Goldenberg and Phillips 2000; Goldenberg et al. 1993; Palmisano 1988; Phipps 1990; Phipps et al. 1995).

Sample sizes for respondent debriefings vary from as small as 20 in qualitative studies, to as large as several hundred in structured RASs. Sample designs vary depending on research goals. They may be randomized, or selection may be purposive to focus on respondents who made reporting errors (Phipps 1990) or to compare early and late respondents (Ware-Martin et al. 2000). For example, respondent debriefings may target so-called "good" (i.e., timely) reporters, using retrospective focused interviewing techniques to identify reporting errors within the context of the survey response process model from Section 7.1.1 (Hak and van Sebille 2002).

In-person interviews are the preferred mode for conducting respondent debriefings, while telephone represents a viable mode for RAS structured interviews. However, self-administered respondent debriefings may be conducted when testing new or redesigned self-administered survey questionnaires (DeMaio and Jenkins 1991). Questions are added to the main survey asking respondents to evaluate the questionnaire in general, and to note problem questions, difficult terminology, and burdensome record retrieval (Jones et al. 2007a). In addition, cognitive-style probes can be modified for self-administration in order to provide indepth evaluation of selected questions (Davis et al. 1995).

Reinterview Studies and Content Evaluations Another variation of respondent debriefing methodologies consists of reinterview studies that may or may not include reconciliation with previously reported figures (Biemer and Fecso 1995; Paben 1998). The primary goal of *reinterview studies* is to estimate bias in summary statistics due to reporting errors; hence, sample sizes and data collection methodologies must be sufficient for this purpose. *Content evaluations* are reinterview studies that examine the components of reported figures and their sources and reliability. Sources of reporting errors are not only identified, as in a RAS, but are also quantified to provide empirical evaluation of data quality (Corby 1984, 1987). However, recommendations for revised question wording and instructions are not considered the primary goals of such studies, although they are a seemingly natural extension of such research. This was emphasized by Van Nest (1987), who piloted a content evaluation study, only to find so much variation in response strategies that a full study was deemed infeasible. Instead, results supported numerous recommendations for changes to data collection questionnaires and instructions.

Validation/Record-Check Studies Measurement error may be directly assessed if there exists a known "true value" that can be compared to respondents' answers to survey questions. Sources for such figures may be individual organizational records or administrative sources. These *record-check studies* may be used to validate reported data (Groves et al. 2004; Nicholls et al. 1997). Since data reported in business surveys are supposedly retrieved from business records, record-check studies would seem to be a common methodology for investigating measurement error. However, we have seen many reasons why requested data may not reside in or be easily retrieved from business records (see Chapter 2). Because of labor-intensive retrieval processes and the proprietary nature of business data, record-check studies are rare in business surveys. Cooperation with selected businesses would need to be cultivated, and confidentiality ensured. Thus, the use of record-check studies to validate reported survey data tends to be reserved for highly sensitive statistics with potentially great policy implications.

7.4.3 Methods for Identifying Questionnaire Problems in Recurring Surveys

Many business surveys, particularly those conducted by NSIs, are repeated with regular periodicity to generate time series (e.g., monthly, quarterly, annually) offering the opportunity for postcollection evaluations of past responses to help identify problematic questions or instrument designs. Any of the methods described in the previous section are appropriate and useful to achieve this purpose. Yet, as described above, the primary goal of postsurvey evaluations of recurring surveys tends to be to identify and correct errors. Results from RASs, reinterviews, content evaluations, or other respondent debriefings conducted within operational data collections are used primarily to evaluate data quality and to suggest limitations in interpreting survey data. Historically, results of these more rigorous postcollection evaluations were used in varying degrees, or may not have been used at all, in revising questions and instructions to prevent errors in the first place. Instead, feedback from staff interactions with respondents, although less formal, has tended to result in revisions to survey questionnaires. We describe a few methods for obtaining such feedback in a structured or semistructured manner, along with other methods that are useful for evaluating questionnaires for recurring surveys.

Feedback from Survey Personnel Evaluation of questionnaires for recurring business surveys often relies on the experiences of survey personnel who have routine contact with respondents. Survey personnel typically convey respondent feedback from the previous collection of a recurring survey early in the questionnaire development cycle for a subsequent iteration. Although these personnel mediate response, they act as proxies for respondents when evaluating the performance of questionnaires. Despite their lack of rigor, such informal evaluations are useful for suggesting areas requiring further indepth investigation using more systematic research methods. Such feedback methods include

- *Interviewer Debriefings.* Borrowed from interviewer-administered household surveys, interviewer debriefings can be conducted for interviewer-administered business surveys, to identify problem questions (Goldenberg et al. 1997; Oppeneer and Luppes 1998; Paben 1998). Other debriefing methods include focus groups, meetings, or conference calls with supervisory interviewers. Having professional survey personnel observe field interviews between professional interviewers and respondents also helps identify problem questions and ensure their correct understanding (Goldenberg et al. 1997; Mueller and Phillips 2000).
- *Respondents' Questions, Comments, and Complaints.* Feedback from staff interactions with respondents is commonly used to informally evaluate questionnaires. These interactions include help desk phone inquiries from respondents (McCarthy 2001), as well as follow-up phone calls to respondents by survey personnel investigating data that failed edits. However, questionnaire development experts must recognize that anecdotal feedback may sometimes be

overstated. Alternatively, respondent feedback may be systematically recorded in an integrated database.

- *Interviews or Focus Groups with Survey Staff.* Interviews or focus groups may be conducted with survey personnel who review data and resolve edit failures. Their feedback can be used to help identify potential candidate questions for redesign or evaluation by other methods (Rowlands et al. 2002; Tuttle et al. 2010a).

Examining Completed Survey Questionnaires Survey questionnaires completed by respondents during production data collection often hold clues that aid survey methodologists in identifying questions or questionnaire designs that may be problematic for respondents. Respondents may provide notes on the questionnaire about data items that were difficult to answer or where estimates were made, such as providing total amounts rather than component parts. Items where answers were entered and later changed (e.g., crossed out) may indicate problems with questions, instructions, or formatting. Respondents may write remarks or explanations in areas where notes are requested, or specify alternative answers when "other" is marked. Trained questionnaire methodologists may recognize and interpret numerous clues left by respondents as potential concerns for response error or added burden, warranting further investigation or suggesting improved designs. (Giesen and Hak 2005; Featherston and Stettler 2011).

For example, Featherston and Stettler (2011) noticed a tendency for respondents to mark through one or more entire pages of questions, often adding "N/A" to indicate that they were not relevant to their businesses. This may have been a way for respondents to note for themselves, their supervisors, or the data collector that those pages had not been skipped over or missed.

Statistical Process Indicators Several indicators typically used to evaluate the quality of collected data and statistical estimates may also be used to assess the performance of individual questions. They may also indicate problematic questions for specific subdomains of the target population, aiding a more indepth analysis of data quality and potential error sources. Statistical process indicators, such as unit nonresponse rates, item nonresponse rates, plausibility indices, and edit failure rates, may be useful, but they require care in interpretation:

- *Unit nonresponse rates* may aid in identifying businesses characteristics, such as size or industry, associated with problematic reporting.
- *Item nonresponse rates* are somewhat less straightforward to calculate and interpret, as a questionnaire item left blank or a zero may mean "not applicable," "applicable but zero," "refuse to answer," or "I don't know." One metric used by Giesen and Hak (2005) was the proportion of blank fields replaced with a value during editing.
- *A plausibility index* is defined to be "an index of deviation from expected values that are computed from [administrative] data, data from previous years and data

from comparable firms" (Giesen and Hak 2005). Plausibility may need to be interpreted with care, however, because its variability may be an artifact of economic conditions.

- *Edit failure rates* may aid in identifying problematic questions. High rates of edit-failing data may indicate which items are difficult to report or that some design issue exists that inhibits recording answers in a correct manner. However, edit failures may also be difficult to interpret as some edits look for associations among multiple items or with previously reported data, making it difficult to isolate error sources (Tuttle et al. 2010a).

7.5 DEVELOPMENT, TESTING, AND EVALUATION OF ELECTRONIC DATA COLLECTION INSTRUMENTS[3]

For our purposes here, we will focus on development, testing, and evaluation of surveys collected over the Internet or the World Wide Web (web), or by other computerized software. For example, at the US Census Bureau, there are two primary electronic reporting options: the Internet and executable software that must be downloaded and installed on respondents' personal computers. The latter is used for large-volume establishment-level data collections, such as the US Economic Census, permitting the respondent to take advantage of additional user options that facilitate reporting.

The volume of research to support web surveys has led to considerable, yet ever-evolving, design strategies and best practices. Those pertinent for business surveys are covered in Chapter 8. The reader is also referred to the text by Couper (2008). Additionally, while Internet-only surveys are becoming more common worldwide, electronic instruments are often administered in mixed-mode designs (Dillman et al. 2009). Chapters 8 and 9 contain discussions of experience and recommendations for mixed-mode business surveys.

Here we will describe some common recommended practices for development, testing and evaluation of electronic survey instruments.

7.5.1 Development Strategies for Electronic Instruments

Development strategies for electronic data collection instruments typically follow some of the same procedures commonly associated with development of software applications. Here we will cover only four of them: (1) user-centered design, (2) requirements gathering/task analysis, (3) expert/heuristic reviews, and (4) functional and performance testing.

User-Centered Design It is important to take a user-centered design approach, by paying attention to the needs of respondents, when developing electronic instruments. This approach focuses on how business survey respondents want to use the survey

[3] Portions of this section were written and contributed by Amy E. Anderson Reimer of the US Census Bureau.

instrument, rather than attempting to get respondents to learn idiosyncratic features of the instruments and alter their behaviors to use them. Growth in the web has generated many opportunities for creating web-based surveys, and for developing and refining user-centered testing techniques (Nielsen and Loranger 2006).

Survey researchers advocate a user-centered design approach to web survey development, ensuring that the instrument is intuitive and easy to use for respondents, in order to reduce potential response burden and, hopefully, increase takeup rates. A user-centered design approach to web survey development utilizes a number of techniques for specifying, creating, testing, and evaluating instruments that meet respondents' needs.

Usability testing, described in Section 7.5.2, should be conducted to ensure that a user-centered design has been achieved. Conducting initial requirements gathering research with input from the actual users of the system, the respondents, helps researchers and programmers ensure that the system being built will begin with a user-centered design.

Requirements Gathering/Task Analysis *Requirements analysis* (or gathering) is a term often associated with the systems engineering or software engineering fields. The underlying goals of this step are also applicable to the development of electronic surveys and probably occur, at least informally, during the development phase. Requirements are specifications for all aspects of the system and come from the goals and purpose of the system, the needs of the respondent, and a functional analysis of the system in which the electronic survey resides (Hix and Hartson 1993). During requirements gathering, members of the design team obtain ideas from various sources and turn them into requirements. Sources for requirements may include stakeholders, respondent complaints, or respondent requests (Unger and Chandler 2009). Requirements should be documented, actionable, measureable, testable, traceable, and related to the needs of the stakeholders or respondents (McConnell 1996).

Task analysis takes a more user-centered approach in determining the design of the electronic survey. Task analysis is the process of identifying respondent tasks and understanding how they might be addressed using the electronic survey. The analysis begins with the question "What does the respondent want to do?" This is followed by learning what respondents need to perform these tasks and what actions they need to perform within the electronic survey in order to achieve this. Additionally, it is important to understand how respondents would evaluate the quality of the electronic response process. Even though the respondent may have a successful outcome, it is important to ensure that the process of getting to that outcome is satisfying.

Task analysis should be accomplished as early as possibly in the development process to provide input into the system design (Nielson 1993). Typically, a researcher meets with respondents to discuss their methods for completing the requested survey tasks. Respondents are asked about their expectations for the electronic survey instrument and what they will need to complete the survey successfully. During this discussion, the researcher can probe about how and why certain actions are taken by the respondent (Anderson and Morrison 2005).

Just as cognitive testing for business surveys is conducted at the business location, so should task analysis, enabling researchers to observe the work environment of respondents. Respondents can access paper or electronic files that may be necessary in gathering information for the business survey in an electronic mode. If it is necessary for respondents to gather information from others within the organization to complete the survey, the researcher's presence in that environment provides a firsthand glimpse of how that communication would occur and how the electronic instrument would help or hinder that process. Debriefing interviews may be conducted with respondents to identify requirements that may help them transition easily from a paper-based environment to an electronic one.

For example, requirements gathering played an important role in preparing to reengineer the electronic data collection software used to collect the 2007 Economic Census at the US Census Bureau (Anderson and Harley 2007). The goal of extensive research was to identify recommended changes to the instrument from the respondent's perspective by conducting a task analysis to determine the processes that respondents undertake when answering an economic census electronically. Respondents then evaluated the effectiveness of the current version of the data collection software in performing the necessary tasks. Areas where respondents felt that a feature (navigation, layout, functionality, etc.) was inadequate or missing became a requirement for the future version. These areas also became the focus of subsequent usability testing after low- and high-fidelity prototypes became available.

Other research techniques that can be used to identify information about the respondent's environment and preferences (Unger and Chandler 2009) include

- *Contextual inquiry*—on-site visits are conducted to observe participant behavior in their everyday environment.
- *Surveys*—the goal is to gather information about preferences rather than performance.
- *Card sorting*—respondents are given items on cards and asked to sort them into meaningful groups to assist researchers in identifying the ideal organization of information online.

Incorporating methods such as requirements gathering and task analysis into the development process of electronic business surveys helps ensure that survey organizations design an instrument that meets respondents' expectations for how an electronic survey should operate. When respondents face an instrument that behaves differently than expected, they will be led down "unproductive paths through the underlying data structure, creating frustration and workarounds to minimize use of the tool" (Murphy et al. 2000).

Electronic Instrument Heuristic Evaluations, Style Guides, and Standards Electronic instruments are also subjected to expert reviews, which are called *heuristic evaluations*, because they are typically assessed relative to human-computer interaction principles or heuristics (Neilsen and Mack 1994). Reviews using heuristic coding schemes may be carried out by individual usability experts, expert panels (Sweet et al. 1997), or members of the development team (Fox 2001).

The additional burden of testing electronic instruments with respondents can be minimized by adhering to a style guide or standards for the user interface when developing electronic instruments (Farrell and Hewitt 2011; Burnside 2000; Harley et al. 2001; Murphy et al. 2001). These rules for navigation, layout, screen design, graphics, error handling, help, user feedback, and other internal electronic mechanisms are based on usability standards and best practices, as well as previous testing and operational experiences with electronic instruments.

Functional and Performance Testing Toward the end of the development phase of electronic surveys, programmers test the system in various ways to ensure that all necessary components are functioning as anticipated. During performance testing, programmers and other internal testers work with the system to troubleshoot any weaknesses related to the design, speed, or responsiveness of the electronic survey. To do this, they may run various data scenarios through the system.

In addition, programmers perform *load testing*, in which they simulate the expected demand on a system to determine how the system will behave under normal and anticipated peak use conditions. This allows programmers to identify the maximum operating capacity of the system and to anticipate any bottlenecks (Meier et al. 2007).

7.5.2 Usability Testing

Computerized self-administered questionnaires and web survey instruments add another layer to the interaction between respondents and the questionnaire—that of the user interface (Morrison and Anderson 2005). *Usability testing* is the broad name given to methodologies for testing software graphical user interfaces, such as navigation and embedded edits. Some examples of usability testing of the respondent interface for business surveys can be found in the following sources: Anderson et al. (2001), Burnside and Farrell (2001), Fox (2001), Nichols et al. (1998, 2001a, 2001b), Saner and Pressley (2000), Sperry et al. (1998), and Zukerberg et al. (1999).

Use of Prototypes Since the timeline for electronic instrument development must include programming, testing may begin with paper prototypes or nonfunctioning screen shots, called *low-fidelity prototypes*, to affect software development. Usability testing is typically conducted with partially or fully functioning, or *high-fidelity*, electronic prototypes. Debriefing interviews may be conducted with business respondents, using existing instruments, to detect usability issues with electronic surveys undergoing redesign (Anderson and Harley 2007).

Observation Usability testing relies heavily on direct observation of business or institutional respondents. Researchers assign tasks typically encountered in using the instrument, and then observe respondents as they perform actions such as entering data, navigating skip patterns, correcting errors, and printing or submitting completed forms. Behaviors of respondents interacting with the instrument may also be coded, supporting empirical analysis of usability problems.

As with cognitive testing of business survey questionnaires, usability testing with business respondents typically takes place on site at the business or institution. Some researchers stress the value of observing the task being completed in the user's natural environment, studying the atmosphere of the workspace and how the environment affects the reporting task. Usability tests may be videotaped, with the respondent's permission, in order to facilitate review of respondent behaviors.

Cognitive Methods Researchers may adapt and incorporate cognitive methods into usability testing, such as concurrent probes, retrospective debriefings, and user or respondent ratings. Vignettes and other hypothetical scenarios have been used to test alternative design options. After interacting with the instrument, users may be asked to complete self-administered debriefing questionnaires or participate in focus groups to provide feedback on their experiences.

Eye-Tracking Researchers have been increasingly using eye-tracking equipment to supplement information learned during traditional usability testing. Eye-tracking consists of following the trail of where a person is looking. Eye-tracking equipment is built into the computer monitor, and eye-tracking software captures the display on the screen along with what the respondent is looking at. The output displays the path and duration spent viewing objects on the screen (Nielson and Pernice 2010). Most eye-tracking equipment is located in usability laboratories, and therefore difficult to incorporate into business survey usability testing performed at the business location. Portable eye-tracker glasses have been developed to enable eye-tracking research outside of laboratory settings.

7.5.3 Evaluating Web Survey Data Collection Instruments

Electronic instrument use can also be assessed empirically once the instrument has been fielded. Design decisions may be informed by examining variables that may be associated with mode selection (Burr et al. 2001; Dowling and Stettler 2007).

Paradata[4] For web surveys, *paradata* refers to process data created as a byproduct of web survey data collection (Couper 2008). These data can be amassed and analyzed to provide information about the process that respondents undergo to fill in a web survey questionnaire. Paradata describes how respondents complete the survey as opposed to the content of their answers (Heerwegh 2003).

[4] Portions of this section were drawn from a literature review prepared by Kristen Cibelli when she was student research assistant at the US Census Bureau in 2010, provided under contract with the Joint Program in Survey Methodology at the University of Maryland.

Capturing paradata consisting of information about page requests, or "visits" to web pages, makes it possible to monitor respondent progress on survey completion, with regard to how respondents initiated the survey, how many completed it, and at what point did respondents break off. It is also possible to examine respondent behavior on completing the survey; for example, whether respondents tend to log in and complete it in one sitting, or return to it multiple times before submitting, it. Examining the point at which respondents break off could help identify problematic questions.

Paradata collected on the respondent's computer make it possible to examine the response process of respondents in their natural setting (Heerwegh 2003). Each "meaningful" action is logged, such that it can be identified and situated in time. Since one of the main challenges with analyzing paradata is dealing with very large datafiles and extracting useful information from them, researchers are advised to decide which actions are meaningful depending on issues of interest. Possibilities include

- Clicking on a radio button
- Clicking and selecting a response option in a dropdown box
- Clicking a checkbox (checking/unchecking)
- Writing text in an input field
- Clicking a hyperlink
- Submitting the page

Some examples of paradata collected and examined by researchers include

1. Response latencies—the time that elapses between when the screen loads on the respondent's computer and the answer is submitted (Heerwegh 2003)
2. Whether and in what direction respondents change their answers (Heerwegh 2003)
3. The order in which questions are answered (Stern 2008)

Survey researchers have analyzed paradata from electronic instrument event logs, which track keystrokes and timing, enabling them to identify user problems with electronic interfaces and functionality (Snijkers and Morren 2010; Saner and Pressley 2000; Sperry et al. 1998).

7.6 ANALYZING QUALITATIVE DATA[5]

Just as in surveys of households and individuals, many of the questionnaire development, testing, and evaluation methods described in this chapter for business surveys utilize qualitative research methods. Qualitative research is typically characterized by purposive or convenience sample designs with a small number of cases.

[5] This section was written and contributed by Jacqui Jones of the UK Office for National Statistics.

We have seen that qualitative methods vary and can include personal contact between researchers and subjects, overt or covert observation, and analysis of texts and documents. Although it is inappropriate to make statistical inferences generalized to a target population, resulting data are rich in detail, providing underlying context and insight for empirical information. We now discuss methods for analyzing, summarizing, and reporting results from qualitative research.

The qualitative data collected during the development and testing of survey instruments will consist of textual accounts of conversations, observations, or documents. Qualitative data provide insight into the "what" and "why," and are characterized as providing rich, detailed information. As in any other data collection and analysis process, errors may occur along the way, including researcher bias during the analysis process. These possible issues lead to the need for systematic methods for qualitative analysis (Miles and Huberman 1994).

To some degree the analysis of qualitative data can be compared to the coding of written responses to classification questions (see Chapter 11), but it is far more involved than this. Unlike classification coding to predefined nomenclature, qualitative data analysis requires identification of themes and associations in the data.

Analyzing qualitative data is no easy task, as vast amounts of qualitative data are generally collected. For example, a one-hour cognitive interview may provide approximately 10 pages of transcribed text; this is our data. So, 10 one-hour cognitive interviews provide roughly 100 pages of data to thematically analyze. The volume of data may seem overwhelming, leaving the researcher unsure of where to begin. For example, interview recordings, notes, documents, videos, or other media might be used to collect the data. The task is to turn the masses of data into an analytical summary, and ensure that the summary is a true reflection of the data.

7.6.1 Data Capture

First, the data need to be captured from the medium in which they are held. Audio and videorecordings of interviews may be summarized or transcribed by the researcher, or they may be sent for professional verbatim transcription. If others will capture the data, then ethical and confidentiality pledges must be ensured and upheld at all times. Additionally, quality assurance methods must be applied to ensure the accuracy of the captured data.

7.6.2 Types of Analysis

Different types of analysis techniques are required for different types of qualitative research.

Data collected from interviews, focus groups, or observations are typically analyzed using *content analysis*. This approach consists in identifying themes from the data, or categorizing information according to previously identified themes. According to Miles and Huberman (1994, p. 9), the analytical steps involved are

- Affixing codes to a set of field notes drawn from observations or interviews
- Noting reflections or other remarks in the margins
- Sorting and sifting through these materials to identify similar phrases, relationships between variables, patterns, themes, distinct differences between subgroups, and common sequences
- Isolating these patterns and processes, commonalities, and differences, and taking them out to the field in the next wave of data collection
- Gradually elaborating a small set of generalizations that cover the consistencies discerned in the database
- Confronting these generalizations with a formalized body of knowledge in the form of constructs or theories

Data collected with the objective of analyzing linguistics will generally use *discourse or conversational analysis*. [See Alba-Juez (2009) for further information.]

Most qualitative data analysis will have three concurrent flows of activity: data reduction, data display, and conclusion drawing/verification (Miles and Huberman 1994).

7.6.3 Data Reduction

One of the best accounts of qualitative data reduction was given by a PhD student, who reported the feeling of drowning in data and not knowing where to start. To overcome this feeling, the student purchased a "clotheshorse"—little plastic bags, pegs, colored pens, and sticky labels. Once immersed in the data, each time the student identified a theme, a color was assigned to it, a plastic bag was labeled with the theme, and the bag was pegged onto the clotheshorse. The relevant text was then cut out and placed in the bag. This continued until all the data had been sorted. Individual bags (themes) were then analyzed further to identify subthemes and associations. This account really sums up the process of data reduction.

One challenge of qualitative data analysis is to take a neutral position, without any preconceptions, in order to ensure validity and reliability. This is vital in qualitative data analysis, as the researcher acts as the analysis instrument. When researchers conduct the qualitative data collection themselves, they may have opinions on the key themes and associations before beginning systematic data reduction. These preconceived ideas must be abandoned, to allow fresh focus on what the collected data may reveal. To accomplish this, researchers become immersed in the data, in order to identify the key themes emerging from the data. This immersion may be aided by automated qualitative analysis tools (e.g., NVivo, ATLAS, Nudist), which distance the researcher from the data (Barry 1998).

During data reduction, contextual details should also be considered. For example, did the cognitive interview take place at the business premises? If so, then the interviewee may have been able to check data held in records against what was required. However, if the boss was in the office at the time of the interview, the interviewee may not feel comfortable speaking openly about the business data, particularly if the boss does not agree with a company policy of complying with business surveys. Context, therefore, can be very important, and it should be noted and considered during analysis.

The process of data reduction also depends on the objective of the research and the method used to collect the data. For example, if survey questions were tested using cognitive interviews, then data associated with individual questions should be collated together before beginning the process of data reduction.

7.6.4 Data Display

Following data reduction, the themes must be examined to identify any associations among them. For example, perhaps many interviewees who could not provide a certain data item according to the specified definitions may have consistently been in small businesses. This brings the theme and context together. This activity may be done visually, using flowcharts.

Without this step, research results may consist only of themes and subthemes, but without the connections between them.

7.6.5 Drawing Conclusions and Verification

Once the analysis has been carried out, conclusions may be reached by focusing on common patterns discerned from the data. As these conclusions are being developed, researchers will often verify them by returning to the raw qualitative data, or they may have detailed discussions with colleagues as a means of checking the validity and reliability of their conclusions (Miles and Huberman 1994).

Section 7.A.1 in the Appendix presents an example of conclusions from a qualitative research project studying public trust in statistics. Qualitative data were collected from nine focus groups across England and Wales. These conclusions were extensively debated among the researchers involved in the project, and validated by returning to the raw qualitative data.

7.7 MULTIMETHOD APPROACHES TO DEVELOPMENT, TESTING, AND EVALUATION

Since many pretesting methods are qualitative, using more than one method to develop, pretest, and/or evaluate a questionnaire can increase both researchers' and sponsors' confidence in the results. The various methods have their strengths and weaknesses, and they tend to be complementary. Some methods help to expose or understand the nature of problems with business survey questions or questionnaires, while other methods suggest solutions, while still other methods compare alternatives or evaluate effectiveness. Bavdaz (2009) provides a review of the tradeoffs in using some common qualitative research methods for collecting and analyzing data from business respondents, and interpreting results, while also providing guidance on using them effectively.

Methods that rely on the judgment of methodological or subject matter experts should be complemented with methods that involve respondents and obtain their feedback. Qualitative methods that obtain indepth descriptive information from respondents may be supplemented with quantitative methods that indicate frequencies or degrees of particular response properties. Methods using small purposive or convenience samples can be supported with hypothesis testing using probability samples that generate estimates with known statistical properties.

Different methods, or the same methods used in different ways, are applied to investigate the different stages of development, pretesting and evaluation. For example, Tuttle et al. (2010a) used an approach consisting of three phases to redesign an existing questionnaire for a quarterly business survey that collected highly technical data on foreign direct investment in the United States:

- Phase 1, conducted before beginning the redesign, consisted of background investigations to aid the researchers' understanding of the subject matter, the data collection forms, the types of data collected, and potential sources of response error. The methods used during phase 1 were (1) focus groups with data analysts to isolate major problem areas, (2) observation of phone conversations between analysts and respondents to identify typical language, and (3) respondent debriefings to discern underlying issues associated with problematic questions.
- During phase 2, questionnaire sections were drafted utilizing the design expertise of trained survey methodologists in the context of phase 1 findings. Then the redesigned questionnaire was cognitively pretested in an iterative manner, with revisions made between each of five rounds of interviews.
- Phase 3 consisted of a pilot test of the final questionnaire, fielded with a subsample of businesses during production data collection. The purpose was to evaluate the effectiveness of the new questionnaire in a production setting. This enabled survey analysts to become familiar with the new questionnaire and compare data with that collected by the legacy form. In addition, a small number of questions appended to the redesigned form were designed to obtain respondents' opinions of some of the features of the new form in comparison to the previous form. Finally, respondent debriefings were conducted with a small number of pilot test respondents, to confirm appropriate response strategies.

Giesen and Hak (2005) used a similar multiphase, multimethod approach for research to support the redesign of Statistics Netherlands' annual Structural Business Survey:

- *Phase 1: Problem Finding Phase.* Data quality and response burden were assessed using the following information or methods:
  1. Previous formal reports about problems with the questionnaire
  2. Review of a sample of survey questionnaires completed by respondents
  3. Respondents' questions, complaints, and comments
  4. Focus groups with survey staff that worked with respondents or with the data

5. Statistical process indicators, such as unit and item nonresponse rates and plausibility indices
6. Assessment of findings with stakeholders to ensure completeness and to create consensus

- *Phase 2: Diagnostic Phase.* The goals were to validate results from phase 1, explore causes of the problems associated with data quality and response burden, and understand the response process for this particular survey to identify how errors might happen. Methods included
  1. Expert review of the survey questionnaire by a design specialist from outside the survey organization
  2. Real-time on-site observation of how respondents actually filled out their survey questionnaires
  3. Retrospective focused interviews with respondents who had already completed the form, reconstructing their response processes
  4. Interviews with different types of respondents to learn about their experiences with the questionnaire
  5. Telephone interviews with nonrespondents to determine whether questionnaire problems contributed to their nonresponse decisions
- *Phase 3: Design Phase.* The researchers worked with the survey project team to review recommendations resulting from the problem finding and diagnostic phases, and to make decisions with regard to redesigning the questionnaire.
- *Phase 4: Testing Phase.* Multiple methods were employed:
  1. Lab tests, with members of the survey organization's staff who completed the form using mock records
  2. Expert interviews and written evaluations from survey staff
  3. Field testing with a small number of companies, using real-time on-site observation and retrospective focused interviews.

The multimethod approach used by McCarthy and Buysse (2010) to revise and test the questionnaire for the 2012 Census of Agriculture included

1. Expert reviews from a variety of stakeholders, including members of professional organizations, advisory committees, commodity groups, data users, and survey staff
2. Evaluation of historical data, including item-level frequency of edits and imputations, and the substance of respondents' phone calls to the help line
3. Cognitive testing of revised questionnaires with a small purposively selected sample of respondents
4. Field testing where questionnaires revised according to results from the previous methods were mailed to a sample of several thousand cases, which also included several split-sample experiments for comparison of alternative designs and strategies

5. Follow-up interviews where a subset of field test respondents exhibiting questionable data were recontacted and debriefed regarding their responses and response experiences

McCarthy and Buysse (2010) explain further how these different methods permit examination of different types of information—frequency of different problems under different conditions, details provided by respondents, narrative opinions, and response rates. Additionally, the different types of information represent different sources and their perspectives—different types of businesses, survey staff in headquarters and in the field, and external professional and trade groups. Different collection strategies are also represented—structured and unstructured self-reports from groups, empirical information from historical data, and interviews using open-ended questions and/or closed-ended questions.

In summary, using a variety of methods permits new complementary discoveries over the course of the research. Qualitative methods seem particularly useful for identifying problems and diagnosing causes in preparation for field studies that use cognitive methods, observation, and indepth retrospective interviews, which are also qualitative techniques. When research findings appear to conflict, subsequent research using alternative methods enables further investigation. Although qualitative methods may not permit statistical inference and generalization to a target population, their use may help identify issues on which to focus when conducting more expensive and elaborate research using quantitative statistical methodologies. As results are replicated and converge across different methods, survey sponsors can gain confidence in the research findings, redesign decisions, and implementation.

7.8 ORGANIZATION AND LOGISTICS

7.8.1 Survey Organizational Structures for Instrument Development, Testing, and Evaluation

In many NSIs, business survey personnel are organized by survey or program area, rather than along functional lines, and so they acquire expertise in the subject area covered by the survey. In most cases, they are integrally involved in all stages of the survey process—development, collection, callbacks, data review and summary, tabulation, and publication—and draw on experiences at later survey stages to suggest modifications to questionnaires. A drawback of this structure is a lack of cross-fertilization, which could lead to innovation in testing methodologies and greater consistency in instrument designs across surveys.

Questionnaire design experts typically work outside these dedicated survey areas, although some organizations assign survey design specialists directly to a survey program. Questionnaire design experts are usually enlisted on a consultancy basis to facilitate development and testing of survey questions. It is important that they become part of the survey team as early in the development process as possible, rather than waiting until after an instrument has initially been drafted. Bringing their knowledge of respondents' perspectives and best practices in questionnaire design, questionnaire design experts can contribute to the processes of conceptualization and operationalization, alongside stakeholders, subject matter experts, and survey managers, helping to resolve common issues and avoid pitfalls early on.

In the decentralized statistical system of the United States, experts in questionnaire development and survey methodology usually specialize in either business or social surveys. Throughout the rest of the world, it is more common for these specialists to work on both types of surveys.

7.8.2 Pretesting Logistics

Procedures for pretest interviews, including cognitive interviews, have been modified to accommodate the characteristics of businesses. First, pretest interviews are conducted at the business site, rather than in a laboratory, so that respondents have access to their records (Giesen 2007; Babyak et al. 2000; Eldridge et al. 2000). While this is more costly and time-consuming for the survey organization, on-site testing provides a richer, more realistic context than conducting comparable interviews in a laboratory (Moy and Stinson 1999). Because of resource constraints, cognitive pretesting has occasionally been conducted via telephone interviews (Cox et al. 1989; Stettler and Willimack 2001), or open-ended cognitive-style probes may be embedded within a self-administered questionnaire. However, comparison of these shortcuts with results from traditional face-to-face cognitive interviews has shown that the latter provided more complete and detailed information (Davis et al. 1995).

Most pretest interviews with business respondents are limited to 60–90 minutes in duration, because taking respondents away from their job responsibilities represents a tangible cost to the business. Since the entire questionnaire may be lengthy and require laborious data retrieval from records, testing is often limited to a subset of questions (Jenkins 1992; Stettler et al. 2000; Ware-Martin 1999). Thus, researchers must be careful to set the proper context for the target questions.

Sample sizes for pretest interviews vary widely. In the United States, collection of data from 10 or more respondents requires approval from the Office of Management and Budget (OMB), so many studies are conducted with samples of nine or less. (OMB allows some federal agencies, including the US Census Bureau, the Bureau of Labor Statistics, and others, a streamlined clearance process for pretesting with larger samples sizes.) Survey organizations in other countries face similar sample size restrictions. Moreover, recruiting business and institutional respondents presents a significant challenge, because they are requested to take time away from their jobs. Studies may utilize as few as 2–5 cases and as many as 60 or more cases, depending on study goals and resources (Babyak et al. 2000; O'Brien et al. 2001).

Although conducting only one round of interviews with a small number of cases is not uncommon among cognitive studies, multiple rounds, or phases, of pretest interviews are preferred, where the questionnaire is redesigned between phases (Goldenberg 1996; Goldenberg et al. 1997; Schwede and Gallagher 1996). An alternative strategy is to use coworkers as substitutes for respondents, particularly during early development (Kydoniefs 1993), or when resources are limited (Goldenberg et al. 2002b).

Typically, formal cognitive pretest interviews are conducted by experts in questionnaire development and survey methodology. Survey personnel trained in cognitive interviewing techniques may also conduct pretest interviews, guided by questionnaire design experts, who debrief survey staff, summarize research findings, and collaborate on recommendations.

Section 7.A.2 presents a case study illustrating the logistics, practices, and experiences typical of US Census Bureau researchers who undertake questionnaire development, pretesting, and evaluation activities for business surveys.

7.9 SUMMARY

In this chapter, we have defined the stages of development, testing, and evaluation of data collection instruments:

1. Question(naire) *development* consists of the activities and processes where concepts are defined, measurable attributes are specified, and survey questions are formulated to collect these measurements.
2. Question(naire) *testing* consists of research conducted prior to production data collection for the purpose of determining the degree to which responses obtained from survey questions meet the intent of the question as specified by the desired measurements.
3. Question(naire) *evaluation* consists of investigations conducted to assess the performance of the survey question in measuring the underlying concept as reflected in the quality of the collected response data.

We placed these steps in a quality framework, linking the purpose of each stage with examination and reduction of nonsampling errors associated with validity, specification, and measurement:

- *Validity* examines questions such as "What can be measured to reflect the concept?" and "How well does the measurement reflect the concept?"
- *Specification error* is associated with the question "How well does the question capture the desired measurement?"
- *Measurement error* is associated with the question "How well does the response match the intent of the question?"

We also identified constraints and limitations in applying questionnaire development, testing, and evaluation research methods in the business survey setting:

- The *nature of the requested data* requires the measurement of technical concepts with precise definitions.

- *Survey response is labor-intensive*, representing a tangible, but nonproductive, cost to the business, which may be exacerbated by the need for multiple respondents and data sources.
- Questionnaire testing, experimentation, and evaluation add to the already extensive *respondent burden* experienced by business survey respondents.

We further described the complex survey response process for businesses, which wraps organizational steps (1, 2a–2c, and 7) around the four traditional cognitive steps (3, 4, 5, and 6):

1. Record formation and encoding of information in memory
2. Organizing the response tasks
   a. Selection and/or identification of the respondent(s)
   b. Scheduling the work
   c. Setting priorities and motivation
3. Comprehension of the survey request
4. Retrieval of information from memory and/or records
5. Judgment of the adequacy of the retrieved information to meet the perceived intent of the question
6. Communication of the response
7. Review and release of the data to the survey organization

We explained how variation in the ways that business respondents perform these steps contributes to measurement error. We demonstrated how this model provides a tool to facilitate the development, testing, and evaluation of data collection instruments for business surveys, in order to identify, assess, and address threats to data quality associated with validity, specification, and measurement. With this as our foundation, this chapter describes various methods listed in Table 7.2, for use in the development, testing, and evaluation of business survey questionnaires and electronic data collection instruments.

We noted that many questionnaire development and testing methods utilize qualitative research methodologies. To demonstrate the need for care in analyzing and interpreting data obtained from qualitative research, we provided some guidance on analysis techniques.

We particularly emphasize the need for attention during the development stage, when efforts spent specifying concepts and operationalizing them in the form of survey questions may improve the efficiency and effectiveness of subsequent testing. Moreover, since many business surveys conducted by NSIs recur with some periodicity, postcollection evaluations of data quality may be especially useful in identifying problem areas and allocating resources effectively for research to make improvements.

Nevertheless, we must remind the reader that there is no single magic bullet for ensuring business survey data quality; no single questionnaire development, testing, or evaluation method will find and solve all potential problems associated with

Table 7.2 Summary of Methods for Development, Testing, and Evaluation for Survey Questionnaires and Data Collection Instruments

Method	Stage		
	Development	Testing	Evaluation
Dimension/attribute analysis	X		
Data user needs assessments	X		
Questionnaire mining	X		
Early stage scoping interviews/focus groups	X		
Concept mapping	X		
Feasibility studies	X		
Accounting/subject matter expert review	X		
Recordkeeping studies	X	X	
Methodological expert reviews/cognitive appraisals	X	X	
Exploratory/feasibility studies and site visits	X	X	
Focus groups with respondents	X	X	
Cognitive interview methods		X	
Vignettes		X	
Ethnographic methods		X	
Pilot studies		X	
Behavior coding		X	X
Split-sample experiments		X	X
Postcollection empirical evaluations; coherence analysis; benchmark comparisons			X
Respondent debriefings		X	X
Response analysis survey			X
"Reinterview"/content evaluation			X
Validation/record-check study			X
Feedback from survey personnel (e.g., survey analysts, interviewers, help desk)	X		X
Examining completed survey questionnaires	X	X	X
Statistical process indicators			X
Requirements gathering/task analysis	X		
Heuristic evaluations	X		
Functional and performance testing	X		
Usability testing methods		X	
Paradata		X	X

collecting economic data from businesses. Rather, we recommend using multiple complementary methods to consider measurement issues throughout the development–testing–evaluation processes, as each method has strengths and weaknesses. Most importantly, as results are replicated and converge across different methods, survey designers can gain confidence in using the research findings to support design decisions.

There is no cookbook for effective development, testing, and evaluation of business survey data collection instruments. There is, however, an extensive toolkit of research methods listed in this chapter, to aid business survey designers in ensuring effective data collection instruments that collect high-quality data while also controlling response burden and reducing nonresponse.

ACKNOWLEDGMENTS

Roughly half of this chapter is drawn from a monograph written by Diane K. Willimack, Lars Lyberg, Jean Martin, Lili Japec, and Patricia Whitridge, previously published as Chapter 19 in *Methods for Testing and Evaluating Survey Questionnaires* (Presser et al., editors, 2004). In addition, with the exception of Section 7.2.1, the remainder of Section 7.2 is extracted from a paper by Ger Snijkers and Diane K. Willimack (2011). We also wish to express gratitude to Alfred D. Tuttle, Amy E. Anderson Riemer, Kristen Cibelli, and Jacqui Jones for sharing their expertise by preparing sections included in this chapter, as noted within. We acknowledge helpful reviews of this chapter by Kristin Stettler, Joanna Fane Lineback, Eric Fink, and Jennifer Whitaker of the US Census Bureau.

APPENDIX

7.A.1 Example of Conclusions from Qualitative Data Analysis: Public Trust in Statistics: A Conceptual Model of Trust Determinants[6]

Four determinants of trust were consistently identified across all of the focus groups. These were perceived independence, perceived accuracy, personal knowledge and experience. All of the determinants were used by people to establish an overall level of trust in specific statistics:

$$\begin{matrix}\text{Level}\\ \text{of}\\ \text{trust}\end{matrix} = \left(\begin{matrix}\text{perceived}\\ \text{independence}\end{matrix} + \begin{matrix}\text{perceived}\\ \text{accuracy}\end{matrix}\right) + \left(\begin{matrix}\text{self-validation}\\ \text{using knowledge}\end{matrix} + \begin{matrix}\text{self-validation}\\ \text{using experience}\end{matrix}\right)$$

However, the emphasis placed on each, and therefore the balance between the determinants, varied. For example, more weight was placed on knowledge and personal experience when information about independence and accuracy was not forthcoming.

Alternatively, participants would substitute for perceived independence and accuracy. For example, since they always knew how they had come across a particular statistic, they would focus their judgment of independence and accuracy on the basis of their perception of the presenter rather than the statistic.

[6] Section 7.A.1 from Wilmot et al. (2005, pp. 25–26). © Crown copyright 2005.

Furthermore, the emphasis placed on each determinant also varied depending on the individual assessing the statistics and their individual knowledge and experience:

- **i.** Where participants were confident in the independence and accuracy of a statistic, and that statistic agreed with their knowledge and experience, then they were more likely to trust it.
- **ii.** Even where perception of independence and accuracy was low, in the current climate, when the statistic agreed with individual knowledge and experience, then it was likely that the statistic would be trusted.
- **iii.** Although participants said that they would trust a statistic more if they could be assured of the independence and accuracy determinants, and that this would make them less reliant on knowledge and experience, because of the hypothetical nature of the discussion, it was unclear whether people would, indeed, then trust the statistic if it was not in accordance with their knowledge or experience.

This conceptual model was developed as a result of the analysis. The determinants are referred to throughout the report. It should be born in mind that the development of the model is still in its infancy, and to date based on nine focus group discussions. It may not be generalizable to the population and requires further investigation. However, the evidence from the groups supports the basic principles.

7.A.2 Case Study: Logistics of Questionnaire Pretesting for Economic Survey Programs at the US Census Bureau[7]

Economic survey programs at the US Census Bureau have integrated questionnaire pretesting into the survey design process. This activity is conducted primarily by members of a staff dedicated to questionnaire development, pretesting, and evaluation. Specially trained in the social and behavioral sciences, such as sociology, psychology, economics, and anthropology, these researchers have expertise in cognitive research and familiarity with statistical procedures.

Their approach to questionnaire pretesting for economic surveys takes advantage of close collaboration with subject matter experts in survey program areas. A typical research team usually includes two researchers trained in survey data collection methodology who work closely with survey analysts and subject matter experts. The survey methodologists contribute their expert knowledge of cognitive research methods, pretesting, and questionnaire designs that reduce response errors. Subject matter specialists contribute their expert knowledge of the concepts to be measured, familiarity with the target population, and hands-on experience with collected data. Together, the methodologists and the analysts identify problem areas to be investigated, define research goals, and develop the project plan.

[7] Section 7.A.2 from Willimack (2008).

Preparation for pretesting consists of developing an interview protocol that addresses the research needs, and recruiting business survey respondents to participate in the study. Research protocols usually start by obtaining some background information about the respondent and his/her role in the company. The protocol includes questions that explore the organizational aspects of the response process. General probes are listed, so that they are readily available to the research interviewer as needed. Question-by-question probes follow, and the protocol may end with questions about what happens after a form is completed.

Researchers use the protocol as a guide, outlining topics to be covered in the interview, rather than as a set of standard probes to be asked verbatim of every respondent. The intent is to get the respondent talking about the issue of interest. If the respondent answers subsequent questions before they are asked, then the interviewer may do no more than verify the information. Researchers must be very familiar with the intent of the protocol, and they are adept at managing the interview so as to ensure the topics are adequately covered.

Participants in cognitive testing are recruited from samples provided by the subject areas. Selection criteria are determined collaboratively with the researchers, and sample sizes vary depending on the number and types of industries represented in the target population and the number and types of questions to be pretested. Since cognitive testing is a qualitative research methodology, sample sizes may range from as few as 8 or 10 to as many as 60 or more and are not meant to be statistically representative of the target population. Because participation in pretesting is voluntary, researchers usually request 5 to 10 times as many cases as the number of completed interviews desired.

Researchers conduct cognitive interviews with respondents onsite at their business locations for two reasons: (1) It is difficult to persuade business respondents to travel to a cognitive lab during work hours; and (2) the respondents' office setting is more realistic, providing access to business records as needed. Researchers try to limit the length of their interviews to 60 – 90 minutes. Sets of questions that cannot be covered during this amount of time may be grouped and distributed across multiple interviews with different respondents, thus requiring a larger sample size to ensure adequate coverage. Interviews are audiorecorded with permission of the respondent, and the recordings are used during summarization to ensure capture of pertinent information provided by the respondent. Subject matter experts participate in cognitive interviews as observers; they help the researchers assess respondents' answers for adequacy vis-à-vis the question's intent.

Researchers prepare written summaries of each individual interview. They also lead meetings to debrief the research interviewers and subject matter specialists who participated in the pretest interviews, looking for findings that were common across multiple interviews, as well as unique or exceptional situations. The researchers then draft a written report summarizing across all of the interviews conducted. The report is structured such that each individual finding is documented, followed by a recommendation for addressing the issue, if appropriate. The draft report is delivered to the survey sponsor and a meeting is arranged to review each finding and recommendation. Decisions regarding each recommendation are negotiated, if necessary, and usually are documented in the final report.

To the degree possible within schedule constraints, researchers conduct multiple rounds of testing for a given questionnaire. Thus, after pretesting of a draft form with roughly 8–15 respondents, the questionnaire is revised according to the results and subsequently retested with another small sample of different respondents. Pretesting projects usually consist of two or three rounds, although as many as five rounds have been conducted for major projects. The minimum turnaround time is 3 months for a single round of interviews with 8–15 respondents. The duration of most pretesting projects is 6–9 months. Large projects with five rounds of interviews, such as those that support a major new or redesigned survey, may last 1–3 years.

CHAPTER 8

Questionnaire Communication in Business Surveys

Gustav Haraldsen

This chapter provides guidelines for designing business surveys questionnaires. We use the term *questionnaire communication* instead of *questionnaire design* in the title to indicate that the request for business information is communicated by the questionnaire. In fact, one of the main arguments in this chapter is that business questionnaires can be improved by replacing series of one-way information on paper with a more dynamic two-way communication in electronic questionnaires. Chapter 9 continues this approach by using the term *survey communication* to describe the communication between the surveyor and the businesses *about* the questionnaire prior to, during, and after data collection. These two kinds of communication, together with a sampling plan, form the core of the data collection design.

In Chapter 3 we tried to identify the main quality challenges encountered in business surveys, and in Chapter 7 we described development and testing methods for learning more about these challenges in a specific survey. In Chapter 3 we also identified the main aspects of a survey instrument; the text elements, the visual elements, and the functional elements. In this chapter we discuss how these instrument elements can be used to address the quality challenges in business surveys. Chapter 9 focuses on how these same challenges are addressed by the survey communication strategy.

8.1 COMMUNICATION MODES

Questionnaires can be either used by interviewers or read and completed by the respondents themselves in self-administered surveys, which are the most common data collection method in business surveys (Nicholls et al. 2000; Rosen 2007).

Designing and Conducting Business Surveys, First Edition.
Ger Snijkers, Gustav Haraldsen, Jacqui Jones, and Diane K. Willimack.

The main focus in this chapter is on self-completion questionnaires, which is the measurement instrument in self-administered surveys. But what is even more important is that in self-administered surveys the whole response process is controlled by the words, structure, and visual design of the questionnaire. Therefore, self completion questionnaires need to be more carefully designed than interview questionnaires, and that is an advantage when the focus is on the measurement instrument. We also think that it is a good procedure first to try designing a questionnaire that can work on its own, and then discuss whether interviewers can contribute with further improvements. Note, however, that even if self-completion questionnaires are more challenging than interview questionnaires, we do not claim that self-administered surveys are more challenging than interview surveys. It is only that the challenges are different and more dependent on the quality of the questionnaire.

Usually two types of self-completion questionnaires are used in business surveys. Until quite recently paper questionnaires were the dominant mode (Bethlehem and Biffignandi 2012; Dillman et al. 2009). Now paper questionnaires are often combined or replaced by a web or other electronic questionnaire (Snijkers et al. 2011b). Most surveyors prefer electronic versions because digitizing normally facilitate their postsurvey activities. The answers do not need to be keyed in or scanned, and edits previously performed in office may to a certain extent be outsourced as error checks that run while the questionnaire is completed. Some survey organizations (e.g., Statistics Norway) currently omit the paper questionnaire at the first contact, and direct those who want to respond on paper to a telephone service where they can request a paper version. Commercial surveyors sometimes choose to omit the paper version altogether. When business surveys are mandatory, however, the surveyor is usually obliged to provide a paper version to those who want it. Thus, two versions normally still have to be designed. We also believe that even if the paper version may play a minor role in the data collection, the questions should be drafted on paper before creating an electronic version. Such a procedure helps the designer focus on one aspect of the design at the time: first on the words and response formats that should be used for individual questions, and next on how questionnaire communication may be improved using the electronic tools available. As long as we have done our best within the paper format, we are less worried about possible mode effects that arise from improvements made in the web version. A questionnaire designer who goes straight to the web version, however, might easily be overwhelmed by the possibilities that computerized questionnaires offer, and forget well-proven phrasing and layout principles that apply to web questionnaires as well as to paper questionnaires. In some instances it is also challenging to transfer these principles to a different communication mode. Therefore preferably the web questionnaire should be based on a well drafted paper questionnaire.

To conclude this introduction, the procedure we propose is to begin with a paper questionnaire and then consider how the questionnaire communication can be enhanced by web tools or by the personal relationship established by an interviewer. A point that we will elaborate on later in the chapter is that these three modes of data collection represent three degrees of two-way communication. When answering a paper questionnaire, the respondents will have no immediate feedback on how they read and answer the questions, while an interviewer-administered questionnaire allows for a continuous interplay between the interviewer and the respondent. Web questionnaires are located somewhere between these extremes. Different kinds of feedback can be implemented in the questionnaire, but only according to a prescribed script. In this way the feedback from web questionnaires is more dynamic than in a paper questionnaire, but less dynamic than in an interview.

The choice between modes should be based on considerations of the relative importance of two-way communication, to ensure that the questions are correctly understood and that the respondents react to them the way we want them to. Then economic and practical considerations, of course, also need to be factored in. One advantage of web surveys is that they are less costly than personal interviews. This is particularly true in surveys where it takes time to collect the information that should go into the questionnaire, which is often the case in business surveys. Self-completion questionnaires are more common than interviews in business surveys because self-administration give business respondents more time to complete the questionnaire, but also because the surveyor cannot afford to pay interviewers to remain with the respondents while they look up information needed and do calculations needed for the answers.

This does not mean, however, that there are no roles for interviewers in business surveys. Instead of being the main mode, we think that interviewers could fill in some gaps in predominately self-administered survey designs. Some tasks are best handled by personal communication. The first task is to gain support from the business management to ensure that they give priority to the survey, choose competent respondents, and allocate the time needed to complete our questionnaire. Qualitative interviews in companies suggest that self-administered business surveys could profit from surveyors taking more active and personal initiatives in order to gain support from the management and motivate respondents (Torres van Grinsen et al. 2011). As mentioned several times in this book, managers set the framework for the response process [see Chapter 2, Figure 3.4 in Chapter 3 and Haraldsen and Jones (2007)]. Interviewers could play an important part in such efforts because they are trained in persuasion and motivation techniques. A more detailed discussion on persuasion and motivation techniques is given in Chapter 9.

One other area where interviewers can be more involved than they were in the past is in help services. As self-administered surveys are computerized and become more advanced, they also become more complicated for respondents to complete. Much can be done to improve usability. But despite all efforts to make web questionnaires user-friendly, we think there also is a growing need for competent help desks where respondents in trouble can seek personal guidance. This kind of assistance comes on top of traditional help services. Generally we also think that personal contacts between respondents and employees in the survey organization can create a feeling of collegiality, which subsequently has a positive effect on motivation (Torres van Grinsen et al. 2011). Interviewers are well trained and experienced in relating to new people, and could also be trained to tackle questions commonly asked by business respondents. We also think that an interviewer working at a help desk could be given the opportunity to change a request for help into an interview if the respondent feels prepared to answer the questions on the phone.

8.2 ELEMENTS IN CONCERT

The survey methodology literature is full of recommendations and checklists on how questions should be worded and self-completion questionnaires should be designed. The most comprehensive list of guidelines is probably found in Dillman et al. (2009). Morrison and her colleagues later used these guidelines as a basis for 18 principles specifically relevant to business surveys (Morrison et al. 2010). Morrison et al. design principles recommend that questionnaire designers:

1. Phrase data requests as questions or imperative statements, not sentence fragments or keywords.
2. Break down complex questions into a series of simple questions.
3. Use a consistent page or screen layout.
4. Align questions and answer spaces or response options.
5. Clearly identify the start of each section and question.
6. Use strong visual features to interrupt the navigational flow.
7. Use blank space to separate questions and make it easier to navigate questionnaires.
8. Avoid unnecessary lines that break up or separate items that need to appear as groups.
9. Use visual cues to achieve grouping between question and answer categories.
10. Avoid including images or other graphics that are not necessary.
11. Incorporate instructions into the question where they are needed. Avoid placing instructions in a separate sheet or booklet.
12. Consider reformulating important instructions as questions.
13. Consider converting narrative paragraphs into a bulleted list.
14. Use white spaces against a colored background to emphasize answer spaces.
15. Use similar answer spaces for the same task.
16. Limit the use of matrices. Consider the potential respondent's level of familiarity with tables when deciding whether to use them.
17. If a matrix is necessary, help respondents process information by reducing the number of data items collected and by establishing a clear navigational path.
18. Use font variations consistently and for a single purpose within a questionnaire.

We will refer to these principles several times in this chapter; some of them more often because we consider them more important than others. But there is also an important difference between the principles listed by Morrison et al. and our discussion about important business questionnaire features. We will concentrate more on how questions are phrased than what is done by Morrison and her colleagues. In addition we will focus more on computerized business survey designs.

In his book about telephone interviews and mail surveys from 1978, Don Dillman introduced the term *total design method* (TDM) (Dillman 1978). He emphasized that all parts of a data collection design and all steps in its implementation should work in unison to achieve good results. For example, a well-designed questionnaire is somewhat useless if there are coverage problems in the sample. Furthermore, even if both the sample and questionnaire are of high quality, a poor contact and communication strategy will normally lead to a low response rate and to nonresponse bias. Later Dillman took this total design principle one step further. It is not only the main elements of the survey design that should work together. The same principle applies to the relationships between different elements in the questionnaire and between different elements of individual questions. At the questionnaire level the total design principle applies to the relationship between instructions, headings, page layout, question order, and dynamics. At the individual question level it applies to the relationship between the text elements, the visual presentation, and functionality as depicted in Figure 3.5. Hence, it is possible to apply a holistic approach for (1) each question, (2) the questionnaire as a package, and (3) the different survey activities (Dillman et al. 2009). This chapter addresses levels 1 and 2 in a survey design. The project plan, the survey communication, and sample management procedures described in Chapters 4, 9, and 10 relate closely to how the different parts of level 3, the most overarching survey level, should be coordinated.

Figure 8.1 shows how an important message can be emphasized in different question elements instead of in a lengthy question text. In this example the instruction to give the amount in thousand euros is first mentioned in the question text, then indicated by the response labels and finally by the three preprinted zeros. It is an important example because the so-called 1000 errors probably is one of the most common errors made in business surveys. Editing to correct this type of error is very time-consuming.

A similar complementary strategy can be applied at the questionnaire level. In the example in Figure 8.2 worded navigational instructions are kept short, but are supported by graphical symbols and by the prominent numbering of the questions.

Redline et al. (2004) provide evidence that the combination of question complexity and the design of branching instructions may affect respondents' ability to navigate correctly through questionnaires with filter questions. They interpret their results as an illustration of how the interplay between verbal and visual messages affects the comprehension, information retrieval, and ultimately respondent performance (Redline et al. 2004).

❶ Please report in thousands of euros the cost incurred for research and development in 2010:

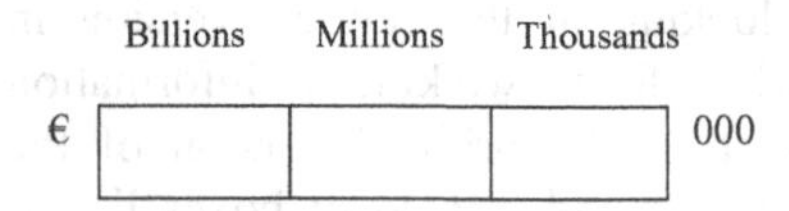

Figure 8.1 Question elements that work together.

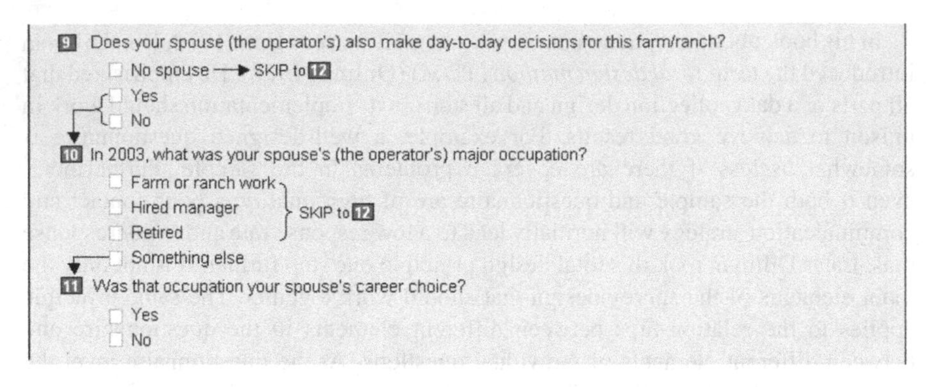
9 Does your spouse (the operator's) also make day-to-day decisions for this farm/ranch?
☐ No spouse → SKIP to 12
☐ Yes
☐ No
10 In 2003, what was your spouse's (the operator's) major occupation?
☐ Farm or ranch work
☐ Hired manager
☐ Retired
SKIP to 12
☐ Something else
11 Was that occupation your spouse's career choice?
☐ Yes
☐ No

Figure 8.2 Navigational instructions communicated in words and by visual means.

We recommend keeping the text short and complementing words with supportive graphical and visual features. This is particularly important when questionnaires are presented on computer screens. Computer screens do not have space for as much text as paper; text on computer screens is of lower quality than text on paper; also, computer screens are probably read more superficially than hardcopy (paper) pages (Eskenazi 2011).

An important prerequisite for this holistic presentation of questions is that the different elements follow a kind of visual grammar that the respondents recognize and understand. In the sequence in Figure 8.2, for example, our attention is drawn to the white numbers on dark background, the question numbers, which consequently help lead respondents from one question to the next.

The Morrison group's list of recommendations for business questionnaires have several references to visualization principles. Furthermore the importance of consistency is underscored 3 times: in point 3, regarding page layout; in point 15, about response boxes; and in point 18, about the use of letter fonts. There is also an implicit reference to consistency when it is argued in point 6 that strong visual features are needed to introduce the respondents to new directions of question flow.

A fundamental issue in the literature on visualization is that figures are the first elements on a page that attract our attention. We see figures before we read text (Hoffman 1997; Johnson 2010; Palmer 1999). This indicates the importance of the combination of text and visual features not only to communicate a common message but also because the combination of text and visualizations appear as units. Next, within these units, the response boxes should attract attention by being clearly designed as figures. This is illustrated in Figure 8.3, which compares old and revised versions of a US agricultural survey (Dillman et al. 2005).

What immediately catches our attention looking at the revised version in Figure 8.3 are the two colored squares that clearly indicate two kinds of information on the page. No such visual clues are present on the original version of the questionnaire. With white spaces on shaded background instead of boxes linked together, the response boxes also clearly appear as figures in the revised version.

NATIONAL AGRICULTURAL STATISTICS SERVICE

2003 AGRICULTURAL RESOURCE MANAGEMENT SURVEY PHASE III COSTS and RETURNS REPORT

ERS

DRAFT 4-23-03 V-4

NOTICE: YOUR REPORT IS CONFIDENTIAL and it will only be used for statistical purposes. Your report CANNOT be used for purposes of taxation, investigation, or regulation. The law also provides that copies retained in your files are immune from legal processes.

OFFICE USE ONLY	VERSION	ID	TRACT	SUBTRACT	SCREENING	BEGINNING TIME
	02		01			

SECTION LAND in FARM/RANCH

Include all cropland, idleland, CRP, pastureland, woodland, wasteland, etc.

In 2003, record for this operation–

1. Farm and Ranch land OWNED ... + ACRES
2. Farm and Ranch land rented or leased FROM OTHERS –
   a. CASH, with the payment being a fixed cash amount? ... +
   b. CASH, with the payment being a flexible cash amount? (Rent paid depends on prices and/or yields, or is otherwise not a fixed amount) ... +
   c. a SHARE of the crop or livestock production?
   d. rented for FREE? ... +
3. Farm and Ranch land rented or leased TO OTHERS– ... −
4. TOTAL ACRES in this operation in 2003 (items 1 + 2a + 2b + 2c + 2d − 3) ... = TOTAL ACRES
   a. Of the total acres in this operation (item 4) how many acres were considered CROPLAND, including land in government programs? ACRES

CROP INSURANCE, CONSERVATION, AND ORGANIC PRACTICES–

5. Of the total acres reported in item 4, how many acres were– ACRES
   a. Covered under a Federal or other crop insurance policy?
   b. Enrolled in the Conservation Reserve (CRP) or Wetlands Reserve Program (WRP)?
   c. Used to raise certified organically produced crops?

NATIONAL AGRICULTURAL STATISTICS SERVICE

2003 AGRICULTURAL RESOURCE MANAGEMENT SURVEY

ERS

U.S. Department of Agriculture

Please make corrections to label name, address, and zip code if needed.

Results from this important data collection will be mailed to all respondents in July 2004

SECTION A: LAND in FARM/RANCH

Please report farm/ranch land owned, rented, or used by you, your spouse, or by the partnership, corporation or organization for which you are reporting in 2003.

NONE NUMBER OF ACRES

1. How many acres of farm and ranch land are owned?
2. How many acres of farm and ranch land are rented or leased from others...
   a) for a fixed cash payment?
   b) for a flexible cash payment?
   c) for a share of crop or livestock production?
   d) rented for free?
3. SUBTOTAL (items 1+2a+2b+2c+2d) SUBTOTAL
4. How many acres of farm and ranch land are rented or leased to others? NONE NUMBER OF ACRES
5. TOTAL ACRES in this operation in 2003 (subtract item 4 from item 3)

Office Use Only | Response Code | Reporting Unit | MM/DD/YY | Name/Address Change

Figure 8.3 Original and revised versions of the 2003 Agricultural Resource Management Survey.

The different questions are separated by white numbers imposed on a dark box. Again the boxes are clearly interpreted as figures.

On the survey form shown in Figure 8.3 the question text and response boxes are connected by a dotted line. This design is a weak visual cue, and the combination of question text and response box hardly appears as a figure. A much more effective presentation of questions and response boxes as parts of the same figure is to shade alternate lines, as shown in Figure 8.4.

Generally, we recommend questionnaire designers to be cautious with visual features that may have unknown effects on respondents' comprehension of questions.

❶ Please indicate your level of agreement or disagreement with each of these statements regarding the store you visited:

	Strongly Agree	Agree	Neutral	Disagree	Strongly Disagree
Stores are conveniently located.	○	○	○	○	○
Store hours are convenient for my shopping needs.	○	○	○	○	○
Store atmosphere and decor are appealing.	○	○	○	○	○
A good selection of products was present.	○	○	○	○	○
Store has the lowest prices in the area.	○	○	○	○	○
Merchandise sold is of the highest quality.	○	○	○	○	○
The merchandise sold is a good value for the money.	○	○	○	○	○

Figure 8.4 Shading questions and response boxes.

In modern questionnaires, this caution particularly applies to colors, pictures, video clips, or other visual media that make the layout more exciting, but that may have unintended effects (Couper 2008). Questionnaire designers should also use caution when considering the amount of visual features to include, as too many may overload the respondent. Visual features that are repetitive or redundant do not enhance question comprehension. Too many visual effects may also lead to a competition between features that result in the features canceling each other out. The Morrison group's warnings in points 8 and 10 about images, graphics, or lines that are not necessary or break up units that should appear as groups echo these cautions.

To summarize, we consider the following visual principles to be the most relevant in self-completion questionnaires:

- Size and contrasts arrange elements from the forefront to the background and draw borders between regions that contain related elements. By changing the size of letters, respondents can more easily differentiate between the headings and the questions under those specific headings.
- By using boldface letters for question texts and lightface letters for instructions, we give the questions first priority.
- White question numbers or response boxes against a darker background set off the question numbers and response boxes. Elements surrounded by a box are perceived as having something in common.
- Simple, regular, and symmetric figures are visually preferred, and are those that are easiest to interpretate and remember.
- Items or graphics that appears similar are interpreted similarly. That is why a consistent layout for different question elements is so important. For instance, if bold letters are initially established as a layout convention for question texts, and then are used to underscore parts of an instruction, this inconsistency undermines the predictability of bold letters as visual feature.
- Visual distance signals what belongs together and what does not. If definitions follow directly after a question text, they are more likely to be read than if a linebreak disconnects them. If the distance between a question text and its response options is shorter than the distance to the next question, this signals that the question and response options are parts of the same information unit.
- Some images have commonly understood symbolic meanings. Arrows, like those used in Figures 8.2 and 8.3, visually guide the reader. The € in front of the response box in Figure 8.1 tells the reader that the amount should be reported in euros. Computers and computer programs use many traditional symbols and have introduced a number of new ones (Dubois 2008). For example, circular response boxes, called *radio boxes*, are used for exclusive response options, while squared response boxes allow multiple responses. Other examples include the following, which activate certain functions:

The following symbols are used to exit the questionnaire or indicate the seriousness of error messages:

With consistent, cautious use of visualization principles, the survey designers should try to present their different information requirements in graphics that work together with the text. But these principles are only the tools used to attract the respondents' attention to what is important. To build good questions, the presentation technique that we have introduced in this section, should to be combined with a clear perception of what we want to communicate, balanced with an awareness of what the respondents perceive as burdensome. When we discussed this challenge in Chapter 6, we recommended that the survey designer should be open about burdensome tasks. To achieve a balance between clear questions and minimum response burden, it is important to consider how to help the respondent retrieve and decide on the correct information. For example, Morrison and colleagues provide some guidance on how to split up complex questions and alternatives to matrix questions (points 2, 16 and 17). We will return to this challenge in Section 8.5.

8.3 TAILORING FOR BUSINESSES

In the second edition of his book, Dillman changed the meaning of the acronym TDM to *tailored design method* (Dillman 2000). He is now more concerned about how the questions and the questionnaire can be tailored to the kind of respondents addressed. This is mirrored in the table of contents where he distinguishes between surveys directed toward households, private persons, businesses, and other organizations (Dillman 2000). According to Dillman, "tailored design involves using multiple motivational features in compatible and mutually supportive ways to encourage high quantity and quality of response to the surveyor's request" (Dillman et al. 2009). The original point about elements that function together is still reflected in this quotation. But now there is an additional concern: tailoring the questionnaire to the respondents that the survey addresses. Applying this approach to business surveys, we will focus on three characteristics that we consider particularly important in business questionnaire design:

1. The questionnaire should be tailored specifically to respondents and response processes typical to business surveys rather than social surveys. As described in Chapter 2, the respondent in business surveys may be the actual manager, but will usually be an employee in that company or in an accounting office or other type of service company that is delegated to complete surveys for their customers. Whatever the case, business survey respondents are usually more professional than social survey respondents and should be treated accordingly.

Respondents in larger firms and accounting offices that provide service to several companies will normally have to complete more than one survey. Also, because many business surveys are panels that ask for monthly, quarterly, or annual reports, one survey may require recurrent response sessions. Hence, participation in business surveys will seem more like a routine, everyday event than a one-time experience. Because business questionnaires often ask for information regarding different activities and parts of the businesses, the response process may also involve more than one respondent. If so, the person who reports to the surveyor will often act as a coordinator of an internal data collection process.

2. It is important to acknowledge both the differences and similarities between business survey and social survey questions. There are two basic types of questions in surveys: (a) those regarding behavior and events, including their outcomes; and (2) those about opinions and attitudes (Schaeffer and Presser 2003). Business surveys are different from social surveys in that attitude questions are rather rare, and hence that the balance between question types is different. The attitude questions asked in business surveys are also more often evaluations on behalf of the company rather than questions about an individual's personal values and attitudes. The behavioral questions also have specific features. They generally focus on the economic input and output (results) of operational activities rather than on the activities themselves. This information will often reside in management system databases rather than in the memory of the respondents. In Chapter 3 we suggested that the most challenging measurement problem in business surveys is that business records are organized differently from business survey questions. Neither the information nor the units in these records may coincide with the definition and units asked for in the business questionnaires. This is what is commonly called *mapping problems*, in contrast to *clarification problems* (Redline 2011). Hence, questionnaire designers need to consider the characteristics of behavioral and attitude questions, the mixture of different kind of questions, and the mapping challenges that are typical to business surveys.
3. The response burden seems to be an even more important and critical issue in business surveys than in social surveys. Time is a tangible cost in businesses. Therefore we must expect the time that businesses set aside for a survey to be limited and that the respondents are expected to complete the questionnaires in an efficient way (Haraldsen and Jones 2007). If the management thinks that the questionnaire looks unprofessional, they might not allocate much time for it. Also, if the respondents find the questionnaire burdensome to read, the questions difficult to understand and time-consuming to respond to, they will probably find their own and less fortunate ways of reducing the response burden. As a result, they may skip important information and offer rough estimates instead of accurate answers. Therefore it is important that the questionnaire present itself as an efficient measurement instrument.

These reflections about survey communication modes, the relationship between survey elements, and tailoring can be summarized as follows:

- Take quality challenges specific to business surveys concerning the roles respondents play, the kind of information we seek, and the vulnerability to response burdens as a starting point.
- Design the questionnaire in three steps:
  1. Start by suggesting how the design challenges should be met in a paper questionnaire.
  2. Then look for improvements possible in a web version.
  3. Finally, consider how the data collection can be further improved by personal contacts between the surveyor and the businesses surveyed.
- Apply a cautious and holistic approach to the presentation tools at hand. This twofold principle should both govern how different sections and individual questions are designed.

From here we move to a more concrete discussion on how business questionnaires should present themselves to the respondents, how business respondents should be given access to the questionnaires they are expected to complete, and how questionnaires should be tailored to a situation with several information sources.

8.3.1 Setting the Scene

Visual properties, such as size, color, and striking text elements, facilitate a basic understanding of what kind of document the respondent faces. A visually pleasant design may also make the respondents have a more favorable view of filling out the survey (Norman 2004). In this way, first impressions set the scene for the relationship between the respondent and the surveyor. This first impression is important because the respondents' initial understanding of the relationship they are invited to take part in and how user-friendly the questionnaire appears, will affect how carefully and seriously they will later respond to the questions posed in the questionnaire.

We think that a questionnaire in the same format as the cover letter provides a neat and professional appearance. The visual presentation and text styles should be guided by cautious use of colors and sans serif typefaces. We recommend starting with a title page, or alternatively reserve page 1 for key information about the survey and for questions that confirm or update information that define the business unit for which the respondent should report. A simple drawing that refers to the title of the survey and questions given a visual design similar to that shown in Figure 8.3 could have a positive effect on the respondent's initial motivation.

We act differently in different social situations (Schober 1998). Because the terms and classifications used in the business questionnaire may differ from similar information that respondents report to their management, it is important that the respondents acknowledge these roles as different. One unfortunate difference, however, is that inaccurate reports to surveyors may not have the same consequences for the business respondent as inaccurate reports to the business management. If the relationship with the surveyor is perceived as a fleeting acquaintance with no lasting implications for the company and no obligations for the future, surveyors cannot expect respondents to always seriously attempt to produce and report accurate figures and responses. Unless they know that the figures will be checked by their management before they are released, or that they errors will be followed up by the surveyor, business respondents may not feel obligated to ensure that they understand the terms and tasks exactly as they are intended. Therefore, setting the scene in business surveys is a twofold challenge: (1) to clearly distinguish between business surveys and management reports and (2) to communicate that this difference is not based on the accuracy of the information. It is important that the business questionnaire present itself as part of a serious business request and as a part of the job that is expected from professional enterprises (Gower 1994).

8.3.1.1 Creating Trust

The relationship between respondents and the survey organization is influenced by the their trust in both the survey industry in general and the survey organization that conducts the current survey. As we recognized when we discussed nonresponse issues in Chapters 3 and 6, business respondents seem to have more confidence in surveys conducted by official statistical agencies than those conducted by nonofficial business surveyors. One reason for this is probably that they recognize and appreciate the fact that official statistical agencies do not have any commercial errand or agenda. In many countries the national statistical offices are also well regarded for maintaining a high quality standard. Moreover, reports presented in official business surveys may be considered similar to other reports that the businesses are obliged to provide, such as for tax authorities and other governmental institutions. Together with this acceptance of official, clearly noncommercial surveys, however, there seems to be a general skepticism toward the survey industry.

Anticipated skepticism toward the survey institute or to the topic of the survey should first of all be addressed during the surveyor's initial contact with the sampled businesses (see Chapter 9 for recommendations on this approach). However, concerning the questionnaire itself, we recommend keeping the questionnaire in a sober, official-like style. Questionnaires with fancy visual effects in striking colors can resemble commercials or "junk mail" and risk ending up in the business' dust bin. If the survey institute is well known and well respected, the logo of the institute printed on the top of each page and the signature of the director in the cover letter are examples of features that communicate seriousness and authority. In his book on designing effective web surveys, Mick Couper (2008) distinguishes between task- and style-related visual elements in questionnaires. While task-related elements relate to the questions, style-related elements communicate something about the surveyor presenting the questionnaire, such as whether it is a commercial or nonprofit agency.

Another trust issue that is more directly linked to the content of the questionnaire is the relationship between the title and the introduction letter and questions that follow. Does the questionnaire deliver what it promises? Consider confidentiality assurances as an example. Businesses may be reluctant to disclose market-sensitive data unless the confidentiality measures described in the cover letter appear reliable. If so, and if confidentiality measures are highlighted in the cover letter, they are not very trustworthy if, at the same time sensitive questions are printed on the first or last pages of the questionnaire, where they are openly visible to everyone. On the other hand, strong confidentiality assurances in the cover letter in a survey that apparently does not contain sensitive questions may also cause suspicion. If confidentiality assurances are underscored, it should also be clear which questions are considered sensitive, and they should not be printed on the first or last page of the questionnaire.

This principle of coherent messages also applies to the relationship between the topic of the questionnaire indicated by the questionnaire title and the first questions asked. Business questionnaires often start with rather long general instructions and questions about mailing addresses, location, main activity, and operational status before the first survey topic question is posed. Except for information that legal requirements stipulate should be on the questionnaire, most general instructions can advantageously be moved to the cover letter or other enclosures.

Questions about the sampled enterprise or establishment are usually of two kinds. Some of them are needed in order to decide whether the business meets the sample criteria or which subunits of the enterprise should be reported for in the survey. Others are used to update the business register. This latter kind could be moved to the end of the questionnaire. We also recommend introducing initial clarification questions in a way that clearly indicates that these questions have to be asked before the respondent proceeds to the core part of the questionnaire.

The information requested should be clearly indicated in the survey title and followed up by subheadings throughout the questionnaire. One source of confusion often seen in business surveys is that the title refers to the statistics that will be produced by the surveyor rather than to the sort of information that the respondent is asked to provide. Terms like *statistics* and *index* should not be used in questionnaire titles because they refer to the aggregated results that the surveyor will produce and not to the information collected in the questionnaire.

One of the main annual surveys run by NSIs is commonly called the *structural business survey* (SBS). It collects a number of detailed items about human resources, production facilities, activities, and financial results. The purpose is to provide an overall picture of the character and performance of the country's industry and commerce, the structure of the business world; and to measure how this structure changes over time. However, similar to the term *statistics* and *index*, the term *structure* probably makes more sense to statisticians than to respondents. Standing alone, the title may rather be associated with questions as to how the business is organized than with questions about revenue, costs, inventories, purchases, and employees. A good tip is to look for a better title in the main purpose of the survey. In this case that could lead to a title such as "Annual survey of business activities and results."

8.3.1.2 Assigning Tasks

Not only are business respondents more "professional" than social survey respondents. Their tasks are also differ from those presented in most social surveys.

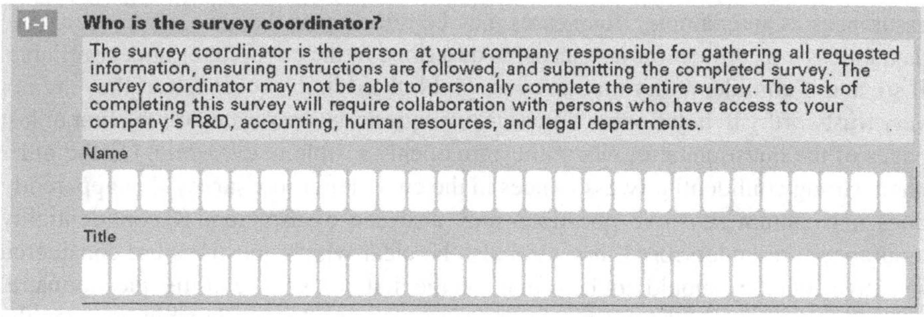

1-1 Who is the survey coordinator?

The survey coordinator is the person at your company responsible for gathering all requested information, ensuring instructions are followed, and submitting the completed survey. The survey coordinator may not be able to personally complete the entire survey. The task of completing this survey will require collaboration with persons who have access to your company's R&D, accounting, human resources, and legal departments.

Name

Title

Figure 8.5 The business respondent identified as survey coordinator.

Business respondents do not report about themselves as individuals, but rather about the company they are working for. Because the questions do not ask for personal information, a personalized style, which is commonly recommended in social surveys (Dillman et al. 2009), may not be appropriate for business surveys. Instead, the respondents should be addressed in a more formal way.

The respondent's role as informant, collecting information from various sources, should be also highlighted. Sometimes a contact person is identified before the data collection begins, with the appellation "contact person" or "survey coordinator" on the first page of the questionnaire. In other instances this person's identity and perhaps signature are provided at the end of the questionnaire. Figure 8.5 gives an example of how the role as survey coordinator is described in the US Census Bureau's Research, Development, and Innovation Survey.

Other practical questions that respondents ask themselves and need to be answered include the following:

- Is completion mandatory or voluntary?
- When is the return deadline?
- How long will it take to complete the survey?

Official business surveys are usually mandatory. By tradition, this is often indicated in the upper right corner of the first questionnaire page. Since that is the least prominent place on pages that normally are read from left to right, this information might not be recognized by first-time respondents. On the other hand, as long as business respondents generally expect official business surveys to be compulsory, it might not matter much if this text is missed. More commonly, respondents will probably not notice when a business survey that is expected to be mandatory actually is voluntary (Willimack et al. 2002). Another, potentially more serious, consequence would be failure to note the return date, which often is printed together with the obligation to respond. Therefore, to ensure greater visibility, we think that both the text indicating that the survey is mandatory and the deadline date should appear somewhere inside the upper left quadrant of the front page.

It is probably unnecessary for the survey organization to estimate how long it will take to complete a paper questionnaire. We believe that respondents will prefer to make their own estimates simply by skimming through the pages. In web questionnaires, clues that indicate how long it will take to complete the questionnaire are less obvious, and thus the need for an estimate made by the survey organization is more important. Filter questions and the wide variation in how much different companies should report make this task quite difficult. If the surveyor has included questions about response time in previous, similar surveys, however, it should at least be possible to suggest an interval that applies to most respondents. This kind of information could also be tailored to businesses of different sizes or in different industries.

A point that we will return to several times in this section is that the questionnaires should be designed to give the respondent the impression of having an ongoing relationship with the surveyor. One way to encourage such a feeling is to add an evaluation question at the end of the questionnaire where the respondents are asked to compare the time estimate suggested by the surveyor with their actual experience, and to comment on differences.

8.3.2 Common Access to Several Web Surveys

When business questionnaires are distributed electronically, businesses that are sampled in a number of surveys or accountancy offices that respond for several companies might be allowed to access the surveys that they are expected to complete from a single point of access, a web portal. Currently, many statistical institutes allow electronic or web-based reporting for their business surveys; there are also a variety of NSI web portals on Internet. In the previous section on survey modes, we indicated that in self-administered surveys, the whole response process is controlled by the words, structure, and visual design of the questionnaire. In web surveys this applies not only for the question–answer process but also for the administrative tasks that are necessary in order to open the questionnaire and mail it when the questions are answered. Presented on Internet, these activities are governed by questionnaire screens with descriptions of tasks and response alternatives that, when activated, initiate different actions. These screens are just as important for the survey as the screens on which the survey questions are posed. Consequently, those screens should be designed according to the same wording and layout principles used in the survey questions. Usability might also be enhanced if the different types of questionnaires in a web survey system have similar designs. Hence, we recommend applying the principle of visual consistency [see Morrison et al. (2010, principle 3)] not only to the survey questionnaire, but also to the relationship between the administrative and the substantial questionnaires in a web survey system.

8.3.2.1 Simple and Advanced Business Web Portals

A *web portal* is a website where businesses can find and access different surveys or other reporting obligations. Business survey portals differ in layout, in how many surveys they give access to, and how the surveys are accessed. In the United States,

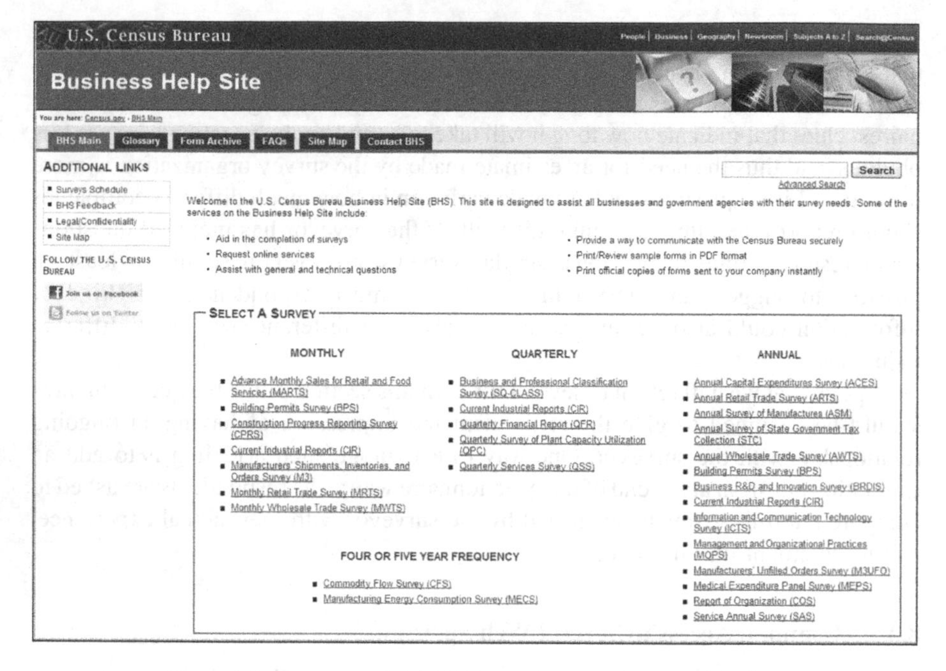

Figure 8.6 The design of the US Census Bureau's Business Help Site.

official business surveys are accessed through different portals owned by different survey organizations. An example of such a website is the US Census Bureau's Business Help Site (`http://bhs.econ.census.gov/bhs/index.html`). Figure 8.6 shows the homepage for this web portal.

The US Census Bureau is only one of several federal institutions that conduct business surveys. In Europe it is more common for national statistical agencies to collect most of the data used in official business statistics, in which case the list of surveys will be longer. An example is Statistics Finland, which lists alphabetically more than 60 data collections concerning enterprises on its website (`http://stat.fi/keruu/yritys_en.html`). From this list participants can look at survey information before they log into individual homepages of the surveys for which they are selected to participate.

Surveys may be accessed individually or by a common password that grants access to all surveys presented in the portal. In the latter case, the business identification can be used to tailor the list of surveys to ensure that only those relevant to the enterprise are listed.

If the list is long, as in the example from Finland, it can be difficult to find the business survey one is looking for. At the Finnish website the surveys are listed alphabetically by the first word in their title. Because the first word seldom is the best identifier, this is not very user-friendly. Examples are that the "quarterly inquiry on international trade in services" is listed under "Q" and not "I" for "International" or "T" for "trade" and that "inquiry of manufacturing commodities"

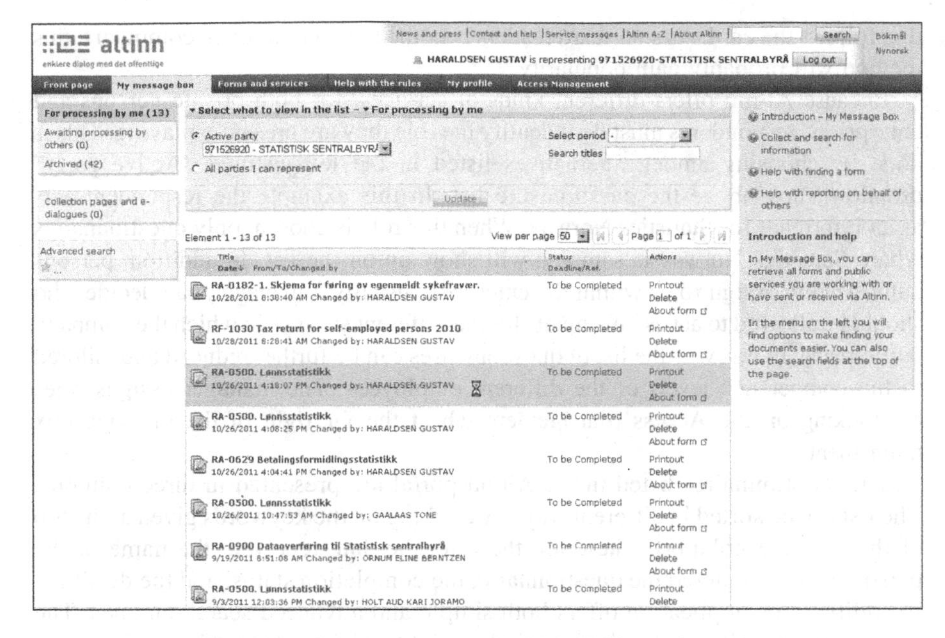

Figure 8.7 List of questionnaires as presented in a web portal.

is listed under "I" and not "C" for "commodities" or "M" for "manufacturing." A list based on significant keywords would probably work better.

More advanced and extensive web portals are found in Norway and Denmark. Altinn.no (`https://www.altinn.no/`) and Virk.dk (`http://www.virk.dk/pid/1`) are web portals that not only lead to business surveys but also give access to a number of other official forms that businesses are obliged to fill out. These portals also serve as a common point of access to governmental services and official forms for individuals and households. We will return to some of the possibilities that this portal structure offers in the next section. Here, we focus on the design of the screens that take the respondent to the questionnaires.

Figure 8.7 lists business surveys conducted by Statistics Norway the way that they are presented in the web portal Altinn. Altinn is the Norwegian authorities' joint solution for reporting and dialog with businesses. The portal also offers services for private citizens, the most important being the opportunity to hand in their tax-return form electronically.

The Altinn user logs into the system directly and not into each questionnaire, as was the case in the US and Finnish examples. Here one fixed password and one temporary password are used. Temporary passwords are commonly ordered in advance and received via mail, sent by SMS (short message service) after a fixed password is entered or generated by a password calculator. Altinn offers all these methods, as well as login with smartcards. Login with a smartcard requires connection of a smartcard reader to the user's computer and installation of the necessary software that comes with the card reader. The user logs in with a pin code that follows the card. As smartcard readers become more common in computers, this method will probably gain popularity.

Because Altinn offers different kinds of services, for both private persons and enterprises, respondents must first identify the role they are presently playing. This is done by choosing among alternatives listed in the left-justified "Active party" dropdown box above the questionnaire list. In this example the respondent acts as an informant for Statistics Norway. When this role is chosen, only questionnaires where Statistics Norway is sampled will show up on the list. In addition, persons authorized to assign roles within the enterprise, usually a manager, can decide who should be allowed to act as informants for the different surveys in which the company participates. In this way the list of questionnaires can be further reduced and tailored to the competence levels of the different employees. The manager assigns roles by clicking on the Access Management tab at the far right of the message box main menu.

The questionnaires listed in the Altinn portal are presented in three columns. The list can be sorted in different ways by clicking on the keywords given at the top of the two first columns. These are the title, the date received, the name of the person who last entered the questionnaire, the completion status, and the deadline. In addition, the application offers both simple and advanced search engines. The simple search criteria are time, period, and title. The advanced search engine provides a number of other options. Without going into more detail, it is obvious that these features make it easier for respondents to find the surveys they are requested to complete.

On the screenshot in Figure 8.7, all items on the list are questionnaires. This is depicted by the pencil-and-document symbol in front of the questionnaire number and questionnaire title. An interesting feature that has not yet been utilized in business surveys, is that the message box also can include emails. Emails are shown with an open or closed envelope symbol in front of the topic name. Emails in the message box could possibly vitalize the relationship between the survey organization and the respondent. Qualitative interviews with business respondents have indicated that insight into and understanding of the data recipient's job may have a positive effect on their own commitment as survey respondents (Torres van Grinsen et al. 2011). One option could therefore be to use the message box to send respondents reports on how the data collection progresses and information about results after the data collection is finalized. In Chapter 9 we will discuss the types of dialog which are appropriate at each step of the data collection process.

8.3.2.2 Visual Design of the Web Portal

The screen in Figure 8.7 is visually divided into a navigation area along the left margin, an input area in the middle, and an information area along the right margin. In addition, there is a header at the top with tabs and hyperlinks to different portal services. This is a rather advanced design, but still quite common on websites. More importantly the design introduced in this portal basically is the same as the one that respondents will find in the individual questionnaires (see Figure 8.11). In the message box the navigation pane offers navigation between different lists of questionnaires, while in the questionnaires it offers navigation between different set of questions. In the message box, the input area is where the respondent chooses a questionnaire; in the questionnaire, this is where the respondent reads questions and responds to them. Finally, the information given in the right-justified pane in the message box provides help with message box operations; information given in the questionnaires provides help with the questions or present error messages. Hence, even if the navigation, input, and information content change, the kind of information presented remains fixed. When respondents familiarizes themselves with this method of presenting information on the survey administration screen, the navigation on the questionnaire screens should also become easier.

Mick Couper distinguishes between the primary and secondary, or supportive, tasks that survey respondents are asked to perform (Couper 2008). On the questionnaire pages, the primary task is to focus on the questions and respond to them, the secondary tasks consist in navigating from question to question, seeking help if needed, and responding to error messages. Even if the tasks the respondents are asked to perform on the portal page are different, they can also be divided into primary and secondary tasks. The primary task is to ensure that the respondents enter the questionnaire that we want them to complete; secondary tasks are to check whether the respondent is reporting for the right unit, to bring up the right list of questionnaires, sort that list, and form a first impression of the nature of the questions that are waiting. The visual design should work together with the text in a way that makes these tasks easy, efficient, and accurate.

We see before we read. Colin Ware has described how people process visual information in three steps, beginning with an overview of the overall structure of the page or screen and ending with focus on the details that seem to be the most important (Ware 2004). If we combine this description with Couper's distinction between primary and secondary tasks, the objectives of the elements on the web portal and questionnaire screens will be to

- Identify the basic visual properties of the page or screen for the respondents
- Help them distinguish between different visual elements and perceive the relationships between them
- Direct their attention toward the primary task
- Help them focus on the primary tasks and alert them to the secondary or supportive information and features available

We will not give a lengthy discussion of how well the web portals presented in Figures 8.6 and 8.7 may serve this purpose, but will mention a few points that we think are worth reflecting on.

Both the US Census Bureau's Business Help Site and the Altinn message box present information about the services they offer and how these services work. However, neither site communicates these offerings clearly. On the US Census Bureau's homepage, the most prominent feature is the box containing survey links and the request to select a survey. Other features on the site are not as noticeable.

On the Altinn portal the shadowed box at the top where respondents can specify their roles, the name of the survey, and the period is perhaps what first catches the user's attention. Then, because providing this information leads to an update of the list below, the next step will obviously be to choose a survey on the list. Respondents who fail to find the survey they are looking for will probably turn to the right or left pane for help. The right pane contains links to a general introduction on how the message box works and what services it offers. A more active promotion of this information and these services could be considered.

Following the principle of caution, elements on web screens that may distract the respondent from the primary task should be avoided, such as a hyperlink that takes the respondent away from the survey to some point of no easy return out in cyberspace (Couper 2008). In a test reported by Haraldsen (2004), one test respondent who was an eager Internet user was able to move from the Norwegian test survey to one of the statistics reported by Statistics Canada. For another example look at the information and login screen for the US Census Bureau's web surveys in Figure 8.8. Even if the icon that the respondent should click on in order to report online is quite prominent on the screen, there will also be a number of hyperlinks leading the respondent away from the questionnaire to statistical presentations and other services.

Common to both the US Census Bureau and the Norwegian Altinn portal is a feature that requires respondents to click on a hyperlink text or button to begin reporting. While this provides a straightforward method for respondents to open a questionnaire, it is still different from the response alternatives presented later in the actual questionnaires, which are radio buttons, dropdown boxes, checkboxes and various textboxes.

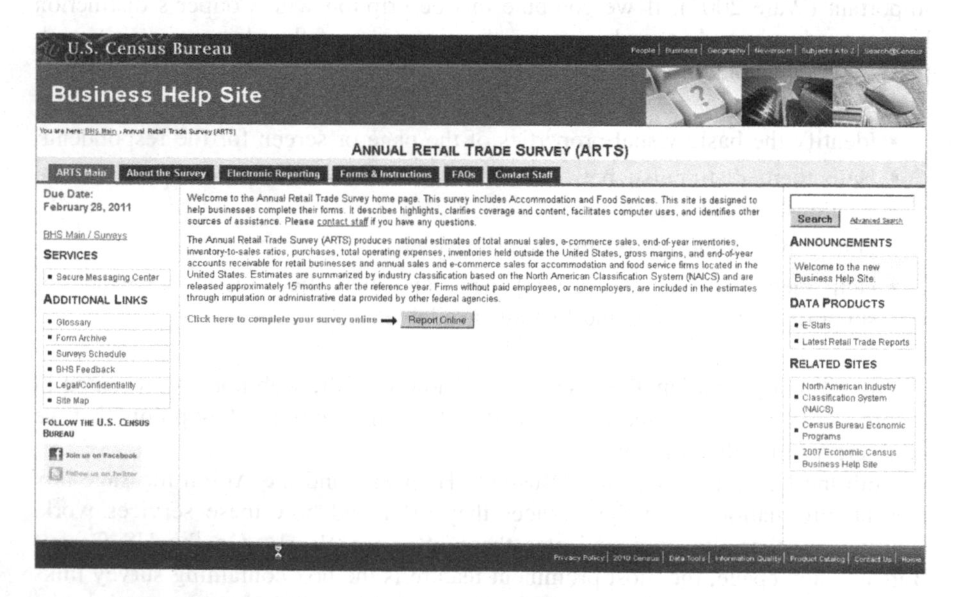

Figure 8.8 Information and login screen for the US Census Bureau's web surveys.

The Altinn portal screen presents a combination of radio buttons and a dropdown box to specify the respondent's role. Two textboxes are used to search for questionnaires. When a specific questionnaire is identified, the response options are to (1) begin reporting, (2) print a paper copy of the questionnaire, (3) learn more about how it should be completed, or (4) delete it from the message box. Option 1 is activated by clicking on the questionnaire title. To select option, 2, 3, or 4, the respondent clicks on one out of three hyperlinks in the "actions" column, which is quite distant from the questionnaire titles. This mix of response formats presented in different places deviates from the principles of grouping related elements and ensuring visual consistency (Dillman et al. 2009). We suggest using radio buttons instead of hyperlinks as response options because the former clearly communicate the questionnaire character of the web portal screen. Moreover, we would suggest placing these four radio buttons together in the actions column. Shading certain lines, as shown in the example in Figure 8.7, help respondents first to see and then read the survey title, note the completion deadline or status, and finally, chose an action option.

8.3.2.3 Data Collection Instruments for Small Companies and Small Questionnaires

Web portals are particularly useful for larger businesses that are expected to respond to a number of questionnaires. Those who probably profit most from a common portal that accommodates business surveys and other governmental reporting obligations are accountancy companies, which serve as respondents and also perform other reporting tasks for a number of firms. Web portals are also best suited for recurrent surveys. At the other end of the scale, we find smaller firms that may be selected for only a one-time survey. For businesses asked to respond to only one or a small number of surveys, going through a web portal seems to be an unnecessary detour. Access management systems, like the one offered in Altinn, will probably also seem to be a hassle for businesses that only have a few employees who are relevant respondents. For these businesses a more direct path and a simpler login procedure is probably preferred. The following is a story suggesting that the advanced facilities offered in a web portal may be either unrecognized or unappreciated:

> Until December 2006, the Quarterly Survey of Self Reported Sick Leave, conducted by Statistics Norway, was offered on paper with an Altinn web version as an alternative. The web pickup rate had been below 10% for a long time. Therefore it was decided to make the survey available in Idun, which is a simpler, low-security web portal also used for business surveys. To access this portal, the user only needs to know a username and password printed on the paper questionnaire. After this web alternative was introduced, the web rate rose to close to 50% within a quarter. Later the web pickup rate jumped to above 90% when a paper version was no longer enclosed with the survey invitation letter (Haraldsen and Bergström 2009).

Because of the wide variety of businesses, professional surveyors should both be prepared to offer their questionnaires in web portals and by more direct access methods. Direct-access methods are useful not only for businesses that are small; the business survey is often small as well. If only a few questions are asked, a complicated login procedure may appear more burdensome than the questionnaire itself.

A self-administered mode that is extensively used and has been successful with short questionnaires in the UK is touchtone (or telephone) data entry (TDE) [Thomas (P.) 2002a]. Touchtone data entry uses the tones of a telephone keypad to make responses and to allow respondents to access a set of recorded messages. With modern smartphones, a dialog could also be presented as a simple questionnaire on the telephone screen. Hence, the traditional TDE moves toward an electronic questionnaire on the mobile phone. Another extension of touchtone data entry is to replace keying with voice recognition; this is called *interactive voice response* (IVR) or *audio computer assisted self interviewing* (ACASI) in survey methodology.

In the UK, surveys containing less than 10 items offer TDE as an alternative to paper questionnaires. As long as TDE was offered as an alternative to an already available paper questionnaire, the pickup rate was rather low (11%). When a paper version had to be ordered, however, as many as 97% responded by TDE (ibid.).

Another interesting design option is to offer TDE to respondents who have nothing to report. In Chapter 3, we discussed a sampling problem with short time statistics on international transactions where only an unknown minority in a traditional business sample have something to report. With a broad sample, this may lead to frustration for the majority of those sampled. On the other hand, if we restrict the sample to business that we are quite certain that they have something to report, we may miss important data. This is an example of a survey for which a TDE option for those who have nothing to report might be useful. In the UK, TDE is used to collect nil responses from surveys such as the "Capital Expenditure and International Trade in Services". Thomas (ibid.) reports that 35% of the nil data are returned by this method. The TDE responses also save processing costs compared to paper questionnaires. Concerns about respondents who would be inclined to report nil response even if they have something to report have also proven unwarranted. Overall, TDE seems to be a cost-efficient mode for respondents who are reporting nil (zeros) or only a few figures (ibid.).

8.3.2.4 Mobile Data Collection Instruments in a Mixed-Mode Design

In Chapter 3, we discussed the Norwegian lorry (truck) transport survey, in which the number of transport firms that had been inactive during the reference period were suspiciously high. The anomalous data were probably due partly to the high response burden of this survey. Compared to other business surveys in Norway, this survey also has an enduring low web response rate. A follow-up study conducted in 2009 indicated that this low response rate may have resulted because respondents found web reporting to be even more burdensome than filling in a paper questionnaire. Drivers were unlikely to bring laptops with them and log into web surveys from their vehicles while they were out transporting products. Instead, drivers probably tended to take notes while they were driving, leaving those notes for someone at the transport office to enter into the web-based survey. If this method was the procedure, it was cumbersome and vulnerable to error. In order to increase the proportion of electronic questionnaire reporting, devices that are smaller or more mobile than laptops might be easier to use. Smartphones and tablets with Internet access represent an interesting alternative.

Initially the lorry transport survey does not present itself as a candidate for presentation on a small smartphone screen. The main part of the original paper questionnaire was a two dimensional diary matrix where transport commissions were listed vertically and a number of details for each commission were listed horizontally. If smartphone reporting were combined with a web survey, however, certain details such as geographic coordinates and timestamps identifying the start and end of each tour, could be entered into the device during driving and then later downloaded into a web survey that was then completed to include other details. To our knowledge, a mixed-mode design such as this has not yet been developed for this kind of survey, but we think it is an interesting design prospect for the future.

8.3.3 Layout and Functional Tailoring for Internal Data Collection

Not only do businesses commonly take part in several business surveys; several employees may also participate in the internal data collection process necessary in order to complete individual questionnaires. As mentioned in Chapter 3, the survey organization may have limited influence on who will be answering the questions; we will discuss this issue in the following sections.

8.3.3.1 Guiding the Questionnaire to the Right Respondent(s)

In traditional paper questionnaires, section headings are one of the most important features. Section headings should indicate what kind of competence or information sources are needed to provide answers, and consequently who the appropriate internal data provider should be. Thus, section headings should be clearly visible, not be confused with other kinds of information, and clearly communicate the information asked for. One of the strongest visual tools is contrasts. An effective way of highlighting topic change is therefore to write section headings in white text against a dark background. In the example in Figure 8.9, the shift from one section to the next is also emphasized by the rounded thematic boxes and the intervening white spaces.

In the example shown in Figure 8.10, the questionnaire starts with a table of contents that presents the topic and information requirements for each section. Each section is distinguished by a different color. This is a good example of text and visual design working together to communicate the same message.

8.3.3.2 Guidelines and Tailoring Tools in Web Questionnaires

Now we leave the paper questionnaire again, and turn to the opportunities in web questionnaires to give an overview of the questions and indicate who should provide what kind of information.

When the questionnaire is on paper, the respondent can easily skim through the pages to get a clearer idea of the topics and the number of questions in each section.

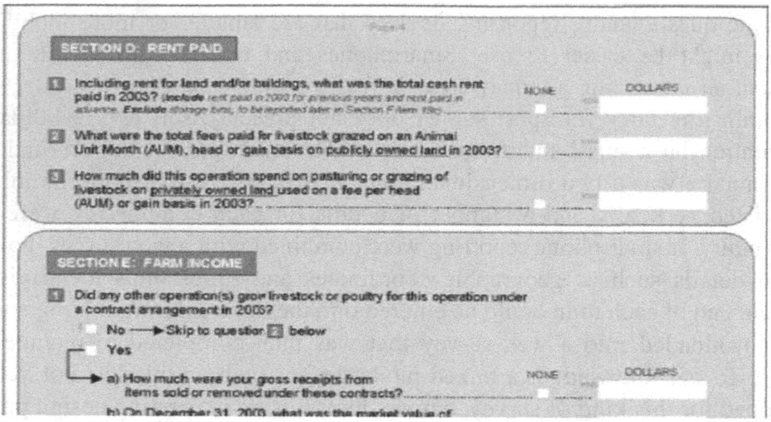

SECTION D: RENT PAID

1 Including rent for land and/or buildings, what was the total cash rent paid in 2003? (Include rent paid in 2003 for previous years and rent paid in advance. Exclude storage bins, to be reported later in Section F item 19c) ..... NONE ▢ ..... DOLLARS

2 What were the total fees paid for livestock grazed on an Animal Unit Month (AUM), head or gain basis on publicly owned land in 2003? ..... NONE ▢ ..... DOLLARS

3 How much did this operation spend on pasturing or grazing of livestock on privately owned land used on a fee per head (AUM) or gain basis in 2003? ..... NONE ▢ ..... DOLLARS

SECTION E: FARM INCOME

1 Did any other operation(s) grow livestock or poultry for this operation under a contract arrangement in 2003?

▢ No ⟶ Skip to question 2 below

▢ Yes

⟶ a) How much were your gross receipts from items sold or removed under these contracts? ..... NONE ▢ ..... DOLLARS

b) On December 31, 2003, what was the market value of

Figure 8.9 Section headings. [From Dillman et al. (2005).]

This task is more cumbersome in web questionnaires, where the different sections are usually indicated either in a Windows Explorer–type navigation menu along the left margin or by tabs along the top of the computer screen. In both cases the section names are normally also hyperlinks that take the respondent to different parts of the questionnaire. It is important that these hyperlinks be named in such a way that they indicate the competence or skills needed to complete the different sections.

Table of Contents

Survey Overview and Table of Contents

The survey is divided into six sections. Each section asks questions about different aspects of R&D or innovation at your company. Due to the specialized nature of each section, it may be necessary to collaborate with colleagues in different departments to answer the questions. The sections are color coded and cover the following topical areas:

Figure 8.10 Questionnaire table of contents. (From the US Census Bureau Business R&D and Innovation Survey.)

The respondents should be able to answer the sections in the order they want. They should also be able to answer some of the questions, exit the web survey without losing any of the answers, and be able to return to their last response when they log in again. When all questions in a section are completed this is often indicated by the section title taking a different color or by a check mark in the navigation menu.

In Figure 8.11 we give an example of a web questionnaire with a left-aligned navigation menu (screenshot from the Norwegian web portal Altinn), followed by an

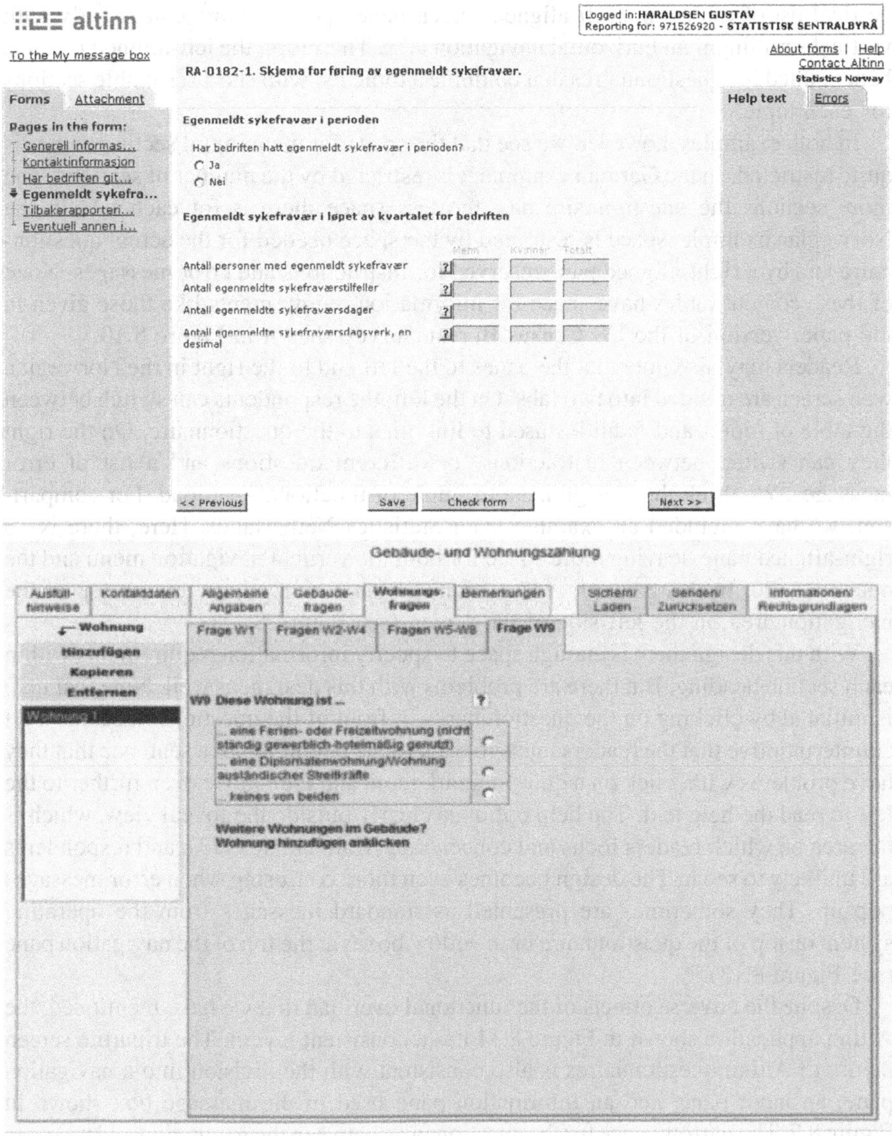

Figure 8.11 Left-aligned and tab-based navigation menus.

example a of tab-based navigation menu (screenshot from a German business survey).

To our knowledge the usability of these two alternatives has not been systematically compared. However, user tests and experiences indicate that the tab-based navigation menu at the top of the screen is quite difficult to understand (Blanke 2011). Several other arguments also point toward the left-aligned menu as being better. Navigation menus at the left margin of the screen are more common in other kinds of websites. This layout is also more similar to tables of content. Finally, a vertical list of topics in a left-aligned screen pane operates more consistently with vertical scrolling than horizontal navigation tabs. Therefore, the left-aligned menu is better suited for questionnaires that combine a topic list with short scrollable sections for each topic.

In both examples, however, we see that the space for meaningful section names is quite restricted. In the German example it is restricted by the number of sections. The more sections the questionnaire has, the less space there is for each tab. In the Norwegian example, space is restricted by the space needed for the actual questionnaire and by a right-aligned pane reserved for instructions and error messages. None of these content tables have space for information requirements like those given in the paper version of the US Census Bureau survey shown in Figure 8.10.

Readers may also note that the panes to the left and to the right in the Norwegian web screen are divided into two tabs. On the left, the respondents can switch between the table of topics and facilities used to link files to the questionnaire. On the right they can switch between instructions for different questions and a list of error messages. We think this design is an example of functional overload. For comparison, we have included an example from Statistics Netherlands. Here, there is no right-aligned pane, leaving more space for both the vertical navigation menu and the questionnaire. In this example, instructions are shown in popup text on top of the navigation area on the left side of the screen (see Figure 8.12).

With this design there is enough space to specify information requirements within each section heading. But there are problems with this design, as well. Note that help is initiated by clicking on the questionmarks in front of the questions. It seems a bit counterintuitive that the readers must go back to the beginning of a sentence that they have problems with, click on a questionmark icon, and then move even further to the left to read the help text. The help option is clearly outside the foveal view, which is the area on which readers focus and concentrate (Kahneman 1973), and respondents are unlikely to see it. The design becomes even more confusing when error messages pop up. They sometimes are presented as standard messages from the operating system on top of the questionnaire or in yellow boxes at the top of the navigation pane (see Figure 8.13).

Despite the adverse effects of the functional overload that we have mentioned, the Altinn application shown in Figure 8.11 has a consistent layout. The tripartite screen layout of Altinn questionnaires is also consistent with the division into a navigation pane, an input pane, and an information pane used in the message box shown in Figure 8.7. The narrow space for the questionnaire between the navigation and message panes could perhaps be broadened by designing in-sliding navigation and message

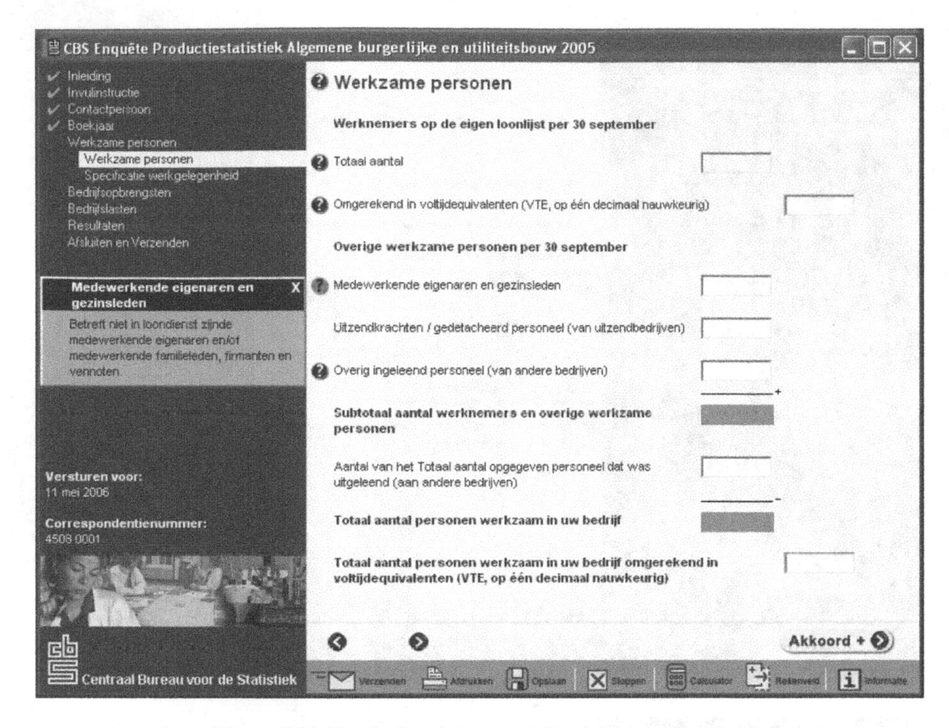

Figure 8.12 Standard web layout of Dutch business surveys.

panes, to show all three columns simultaneously or only the navigation and questionnaire pane, only the questionnaire pane, or only the questionnaire and message pane as needed. The movement of the panes sliding in from the left or right might also attract the respondent's attention to navigation issues, or help messages, respectively. In this way, a consistent layout can be combined with effects that clearly interrupt normal reading flow when necessary [see Morrison et al. (2010, principle 6)].

As questionnaires tend to be divided into sections that call for different kinds of knowledge, one major advantage of the web survey is its immediate availability to all employees who are granted access, unlike the paper version, which needs to be copied and distributed internally to collect information from different informants. The study by Bavdaz (2010b) referred to in Chapter 3, revealed that survey coordinators who needed assistance with a paper questionnaire sometimes copied and distributed it without the enclosed instructions. At other times they called colleagues and carried out a kind of telephone interview. Some also simply made their own estimate without asking anybody (Bavdaz 2010b). A web questionnaire that can be accessed by different reporters should make it easier for employees with relevant competence to contribute to the questionnaire in a uniform way. Time measurements made by Statistic Netherlands before and after the introduction of their web version of their structural business survey (SBS) also indicate that the internal data collection is carried out in a more efficient way with a web solution

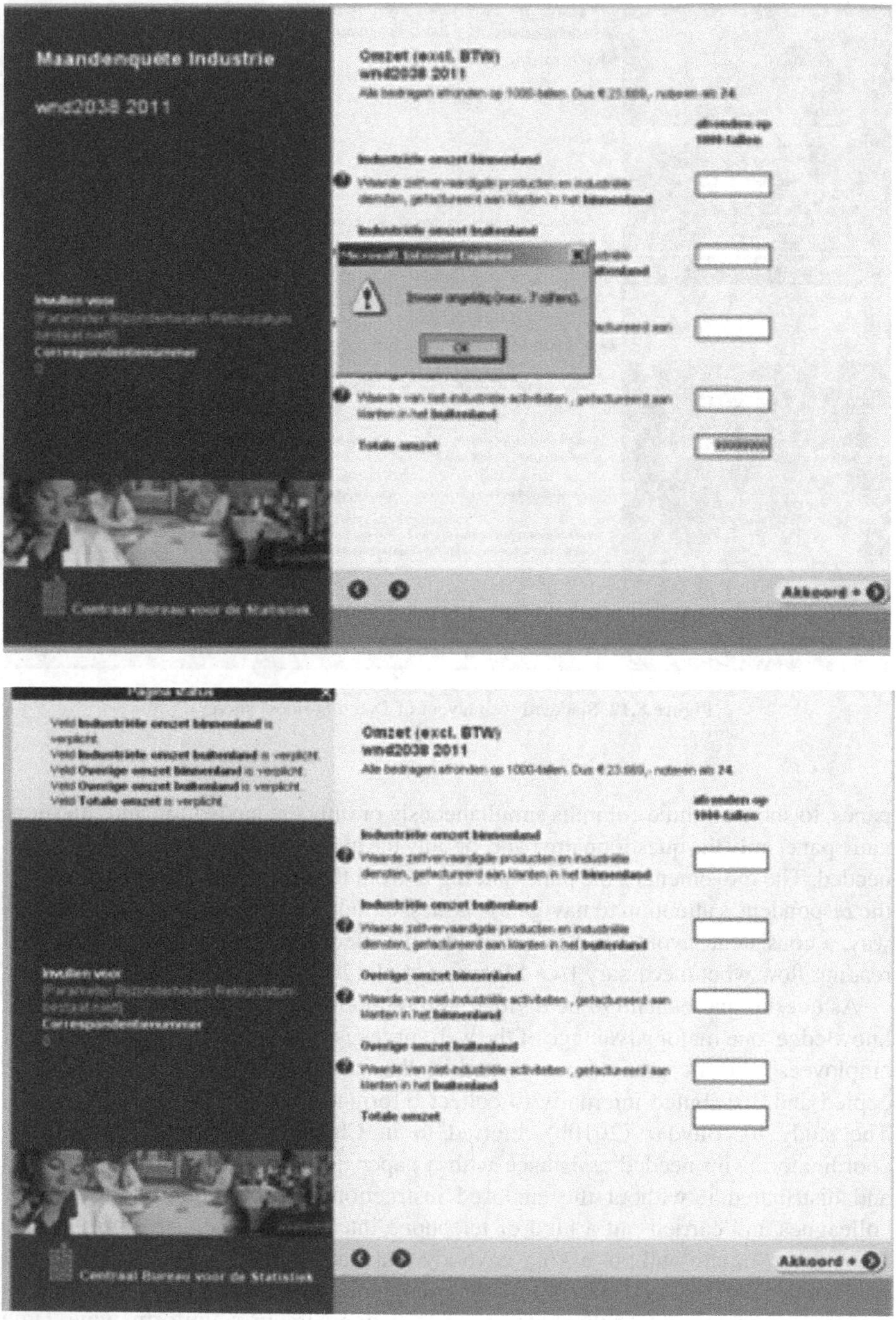

Maandenquête Industrie

wnd2038 2011

Omzet (excl. BTW)
wnd2038 2011
Alle bedragen afronden op 1000-tallen. Dus € 23.689,- noteren als 24.

afronden op 1000-tallen

Industriële omzet binnenland

Waarde zelfvervaardigde producten en industriële diensten, gefactureerd aan klanten in het **binnenland**

Industriële omzet buitenland

Microsoft Internet Explorer

Invoer ongeldig (max. 7 cijfers).

OK

Waarde van niet-industriële activiteiten, gefactureerd aan klanten in het **buitenland**

Totale omzet

Invullen voor

Correspondentienummer

Centraal Bureau voor de Statistiek

Akkoord

Pagina status

Veld **Industriële omzet binnenland** is verplicht
Veld **Industriële omzet buitenland** is verplicht
Veld **Overige omzet binnenland** is verplicht
Veld **Overige omzet buitenland** is verplicht
Veld **Totale omzet** is verplicht

Omzet (excl. BTW)
wnd2038 2011
Alle bedragen afronden op 1000-tallen. Dus € 23.689,- noteren als **24**.

afronden op 1000-tallen

Industriële omzet binnenland

Waarde zelfvervaardigde producten en industriële diensten, gefactureerd aan klanten in het **binnenland**

Industriële omzet buitenland

Waarde zelfvervaardigde producten en industriële diensten, gefactureerd aan klanten in het **buitenland**

Overige omzet binnenland

Waarde van niet-industriële activiteiten, gefactureerd aan klanten in het **binnenland**

Overige omzet buitenland

Waarde van niet-industriële activiteiten, gefactureerd aan klanten in het **buitenland**

Totale omzet

Invullen voor

Correspondentienummer

Centraal Bureau voor de Statistiek

Akkoord

Figure 8.13 Error messages in a Dutch web questionnaire.

(Giesen et al. 2009b). Norwegian results indicate that this is particularly true for multibusiness enterprises, which typically need to collect information from different sources (Haraldsen et al. 2011).

In a reporting system for regular municipality surveys established by Statistics Norway, one extra step is taken to facilitate internal data collection. In this system, it is possible for the local coordinator to cut and forward different parts of a questionnaire to different information sources within the municipality. Then, as different data providers respond, their data enters the original questionnaire (Dale et al. 2007b). This system encourages a more active data capture strategy by the survey coordinator than a system that simply makes the questionnaire available to different data providers.

An even more radical solution would be to change the content of the questionnaires so that they no longer reflect the information needs of specific statistics, but rather the reporting competences of different business respondents. The current division of questionnaires by statistical topics is perhaps the most cost-efficient way of distributing and collecting paper questionnaires. In an electronic system, however, the cost is not affected by the number of questionnaires in the same way. Including all questions that belong to a certain statistics in the same questionnaire ensures that all the needed information is collected at the same time and within the same deadline. From the respondents' perspective, however, questionnaires addressed to different professions within the company still seems to provide a better way to ensure valid and reliable information.

8.3.3.3 Intercomputer Communication

In addition to knowledgeable colleagues, an important source of information for business surveys is the companies' administrative records. When the information asked for is available in an accountancy or personnel system, it seems to be a cumbersome detour for respondents who report in a computerized questionnaire, first to log into the company's administrative system, then look up figures there, copy them, and finally paste them into the relevant response box. A more efficient procedure, particularly in recurrent surveys, would be to link the response box to the relevant post in the administrative system so that it will be automatically updated each time figures are to be reported.

This method can be realized in two ways; by

1. Implementing a procedure in the administrative software that transports data from this system to the survey questionnaire, and from there further on to the survey organization. In Norway, major accountancy and personnel software suppliers have implemented such a system for transferring data on wages and sick leave to the national statistical office. When the administrative system user clicks on the option "Statistical reporting," the data are first transported to a questionnaire that resides in the Altinn web reporting system. There the data are checked in the same way as if they were entered manually. Any errors are reported back to the administrative computer system and can be corrected before the data are resubmitted (Vik 2007).

2. Initiating and controlling the data capture from the web questionnaire. In this way, the software companies are not responsible for providing the tool used to control the mapping process and establish a link between the administrative system and the survey system; this is done from the survey questionnaire. This affords the surveyor, who has designed the web questionnaire, better control of the mapping. Mapping tools designed by software companies may differ in quality and usability from one software producer to another, and between different versions of the administrative computer program. If the mapping is done from the web questionnaire, all respondents are working on the same application. Working from the survey questionnaire is also better when automatically captured data are combined with questions that need to be answered by the respondent. Moreover, we believe that it is easier to correct answers that trigger error messages when the respondents also see the questions and the context within which the questions are posed. Finally, working in the web questionnaire may also give the respondent an opportunity to modify captured data that do not completely fit the survey definitions. This is particularly important because data requests rarely map 100% with information stored in administrative records (Roos 2011).

There is a noteworthy similarity between this data capture design and the data collection designs that we described for the Norwegian municipality reporting and the design suggested for mobile respondents in the lorry transport survey (see Section 8.3.2.4). The common denominator for all these designs is that data from different sources and subquestionnaires are collected into a "mother questionnaire." Here, still missing data are added and quality controls are run before the information is sent to the surveyor. We believe that this is a kind of mixed-mode design that warrants further study and development.

8.4 PHRASING QUESTIONS IN BUSINESS SURVEYS

Setting the scene and tailoring the questionnaire to how businesses operate is important for motivating the respondents to address the questions in a serious way. Still, the primary task is to read, understand, and respond adequately and correctly to the questions that are posed in the questionnaire. The key model behind question wordings is Tourangeau's four-step cognitive model (Tourangeau 1984; Tourangeau et al. 2000). With this model we see two main challenges: on one hand to avoid misinterpretation; on the other to avoid unnecessary response burden (see also Figures 3.4 and 3.5).

8.4.1 From Requirements to Inquiries

A striking feature of business questions is that they often are not real questions, but only keywords or imperatives combined with response labels that indicate the information that the survey organization is requesting. The following examples are from two Norwegian business surveys:

Q1. Merchandise. By *merchandise*, we refer to goods that are bought and sold again without being processed, that is, pure merchandise goods. Income and cost for services and self-produced goods should not be included here.

Q2. Total turnover and withdrawals *within* and *outside* the VAT regulations. See instructions.

Sentence fragments like these render the questions tail-heavy (end-loaded). The first words only indicate what the topic of question is. To grasp the full meaning, the respondent needs to read more or even look up definitions in a separate manual. Cognitive research indicates, however, that readers concentrate more on the initial words than on what comes later (Gernsbacher 1990) and that they seldom read separate instructions. It may even be difficult to make respondents read definitions that directly follow a question text (Couper 2008), especially respondents who feel that they already know the terms used, as will often be the case in business surveys. Hence, the principle that business survey requests should be phrased as full questions is aptly listed as first priority by Morrison et al. (2010, principle 1). Some methodologists have even suggested that important clarifications should precede the information inquiry (Redline 2011). We think this unfamiliar order should be used only in cases when it is vitally important that the respondent read definitions or delineations. Normally the question should precede clarifications. Another option would be to rephrase questions to avoid the need for additional clarification. This can often be achieved by replacing terms with specifications used to define the terms. Instead of constructions such as

> "How much money did the enterprise invest in renewable power sources?" With "renewable power sources," we think about sources such as hydroelectric power, wind power, power from sea waves, solar energy, or bioenergy."

the following version is clearer:

> "How much money did the enterprise invest in hydroelectric power, wind power, power from sea waves, solar energy, bioenergy or other renewable power sources?"

Furthermore, we think that an inquiry phrased as a question or polite request rather than an imperative requirement is a more respectful way to address respondents; the latter tone could lead to irritation, reluctance, and eventually low response quality.

Survey questions point toward a response field. Thus, questions should clearly indicate what sort of answer is required and at what detail level. Two very important quality criteria for questions are (1) a clear and obvious link between the question and the response task and (2) that the response options match the information requested (Fowler and Cosenza 2008). This is often best achieved if one starts working with the response options instead of the question text. When you know what kind of response field the question should lead up to, it is easier to phrase good questions.

The questions posed in business surveys often cover quantities such as number of employees, investment figures, energy consumption, production volumes, and economic results. The response format used for such questions are open-ended, but should not be open to all kinds of answers. There are three ways of indicating that we want a numerical response what measurement unit and what amount of detail the response field should contain (Dillman et al. 2009):

1. The unit desired should be specified in the question text. Unspecified expressions like "How much" and "How many" should be avoided and replaced by specific phrases such as "How many US dollars" or "What is the number of employees?"
2. The layout of the response box should indicate the maximum number of digits that we expect. Sometimes, as in the example in Figure 8.14, we also group response boxes in order to indicate where decimals, tens, hundreds, thousands, and millions should be written.
3. The measurement unit should be repeated together with the response box. Sometimes we also label boxes that should contain numbers at different levels and write in zeros when numbers should be rounded.

This is what is called "question elements in concert," earlier in this chapter. More illustrations are given in Figure 8.14. These examples are taken from Statistics New Zealand's layout standard for self-administered paper questionnaires (available at `http://www.stats.govt.nz/surveys_and_methods/methods/survey-design-data-collection/layout.aspx`).

The risk of superficial reading and taking definitions for granted is particularly high when questions are changed in panel surveys. Respondents who are familiar with earlier versions of the questionnaire do not expect changes and may therefore skip announcements about new or revised questions (Bavdaz 2006). Therefore, highlighting changes in questions or requirements may be of utmost importance in recurrent business surveys. Information about changes is commonly given in

35	Question text	
	dollars and cents	$.
	whole dollar	$ 00
	rounded to thousands	$ 000
	percentage	%
	time in hours	hrs
	time in hours and minutes	hrs mins

Figure 8.14 Visual layout and labels in open-ended, numerical response boxes.

separate flyers or by indicating the changes in the beginning of the questionnaire. This may work well, but does violate the general principle that instructions should be presented where they apply (Morrison et al. 2010, principle 11). If changes are announced in a separate flyer, it might not be distributed together with the questionnaire when several employees are involved in filling in the information. An explanatory note together with the question or definition that has been changed will normally work better. Statistics New Zealand suggests that notes be distinguished from other kinds of instruction by the keyword "Note" followed by colon, as shown in Figure 8.15.

Another option that perhaps should be considered in questions where notes compete with other kinds of questionnaire instructions is what Statistic New Zealand calls *reminder instructions*, which are presented in bubbles to the right of the question,[1] as shown in Figure 8.16.

If bubble reminders are reserved only for this kind of purposes, we believe that their unconventional shape will catch the respondent's attention.

8.4.2 Clarifying Terms and Tasks

Even if the survey inquiries are formulated as polite questions and are designed in a way that attracts the respondents' attention to what is important, respondents may still misinterpret terms or misunderstand which tasks they have to perform in order to provide valid and accurate answers. Redline (2011) distinguishes three weaknesses that may cause misunderstanding among readers.

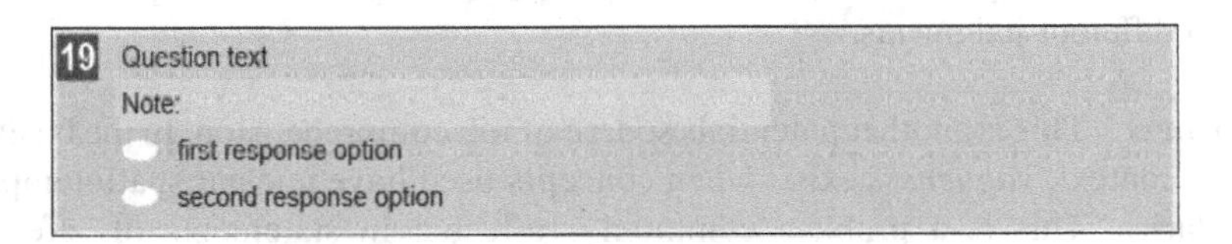

Figure 8.15 Visual layout of question note.

Figure 8.16 Visual layout of bubble reminder.

[1] Information is available at http://www.stats.govt.nz/surveys_and_methods/methods/survey-design-data-collection/layout/instructions.aspx#general.

Ambiguity This can result when terms have different lexical meanings or definitions in different linguistic or social contexts. If the respondents do not recognize this and do not recognize the role they are asked to play in the survey, this may lead to miscomprehensions or what Tourangeau and his colleagues call "misalignments" (Tourangeau et al. 2006). We have already mentioned that business surveys often define everyday business terms in survey contexts slightly different from those in routine daily business activities. Therefore business survey respondents should understand that they are performing a different tasks when they interact with survey institutes than when they interact with their business colleagues. Moreover, business respondents may also play different roles within the company and may erroneously answer the survey questions from another perspective than that intended. A typical example is when business respondents are asked to express their own opinions about conditions that also are addressed in the companies' policy declarations. Even if the questions clearly ask for personal assessments, employees may still feel obliged to report the official company position rather than their personal opinions.

The issue of which role the respondents are expected to take is particularly relevant in *business tendency surveys*. In these surveys, the participating companies are asked to judge the current situation of their business and are questioned about their plans and expectations for the near future. A number of countries are conducting business tendency surveys both monthly and quarterly; the quarterly surveys are longer and cover more aspects. There are also different versions of the questionnaire tailored to companies within manufacturing industries, construction, retail trade, and services (OECD 2003). The survey is addressed to the management, but there may be different professions represented in the management or the survey may still be answered by subordinates. Hence different professional perspectives by different respondents may lead to different judgments.

Vagueness This is another potentials source of miscomprehension. In the business survey context, *vagueness* exists when concepts used have unclear spatiotemporal boundaries. This is a problem commonly reflected in statements of reference periods and in lists of what to include and exclude (do/don't lists) in survey questions. But even if this is a well-recognized problem, it has challenging aspects. Often the delineations of everyday terms are not questioned and thus the need for definitions not recognized. On the other hand, when it is recognized, it sometimes leads to do/don't lists that are so long that respondents are unlikely read them. You will find many examples of such lists in instruction manuals that accompany business questionnaires. Hence, when definitions seem intuitive, respondents often consider them unnecessary, and when they are not intuitive, respondents tend not to read or follow them (Gerbner et al. 1996). Cognitive research has shown that if distinctions are not observed or well communicated to the respondents, they tend to form their own interpretations relative to their own context (Clark and Schober 1992). Finally, an often overlooked problem with reference periods are that they treat conditions that vary as if they were stable. An example is when we ask questions like "How many employees did your company have in 2011?" How should a company with seasonal variations answer such a question?

Another vagueness problem more specific to business surveys and much more difficult to address is when questions regarding activities cannot be related to a particular unit, but rather is a part of employees' job or is taking place in several production units. In these cases the questions are vague because they ask about a unit that can't be seen either on the organization map or in reality. We discussed this delineation problem in Chapter 3 (see Figure 3.8), where we mentioned that some business questions may not have exact answers but need to be estimated. Common methods that aim to help respondents to remember and quantify actions or events (Bradburn et al. 2004) seem irrelevant when the units are abstract. Instead, the best advice we can give is probably to acknowledge that respondents need to make estimates and then lead them through the estimation process step by step. This may, for instance, mean that we start with questions about where the production in question takes place, and then proceed with questions about how many employees are involved in this kind of production and how much of their time is spent on this kind of activity.

Mapping Discrepancies Another potential source of comprehension errors is an inexact fit between the survey definitions and the situation experienced by the respondent. This is known as a *mapping problem* (Redline 2011; Tourangeau et al. 2006). Even if the definitions are clear, the respondent may still have difficulty classifying borderline cases. The employment question from above could be used as an example. Full-time workers in permanent positions should obviously be counted as employees. But what about part-time workers, people who hold temporary positions or are hired in from an employment agency for a limited period of time? These and other borderline cases may cause severe classification problems. What this also should tell us is that mapping problems typically will vary between industries and between enterprises of different size and complexity. Hence, because the business population is so heterogeneous, measures taken in order to resolve mapping problems may be an unnecessary burden for some, but not good enough for others.

Discussion We will discuss how various miscomprehensions can be avoided by studying an example. Assume that the respondent is asked to divide into percentages the company's turnover by six commercial areas, as shown in Figure 8.17. The six terms used should be well known to business professionals. Still, the first three are explained in the question. In addition, more definitions and specifications were available in a separate instruction manual. The example is taken from the Norwegian Structural Survey of Wholesale and Retail Trade.

The question posed in Figure 8.17 refers to figures already reported to the Norwegian tax office. In the original, Norwegian version of the question, the respondent is asked to summarize a number of specific posts from this report and then split the result into the categories listed in the question. Whether the respondents have this document in front of them is, however, unknown. The respondents could either add posts from the tax office report as instructed or assume that they know what is included in the total turnover and base their answers on their own calculations. Because the

❶ Divide the total turnover reported in the Tax Office Trading Statement by percentage on the different commercial areas listed below.	Percent with one decimal
Retail. Sales of merchandise primarily for the purchaser's personal use or to private households. Merchandise is for goods sold on without being processed	___ , ___ %
Wholesale. Sales of merchandise to commissioners who sell on to resellers and companies	___ , ___ %
Commission trade. Turnover from the sale of merchandise on commission from other companies.	___ , ___ %
Primary industry products	___ , ___ %
Industry products	___ , ___ %
Services	___ , ___ %
The percentages should total or sum to 100	100 , 0 %

Figure 8.17 Questions from a Norwegian survey of wholesale and retail trade.

respondents are not asked what the total turnover was, it is also impossible to detect inconsistencies during editing that may relate to personal perceptions of what should be included in the terms. So, even if the definition used in the tax office trade statement is brought to the respondent's attention, no action is taken to ensure that the correct definition actually is used. If the respondents were asked to state what they have reported to the tax office, that would anchor the following questions in a more secure way.

There is a similar problem with the preprint that indicates that the percentages should add up to 100. Both the text and the preprint try to bring this to the respondents' attention, but there is no check as to whether this instruction is recognized and followed. A respondent can easily fill in percentages that don't add up to 100%. Instead of the preprint, we think that the respondents should total their figures and check their answers if the figures did not total 100%.

The basic difference between retail, wholesale, and commission trade is probably known to the respondents. Some of the definitions included in the question, for example, that commission trade is turnover from sale of merchandise on commission from others, are probably superfluous, and could even be perceived as slightly insulting to the respondent's intelligence. The instruction document, however, contains other specifications that are not so obvious. One is that commissions for the provision of betting, lottery, or other games should not be treated as retail or commission trade, but as services. This specification is repeated 3 times in the instruction document, but is not mentioned in the questionnaire. Because "services" is last on the list, there is a high risk that respondents who do not read the instructions will include results from this kind of activity in one of the previous categories.

The commercial areas listed in the question in Figure 8.17 fall into two groups: merchandise and own production. *Merchandise* is defined with the first item, *retail*, but the shift from merchandise sources to turnover from own production, starting with primary industry products, is not indicated. We think it would be easier to distinguish between these types of turnover by splitting the question into two sections. Moreover, *Merchandise*, and the definition, should be the title of the first section.

Three types of merchandise are listed first, obviously because the questionnaire is addressed to enterprises within the wholesale and retail trade industry and hence that merchandise is their major source of revenue. From a questionnaire design perspective, however, misclassification of commissioned betting and lottery activities would be easier to avoid if the list had begun with turnover from self-production. Then one could have specified turnover from services as "services, including commissions for the provision of betting, lottery, or other games" before the respondent reached other items where it could be tempting to believe that this kind of turnover belongs. This is an example of the importance of the order of question and answer categories for question comprehension.

Another way to avoid misinterpretation would be to consider whether terms than need to be explained can be simplified by using the explanation as the basis for the question. Earlier we gave an example of this with the term *renewable power sources*. Here the specification of primary industry products is another example. In the instruction document enclosed with the questionnaire, this is specified as goods or services that the enterprise produces within the area of farming, forestry, hunting or fishing. So, why not abandon the term *primary industry products* and simply ask for turnover that stems from these specific activities?

So far we have discussed sources of ambiguity that are related only to which activities or products should be included or excluded from the reports that the respondents give. Another common source of ambiguity in business surveys is how numeric answers should be calculated and reported. All figures should be reported except VAT. This is repeated several times in the instruction document. There one can also read that public taxes included in the price of retail products should be included, while public subsidies should be excluded. It is rather confusing when definitions of terms are mixed with calculation rules. Calculation rules are not needed before the respondent has understood what services or products should be considered. What is to be included and excluded in the reported figures should therefore either be stated after the terms are defined or even not before the respondents focus on the response boxes. An option is therefore to state that VAT should be excluded above the response boxes. The present instruction that figures should be given with one decimal seems less important and could probably be dropped.

8.4.3 Listening to Questionnaires

Now we turn from the terms used in survey question texts to the full-question formulations. We have already underscored the importance of recognizing the distinction between the business survey scene and normal business management. If we have succeeded with setting the scene, the environment that the respondents will relate to is the questionnaire. From a communication perspective, the questions in the questionnaire should be regarded as parts of a conversation. It has been argued that the relationship between the respondent and the questions posed in questionnaires is similar to any relationship in that it is subject to common conversational norms (Schober 1998). But at the same time the respondent's relationship to the questionnaire also differs from daily-life conversations in that the opportunities for two-way clarifying negotiations between the surveyor, talking through the questionnaire, and the respondent are limited (Schwarz et al. 2008).

We build our understanding of the norms that govern communication acts on Grice's (1975, 1989) four conversational maxims of quantity, quality, relation, and manner. Because the communicating actors are involved in an effort to cooperate, Grice asserts that the words and sentences shared are expected to be informative (quantity maxim), truthful (quality maxim), relevant (relation maxim) and clear (manner maxim). Because of the limited opportunities to ask for clarification and to test tentative answers, however, the respondents are virtually unassisted in making sense of the survey questions. Some important consequences of taking a conversational view are described below.

8.4.3.1 How Question Order Affects Comprehension

The maxim of quantity suggests that because respondents will expect every question to be informative, they will not expect the same question to be asked several times. This way of thinking may lead to unexpected answers under two main conditions: (1) when a specific question is asked before a general one and (2) when two similar questions follow each other. In both cases respondents tend to read the second question as "aside from what you were thinking about in the previous question" (Schwarz et al. 2008). Hence, unless we do not clearly state that we want the respondent to summarize answers to specific questions into an overall assessment, we run the risk that the answers to the specific questions may be excluded when the respondent answer the general one. When similar questions are placed close to each other, there is also the risk that minor differences will be given more importance than intended. An important consequence of this latter phenomenon is that correlations between questions that are formulated to measure the same concept (e.g., in validity tests) may be surprisingly low.

When we suggested that the item in which betting commissions should be included should come first in the list of different commercial areas in the question presented in Figure 8.17, we turned the quantity maxim to our benefit. By assigning this source of revenue in a commercial area early in the list, we prevent it from being included in items further down.

1 Divide the total turnover reported in the Tax Office Trading Statement by percentage on the different commercial areas listed below.

		Percent with one decimal
Primary industry products	☐ None	☐ , ☐ %

Figure 8.18 Making "none" a legitimate response alternative.

8.4.3.2 Implicit Expectations

The maxim of quality suggests that because respondents expect surveyors to be honest and know what they are asking for, the fact that a question is asked presupposes that the issue exists. Hence, the respondents do their best to make sense of and answer even obscure or hypothetical questions. There are numerous examples from TV entertainment programs where people on the streets answer fictitious questions the best way they can. Business questionnaires do not try to fool people, but questions may sometimes seem tricky or obscure. These may be questions that have no obvious link to the topic of the questionnaire or that are uncommon to ask in a business setting. They can also be questions asking for answers reported in an uncommon measurement unit. What we consider the most common and important example in business surveys, however, is the tendency to exclude filter questions and expect that respondents who have no numeric values to report will report a zero. Again, the question in Figure 8.17 can serve as an example. Here respondents are expected to fill in nil percent for items that did not contribute to their total turnover. But according to the quality maxim, the lack of a filter question, in this case a response box for "none," seems to imply that most business have something to report (Yan 2005). Combined with slightly unclear terms, this may lead to over-reporting marginal revenues. A question design like the one in Figure 8.18 would prevent this.

8.4.3.3 Looking for Hints

The maxim of relation suggests that in order to make sense of ambiguous questions, respondents utilize different kinds of information in the questionnaire (or communicated by the interviewer in case of interview surveys) that they consider related and relevant to the question text that they are struggling with. We have already given some examples of how this can work. Respondents tend to use previous questions to define the present topic and similar questions nearby to decide what the present question does not ask for. Another example is where the size and shape of the response box indicate the kind of response expected. Figure 8.14 gave some examples of how the visual layout of response boxes could be designed in order to substantiate what figures the questions ask for. In web surveys radio buttons tell the respondent that the response options should be considered mutually exclusive, while checkboxes create opportunities for more than one answer. We recommend that the same convention be used in paper questionnaires, in particular when the respondents can chose between paper and the web.

The most obvious place to look for cues is in the response categories. There is a wide body of literature on how verbal or numerical response labels used with rating or frequency scales can affect the respondent's comprehension of questions. It has been shown that numbers as well as text labels are interpreted for instance, that a rating scale ranging from −5 to +5 is interpreted differently from one ranging from 0 to 10. While response options labeled with negative numbers tend to be considered appropriate only for those who have specific complaints, a rating value of 0 or near 0 will also be considered appropriate for those who are only not satisfied (Schwarz et al. 1991). Another example is that frequency scales often are understood as reflecting population distributions, with uncommon values at both ends of the scale and the most common value in the middle. Read this way, the respondents, instead of actually considering the response alternatives, may position themselves as less frequent, in line or more frequent than the average (Schwarz 1996).

Ratings and frequency scales are not as common in business surveys as in social surveys. As we have shown several times in this and previous chapters, the most common response formats in business surveys are labeled, open-ended response boxes. Still different kinds of response scales are found in business surveys of working conditions and in business tendency surveys. In addition, rating scales are used in customer evaluations.

The European Working Condition Surveys is a series of interview surveys that have been conducted several times since 1990 in a number of European countries.[2] During the interview a card with questions like that posed in Figure 8.19 is shown to the respondents.

Strictly speaking, the questions shown in Figure 8.19 are yes/no questions with a follow-up frequency question baked into the different "yes" alternatives. A question construction like this both violates the rule of thumb which says that you should ask one question at the time and that the response options should match the question text

Q23 Please tell me, using the following scale, are you exposed at work to . . . ?

	All of the time	Almost all of the time	Around ¾ of the time	Around half of the time	Around ¼ of the time	Almost never	Never	DK	Refusal
Vibrations from hand tools, machinery, etc	1	2	3	4	5	6	7	8	9
Noise so loud that you would have to raise your voice to talk to people	1	2	3	4	5	6	7	8	9
High temperatures which make you perspire even when not working	1	2	3	4	5	6	7	8	9

Figure 8.19 Extract from the European Working Condition Surveys.

[2] Information is available at `http://www.eurofound.europa.eu/surveys/ewcs/index.htm`.

(Fowler and Cosenza 2008). We also notice that no time reference is specified. Thus it is not clear whether the response options refer to a typical work day or to a longer period of time. Verbal labels such as “all of the time” or “around half of the time” may make the respondent think about what happens daily, while the label “never” indicates a longer time period. What concerns us most here, however, is that while the verbal labels range from high to low frequencies, the numbers range from low to high values. The card used in the survey lists nine descriptions of working condition problems, so it is quite long. As the respondent moves down the list, the distance to the original verbal labels increases and there will consequently be an increasing risk that the numerical labels will predominate and turn the scale around.

Business tendency surveys ask about current and future prospects of businesses. Core aspects that the managers are asked to evaluate are the inventory and order books status, production capacity, prices, and market prospects. For most of these questions, a three-option ordinal scale, distinguishing between upward, stable, or downward trends, is recommended (OECD 2003). An alternative that is used in evaluation of production capacity is to ask the respondent to estimate the percentage of how much of the capacity is utilized.

Two examples of business tendency questions gathered from the Canadian business condition survey for the manufacturing industries (English version) are presented in Figure 8.20.

As with the example from the working condition survey, the questions listed in Figure 8.20 are conducive to numerous comments. One is the mixture of numbering systems, with ordinary Hindu-Arabic numerals for the question, Roman numerals for of the different statements, and letters for the response alternatives. This mix clearly breaks with the common recommendation for numbering the questions in a simple, consistent style (Dillman et al. 2009). But let us return to the scales, which is the main topic here.

Scale length should reflect how people think about the topic in question. We do not know how managers view the current status and prospects of their business, but recognize that a three-point scale does not allow for any considerations about the strength of decline or progress. For this reason, five- or seven-point scales are often preferred over three-point ones (Krosnick and Presser 2010). In the guidelines for business tendency surveys given within the framework of the Joint Harmonised EU

---

2. ALLOWING FOR NORMAL SEASONAL CONDITIONS, it is the opinion of the management of this establishment that:

I The inventory of finished goods on hand in this establishment is:

(a) too high ☐ (b) about right ☐ (c) too low ☐

II Orders received by this establishment are:

(a) rising ☐ (b) about the same ☐ (c) declining ☐

Figure 8.20 Extracts from the Canadian business condition survey.

Q1 How has your production developed over the past 3 months? It has. . . .

+	increased
=	remained unchanged
–	decreased

Q1 How has the financial situation of your household changed over the last 12 months? It has. . . .

++	gotten a lot better
+	gotten a little better
=	stayed the same
–	gotten a little worse
– –	gotten a lot worse
N	don't know.

Figure 8.21 Recommended visualization of response alternatives.

Programme of Business and Consumer Surveys, a five-point scale is recommended for investment judgments and expectations, but not for other assessments. We also recognize that a five-point scale is recommended for consumer surveys asking household members the same kind of questions as those posed in business tendency surveys (ECFIN 2007).

The Joint Harmonised EU Programme of Business and Consumer Surveys also recommends that the response alternative be supplemented with simple mathematical symbols as shown in Figure 8.21. As we see it, these visual labels coincide with the verbal labels in a simple, intuitive way. Compared to this scenario, the letters used in the Canadian example appear as visual clutter without any clear message.

For most questions in business tendency surveys, the first response alternative indicates an advantageous situation or optimistic view of the future. Because success is what is valued in business, one may wonder whether putting the optimistic option first does not invite socially desirable answers. Particularly in times when most businesses are prospering, the respondents may feel that the first response option is expected and that it might be slightly embarrassing to admit that one's company differs from the majority. By saying this, we also suggest that the evaluation scales may work differently in recession periods. Generally business tendency surveys seem to predict future economic trends quite well [see, e.g., Lemmens et al. (2004)], so this may not be a major problem in this survey. Still we think it is important to mention that these and other business survey questions may be more vulnerable to social desirability effects than expected.

A technique that is used to obtain honest answers to questions where the respondents may be tempted to embellish realities is *bogus pipeline procedures*—this is a general term hinting that survey questions will be checked against other sources, and consequently that white lies will be exposed (Tourangeau and Yan 2007). Because much of the information reported in business surveys is equivalent or similar to that in reports given to other authorities, it may have a positive effect on the response quality if respondents are told that other information sources also will be used as input to the statistics. Business respondents who participate in recurrent surveys should also sometimes be contacted by the survey organization for clarifications or receive other kinds of feedback from the data collection.

The last comment that we will make about business tendency questions is that even if the first response alternative generally indicates that the business is currently doing well and that its prospects are good, there are a few exceptions that may cause confusion. The first question in the Canadian survey depicted in Figure 8.20 is one example. Contrary to what perhaps is the most obvious association, a higher scale score in this context does not mean improvement, but the contrary. When the same scale is used for a number of similar questions, as in this questionnaire, the designer should be aware of the risk that the respondent may fail to notice that a response option that was previously positive has become negative. To avoid this risk, we recommend separating questions that ask about positive and negative trends and clearly indicate when the impact of different scale alternatives change.

8.4.3.4 Talking with Web Questionnaires

Traditionally the recommended method for avoiding measurement error in surveys has been to present exactly the same stimuli to all respondents; this is called a *mass communication strategy*. Interviewers were, for example, thought to read questions word by word as they were written in the questionnaire and to follow standard probing rules (Fowler and Mangione 1990). However, uniform wording does not necessarily lead to uniform understanding. Because the meaning of terms and messages are not fixed, but to a certain degree dependent on the circumstances in which they operate, there may be a need for a dialog that establishes a common ground of mutual understanding before the questions can be properly answered (Conrad and Schober 2010). We think that this may be particularly important in business surveys because the distinction between how terms are used in the production environment of the businesses and in business surveys may be unclear.

As we pointed out at the beginning of this chapter, mailed paper questionnaires are basically a one-way communication. There is no room for a two-way interaction to adjust for misunderstandings as there is in a conversational (in-person) interviewing approach. In web surveys, however, a kind of conversation similar to the collaborative interaction that we know from personal communication can be built into the questionnaire (Conrad et al. 2007). In fact, to a certain degree this has already been done, even if the interaction may not be very dynamic and flexible yet. In interview surveys the interviewer speaks on behalf of the surveyor. In web surveys the computer program has a similar role. If web respondents are unsure of how a question should be understood or responded to, they can click on an icon that brings them to a more detailed explanation or instruction. The questionmarks in the Norwegian, German, and Dutch web questionnaires shown in Figures 8.11 and

8.12 are such icons. Moreover, if respondents skip questions or provide inconsistent answers, error messages that suggest or confirm errors may be activated. For example, a turnover rate that should be divided into different commercial areas in a web version of the question presented Figure 8.17 might not add upto 100%.

Help that is initiated by the respondent clicking on an icon is an example of *obtainable clarification*, while activated error messages are an example of *unsolicited clarification* initiated by the computer program (Conrad et al. 2007). Other combinations and more sophisticated help and error messages are possible and should be considered. The first step could be to replace passive help and error messages with *questionnaires within the questionnaire*, which investigate what the problem is and then tailor help texts to the respondent's needs. The next steps could be to peronalize the dialog by offering direct mail, telephone, or video contact with a virtual or real help desk during the completion process. In additon to help initiated by the respondent, help offers could also be activated by computerized risk analysis based on client-side paradata. Paradata are registrations that are carried out behind the scenes while the respondents fill in the questionnaire (client-side paradata) or communicate with the web server (server-side paradata) (Heerwegh 2002). Examples of client-side paradata that could be included in a risk analysis and trigger help functions could be temporal measures indicating a suspiciously long or short time spent on certain questions, the number of times that presented answers are corrected, or error messages that are activated several times (Haraldsen et al. 2006).

We will return to the topic of how web technology can facilitate tasks initiated by the survey questions in Section 8.5.2.

8.5 MINIMIZING RESPONSE BURDEN

In Chapter 6 we distinguished between four kinds of response burden reduction measures: reductions of sample, variables or contacts, sample coordination, survey communication measures, and questionnaire improvements. The present section can be seen as an extension of the section on questionnaire improvement measures.

Minimizing response burden in business questions first of all entails minimizing the burden of information retrieval and judgments. Collecting information that should be included in the questionnaire is normally more time-consuming than answering the questions (Haraldsen and Jones 2007). Moreover, judgments when the collected information does not exactly match with what the questionnaire asks for and calculations before the answers can be entered are among the tasks most often perceived as burdensome by business respondents (see Figure 3.6). We think that some of these tasks, such as some calculations, can be removed from the questionnaires and left for the survey organization. Others, such as capturing data from the company's administrative systems, can be automated. But even with these measures implemented, there will normally be a number of tasks, including burdensome ones, left for the respondents to complete. As we mentioned in the beginning of Chapter 6, burdensome tasks in questionnaires cannot be hidden, and can only to a certain degree be balanced against more gratifying tasks. Contrary to what might be the case during recruitment, in the questionnaire we may need to focus on demanding questions.

It takes time and effort to produce valid and accurate data. As we saw in the analysis referred to in Chapter 6, respondents who find the questionnaire easy to complete may make just as many errors as those who find it difficult (Fosen et al. 2008). Thus, on one hand, excessive response burdens should be avoided. On the other hand, the respondents should also be motivated to invest the effort necessary to produce proper answers.

Section 8.5.1 discusses time, and thereby also cost saving measures in business surveys. Section 8.5.2 discusses methods for reducing cognitive burden. Morrison and her colleagues recommend deconstructing complex questions into series of simpler questions (Morrison et al. 2010, principle 2). From a quality perspective, that seems to be good advice. One should, however, be aware that more questions, even if simple, may take more time to complete than a composite version; this indicates a potential conflict between quality and cost considerations.

8.5.1 Time Saving Measures

The most obvious way to reduce the response completion time is, of course, to reduce the number of questions. As shown in Figure 6.6, replacing surveys by utilizing secondary sources is the most common way for NSIs to reduce the business' response burden (Giesen and Raymond-Blaess 2011). But there are also ways of reducing the number of questions without replacing the entire questionnaire. We will stress the importance of a critical review of the relevance of each question included in the questionnaires. This review should be done in two rounds:

1. In the first round, ideally at the beginning of the survey project, specification of the questionnaire content takes place. In business surveys, which commonly were established years ago, regular meetings should be held to update these specifications. Updates should include critical review of old topics as well as addition of new ones. Specification methods like those described in Chapter 7 should be used [see also Snijkers and Willimack (2011)].
2. In the second round, the clients and stakeholders should be able to review the questions included in the questionnaire in the light of results from feasibility studies, cognitive interviews, or pilot studies. Seeing how the questions actually work, and in particular identifying those that do not, this may lead to fruitful discussions about the relevance of the questions. An example of the progress that can be achieved by confronting stakeholders with pilot test results is found in Ballou et al. (2007), who report the developments of a business survey on entrepreneurship. This survey underwent a meticulous planning procedure before it was tested in a pilot study. Still a discussion about the analytical contribution of each question based on the pilot results led to a reduction in the number of questions and the time it took to complete the questionnaire. This was a telephone survey. According to Ballou and her colleagues, the average interview time was reduced from 50 to 20 minutes.

When reviewing the relevance of business questions, those asking for calculations should be studied with some degree of skepticism. Calculation-related questions, often involve complicated and time-consuming tasks. Often a sequence of simpler questions covering factors to be included in the calculation will be quicker to answer and sufficient to enable the survey organization to perform the calculation later. For example, the respondents may be asked to summarize items like those listed in the following question:

"How much money did the enterprise spend on the following IT [information technology] and communication products: computers; accessories such as scanners, printers, and external storage equipment; landline telephones, mobile phones, wide-area networks (WANs) and local-area networks (LANs)?"

Unless this question refers to a post in the business' accountancy system, reporting the money spent on the individual IT items listed here would probably take less time and give more accurate figures than a calculation made by the respondent. Another example of calculations that can be simplified is as follows:

"Please report the sales of merchandise, VAT excluded." Government charges included in the price should be subtracted. Government subsidies should be added. Revenue from self-produced services and goods should not be included.

In this case we assume that the value of sales initially is collected from the accountancy system but may or may not include the elements that the question asks to include or exclude. Then it would probably be easier and quicker for the respondent to first report the figure residing in the accountancy system and then answer a sequence of follow-up questions such as those listed in Figure 8.22.

The example with expenditure on IT and communication products and this last example are illustrations of how instructions can be transformed into questions Morrison et al. (2010, principle 12).

8.5.1.1 IT Tools

Given that the number of questions is reduced to those strictly relevant, the most effective way of reducing the time it takes to answer the questions included in the questionnaire is probably to utilize time saving tools offered in electronic questionnaires. It is important to stress that IT alone does not reduce the time burden. However, we are confident that IT will affect survey questionnaires increasingly in the future.

We have already mentioned different ways of utilizing IT to streamline the internal data collection that precedes questionnaire completion and questionnaire completion itself. When business questionnaires are made available in a common web portal, internal distribution is facilitated and improved. If the respondents are able to copy and paste or even better, link response boxes to administrative information sources, the time needed to answer the questions should decrease.

A computer is a general information processing tool that can be programmed into a range of more specific tools. For examples, computers can be programmed to

1. Was VAT included in this figure?

○ Yes → **Please specify the VAT** ☐ Euros

○ No

2. Were government charges included?

○ Yes → **Please specify these charges** ☐ Euros

○ No

3. Were government subsidies included?

○ Yes → **Please specify these subsidies** ☐ Euros

○ No

4. Were self-produced services or goods included?

○ Yes → **Please specify the value of these services of goods** ☐ Euros

○ No

Figure 8.22 Sequence followed to adjust reported figures.

work as word processors, number crunchers, cameras and image processing tools, audio videorecorders, and editing tools. One source of response burden that is quite high on the business respondents' complaint list is the need for calculations before the questions can be answered (see Figure 3.6). A simple method for facilitating calculations is to provide a calculator together with the questionnaire. If you look at the bottom menu of the Dutch questionnaire for structural statistics in Figure 8.12, you will see a clickable icon that opens such a calculator. A comparison with the previous version of this questionnaire showed that the time it took to collect data decreased by 35 minutes and the time it took to fill in the questionnaire decreased by 19 minutes. Interviews with respondents indicate that this time reduction was due in part to use of the calculator (Giesen et al. 2009b).

The calculator offered in the Dutch questionnaire is a standard one. Taking this idea a step further, however, it is also possible to offer calculators that are tailored to the calculation task asked for in the survey. One example is the calculations asked for in the Norwegian Quarterly Survey of Self-Reported Sick Leave. In one of the questions in this survey the respondent is asked to calculate how many workdays are lost because of self-reported sick leave (in contrast to sick leave imposed by a physician). The number of workdays should be calculated by multiplying the employee's job share by the number of days away from work. Because the job share can be described in different ways, this calculation can be performed in different ways. The following two examples were given in the instructions:

1. One employee works 25 hours per week. He is sick for 2 days. If we assume the standard working hours for one week to be 37.5, which it is in Norway, the total workdays lost are

$$\text{Lost workdays} = 2 * \frac{25}{37.5} = 1.3$$

2. Another employee who works 60% of the work week has been sick for one day. The lost workdays are

$$\text{Lost workdays} = 1 * \frac{60}{100} = 0.6$$

Instead of giving these examples and leaving it to the respondent to calculate lost workdays for different employees with different employment contracts, one could offer a calculator that asked for the factors included in these formulas and then perform the calculations.

Other examples of tailored calculators that can be useful and save time in business surveys are those that sum up totals as details are added (running tallies), those that include or exclude VAT from prices reported, and those that convert production figures reported in different ways into a common measurement unit for easier comparison. Other tools that could save time in electronic business questionnaires are maps specifying travel routes in transport surveys, calendars indicating events or periods, search engines assisting respondents in searching for information to include in the questionnaire, and email services that link the respondent to a support center during questionnaire completion.

8.5.1.2 Presentations

Finally, some time can probably also be saved by streamlining the layout of the questionnaire to facilitate reading. As we mentioned at the beginning of this chapter, consistent use of visual elements contributes to an efficient, user-friendly layout. The combination of font and typeface that is used for questions, instructions, and labels should be consistent throughout the questionnaire. Similarly, response boxes containing the same kind of information should have a consistent layout. Questions and response boxes should be graphically grouped in a consistent way and consistently be aligned to margins or tabulators (Morrison et al. 2008).

One of the most striking differences between business questionnaires and questionnaires used in social surveys is that the former often have a ledger-type design with wide, one- or two-dimensional matrices and open response fields for the respondents to fill in figures. In social surveys a booklet format is commonly recommended for questionnaires to emphasize their similarities to books (Dillman et al. 2009). This format is uncommon for accountancy books often copied in business surveys. For ledger-type questions in paper questionnaires, we believe that a double-page portrait format or a single-page landscape format will work better. However, for web questionnaires presented on small computer screens, the ledger format is difficult to maintain.

Time can also be saved by an effective flow of questions together with skip-instructions that are easy to see and follow. Figure 8.2, which showed how navigational instruction can be communicated both verbally and visually, illustrates how this can be done. The most effective way of determining sequence is to number the questions. Contrary to common recommendations, business surveys tend to use complicated numbers, often with several subnumbers. Often each section is also numbered separately. In Figure 8.20 we showed a glaring example of mixed-numeral numbering. Simple numbering running through the whole questionnaire communicates question order more effectively. If paper questionnaires are combined with electronic questionnaires, however, one should consider using two-digit numbering when filter questions lead respondents past follow-up questions. In electronic questionnaires bypasses such as these can be automated so that the respondents do not see questions that do not apply to them. With simple one-digit numbering, this can be confusing because some questions seems to be missing. One way to handle this problem is to reserve overall numbering for questions that all respondents should answer, but number follow-up questions with a second digit.

8.5.2 Reducing the Cognitive Burden

Clear, well phrased, and user-friendly questions should make it easier to understand the business survey questions and complete the information tasks. Hence, response burden reduction has been a recurrent, underlying topic of this chapter. The two elements that will be discussed in this last section come in addition to those already introduced.

8.5.2.1 From Preprinting to Dependent Dialogs

Preprinting is a frequently used technique in business surveys when previously reported production or sales figures need to be verified or updated. The technique is also often used in agricultural surveys where the respondents are asked to report the size of areas cultivated by different kinds of crops (Mathiowetz and McGonagle 2000). There are several arguments for the merits of preprinting in reducing cognitive burdens and enhancing response quality:

1. Some administrative data, like addresses, contact persons, and industry codes, seldom change. When the same enterprise receives the same questionnaire several times within a short time period, as they commonly do in surveys collecting monthly or quarterly statistics, it appears unnecessary to give the same answers to questions like these time after time. Instead, confirming previously reported information or occasionally correcting them seems to be more effective. When open-ended responses are used to assign codes, there is

also a risk that slightly different answers to the same question at different points in time will lead to different codes that actually do not reflect real changes. The allocation of industry codes is an example of this.

2. Preprint is also used to maintain and update lists of goods, services, or other items that respondents report in recurrent surveys. Price index surveys are a typical example. The advantage of a preprinted list of items is that items previously reported are not missed and that the respondents can check how reasonable the new figures that they are about to enter, are.
3. Seeing what has been previously reported may also help respondents better recognize what type information is requested. This may be particularly important when the respondent changes from one survey round to the next.
4. When survey participants are asked to count events or give figures that belong to a certain time period, information belonging to the previous period may be incorrectly included or information specific to the period in question may be missed. The first kind of error, called *forward telescoping*, is most common. In recurrent surveys that ask for information about consecutive periods, forward telescoping will often lead to *seam effects*, that is, double reporting of events or figures that pertain to the end of the previous period. Preprint of results from the previous period may prevent this because it reminds the respondents of information that they previously reported and thus should not be included in the new figures. This aiding technique was originally developed in household expenditure surveys and is called *bounded recall* (Neter and Waksberg 1964). As the term indicates, it is tailored for situations where numeric values cannot be collected from available records, but need to be memorized or estimated. In business surveys the challenge will often be to get respondents to collect information from relevant business records instead of doing their own estimates. For this reason, the preprint method should perhaps be used cautiously when the survey asks for information that is a part of normal bookkeeping.
5. An interesting kind of preprint is when individual answers to survey questions are summarized and presented for final approval before they are posted to the survey organization. Giving the respondents this kind of overview may expose inconsistencies or inaccuracies not noticed when the answers were initially entered. We consider this type of summary particularly important in web questionnaires, which are much more complicated to review than paper questionnaires. Summaries in web questionnaires should also be printable.
6. Finally, preprint of previously reported data is a close relative to feedback of statistical results based on previous data collections. Both can have a positive effect on respondents' motivation. Feedback signals a two-way contact that may be appreciated. Moreover, a comparison between one's own figures with statistical results may be considered as an interesting kind of benchmark measurement (Bavdaz et al. 2011; Torres van Grinsen et al. 2011).

While previously reported data have not been shown to reduce respondent burden in terms of faster survey completion, it is generally assumed that cognitive burden is reduced (Holmberg 2004; Hoogendoorn 2004). The possible advantages of preprinting need, however, to be balanced against potential drawbacks.

The most common concern is probably that preprinting facilitates shortcuts. Skeptics fear that the respondents will acquiesce to the preprinted data and that this may lead to incomplete updates of item lists and of underreporting of change. We have already pointed out that by showing previously reported data, we may invite respondents to guess what the updated data are instead of looking up figures from the company's administrative sources. Research results indicate that the presence of previously reported data can enhance response quality (Holmberg 2004; Hoogendoorn 2004; Pafford 1988) but in some cases also lead to a certain degree of acquiescence (Pafford 1988; Phillips et al. 1994; Stanley and Safer 1997). For questions that are vulnerable to measurement errors and have high response variability, the quality improvements can be considerable. When this is the case, preprint is recommended, but should be presented in a way that prevents acquiescence. We think that conditional follow-up questions, which can be implemented in web questionnaires, offer a possible solution.

In interview surveys, instead of talking about preprint, we use the term *proactive dependent interviewing* (PDI) in contrast to control questions which are called *reactive dependent interviewing* (RDI). As we have repeatedly pointed out, web technology offers an opportunity to imitate interviewing. Within a web framework, preprinting is a proactive help function, while error checks correspond to reactive dependent interviewing. In interviews and web questionnaires previously reported data can go into a more dynamic dialog than that possible with paper questionnaires. Looking at previously reported data may be an option instead of a fixed preprint, and we may offer a choice as to which data to look at. If answers indicate acquiescence, they may be followed up by a set of investigative questions. In an experiment with a web questionnaire that asked respondents to update information about assets and liabilities, Hoogendoorn (2004) demonstrates how such follow-up questions posed to those who initially report no change may lead to more plausible results. We recommend this as a strategy for preventing acquiescence.

Another serious risk when questions are anchored to previously reported data is that errors in previous reports may not be recognized, and consequently that errors are perpetuated in current and future reports (Pafford 1988). Therefore, one should be particularly careful to pretest and evaluate the quality of questions included in such a preprint design. Even if the answers were originally given by the respondents, feedback of obviously erroneous data may damage confidence in the survey institute. Likewise, the respondents may react negatively to edited results that they recognize as inconsistent with what they reported earlier.

8.5.2.2 Manageable Matrices

Questions combined in a matrix constitute a very common question format used in business surveys. Although efficient in the sense that a number of questions are compressed within a small space, this format has been criticized for being cognitively burdensome (Couper 2008; Dillman et al. 2009). Respondents must relate to the information presented in the rows and in column headers at the same time. In addition, the questions may contain references to separate instructions. If this is too much to keep track of, it may lead to superficial response strategies and information missed (Tourangeau et al. 2000). Even if matrices are space-efficient, they are usually too wide to be presented within a single computer screen.

We can distinguish between two main kinds of matrices used in questionnaires: (1) Questions with common response alternatives grouped together with one question in each row and the common response alternatives in the column heading of the matrix. A set of statements combined with a common evaluation scale is a typical example of this format.
(2) Different items, such as different products or services, listed vertically and a sequence of questions that should be answered for each item presented horizontally. This means that each line and each column ask for different kinds of information. The response format may either be common for all combinations of items and questions or change between matrix cells. It is this two-dimensional, most demanding kind of matrix that is most common in business surveys.

Morrison and her colleagues recommend limiting the use of matrices, but at the same time admit that eliminating matrices by converting each line to a set of individual questions might not always be feasible in business surveys (Morrison et al. 2010, principle 16). Business respondents often express a preference for matrices over tedious repetitions of identical questions. Spreadsheets are a very common tool in businesses, and it seems awkward to split up data that reside in spreadsheets and report them item by item.

There seem to be two basic situations when matrices should be retained in business surveys:

1. When the business questionnaire asks for information that can be copied directly from business spreadsheets. In this case the survey organization may even offer an opportunity to enclose a copy of the relevant spreadsheet with the questionnaire. Remember, however, that business surveys often use slightly different definitions and ask for slightly different figures than those residing in the company's personal and accountancy systems. If so, offering a matrix response format may encourage the respondents to copy and paste inaccurate figures from their administrative spreadsheets. Unless the matrix question is followed up with control questions about what is included and excluded in the figures reported (see Figure 8.22), this kind of error will seldom be detected. In these cases individual questions may work better.
2. When the number of questions asked for each item are so numerous that repeated question sequences may be too burdensome for the respondents. This burden might pose a greater threat to the response quality than the cognitive burden imposed by the matrix format. In cases like these, we recommend combining a question-by-question format with a matrix format as shown in Figure 8.23. Here the first line of the original matrix is replaced by individual questions. By doing this, the respondent becomes familiar with the questions that later appear in the matrix columns. The question-by-question format leaves space for complete question

Original version

1 Investments, maintenance and sale

	Investements in 1000 euros	Maintenance in 1000 euros	Selling price in 1000 euros
Buildings and installations			
Building sites			
Private houses			
Ships, rigs, aircrafts, etc.			
Vans and other vehicles			
Office equipment			
Furniture, etc.			
Other deallocated resources			

Revised version

1 How much, given in 1000 euros, did the enterprise invest in buildings and installations in 2011?

000 euros

2 How much did the enterprise invest in buildings and installations in 2011?

000 euros

3 How much did the enterprise earn from selling buildings and installations in 2011?

000 euros

4 Please fill in the same kind of information for the investement objects listed below.

	Investements in 1000 euros	Maintenance in 1000 euros	Selling price in 1000 euros
Building sites			
Private houses			
Ships, rigs, aircrafts, etc.			
Vans and other vehicles			

Figure 8.23 Combination of individual and matrix questions.

formulations and time for reflections on how the questions should be interpreted. When the respondent enters the matrix, it is with this understanding in mind. The only challenge added is to relate to the item list. We may call this a *hybrid* question format.

8.5.2.3 Computerized Matrices

Finally, we present an example of how computers can generate matrices that are tailored to the information business respondents have to give. This example is taken from a prototype of a web questionnaire on the value of services delivered abroad. On paper, this kind of question commonly is presented in a large matrix that offers a large number of of service–country combinations. The respondents should report values for each relevant combination. Businesses that have little to report will leave most of the matrix cells empty, while businesses that have a much to report may not find the matrix large enough.

In the web version of this question, the respondent is first asked to identify which services the company offers. Based on the answers to this question, a set of small matrices are generated for each relevant service. The size of each matrix is decided by the number of countries to which the services are delivered. They are small enough to be presented on a computer screen. The respondents are asked to choose countries from a dropdown list and can add as many lines with country names as needed. In the example shown in Figure 8.24, the respondents delivered installation and assembly services to two countries and performed reconstructions by/for others in two countries.

If this survey is repeated among the same respondents, the list of services and matrices from the last report can be preprinted. The respondent should then have the opportunity to add new services or new countries and delete services that the company does no longer delivers or countries to which the company no longer deliver.

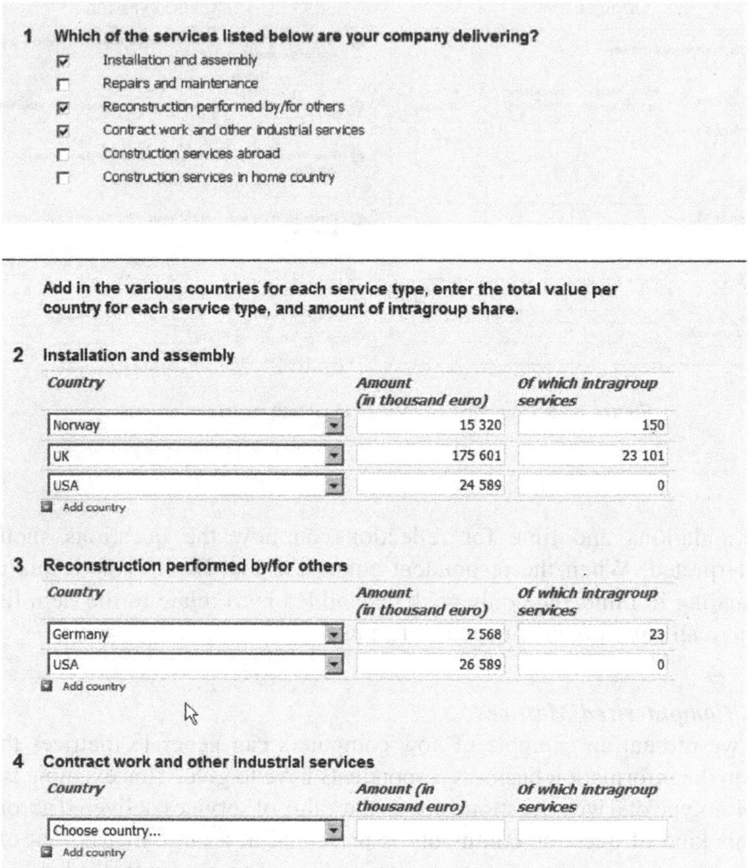

1 Which of the services listed below are your company delivering?

- [x] Installation and assembly
- [ ] Repairs and maintenance
- [x] Reconstruction performed by/for others
- [x] Contract work and other industrial services
- [ ] Construction services abroad
- [ ] Construction services in home country

Add in the various countries for each service type, enter the total value per country for each service type, and amount of intragroup share.

2 Installation and assembly

Country	*Amount (in thousand euro)*	*Of which intragroup services*
Norway	15 320	150
UK	175 601	23 101
USA	24 589	0

Add country

3 Reconstruction performed by/for others

Country	*Amount (in thousand euro)*	*Of which intragroup services*
Germany	2 568	23
USA	26 589	0

Add country

4 Contract work and other industrial services

Country	*Amount (in thousand euro)*	*Of which intragroup services*
Choose country...		

Add country

Figure 8.24 Example of generable services.

Figure 8.24 shows a somewhat split-grid format, not so different from an experiments that Couper and his colleagues have done with a diet history questionnaire where respondents were asked to indicate what fruits they ate during the last 12 months, and then specify the frequency and amount of consumption of those fruits selected (Couper et al. 2011). This two-step procedure limited the number of questions asked at the same time, thereby improving performance. The original paper version of the service delivery question in Figure 8.24 had a single-grid layout with all questions presented at the same time, while the web version first scans which services the business has delivered and then presents a series of double-grid questions.

The striking difference between the fruit consumption experiment and the service delivery design is, that the matrices that follow the scanning question about services does not have a fixed size but is generated for each service delivered. The button that generates a new line and the dropdown box used to identify countries requires a deliberate action that may motivate underreporting. In one of their experiments, Couper and colleagues found that an explicit yes/no question about fruits eaten during the last 12 months seemed to discourage respondents from answering "yes" and consequently led to fewer fruits reported than in another version where "never" was a response option only in the frequency question. The "add country" question in the service delivery design may have a similar effect on the number of countries listed.

8.6 CONCLUDING REFLECTIONS

Quite early in this chapter we recommended drafting a questionnaire on paper before considering the functionality possible in an electronic version. Later we discussed how challenges in business surveys can be met by computer technology. Indeed, even if proven wording and design principles are fundamental in every kind of questionnaire, we think that future opportunities are in computerized questionnaires. On the other hand, technology can hardly solve what has been pointed out as the most burdensome and important quality problem in business surveys; that we ask for information that does not match with business records. On several occasions we have implied an alternative approach to this problem. Instead of asking business respondents to adjust their information to our definitions, we envisage survey designs that collect available data as they are together with surveys questions that map definitions and other characteristics that are necessary to adjust the data for statistical purposes. Jelke Bethlehem has written an article about data capture from secondary sources that he calls "Surveys without questions" (Bethlehem 2008). We think that more data will be collected from secondary sources in the future, but that surveys will not disappear, but may rather play a different role in the data collection designs.

ACKNOWLEDGMENTS

The author is grateful to Jennifer L. Beck, Kristin Stettler, and Heidi M. Butler, all of the US Census Bureau, for their kind and detailed review of the manuscript; and to Solveig Gustad, Statistics Norway, for providing examples of web questionnaires.

CHAPTER 9

Business Survey Communication

Ger Snijkers and Jacqui Jones

9.1 INTRODUCTION

Surveys basically concern communication. In Chapter 8 we discussed communication associated with the questionnaire. In this chapter we discuss all other aspects of business survey communication. As we shall see, there are many aspects. Features of business survey communication include (Cox and Chinnappa 1995): establishing contact with sampled businesses, bypassing gatekeepers, informing respondents about the survey, dispatching the questionnaire to the right person at the right moment in time, reminding and nonresponse follow-ups, and dealing with questions from businesses, using various means such as advance and reminder letters, telephone reminders, websites, and help desks. In some cases an actual appointment is made (e.g., a time deadline for returning the questionnaire), but traditionally (in case of self-completion) this "appointment" comes down to informing the business that it is selected and that the data are due by a specific date. This is a one-sided appointment. The fact that there is considerable use of self-administered data collection in business surveys renders these communication features even more prominent.

Such a one-sided approach receives much criticism and resistance from businesses; they don't appreciate the value of participation, cannot prepare themselves, would like to obtain more information about the background of the survey, don't get any feedback, don't know what happens with their data, and so on. These reactions indicate that with such an approach businesses (or should we say, the people within businesses) are neither motivated nor enabled to cooperate; and this leaves businesses with a feeling of irritation and an unnecessarily high perceived response burden. The attitude that is evoked by such communication can also have a negative

Designing and Conducting Business Surveys, First Edition.
Ger Snijkers, Gustav Haraldsen, Jacqui Jones, and Diane K. Willimack.

effect on the quality of the collected data. This is why an effective business survey communication strategy is important.

The basic aim of business survey communication is to get the questionnaire to the sampled units and get the data back. To do that, a one-sided approach can be applied, or a communication strategy that deals with the known associated problems with a one-sided communication approach can be designed and implemented. Hopefully this latter approach will increase the likelihood of receiving quality responses (i.e., timely, accurate, and complete) and minimize costs and perceptions of burden. This is the topic of this chapter: how to design an effective business survey communication strategy.

To design an effective business survey communication strategy, several factors need to be considered: the role, objectives, and components of survey communication, motivation and facilitation strategies, the business context (as discussed in Chapter 2), internal survey organization policies and procedures (as we discussed in Chapter 4, when planning a survey), and external conditions. In this chapter an holistic approach is applied, in which these factors are combined. In doing so Cox and Chinnappa's (1995, p. 10) advice should be remembered: "minimizing nonresponse and measurement error in business surveys requires advance planning and creative data collection approaches." In addition, we will discuss "creative and flexible" approaches. As we will see, there are four overarching principles for designing an effective communication strategy: objective-driven, coherent, tailored, and planned.

The structure of this chapter is as follows:

Section 9.2: the role, objectives and process of business survey communication
Section 9.3: tailoring to the business context
Section 9.4: the survey organization context
Section 9.5: designing an effective business survey communication strategy
Section 9.6: communication measures, practices, and guidelines
Section 9.7: summary and example of a communication strategy

To design or redesign a complete business survey communication strategy, you may wish to start at the beginning of the chapter. If looking for guidance in relation to specific practices and guidelines, Sections 9.6 and 9.7 are a good place to focus.

9.2 THE ROLE, OBJECTIVES, AND PROCESS OF BUSINESS SURVEY COMMUNICATION

To effectively design a communication strategy, an understanding of the role, objectives and the process of business survey communication is required.

9.2.1 Role

The primary role of survey communication is to get questionnaires to the sampled units and get the data back, that is, to receive responses. Traditionally this primary role has involved communicating relevant instructions and procedures with regard to the completion tasks. It has not focused on motivating and facilitating business cooperation.

We know that some business survey participants are not motivated to comply with a survey request, while others may be motivated but unable to do so because of, for example, competing work commitments or difficulties in understanding or using aspects of the survey components. Extending the traditional role of survey communication, to include motivation and facilitation strategies (i.e., getting respondents in the right mood for survey participation and facilitating their survey-related work) can improve the benefits of business survey communication. Ideally, since not all businesses are the same, the communication would be more effective if it were tailored to the business context (Dillman et al. 2009; Snijkers and Luppes 2000).

Survey communication potentially influences knowledge, attitude, and norms and subsequently behavior (Pol et al. 2007; Vonk 2009). In business surveys, we need to influence the behavior of managers and employees involved in the response process so as to obtain quality responses. (In Chapter 2 we discussed the often complex response processes in businesses in which gatekeepers, managers, response coordinators, respondents, and data providers play a role.) This can be done by influencing knowledge, attitude, and norms:

- *Influencing knowledge* can be assisted by providing relevant insights and background information to the survey in such a way that it helps to achieve a response, such as by providing information on what the survey is used for and what will happen to the data.
- *Influencing attitudes* refers to affecting the attitudes of businesses, gatekeepers, and respondents in such a way that they will have a more positive attitude toward the survey.
- *Norms* refer to general norms outside and within a business with regard to surveys, for example, business policies on survey participation, and the survey-taking climate.

In an attempt to influence knowledge and attitudes, motivation and facilitation strategies should be included in business survey communication to encourage cooperation. Norms are hard to influence; they need to be factored in.

The extended (and complete) role for business survey communication is to be a tailored and persuasive strategy that communicates the survey request and related instructions and procedure, and motivates and facilitates participants to comply with the survey request. To activate this role, the survey communication strategy should be an objective-driven and coherent plan of communication measures, activities, and actions.

9.2.2 Objectives

The ultimate objective of all survey communication between the survey organization and sampled reporting units is to increase the likelihood of receiving quality survey responses: timely, accurate, and complete data. When relating the communication strategy to the total error framework (as discussed in Chapter 3), the objective is to minimize nonresponse and measurement errors by, for example, obtaining a high response rate and representative response distribution.

Survey communication affects both bias and variance of the estimates. They are affected by nonresponse errors. If the strategy is not effective, we will have low response rates, and thus a high variance. If the response is selective, this has an effect on bias. Also, bias and variance are affected by measurement error. How the communication strategy motivates respondents and facilitates the response process has an effect on bias, for example, if respondents satisfice (Krosnick and Presser, 2010). This may also lead to item nonresponse for specific variables, and thus increased variance for these variables.

To achieve this ultimate objective in communication with selected businesses, three objectives can be identified:

- Establishing contact
- Seeking and gaining cooperation
- Communicating information, instructions, and procedures.

These objectives will be discussed in association with the process in the next section.

9.2.3 Process

Communication activities aimed at the three abovementioned business survey communication objectives should ideally be implemented at different phases in the business survey communication process, which typically begins before the collection phase and finishes in the process phase of the survey production process (see Chapters 1, GSBPM, and 4). To assist in increasing our understanding of this process and to enable mapping between communication objectives, activities and the communication process three stages of the business survey communication process are identified (see Figure 4.2; Jones et al. 2008b; Groves et al. 1997):

- *Prefield stage*—prior to the survey field period commencing,
- *Field stage*—during the survey field period, when data are being collected and returned, and
- *Postfield stage*—commencing as data are returned to the survey organization.

Figure 9.1 provides an initial mapping of the communication objectives to the communication process stages and associated activities. At all stages survey communication should focus on minimizing actual and perceived response burden. The shaded activities in the prefield stage may be carried out in either the prefield or field stages depending largely on whether the survey is recurring and cost considerations. Prefield activities may be amalgamated with field activities. Even if a prefield stage is carried out, some of the prefield communication activities may

Survey process phase	Stage in communication process	Survey communication objectives	Communication is focused on
Collect	Prefield	Establishing contact	- Updating contact information, and identifying the appropriate contact person (management level) - Sample frame refinement: identifying ineligible survey units
		Establishing contact and cooperation	- Introducing (changes to) the survey - Contacting the appropriate person and effecting survey cooperation by motivation and facilitation - Building a relationship
		Communicating information, instructions, and procedures	- Presenting information on the survey - Answering questions/queries - Dealing with complaints
	Field	Establishing contact	- Updating contact information, and identifying the appropriate contact person (management and respondent level) - Checking eligibility
		Establishing contact and cooperation	- Contacting the appropriate person and effecting survey cooperation by motivation and facilitation - Completing and returning the questionnaire - Reminding and nonresponse follow-up - Building and maintaining a relationship - Starting data validation
		Communicating information, instructions, and procedures	- Completing and returning the questionnaire - Answering questions/queries - Dealing with complaints
Process	Postfield	Establishing contact and cooperation	- Data validation - Enforcement of nonresponding units (for mandatory surveys) - Collecting late responses (after survey deadline) - Maintaining a relationship

Minimizing response burden

Figure 9.1 Survey communication objectives and associated activities.

continue into the field stage, such as checking for ineligible units and updating contact information.

In each stage the survey organization communicates with the selected businesses. Communication in the prefield stage is in the direction of the business; that is, it is initiated by the survey organization, but may also be two-way, when businesses have a question (e.g., asking for more information) or a complaint. The field communication is two-way, while the postfield stage is one-way.

From both the business and survey organization perspectives, many people will be involved in the survey communication process (see Sections 9.3.1 and 9.4.1). As a consequence, the survey organization will communicate with different people in the business, and business participants will often deal with many different people in the survey organization. Therefore, depending on the size of the business, there will be a many-to-many relationship between the business and the survey organization. This is important to keep in mind when reading the next sections. It is the role of the survey communication strategy designer to organize the survey communication process and activities to achieve quality responses, while ensuring that the strategy's communication measures and actions are coherent across the stages.

A high-level model of the business survey communication process is shown in Figure 9.2, which is an adaptation of Figure 2.1 on business survey communication. In the following sections we will discuss each stage and the associated communication objectives and activities in more datail. The basic focus for the communication process is from the survey organization perspective, with the business perspective brought in.

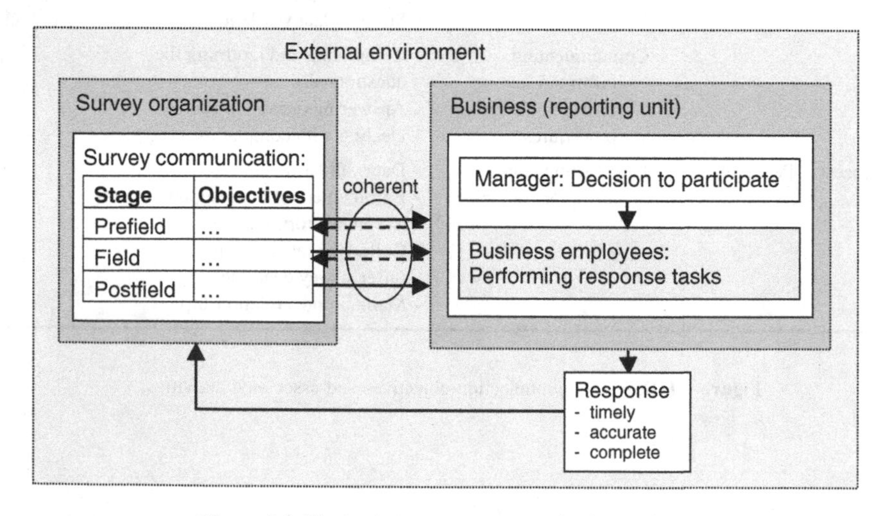

Figure 9.2 The business survey communication process.

9.2.3.1 Prefield Survey Communication Stage

In prefield communication, three activities initiated by the survey organization can be identified (see Figure 9.1):

1. Establishing contact to introduce (changes to) the survey, gaining cooperation and building a relationship, and communicating procedures for businesses asking questions/queries or lodging complaints.
2. Sample frame refinement activities to check for eligibility, update contact information, and identify the appropriate contact person.
3. Supporting communication using websites (with FAQs) and help desk facilities.

Prefield survey communication is therefore aimed at optimizing contact and increasing the likelihood of participation by building a relationship. It focuses on sample frame refinement (to identify ineligible survey units, update contact information, and identify the appropriate contact person, that is, someone in the business who is in a position to approve and authorize the survey request), contacting that person to introduce the survey (or changes to the survey) and seeking survey cooperation, and informing businesses on procedures in case they have questions, queries, or complaints (Jones et al. 2008b; Groves et al. 1997; Ramirez 1997).

A map of the prefield process is presented in Figure 9.3. If no prefield activities are scheduled, these activities are amalgamated with the field stage communication process.

Making Contact This consists in contacting the appropriate contact person(s) in the business. In business surveys there can be up to three contact levels (as is shown in Figure 9.2): the business unit, the management level, and the respondent level. In a business survey, information from the right business unit or cluster of units (e.g., from a large enterprise or from local production units) is required. For that unit a reporting unit is identified, and contact needs to be made. In Chapters 3 and 5 we saw that the relationship between sampled or observational units and reporting units can be quite complex. At the management level, within the reporting unit, contact needs to be made with the manager of the correct department (depending on the data requested), who has the authority to decide on survey cooperation and, where necessary, to delegate the task of completing the questionnaire and also bypassing gatekeepers if necessary. The third level is the respondent level. For actual completion of the questionnaire, contact is needed with a competent person (or persons) with knowledge of the requested data, and where to find them. This person (or persons) should be authorized to complete the questionnaire and have the time to undertake this task (Willimack et al. 2002). Establishing contact is an essential aspect in business survey communication that needs careful attention.

In the prefield stage the contact person is very likely a manager who can decide on survey participation, but it can also be a response coordinator or a competent respondent. In recurring business surveys with overlapping or panel sample designs, the contact person should be known, in which case prefield contacts will involve checking the contact information and introducing any changes to the survey, such as a new data collection mode, or new questions.

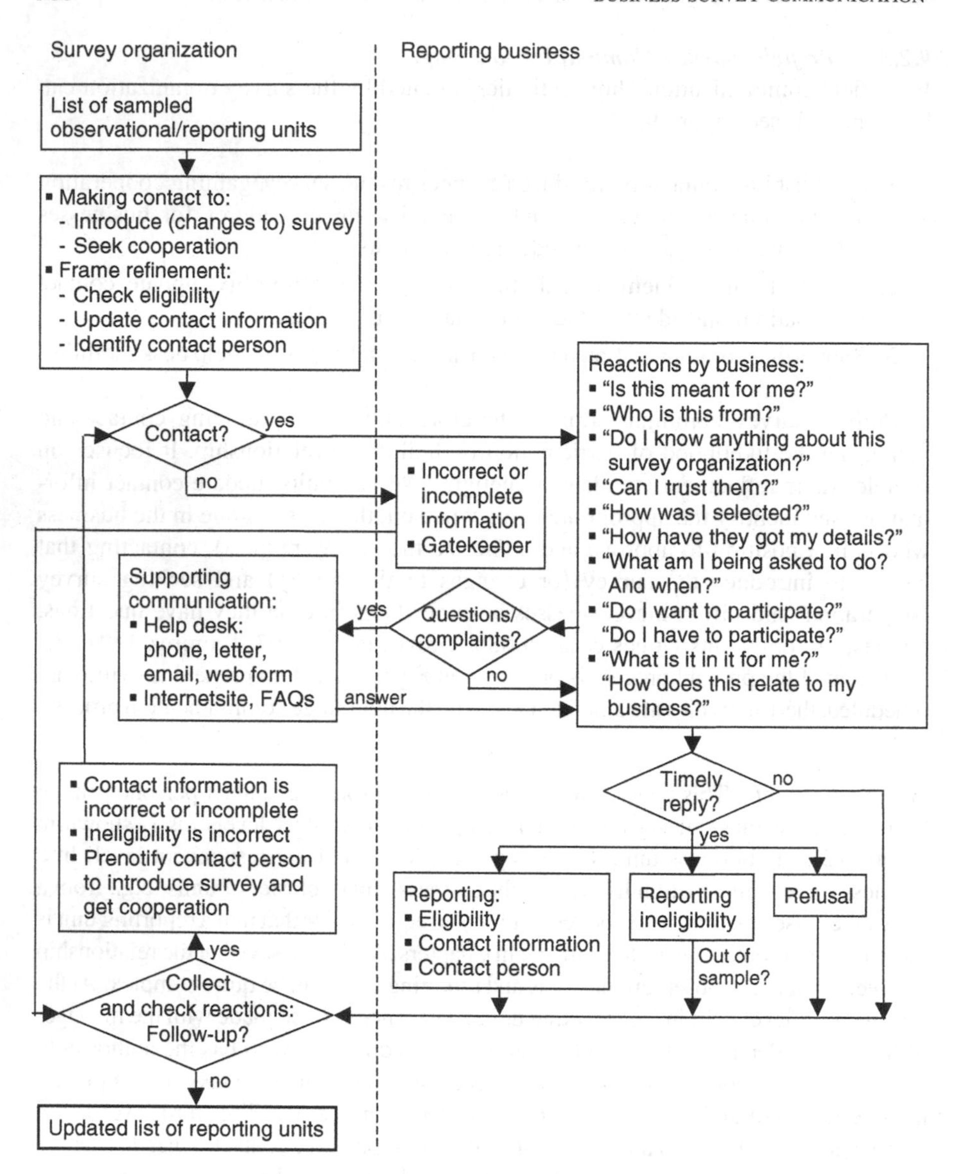

Figure 9.3 The prefield communication process.

Establishing contact requires contact information. Contact information for reporting units may be listed in the sample frame. However, this information may change; businesses change their names, move to new addresses, change their Internet address, and/or email addresses, or change their phone numbers. Reaching the contact person is harder if letters or calls lack an accurate contact name, and places the likelihood of response at risk. In addition, addressing businesses inaccurately can be seen as unprofessional (Ipsos MORI 2010).

Checking Eligibility Businesses may also change in such a way that they become ineligible for the survey. In the business population changes take place all the time. New businesses emerge, old businesses close down, and other businesses startup again (with or without a new name). Businesses merge or are takenover, and businesses breakup. Also, the characteristics of businesses that are important to the survey may change; businesses may change their legal structure or may grow or shrink in size, and their economic activity may change. [See Chapter 5; see also Struijs and Willeboordse (1995); and Ritzen (2007)]. Because of these changes, eligibility needs to be checked and contact information updated. The whole communication process starts with refinement of the selected or observational units, that is, frame cleaning and enrichment (Groves et al. 1997), to increase the chances of establishing contact.

Sample frame refinement is important in the prefield stage, especially if the frame is not periodically updated. Willimack and Dalzell (2006) state that for the industrial research and development survey conducted by the US Census Bureau, which is a telephone survey, major reasons for noncontacts were because a business was no longer active (resulting in disconnected phone numbers), or inability to identify the appropriate respondent or contact person. They conclude that the efficiency of contacting businesses is improved if closed and other ineligible businesses can be removed from the sample frame list prior to the fieldwork. Out-of-date sample information and inaccurate or incomplete contact details delay the process of establishing contact (Ipsos MORI 2010; Barton and DesRoches 2007).

Survey Cooperation Prefield communication is also aimed at introducing the survey, getting cooperation through motivation and facilitation, and building a relationship. In the prefield stage businesses may be notified in advance that they will be selected for a survey and contacted soon. In this way, they can be prepared.

From both the management and respondent levels, we want to obtain cooperation (see Figure 9.2). From the management level we want to get a "yes," indicating survey compliance and commitment. From the respondent level, we want cooperation with regard to the physical action of collecting the necessary data, completing the survey questionnaire, and returning it to the survey organization. For many business surveys, this also includes cooperation in any future data validation queries from the survey organization, such as ensuring that the returned data are correct. Seeking and achieving cooperation therefore involves three aspects (of the survey response process, discussed in Chapter 2):

- Decision to participate
- Performance of the questionnaire completion and return tasks
- Future queries on returned data.

Seeking and achieving survey cooperation can be enhanced by influencing the social behavior in businesses that is pertinent to the survey response process. This involves motivating and facilitating businesses with regard to the survey request (see Section 9.5.3) and questionnaire completion and return. Establishing contact and seeking cooperation are closely linked.

Communicating Information, Instructions, and Procedures The remaining communication activity in the prefield stage involves instructing participants and communicating information and procedures that are related to the survey request and the questionnaire completion tasks. This includes helping respondents find additional information on the survey, providing details on available collection modes and how to ask questions or make complaints, and setting the return date for the survey data.

As a result of this prenotification, businesses may have questions about the survey or the survey organization. They might also have complaints or they might want more information. These questions and complaints are listed in Figures 9.3 and 9.7. In the communication process this needs to be facilitated with supporting facilities such as a help desk and online help site.

Outcome of This Stage For some of business surveys (especially monthly and quarterly) with overlapping or panel sample designs, the prefield stage might take place only when the business is initially selected to participate in the survey. Following this the prefield communication is often combined with the request for data in the field stage. However, any changes to the survey may be communicated in the prefield stage. Because of cost considerations, some organizations might not send out prefield survey correspondence and again combine the prefield communication goals with the request for data in the field stage.

If the prefield stage (Figure 9.3) does take place, the result of this stage will be an updated list of reporting units in which ineligible units are removed, contact information is updated, and commitment possibly made for survey participation (which is registered). The goals may not be fully achieved, which means that establishing contact, and seeking and obtaining cooperation continues in the field stage.

9.2.3.2 Field Survey Communication Stage

In the field stage, data collection begins, and includes the following communication activities:

- Dispatching the questionnaire—making contact to introduce (changes to) the survey, gaining cooperation and building a relationship, and communicating procedures for businesses to use if they wish to ask questions/queries or make complaints
- Issuing reminders and doing nonresponse follow-up
- Supporting communication using websites (with FAQs) and help desk facilities
- Collecting the data and maintaining a relationship by thanking the respondent and providing feedback

These are the field communication activities as listed in Figure 9.1. The field communication process is outlined in Figure 9.4.

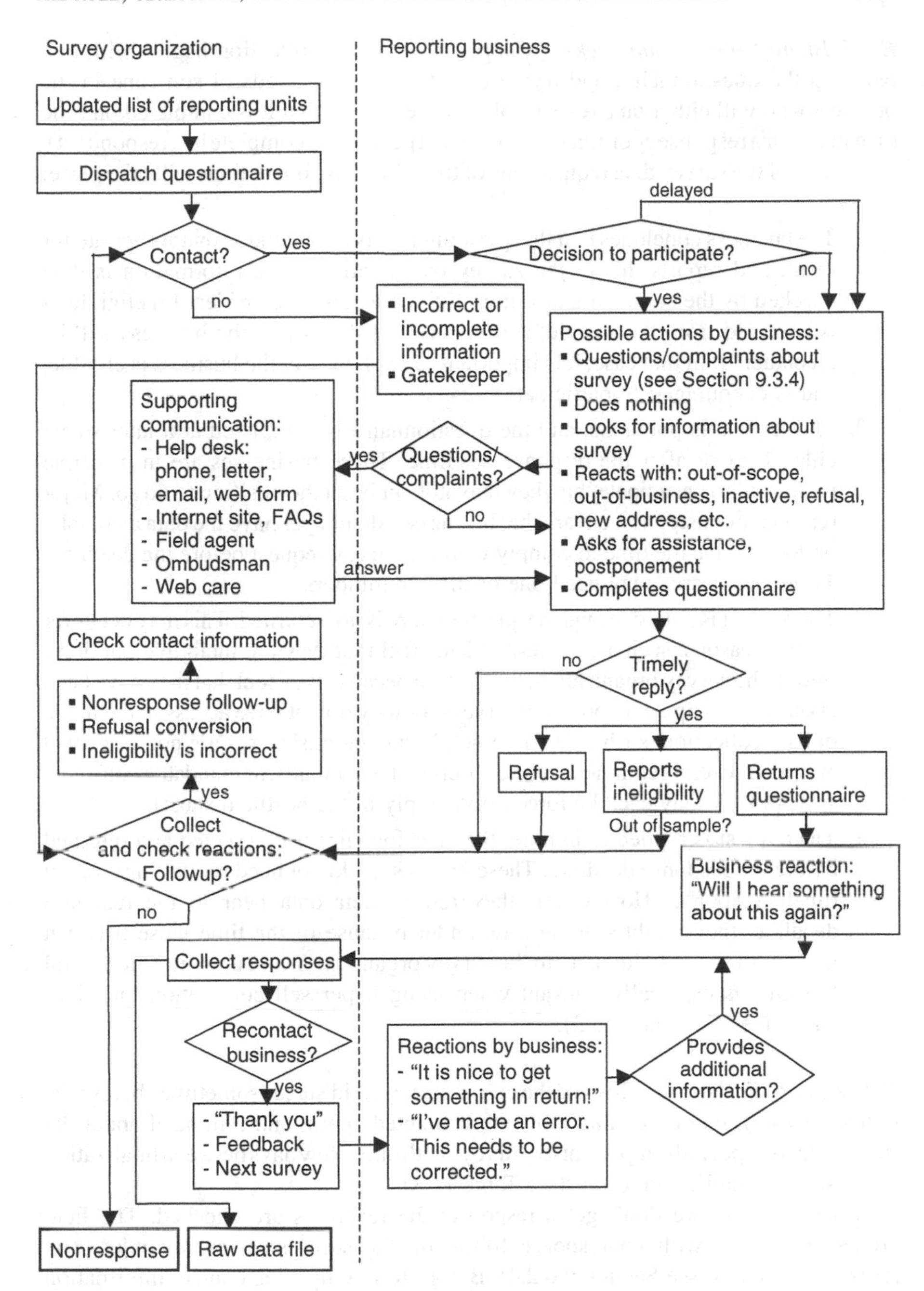

Figure 9.4 The field communication process.

Establishing Contact and Seeking Cooperation Data collection begins with dispatching the questionnaire, and trying to get it into the hands of someone in the business who will either take responsibility for ensuring that it is completed on time (a manager or a response coordinator) or who will personally complete it (a respondent). On receipt of the survey data request, one of the following four actions will take place:

1. The business concludes that the questionnaire is not intended or appropriate for them, and reports this to the survey organization. This information is then checked by the survey organization. If the business is considered ineligible, it is removed from the sample; if it is considered eligible, the business will be recontacted. In that case, it is important to explain why the business is eligible, and is encouraged to participate.
2. The request is put aside, and the questionnaire is completed at a later stage, either near or after the response deadline. These businesses are in principle motivated to participate, but they may not yet be in the position to do so. Major reasons for postponement are that businesses do not yet have the data available, or do not have the time to comply with the survey request before the deadline. These businesses may need one or more reminders.
3. The request is put aside, and the questionnaire is not returned. This may occur for several reasons, such as: the respondents feel that they are ineligible but don't inform the survey organization; they refuse because they feel that they have been asked to participate in too many surveys; or they do not have access to the mode of data collection, such as the Internet. Reluctant businesses are not motivated and need several reminders to be encouraged; eventually in mandatory business surveys they may even be forced (to comply in the postfield stage).
4. The request is replied to in time; the questionnaire is completed and returned before the response deadline. These businesses do not need reminding: we get timely response. However, if they return their data near to the response deadline, they might still get a reminder because of the time lapse between transfer of the questionnaire to the survey organization and updating the central file (this is especially relevant when using paper self-completion questionnaires (see Section 11.2.3).

Nonresponse Follow-up Toward the end of survey field stage (sometimes before the return duedate), nonresponding units are contacted again, and reminded about the survey; this is especially important for surveys with only a few days between finalization of the survey datafile and results (see Table 11.8).

In cases where we don't get a response, the reactions are checked. The field process continues with nonresponse follow-up measures such as reminders and refusal conversion (see Section 9.6.2.4). Before follow-ups, the contact information may be checked again (as in the prefield stage), if time and resources are available.

Communicating Information, Instructions, and Procedures With the questionnaire, survey information, instructions and completion procedures are communicated to help the coodinating person or business respondent. In reaction to this, they may contact the survey organization. This might be done for several reasons: to report that they received the wrong questionnaire (according to them); that they were out of business; or to ask for postponement, more information about the survey, or clarification and explanation. They might also ask for assistance, such as requesting a field agent to visit them and help them complete the questionnaire. They might also contact the survey organization to complain about the survey, or contact the ombudsman (mediator) for complaints on business response burden (Sear 2011). An overview of possible questions and complaints is given in Figure 9.7.

All questions and complaints directed at the survey organization need to be followed up. Businesses might also promulgate their questions and complaints in the open using social media (e.g., Facebook, Twitter; Torres van Grinsven and Snijkers 2013). Dealing with social media requires web care.

Maintaining a Relationship After the responses are sent back to the survey organization, the communication process is not yet finished; the respondent should ideally be thanked for participating in the survey. In addition, feedback can be provided, or information given as to when to expect the next questionnaire. At this point in time the process ends from the perspective of the business: they have complied with the survey, and perhaps have received something in return (as they might expect).

With regard to future contacts in recurring surveys, and in case these businesses are selected for other surveys (which very likely is the case for large businesses), it is important to build and maintain a relationship, such as by giving feedback on returned data and thanking the participants. In fact, building and maintaining a relationship should be attempted from the very first contact in the prefield stage, and also during postfield communication. Feedback to businesses may result in the businesses recontacting the survey organization if they discover errors in the reported data. Responding to a business' queries or complaints is important for building and maintaining a strong relationship. If respondent communications are not followed-up, respondents are trained to become nonrespondents.

Outcome of This Stage The intended outcome of this stage is achieving timely, accurate, and complete response; the unwanted outcome is nonresponse. On receipt of the data, they are collected in a data file and data validation starts, which is continued in the postfield stage; final nonresponse may be followed by enforcement.

9.2.3.3 Postfield Survey Communication Stage

The postfield stage starts as soon as data collection has ended, after the final survey deadline. In this stage there are two parallel processes: data validation, and enforcement: (1) recontacting businesses to validate data that have failed edit checks and requires manual intervention and (2) enforcing compliance of units that have not responded to a mandatory survey.

In addition, data are collected from businesses that respond after the survey deadline has passed. These data can be used if received prior to finalization of the survey datafile, or if part of a time series will be used in the next release of the statistics.

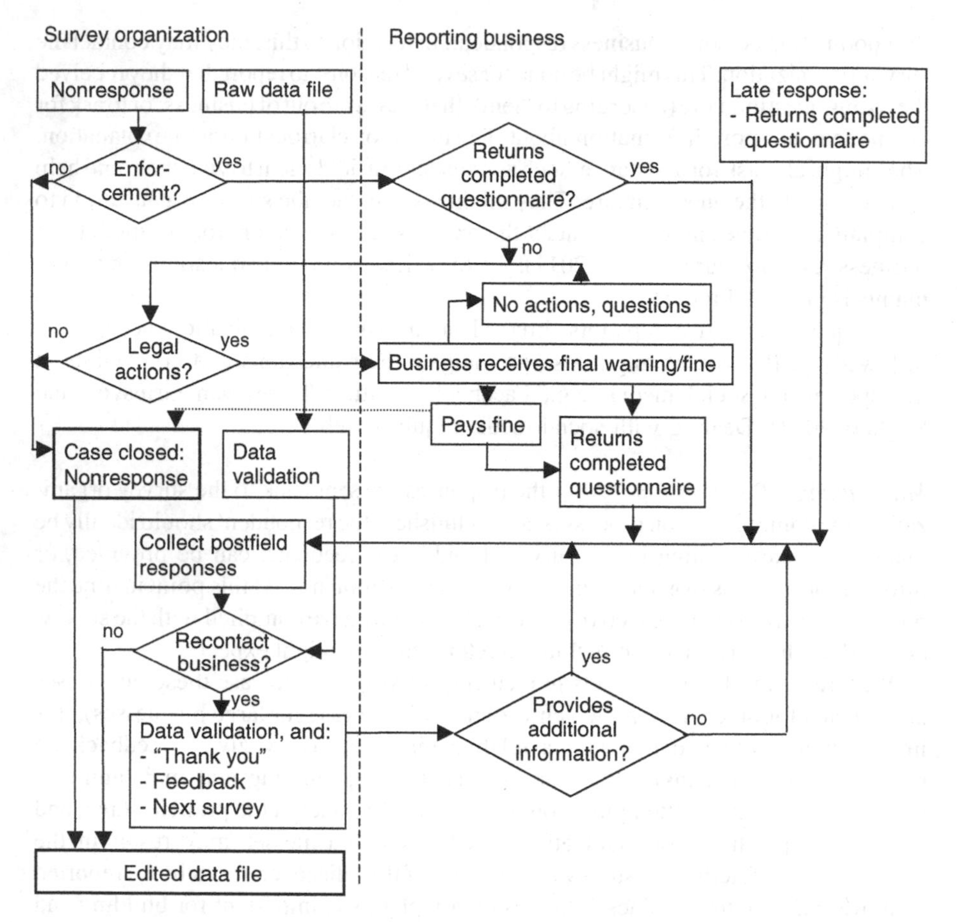

Figure 9.5 The postfield communication process.

Again, these activities are objective-driven, following the activities as specified in Figure 9.1. A map of the postfield process is provided in Figure 9.5.

Communication to Validate Data Businesses may be recontacted to validate the data. This can be done, for example, when answers to questions fail edit checks or if data are internally inconsistent. Businesses may also be recontacted for reasons of item nonresponse, or if they have returned a blank questionnaire [which may be counted as response (in the return rate) but basically is nonresponse]. The registered contact person is recontacted to verify the answers or to obtain additional information. However, business are not recontacted in all cases, such as when the data are edited automatically. Automated editing runs the risk that businesses will not trust the results of the survey, when they have applied a suboptimal completion strategy [such as satisficing; see Krosnick and Presser (2010)]. Since they are not familiar with the data validation procedure, and receive no feedback (which they expect), they think that the provided data are used for analysis resulting in lack of trust. If businesses are recontacted, they may not remember their answers if there is a long gap between responding and being recontacted. Remember that for businesses, the field stages ends as soon as the questionnaire is returned. (The data validation–editing process is discussed in Chapter 11.)

Enforcement If participation is legally required by law, survey compliance is mandatory. Nonrespondents are recontacted after the survey deadline. Enforcement procedures differ for each country. These procedures need to be checked when designing the communication strategy. Enforcement has a positive effect on the response rate, as we will see in Section 9.4.2.2. This does not, however, result in a 100% response rate.

Maintaining Interaction and Outcome of This Stage For all actions, maintaining a relationship by thanking the respondent and providing feedback is an issue. The outcome of this stage is additional response and an edited datafile that is ready for analysis (see Chapter 12).

9.3 TAILORING TO THE BUSINESS CONTEXT

Survey communication is basically a kind of mass communication: we communicate with a large number of groups at once. By tailoring the communication to the business context of specific groups that are more or less alike, we can ensure that the strategy will become more effective. The business context was discussed in detail in Chapter 2. In this section we discuss a number of factors that need to be accounted for when tailoring the communication strategy to the business context, including

- the actors and their roles in the response process,
- the size and economic sector of the business,
- factors affecting the response process in businesses, and
- questions and complaints.

9.3.1 Actors and Their Roles in the Response Process

Actors in the business response process are the recipients of the survey request. These are the people we want to make contact with, and obtain survey cooperation from. As we saw in Chapter 2, business survey participants do not act as private individuals, but as representatives of a business. They act on various levels, and have various roles in the response process. We identified five different roles in the business response process: gatekeepers; managers, authorities, or opinion leaders; response coordinators; respondents or informants; and data providers (Bavdaz 2007, 2010a, 2010b; Tomaskovic-Devy et al. 1994; Groves et al. 1997; O'Brien 2000b; Dillman 2000; Willimack et al. 2002; Fisher et al. 2003; Dillman et al. 2009).

Gatekeepers act as liaison between the business and the outside world, like a receptionist, or a personal assistant. Typically, initial survey contacts, like prenotification or advance letters with the questionnaire, end up on the desk of a gatekeeper, who then, hopefully, transfers it to the appropriate person within the business, preferably a manager or a response coordinator. The manager authorizes survey cooperation, and grants permission to employees to perform the response tasks, to support this process, and at the end signoff the completed questionnaire (see Figures 9.2 and 2.1). The manager may also authorize a response coordinator to ensure the survey completion. Their job is to internally collect and coordinate all needed data, complete the questionnaire, have it signed off, and send it back. To do this they may transfer the questionnaire (or parts of it) to respondents or informants. The respondents will actually complete the questionnaire, and retrieve the actual data from the business records. Here it is important that the right employees be selected: those who are knowledgeable about the data the survey asks for, know where to find the data, and have time to participate in the response process. They may need help from data providers, whom they might contact if needed. These colleagues might be asked to provide specific data that are needed to complete the questionnaire.

In the various stages of the business survey communication process, the surveyor communicates with various contact persons within the businesses, at various levels, having one or more roles in the response process. With regard to the initial contact in the prefield and field stage, the managers are in an essential position. From a communication perspective, they are so-called opinion leaders within the business. They are highly influential with regard to organization of the work, as well as internal policies, opinions, attitudes, and norms. Managers may also affect the work of the gatekeepers, such as by instructing them on how to handle specific information (e.g., letters from a survey organization). Willimack et al. (2002) report that for large businesses this authority often resides with midlevel financial managers; for small and medium enterprises, this position is often assumed by the owner/manager.

A manager may be the contact person, but employees in their roles as gatekeeper, coordinator, or respondent could also assume this position. In many cases the gatekeeper will be the survey contact person; in other cases, respondents. A gatekeeper who is the survey contact person will act as a mediator between the survey organization and the respondent(s). This can also mean that survey correspondence, other than the questionnaire, is never received by the actual respondents. Who will be the contact person for each stage in the communication process depends on whether the survey is a one-time cross-sectional or a recurring survey, and also on the size of the business (see next section). The one-time cross-sectional situation also holds for the first time a business is contacted for a recurring survey.

9.3.2 Business Size and Economic Sector

Size is an important background characteristic of businesses for tailoring the communication strategy, as we noted above with regard to the contact person. A second relevant background characteristic is the industry or sector of the economy.

With regard to size, we need to differentiate between small, medium-sized, and large businesses (O'Brien 2000b; Rivière 2002). With increasing size, the business structure becomes more complex, in terms of the number of departments, vertical levels and hierarchy, role differentiation and specialization, rules and regulations, and accounting systems used (Robbins et al. 2010; Bavdaz 2007; Edwards and Cantor 1991). As a consequence, with increasing size, the internal communication becomes more complex: more departments are involved, more levels are involved, and as a consequence more people in various roles are involved.

For small businesses, with a simple, flat structure, the roles of gatekeeper, manager/coordinator, and respondent/provider may be assumed by one person or different people. What is essential here from a survey communication perspective is that we are dealing with a small community: people know each other and their roles within the business. Communication lines are short, with authority centralised in one (or two) people with a wide span of control. This means that it is easy to establish contact and deliver the survey requests; getting response, however, may be quite difficult.

For medium-sized businesses, the structure and the accounting system are more complex than in small businesses and can vary from business to business. For these businesses the response process is diverse: for some business the various survey roles can be linked to different participants, but for other businesses some roles may be concentrated within one or two persons. This depends on the topic of the survey and also on the business structure. This higher level of complexity as compared to small businesses may affect survey communication (as well as response burden, as we saw in Chapter 6).

For large businesses the situation is quite different. Large businesses may markedly differ in terms of structure, but can be characterized by various production locations, many departments, many vertical levels, high degree of job specialization, and many internal rules and regulations. For these large businesses, the response process can be very complex: many people, in various positions, may be involved in various roles. Because of this increased complexity, for large businesses it is more difficult to contact the appropriate people, but obtaining survey cooperation may be easier. Many national statistical institutes (NSIs) have established special procedures for communicating with large businesses; these are discussed in Section 9.6.3.1.

Another background characteristic that is relevant for tailoring survey communication is the economic sector. For instance, businesses in various industries may have different vocabularies, or when using business organizations to endorse the survey, the organization with which the business identifies may need to be tailored. Tailoring to economic sector, as well as to size, is also relevant to ensure a representative response distribution and unbiased estimates.

9.3.3 Factors Affecting the Response Process

In Chapter 2 we discussed many authority, motivational, and ability (or capacity) factors that affect the response process in businesses (Willimack et al. 2002; Snijkers 2008). We discussed the external environment, business/management, and survey and response burden–related factors that affect the decision to participate. We also discussed factors at the level of the respondent that affect the performance of the response tasks. An overview of these factors is presented in Figure 9.6. These factors need to be considered when designing a business survey communication strategy.

Layers	Authority	Motivation	Capacity
External environment	• Legal and regulatory requirements	• Public goodwill • General norms • Survey-taking climate • Political climate	• Economic conditions
Business	• Policy on survey participation • Reporting priorities	• Cost-benefit analysis • Trust: familiarity with survey organization • Past behavior	• Expectations about response tasks • Record formation and data availability • Organizational structure and complexity: infrastructure • Availability of resources • Established routines • Respondent selection
Respondent	• Authority/mandate to perform tasks	• Employee attitude • Organizational and management priorities • Job performance evaluation criteria • Professional standards	• Employee's role in response process • Routine • Capacity: time, knowledge of data, competence/skill • Workload: competing activities • Data: availability and access • Timing of survey request • Survey-related factors
Survey organization and survey design	• Legal obligation • Reputation and legitimacy	• Survey topic • Confidentiality • Image of the survey organization • Actual and perceived response burden - Sample coordination - Recurring surveys • Questionnaire design • Data collection mode • Communication strategy	• Questionnaire design: - Level of detail - Visual layout • Data collection mode(s) • Communication strategies - Contacts - Coordination - Help desks

Figure 9.6 Overview of factors affecting the response process (see also Table 2.3).

9.3.4 Questions and Complaints

Associated with the factors listed in Figure 9.6, businesses will very likely have a number of questions and complaints, which depend on the stage in the process and the level and role of the contact person. When being addressed with a prenotification or a survey request in the prefield stage, managers look for different kinds of information compared to employees. They are concerned with running the business. Response coordinators, respondents, and data providers are more concerned with their own jobs, the performance of the response tasks, and how this fits in there personal agenda.

Managers or respondents might also contact the survey organization to complain about the survey, the number of questionnaires received, the difficulty in completing a questionnaire, how to find it on the Internet, how to open it and transmit the data, the time it takes to complete, the costs for compliance (response burden), the usefulness and relevance to the business, and mandatory issues and trust (Snijkers et al. 2007a; UK ONS 2010).

Questions and complaints are associated with authority ("Do I have to do this?"), motivational ("Do I want to do this?"), and ability factors ("Am I able to do this?"), as shown in Figure 9.6. The survey communication should address these questions, since these factors influence the decision to participate and performance of the completion tasks. Depending on the contact person (level and role), the stage and goal of the communication different measures are needed. An overview of these questions and possible measures, related to the factors in the response process, is presented in Figure 9.7 (Jones 2011; Snijkers et al. 2007a).

9.4 THE SURVEY ORGANIZATION CONTEXT

Apart from tailoring the communication strategy to the business context, the internal survey organization context also needs to be factored in. This sets the constraints for the strategy. First the business context needs to be considered, followed by matching the design to the survey organization context. In this section we will discuss the survey organizational factors: (1) staff and their roles in the communication process; (2) organization policies, procedures, standards, guidelines, corporate communication, and so on; and (3) survey communication in perspective.

9.4.1 Staff in the Communication Process and Their Roles

In the design and execution process of a communication strategy (see Section 9.5.1), a numerous survey organization personnel in many roles is involved. An overview of the actors and their roles is presented in Figure 9.8, related to phases in the survey production process (GSBPM, as discussed in Chapters 1 and 4).

Those actually involved in two-way communication with businesses are fieldwork staff, help desk staff, mediator (ombudsman) for business response burden, data

Survey factors	Possible question considered by respondents	Possible measures and actions to deal with these questions
Knowledge of and trust in the survey organization	- "Do I know anything about this organization?" - "Can they be trusted with my information?" - "Will the organization keep responses secure and ensure that results are not disclosed?"	- Branding and marketing of the corporate image (see Section 9.4.3) - Try to establish trust (see Section 9.5.3.1—Authority, and Section 9.4.2.1—Informed consent)
Obligation to respond	- "Is it a legal requirement to respond?" - "What will happen to me if I don't respond?" - "Since I have to comply, at least make it easy for me to do so!"	- Inform business that participation is legally required in both prefield and field stages; in postfield stage: enforcement procedures (see Section 9.4.2.1)
Past experiences in surveys participation	- "How did I feel when responding to the last survey?" - "I responded last time, so I will also do so now!" - "I have no experience in responding to a survey!"	- Try to build a relationship with newly surveyed businesses. Past behavior is a very strong predictor for survey participation (see Section 9.5.3.1—Commitment) - A first impression usually counts
Prefield communication (see Figures 9.1 and 9.3): - Sample frame refinement - Introducing (changes to) the survey	- "Is this meant for me?" - "Who is this from?" - "Do I know anything about this survey organization?" - "Can I trust them?"	- Be clear about the objective of each item of communication; mode, message (content, layout, language), recipient, and timing of communication are relevant (see Sections 9.2.3 and 9.5)

- Contacting the appropriate person and getting survey cooperation - Building a relationship - Communicating information, instructions, and procedures	- "How was I selected?" and "How have they got my details?" - "What am I being asked to do? And when?" - "Do I want to participate? Do I have to participate?" - "What is it in it for me (perceived costs and benefits)?" and "Is this relevant for me?" - "How does this relate to my business?"	- See Section 9.2.3.1 for process map and Section 9.6.1 for prefield measures and guidelines - Tailor to the business context - Be flexible - Adopt a proactive, service-oriented marketing strategy
Field communication (see Figures 9.1 and 9.4): - Sample frame refinement - Introducing (changes to) the survey - Contacting the appropriate person and getting survey cooperation - Completing and returning the questionnaire - Reminding and nonresponse follow-up - Data validation - Building and maintaining a relationship - Communicating information, instructions, and procedures	On receipt of the questionnaire: - See prefield questions above - "Do I remember previously receiving any information about this?" - "What is this survey about?" - "What will happen with the data?" - "What will the data be used for? - "Do I have the time to participate?" - "Can I get assistance?" Nonresponse follow-up: - "I know I have to do this, but I am not able yet! I told you that! - "I have always responded in time, for once I am late, and I am being stalked." When submitting the data: - "Do I get something in return?" - "Will I hear anything about this again?" - "Why am I being thanked? - In case of feedback: "It is nice to get something in return!"	- Be clear about the objective of each item of communication; mode, message (content, layout, language), recipient, and timing of communication are relevant (see Sections 9.2.3 and 9.5) - See Section 9.2.3.2 for process map and Sections 9.6.2 and 9.6.3 for field measures and guidelines - Be coherent with prefield communication and over all communication measures - Tailor to business characteristics, past behavior, and availability of data - Be flexible - Adopt a proactive, service-oriented marketing strategy - Use motivation and facilitation strategies - Thank the respondent and give feedback

Figure 9.7 The business perspective on business survey participation. [Adapted from Jones (2011).]

Postfield communication (see Figures 9.1 and 9.5): - Data validation - Enforcement - Collecting late responses - Maintaining a relationship	Data validation: - "I haven't heard anything. I don't trust the outcomes." - "After a long time I am being recontacted. I don't remember this, and I don't remember my answers." - "Why am I being recontacted? Did a make a mistake?" Enforcement: - "Why am I being fined? Do I need to pay this? - "I have always responded in time, for once I am late and I am being threatened."	- Be clear about the objective of each item of communication; mode, message (content, layout, language), recipient, and timing of communication are relevant (see Sections 9.2.3 and 9.5) - See Section 9.2.3.3 for process map and Chapter 11 for postfield activities - Be coherent with prefield and field communication and over all communication measures
Survey topic	- "What is this survey about?" - "Is this a topic that I am interested in?" - "Do I want to respond to questions on this topic?" - "What will the data be used for?"	- Have an appropriate survey title that reflects the survey topic(s) - Inform businesses about the topic of survey and what will happen with the data - Try to establish trust: informed consent (see Section 9.4.2.1) - Try to make the topic salient to the business
Design and navigation of the data collection instrument	- "Does it look as if it will be easy to respond?" - "Do I need to read or listen to lots of instructions to understand how to respond?"	- Have user-friendly questionnaires: see Chapters 7 and 8 (on questionnaire design and pretesting) - In web surveys: make the login–completion–sending procedure easy; test for usability (see Section 9.6.3.2)

Interviewers' approach toward respondents and their knowledge of the survey organization and survey	- "Do I know why this person is contacting me?" - "Do I know who they work for?" - "Do I trust this person?" - "Is their request legitimate?"	- Have interviewers trained in calling strategies (see Section 9.6.2.4)
What information is being requested	- "Why do they want this information?" - "Do I want to give someone this information?" - "Do I feel that the requested information is sensitive?" - "Is it easy for me to respond to the questions?" - For opinion questions: "Do I actually have an opinion on these items?"	- Inform managers and respondents about the kind of information that is requested - Have this information printed on the envelope, to increase the chances of establishing contact with the appropriate respondent(s) - Establish trust
Mode of data collection	- "Can I actually access this mode (especially relevant for web and telephone)?" - "This is OK as I am used to using this mode," "This is not good, as I have used this mode only a few times before," "This is not good, as I have never used this mode before," or "This is not good, as I cannot use this mode". - "Do I want to respond using this mode?"	- Select modes of data collection that are accepted and an available mode that enables the particular respondent to interact (e.g., paper, email, web) - See Section 9.6.3.2 for measures to introduce web questionnaires in a mixed-mode design

Figure 9.7 (*Continued*)

Required process of performing response tasks: - Performance of response tasks: • Comprehension • Retrieval • Judgment • Communication - Who is involved: roles and actors?	- "Can I understand the questions?" - "What do I need to do to respond?" - "Will I need to consult records to find the answers?" - "Do I have a response to the questions?" - "Am I prepared to give an honest answer?"	- Study the business context and the response process before starting the design process (see Section 9.3) - Develop and test the questionnaire to minimize errors, costs, and response burden (see Chapter 7)
Actual response burden: - Time needed to complete a questionnaire	- "How long will this take?" - "How much work is involved?" - "Apart from this questionnaire, I have many more regulations to comply with and forms to complete!"	- Assess the time needed for participation in the data collection process - A combination of measures to minimize the actual burden with regard to sampling, mode, questionnaire; survey communication
Perceived response burden (this is a combination of all of the factors outlined in Figure 9.6)	A combination of all or some of the questions above	- Provide a mediator (ombudsman) for response burden - A combination of measures to minimize perceived burden with regard: sampling, mode of data collection, questionnaire, survey communication

Figure 9.7 (*Continued*)

editing staff, and compliance enforcement staff. In the design and implementation stage the actors in the communication need to be identified and listed, informed, and trained with regard to their roles in the process. We also recommended that these individuals participate in the design stage to factor in the survey organization policies and procedures in the communication strategy.

9.4.2 Survey Organization Policies and Procedures

For all actors in the communication process, specific procedures may be active; these survey organization policies and procedures need to be checked when designing a communication strategy.

They include

- Guidelines on communication modes to be used (see Section 9.5.2)
- Guidelines on communication design [e.g., corporate communication strategies, web design guidelines, standards for envelopes, letters, and standard sections in letters (e.g., informed consent or the mandatory status of a survey)]
- Guidelines on ethics (see next Section)
- Procedures for reduction of response burden [see Chapter 6; also, e.g., UK ONS (2010)]
- Procedures and practices that are used by departments that are involved in the communication process, such as the help desk, reminder procedures for telephone unit, recontacting during data editing, and enforcement procedures
- Systems and tools used in designing and building communication items (e.g., printing options and the use of color)

These policies and procedures will differ for each survey organization. Two aspects that are more broadly relevant with regard to business survey communication are (1) informed consent and confidentiality and (2) mandatory surveys. Informed consent and confidentiality is associated with ethics and eliciting trust; mandatory surveys, with the obligation to respond. Survey organizations may have specific procedures on how to deal with these aspects. Here we will discuss these aspects more generally.

9.4.2.1 Informed Consent and Confidentiality

Trust is key in obtaining cooperation; businesses should trust the survey organization with their data. This can be done by showing professionalism (as will be discussed in Section 9.5.3: authority principle), and by referring to informed consent and confidentiality protection (Singer 2008). The signals that the communication items show and the message should be consistent! For instance, a messy letter doesn't evoke trust, even if it states that the use of the data is strictly restricted to statistical analysis.

Stage	Actors	Role
Design, build, and test	• Project manager	• Setup and manage project • Define objectives (step 1 in Section 9.5.1)
	• Project staff: - Staff involved in prefield, field, and postfield communication and implementation (see below) - Data collection methodologists - Pretesting methodologists - Communication designer - Web designer	• Design, build, and test the communication strategy (steps 2–5 in Section 9.5.1): - Decide on the communication measures - Write texts for letters, brochures, websites, and so on - Design and build webpage - Write procedures, training materials, and so on (linked to the measures) - Define paradata
Collect: implementation	• Staff involved in preparing prefield, field, and postfield communication • Printing department (paper items) • Web IT staff	• Implement the strategy in a data collection schedule (step 6 in Section 9.5.1) • Have all items, procedures, workflows, and so on ready • Train staff • Print communication items • Upload websites

Collect: prefield and field communication	• Process managers and staff	• Control and monitor the process according to the data collection schedule • Collect and analyze paradata
	• Printing departments (paper items) • Fieldwork staff (interviewers, field agents, business profilers, large-business managers) • Data processing staff • Help desk	• Sending out paper items • Contact businesses and receive responses • Follow up on nonresponse • Capture data • Handle questions and complaints • Address technical queries
	• Ombudsman (mediator) for business response burden	• Address questions and complaints on response burden
	• Data editing staff	• Recontact for data validation
Process: postfield communication	• Data editing staff • Enforcement staff	• Recontact for data validation • Enforce mandatory surveys
Evaluate	• Process managers and staff	• Analyze the communication process and seek improvements (step 7 in Section 9.5.1)

Figure 9.8 Survey organization actors and their roles in the communication design and execution process (see Sections 9.2.3 and 9.5.1).

Informed consent refers to ethics in survey methodology where respondents as well as their data, are respected (Groves et al. 2004). In the United States, informed consent is defined in the *Federal Code of Regulations* as the "Knowing consent of an individual or his legally authorized representative . . . without undue inducement or any element of force, fraud, deceit, duress, or any other form of constraint or coercion" (Singer 2008). In surveys the principle of informed comes down to "informing respondents about 'routine planned uses' of the data they are asked to submit" (Tuttle and Willimack 2005).

According to Singer (2008, p. 85), the implementation of informed consent requires four conditions: (1) providing enough information about potential benefits and risks of harm to permit respondents to make informed participation decisions, (2) ensuring that the information is understood correctly, (3) creating an environment that is free of undue influence and coercion, and (4) the fact that survey organizations need some assurance that respondents have been adequately informed and have agreed to participate. As a comment on these requirements, Singer asks herself: "How easy is it to create these conditions in the context of real-life research?"

These conditions provide guidelines as to the information that should be communicated to business respondents (Tuttle and Willimack 2005). The message should include

- The duedate
- Purpose and use of the data
- Types of information the survey asks for
- How the business is selected
- Information on confidentiality protection
- The (legal) basis of a survey, including the obligation to respond

This informs businesses about the reasons for a survey, and what to do; it is related to information businesses expect to receive (as shown in Figure 9.7).

Confidentiality protection relates to assurance of anonymity; it refers to the ethics in surveys stipulating that from survey outputs no data of individual business may be disclosed. A typical paragraph in letters from the U.S. Census Bureau to inform respondents that anonymity is protected by law is (Tuttle and Willimack 2005):

> *YOUR CENSUS REPORT IS CONFIDENTIAL BY LAW.* It may be seen only by persons sworn to uphold the confidentiality of Census Bureau information and may be used only for statistical purposes. Further, copies retained in respondents' files are immune from legal process.

A few additional statements include

- No individual establishment, firm, business (etc.), or its activities may be identified.

- US Census Bureau publications summarize responses in such a way that the confidentiality of respondents and their business activities is fully protected.
- Your reported data are exempt from requests made under the Freedom of Information Act.

These statements may change from one country to another, but similar statements are included in letters from other statistical agencies, like Statistics Netherlands, UK ONS, and Statistics Norway. How to ensure disclosure protection in statistics will be discussed in Chapter 12 (Bethlehem 2009; Groves et al. 2004).

9.4.2.2 Mandatory Surveys

For mandatory surveys, businesses should be informed about the obligation to respond, the legal basis of that obligation, and the consequences of not responding. Following informed consent, this message should be conveyed, even though businesses don't like being reminded about that obligation.

A typical paragraph from the US Census Bureau is (Tuttle and Willimack 2005):

> *YOUR RESPONSE IS REQUIRED BY LAW.* Title 13, United States Code, requires businesses and other organizations that receive this questionnaire to answer the questions and return the report to the U.S. Census Bureau.

Mandatory authority is exercised in late nonresponse follow-ups via the threat of prosecution. In addition, selected large businesses receive courtesy calls before the Commerce Department's Office of the Inspector General (OIG) is informed of the company's noncompliance. If this still doesn't result in compliance, an OIG letter is sent, threatening prosecution (Petroni et al. 2004a).

A similar statement was issued by Statistics Netherlands,which applies a stepwise approach with stronger appeals in follow-up letters:

- Advance letter and "remember" letter (in the lower part of the letter, and in the left margin "Cooperation important"): "We are aware of the fact that we ask a lot from you. However, your cooperation is important to timely publish reliable information. It concerns information about businesses in the Netherlands in general, but also information concerning your own sector of the economy. The government has made your contribution mandatory, since this is considered important input for this information. Therefore, we ask you to ensure that we have received your data no later then [date]."
- First reminder letter (as second paragraph at the top of the letter, and in the left margin "Delivery mandatory"): "As you know, you are obliged to timely deliver your data to us. If we haven't received your data before [date], this may result in sanctions. You can find more about these sanctions on the Internet: `www.cbs.nl/handhaving`. Please make sure that we have received your data before [date]."

- Last reminder letter (as second paragraph at the top of the letter, and in the margin "Submit at once"): "As you know, you are obliged to timely deliver your data to us. This is according to the Statistics Netherlands Act. You can find this act on the Internet: `www.cbs.nl`. Now, you may be sanctioned: your fine can be as high as € 16,000. You can read more about this sanction on the Internet: `www.cbs.nl/handhaving`. You can prevent this from happening by submitting your data at once."

The fine depends on past behavior and the size of the business. If no response has been received after the fieldwork, a team of managers will ensure enforcement. This will be done in the postfield stage. Research in the Netherlands has shown enforcement to have a substantial effect, it can lead to a 7–8% increase in the response rate (Smeets 2006; Roels 2010). Also, it has a positive effect on the timeliness of the response.

It is important to follow-up on the threats of prosecution, in order to be trustworthy. If this is not done, the threats are without authority, and the survey organization is considered a paper tiger.

9.4.3 Survey Communication in Perspective

A final aspect of the survey organization context that needs to be considered when designing a survey communication strategy is how survey communication relates to other forms of communication by the survey organization. For large survey organizations, like NSIs, survey communication is not the only form of communication with the outside world. Large survey organizations also communicate information about their output—the statistics and research outputs they have produced; these are typically communicated via press releases, press conferences, paper and electronic publications, their website, and more recently also via YouTube, Twitter, Facebook, Google Earth, and other media (see Chapter 12). Target groups such as policymakers, researchers, or industry organizations may be reached through specific publications and conferences. For example, the US Census Bureau and the Bureau of Labor Statistics "conduct outreach and survey promotion through trade shows and contact with industry organizations" (Petroni et al. 2004a). This type of communication might also reach actual or potential business respondents. The way in which survey organizations communicate about their outputs is assumed to affect the general perception of these organizations (Giesen and Snijkers 2010).

Specifically, the use and discussion of NSI statistics in the media may have a large impact on the general public's attitudes concerning the relevance and trustworthiness of NSI statistics. Wilmot et al. (2005, p. 57) conclude from an extensive qualitative study in the U.K. about public confidence in official statistics that "the main public source of information on official statistics seemed to come from the media." NSIs have many reasons to be concerned with their corporate image (Voineagu et al. 2008), including trust, and the effect on respondents and their motivation.

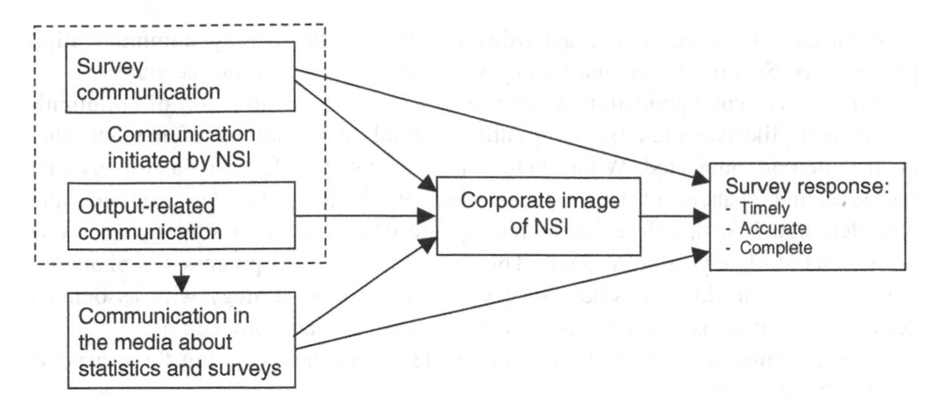

Figure 9.9 Survey communication in perspective. [Adapted from Giesen and Snijkers (2010).]

Roy 2001 states:

> Statistics Canada places a high importance on being identified as the source of its information whether it is in the news media, publications or on the Internet. This relationship is important because it helps to establish a public perception of 'relevance' that engenders strong support for data collection activities.

Thus, corporate communication strategies—on which survey communication, and the communication related to the dissemination of statistics may be based—may affect the attitude toward surveys in general, and the decision to participate in a particular survey.

The relationship between the various communication flows and survey response is shown in Figure 9.9. The direct relationship between survey communication and survey response is obvious, since this is what we aim for. For all forms of communication, we assume a direct effect on corporate image, which, in turn, affects survey response. Also, we assume a direct effect of communication in the media about statistics and surveys among owners/managers/employees of sampled businesses on survey response. Through discussions (e.g., in social media via Twitter, chat sites, Facebook, etc.), people may influence each other's attitude on surveys, and consequently their response behavior. [This effect relates to social validation, as one of the compliance principles (see Section 9.5.3.1).]

9.5 DESIGNING AN EFFECTIVE BUSINESS SURVEY COMMUNICATION STRATEGY

So far we have discussed the business survey communication process and contextual factors related to the business and the survey organization. A high-level model of the business survey communication process was shown in Figure 9.2. A business survey communication strategy should be designed within the context shown in that figure. The strategy should be an object-driven, tailored, and coherent plan of communication measures and activities mapped to the survey communication process (see Section 9.2.3), and taking the contextual factors into account.

An effective communication strategy is not merely a compilation of communication items, like paper letters, phone calls, or emails, and a number of messages that need to be communicated. What, then, makes a strategy effective—a strategy that increases the chances of being successful, of obtaining timely, accurate, and complete data? Also, before implementing the communication strategy, how can we predict its likely effectiveness? This section outlines a process for planning, designing, and testing a business survey communication strategy with associated communication measures, actions, activities, materials, and procedures.

As the communication strategy is being designed, the following three aspects need to be considered:

1. The planning and design steps, and quality assurance of the strategy (Section 9.5.1)
2. Communication modes and their effectiveness (Section 9.5.2)
3. Motivation and facilitation strategies (Section 9.5.3)

In the following sections we will look at these three aspects in turn, taking a multidisciplinary approach to integrate approaches from survey methodology (Cox et al. 1995; Groves and Couper 1998; Groves et al. 2004; Dillman et al. 2009; Church and Waclawski 1998), organizational behaviour theory (Robbins et al. 2010), social psychology (Vonk 2009), persuasion theory (Cialdini 2001; Perloff 2010; Gass and Seiter 2011; Hoeken et al. 2009), motivation theory (Torres van Grinsven et al. 2011; Wenemark et al. 2011), and mass communication and marketing (McQuail 2010; Pol 2007).

9.5.1 Planning and Designing the Communication Strategy

Just like any other process, the communication process needs to be planned, designed, built and tested, implemented, and evaluated. In general, planning an effective survey communication strategy consists of the following five steps, followed by implementation and evaluation:

1. Define the objectives
2. Study contextual factors
3. Define paradata
4. Design, build, and test the communication strategy
5. Revise and finalize
6. Implementation
7. Evaluation

This approach is based on *Intervention Mapping*, a method for designing effective communication strategies in health education programs aimed at behavioral change, as developed by the Universities of Maastricht and Texas (Pol et al. 2007; Schaalma and Kok, 2007). It turned out that that many (re)design projects as reported in the

Activities and actions	Objectives of each activity (Why?) (Figure 9.1)	Communication modes (How?)	Messages to comunicate (What?)	Actors (Who?) Survey organiza-tion staff	Contact person(s) in business	Timing of communication (When?)
Prefield	- Contact	Item 1	—	—	—	—
Measure 1	- Cooperation	—				
—	- Information	Item n				
Field	- Contact	Item 1	—	—	—	—
Measure 1	- Cooperation	—				
—	- Information	Item n				
Postfield	- Contact	Item 1	—	—	—	—
Measure 1	- Cooperation	—				
—		Item n				

Figure 9.10 Components of a business survey communication strategy.

literature (implicitly) follow these steps [see, e.g., Tuttle (2009), Tuttl et al. (2010a), Ipsos MORI (2010), and Petroni et al. (2004b)].

We will discuss each of these steps in turn.

Step 1: Define the Objectives Define the objective of the strategy as a whole, the role and objectives of the prefield, field, and postfield stages, and the associated activities and procedures. This involves an initial completion of Figure 9.10.

The strategy design process starts with defining the survey communication objectives. In general, this includes a specific response rate level as negotiated with the stakeholders, but more specific goals may also be defined, such as reducing perceived response burden by communication measures, or increasing web takeup rates in mixed-mode designs (see Section 9.6.3.2).

The output from step 1 is the initial completion of Figure 9.10. This then provides the basis for developing, testing, and implementing the business survey communication strategy and associated activities and procedures. The columns in Figure 9.10 represent the six components of a communication strategy [partly based on the content–processes–roles model by Church and Waclawski (1998)]:

- Activities and actions
- Objective of each activity
- Communication modes
- Messages to communicate
- Actors: (a) survey organization, and (b) business
- Timing of the communication

To complete Figure 9.10, a number of decisions are necessary. To facilitate this process, questions related to each component are listed in Figure 9.11. At all

Component	Questions
1. Activities and actions	What measures, activities, and actions will be applied at each stage in the communication process? For instance, how will we establish contact, introduce the survey, seek cooperation, support businesses, and remind them?
2. Objective(s) of each activity	What is the objective of each activity (as listed in Figure 9.1)?
3. Communication modes	How will we communicate? What means, materials, or items will we use: letters, personal contact by telephone or face-to-face, emails, brochures, Internet sites, the help desk, and so on, or a combination of these? What is the best mode (or channel) in a specific situation, such as in nonresponse follow-ups? Note that these communication modes are not necessarily the same as the questionnaire mode. (Communication modes are discussed in Section 9.5.2.)
4. Messages to communicate	What message will be communicated with each communication activity? What are the message(s) we want to put through? What should be in a letter? What instructions and procedures need to be communicated? How will we motivate, facilitate, and help respondents (see Section 9.5.3)? Presentation of the message (language, tone of voice, and layout) is as important as the message, and need to be considered as well.
5. Actors	Who are the actors in the communication process?
a. Survey organization	What staff from the survey organization will be involved in the communication process? This includes methodologists, text writers, communication experts, printing departments, Internet designers, interviewers/field agents, help desk staff, editing staff, and enforcement staff.
b. Business	Whom will we contact? Who will be the contact person within the business (i.e., the recipient)? Whom will be addressed, and who needs to process the message?
6. Timing	When are we communicating? When is the best time to communicate, for each activity/item? When are we sending out advance letters with questionnaires, reminder letters, and so on?

Figure 9.11 Questions to answer to initially complete Figure 9.10.

times, remember that the strategy should be an object-driven, coherent plan of activities and actions; you may even return to the objectives during the design process!

Step 2: Study Contextual Factors To tailor the strategy, analyze the contextual factors for the communication strategy: the business context, survey organization context, and external context.

A business survey communication strategy needs to be put into context to be effective. It must be tailored to the business context, the response process within businesses, and response burden issues. Also, external factors, survey organization policies, and procedures as well as project constraints and the survey communication process need to be factored in. These contextual factors have been discussed in Sections 9.3 and 9.4.

In studying the business context and the response process, the objective is to identify potentially effective motivation and facilitation strategies. A starting point for such a study can be a review of literature describing and evaluating strategies that have been used for equivalent surveys. Also, a background research using Google or another search engine can be conducted to obtain information on the businesses in the target population. Surveyors who will be studying the business context and the response process themselves can apply the research methods discussed in Chapter 7, including conducting debriefing sessions with interviewers and help desk staff or a small number of indepth interviews with business representatives. For examples of such studies, we recommend Willimack et al. (2002), Tuttle (2009), Tuttle et al. (2010a), Bavdaz (2007), Fisher et al. (2003), Hales and Webster (2008), and Ipsos MORI (2010; this research report, published by Her Majesty's Revenue & Customs in the UK, includes a discussion guide (for conducting indepth interviews) that can be used as an example for a small-scale study with businesses).

The business context and the response process are also studied to identify subgroups of businesses with similar characteristics, such as businesses that have data available early and those who are late, and their background characteristics. For recurring surveys, this can be done, for example, by a detailed response analysis of past behavior (Morren 2008), or by applying data mining techniques using decision trees to cluster businesses (McCarthy 2008; McCarthy et al. 2009). The communication can be tailored to these subgroups. The communication can also be tailored to predefined sample strata to ensure a representative response distribution.

The survey organization context puts the business survey communication in a larger context from the survey organization's perspective. Relevant policies and procedures are identified, including production systems conditions, corporate communication guidelines, and standardized sections to be included in letters (e.g., informed consent and enforcement of mandatory surveys). Also, survey characteristics (e.g., a mandatory survey, relevant sponsors for endorsement), and project conditions (money and other resources, deadline for delivering the survey data) are identified. The communication process is analyzed to provide a map of the process, identifying the actors and their roles in the process, and critical moments in the process. A study of the external conditions should identify factors that are outside the control of the surveyor but need to be factored in, including general norms and the survey-taking climate (see Figure 9.6).

Step 3: Define Paradata Define paradata to actually measure whether the strategy is effective when used in the field. Plan for an active fieldwork management approach/ responsive design (as discussed in Chapters 4 and 10).

To monitor and evaluate the communication strategy, paradata need to be defined in advance, and built into the strategy. In practice, the paradata will be planned and designed simultaneously with step 4. Paradata, or process data, will be discussed in more detail in Chapter 10, along with fieldwork monitoring.

Paradata can be collected to monitor the effectiveness of the strategy in real time (Groves and Heeringa 2006; Kreuter et al. 2010; Gates 2008; Tutlle et al. 2010a). These paradata and indicators include

- *Prefield paradata*—number of ineligible units relative to total sample (split up for relevant strata); number of noncontacts and reasons for noncontacts; number of questions by businesses and classification of questions; number of complaints and classification of complaints
- *Field paradata*—in addition to prefield paradata; number of advance letters, reminder letters, phone calls associated with timing of communication (including costs); as for phone calls—time of call and outcome (e.g., non-contact, gatekeeper, voicemail, refusal, appointment, response); return rate associated with timing of communication (split up for relevant strata); response rate associated with timing of communication (for relevant strata) and reasons for nonresponse; number of refusals (for relevant strata) and reasons for refusing; number of and reasons for postponement (for relevant strata); audit trails and keystroke data (for electronic questionnaires) such as date of starting to work on questionnaire, peaks in completion activities, and date of response
- *Postfield paradata*—recontact rate (for relevant strata) and reasons for recontacting; enforcement rate (for relevant strata) and reasons for nonresponse

Step 4: Design, Build, and Test the Communication Strategy Communication measures, activities, and procedures are designed, built, and tested. The quality of the strategy is assessed. Figure 9.10 is completed.

Steps 1 and 2 are foundation steps that now allow us, in step 4, to actually design, build, and test the communication strategy, communication measures, activities, and procedures; tailored to the business, survey organization, and external contexts. When designing the strategy, and specifying activities and procedures, you need to ensure that they will be as effective as possible by following considerations on communication modes (see Section 9.5.2), motivation and facilitation (see Section 9.5.3), as well as quality assessment (this step).

Figure 9.10 will be completed during this step. The order in which we study the components is not necessarily the same as in this figure, and may be iterative. While going back and forth in Figure 9.10, we make decisions about the communication activities, materials, modes, messages, and timing; the actors follow from these decisions. Measures and practices are discussed in Section 9.6.

Once we have a clear picture of these activities and processes, we have a draft of a communication strategy. Now we can undertake a quality assessment of the planned communication activities by noting the topics in Figure 9.12 and answering the related questions. In this quality assessment we differentiate between the assessment of individual activities and the combination of all activities, that is, the strategy as a whole. In addition, we also look at related strategies. Working through these questions will help to ensure an objective-driven, coherent, motivating, facilitating, and tailored communication strategy.

When designing the communication strategy, you may find that contextual factors contradict or are too restrictive. Consider, for instance, the timing of launching the questionnaire, the business context analysis may reveal an optimal date, but this may come too late for delivery of the data. Also, you may find that an optimal strategy will be too expensive. Although these survey organization factors need to be considered, the effectiveness of the communication strategy can easily be jeopardized when they

Survey communication component	Related questions
1. Activities and actions	Individual activities and actions: **a.** Will the activity be successful, and achieve the intended outcomes? **b.** Will there be unintended and undesirable outcomes, such as irritated business? **c.** What will be the effect on perceived and actual response burden? Does the activity have low response burden? The strategy, that is, the combined activities and actions: **d.** Do the communication activities fit the purpose and meet the overall objectives stated in step 1? **e.** Are the activities linked together? For example, are they working together in a coherent way to ensure a timely, accurate, and complete response? **f.** Are any measures and activities missing to achieve the objectives?

Figure 9.12 Quality assessment topics and related questions.

2. Objective of each activity	**a.** Is each communication activity clearly linked to an objective in the prefield, field, or postfield stage (see Figure 9.1)? **b.** Does each activity have a clear role in the strategy?
3. Communication modes	**a.** Are the most effective communication modes being used to achieve the objectives, namely, to contact businesses, convey the message, and accounting for the survey costs, response burden, and the business characteristics? **b.** Can the modes be applied flexibly, considering survey costs and business characteristics?
4. Message	**a.** Is the message conveyed in the right way? This includes: - *Content*: Is the right information (including instructions and procedures) included? Is information missing? - *Presentation*: Is the message presented in a readable, comprehensive way, with appropriate wording, tone of voice, and visual design? - *Motivation*: Is the recipient being motivated? Is the motivation strategy tailored to the businesses context? - *Facilitation*: Is the recipient being facilitated? Is the facilitation strategy tailored to the business context? Is a service-oriented approach adopted? **b.** Are all messages in the strategy coherent with regard to each other, taking content, presentation, motivation, and facilitation into account?
5. Actors	**a.** Are the actors who are involved in each measure identified? - *Survey staff*: Are staff informed about the strategy, activities, and procedures? Are they skilled and trained to do the job? - *Business contact person*: Is the appropriate person in the business addressed? **b.** Does the strategy leave a coherent impression? **c.** Is the strategy flexible with regard to the contact person?
6. Timing	**a.** Is the timing of the measure right? - *Survey process*: Will the timing enable us to deliver the survey data on time? - *Response process*: Is the timing proactively tailored to the internal processes of the businesses? **b.** Is timing flexible?
Consistency with other strategies	**a.** Is the strategy consistent with other survey communication strategies? (This is relevant for multisurveyed businesses.) **b.** Is the strategy consistent with output-related communication? **c.** Are there potential areas of good practice to share?

Figure 9.12 (*Continued*)

are taken as a starting point. Therefore, we recommend the following procedure in dealing with these restrictions and conditions in order to achieve an effective strategy:

1. Design an optimal communication strategy based on the analysis of the business contextual factors in order to achieve the objectives, without restricting yourself too much (by asking "How do I want the strategy to be?" instead of saying "This is not possible!").
2. See where the communication strategy should be adapted if the survey organization constraints are taken into account. This will reveal the tradeoffs, that is, where the strategy will lose its effectiveness.
3. Begin negotiations within the survey organization or with the stakeholders in an attempt to optimize the strategy, such as, by negotiating a later delivery date of the survey data or an increased survey budget.

In this way you can make the tradeoffs transparent, identify the risks of not reaching the objectives, and take mitigating actions.

Following the quality assessment, and when you are satisfied with the proposed strategy, activities, and procedures you can proceed with building and prefield testing. The objective of the testing is to assess the effectiveness of each communication activity and procedure; as do questionnaires, they need to be tested prior to field implementation. Testing methods range from small-scale pretest methods to field pilots, as we have seen in Chapter 7. Letters, websites, and other media can be tested with the help of actual respondents, as in usability and cognitive pretests. Procedures, like telephone scripts, can be tested in role plays. To investigate the coherence and effectiveness in practice, pilots (or dress rehearsals) should be conducted. The ultimate goal of the strategy is to ensure an accurate, complete, and timely response. Therefore, the final effectiveness of the communication strategy is assessed by paradata-based quality indicators as defined in step 3.

Step 5: Revise and Finalize Revise the strategy on the basis of the pretest results and finalize it.

The results of testing will provide evidence to inform any subsequent revisions to the design of the strategy, activities, and procedures. Any revisions ideally should be reassessed and tested. In reality, however, this is rarely done, given time constraints. In planning of the design of the communication strategy, time for testing and revising should be scheduled.

Any revisions should be updated in Figure 9.10. Before finalizing the strategy and agreeing to it with relevant stakeholder, it is advisable to undertake a final quality assessment using the topics and related questions outlined in Figure 9.12. You should now ensure that the strategy (in Figure 9.10) is updated and agreed with all relevant stakeholders.

Step 6: Implementation Implement the strategy in a data collection schedule, and run and monitor the fieldwork.

When the communication activities and procedures have been tested, they are ready for implementation in the prefield, field, and postfield stages of the survey. To implement a survey communication strategy, you should map the communication process for your survey.

The processes as shown in Figures 9.3, 9.4, and 9.5 are general process maps that may serve as examples; the actual communication process will very likely differ from these maps. On the basis of the strategy, a data collection schedule is prepared, that is, a detailed timetable of the data collection process with associated activities and actors. Implementation involves the necessary preparations for all communication measures and actions to be carried out, such as printing letters and other paper materials, updating websites and uploading information, training staff, and informing everyone involved.

During these stages, the fieldwork needs to be monitored, to determine whether the objectives (as defined in step 1) are met, for instance, whether the response is coming in as planned or is falling behind. This can be done by the use of paradata-based process and quality indicators, which requires a responsive design (see Chapter 4). Monitoring of the fieldwork is discussed in Chapter 10.

Step 7: Evaluate the Strategy The evaluation results can be input for improving the strategy, thus starting a new cycle: existing business survey communication strategies are redesigned and improved.

After the fieldwork, the communication strategy needs to be evaluated, in order to improve it for a next time (following the PDCA cycle, as discussed in Chapter 4). Evaluation of the strategy requires advance planning to ensure that all required information (paradata in step 3) for the evaluation is collected. Methods used to study the business context and in the pretesting can also be used to evaluate the strategy (these methods were discussed in Chapter 7). For example, a small number of businesses can be recontacted to retrospectively study their internal response process [see, e.g., Bavdaz (2007)].

For the redesign of a communication strategy, a project may be defined. A (re)design project focuses on steps 1–5, that is, from setting the objectives to revising the tested strategy. The planning and management of such a project was discussed in Chapter 4. Step 3 is included in the (re)design project, as it is a necessary step to monitor the results of the strategy during the fieldwork (response rates) and to evaluate it afterward.

9.5.2 Effectiveness of Communication Modes

Making a strategy effective involves considering effective communication modes in relation to the objectives and the other components. Various of communication modes (sometimes called communication channels) can be used in a communication strategy, including paper, email, the web, and the telephone. These may be different from the modes used for the questionnaire. For self-completion questionnaires, like web or paper questionnaires, paper letters or emails may be used to contact businesses and seek cooperation, and the phone may be used for reminding. For a telephone or face-to-face interview, a letter may be used to contact businesses about the interview. The phone may be used to make an appointment. But in what circumstances are these modes effective?

We distinguish between three kinds of communication, which are associated with specific communication modes: one-way communication, two-way communication, and publication. One-way communication includes letters and emails, two-way communication includes telephone contacts and personal visits, and publication includes websites and brochures. Each way of communication has specific characteristics, which makes it useful for specific situations, but unfit for use in others.

One-Way Communication This is outward-focused communication (Erikson 2010): initiated by the surveyor, and directed toward the sampled business. It can be used to inform people about the survey, send out the questionnaire, encourage participation, communicate instructions and procedures, and send survey reminders. Initially, it is a low-cost way to communicate, and many businesses can be reached at the same time, with the same message. That is why it is traditionally and frequently used in prefield, field, and postfield communication. However, it can increase costs further down the survey process, such as by increasing the number of nonresponse follow-ups and treatment of data errors.

As there is no two-way interaction, the persuasive power of one-way communication is limited. There are no guarantees that the communication will reach the intended respondent, or that cooperation will be encouraged. The mode used may have some impact; this is why mandatory surveys tend to use official letters. However, if contact is effected, not everyone will read the letter, brochure, or email. Letters and brochures can get lost on a desk; emails can quickly disappear up the screen, especially if the respondents receive large numbers of emails. When the contact person is out of the office, the request might be left unread in the email inbox, while a letter may be handed over to a colleague, thus increasing the chances of participation.

Two-Way Communication This refers to person-to-person communication by telephone and personal visits. Nowadays, two-way communication is also achieved via the Internet, using options such as chat sites. This kind of communication is characterized by interactive personal contact, which is often appreciated by people; it makes people feel valued and gives them the opportunity to ask questions, and utter their frustrations (Ipsos MORI 2010). Two-way communication can assist in persuading and motivating businesses to comply with the survey request. It should result in tailored communication as the interviewer responds to the respondent.

With appropriate interviewer training, two-way communication modes should be more effective than one-way communication. Compared to one-way communication, the upfront costs are higher (note that it may reduce later costs for nonresponse follow-up and correction of data errors). In surveys with important and large businesses, the telephone is the dominant mode for two-way communication. It is used, for instance, in the (pre)field stage for contacting respondents or reminding respondents in the fieldstage and, of course, when businesses call the help desk. In postfield communication the telephone is used to validate data with respondents. An alternative two-way mode is face-to-face communication. This is the most expensive option. Face-to-face visits are often used in business profiling (see Chapter 5) and can be used to instruct and assist businesses in self-completion surveys. Personal visits are effective in building a relationship with businesses. Both telephone contacts and personal visits require staff to be well trained in gaining cooperation, avoiding refusals, and negotiation skills.

Publication Published communication can include for example, general information about the survey, frequently asked questions (FAQs), and survey calendars. The information can be published on websites, social media (e.g., YouTube, Twitter, Hyves, Facebook) and in brochures. Businesses initiate accessing this kind of information as and when they need it. As such, it can be seen as inward communication. Websites and social media are an efficient and cost-effective way of enabling open and constant access to everyone (with a computer). In contrast, brochures are available only to those who receive them. One issue with website and social media communication is that its availability needs to be drawn to people's attention. This can be done in several ways, such as by mentioning the website in letters or on the questionnaire, providing a link in an email, or having a prominent respondent section on the survey organization website. If people don't find the information they are looking for easily, they will telephone or give up.

9.5.3 Effectively Motivating and Facilitating

We mentioned above that when seeking survey cooperation, we need to influence the social behavior in businesses, pertinent to the survey response process. We will now provide guidelines on how to effectively influence this behavior, by briefly discussing general persuasion principles.

According to persuasion theories [e.g., the Elaboration Likelihood Model and the Heuristic–Systematic Model (Perloff 2010; Gass and Seiter 2011; Hoeken et al. 2009)], organizational survey nonresponse theory by Tomaskovic-Devey et al. (1994), and the response process model as discussed in Chapter 2 [as also discussed by Ramirez and McCarthy (2007)], we can identify three main factors that influence the behavior of people in businesses in order to get survey cooperation:

Authoritative factors	Authority can be seen as a special argument to motivate people (i.e., external motivation) and relates to the question "Do I have to do this?"
Motivational factors	Motivational factors are associated with internal motivation and relate to the question "Do I want to do this?"
Ability or facilitating factors[1]	Ability factors relate to the question "Am I able to do this?"

These factors apply for both the management and the respondent level (as we have seen in Figure 2.1) and generally emerge in the following order. When asked to participate in a survey, business managers and employees will first ask themselves whether they have to participate? If "no," they will ask whether they want to do it. Only if motivated they will check to see if they are actually able to do it. So, to obtain survey cooperation, businesses first have to be motivated (externally, internally, or both) and then be able to comply with the survey request. What irritates businesses is when they are willing to do the job, but for some reason (that is out of their control) cannot, possibly because the response process is not facilitated by the survey design. For example, they might not be able to find or open a web questionnaire, the web questionnaire might stop halfway through the completion process, or the requested data might not yet be available and they have to ask for postponement. Following this dichotomy, both motivation and facilitation strategies have to be applied to achieve a response.

9.5.3.1 Motivation Strategies

Motivation strategies are aimed at influencing the behavior of a manager or respondent by means of communication to obtain survey participation. We previously mentioned that motivating people is generally done by influencing knowledge, attitude, and norms (Section 9.2.1). Now, surveyors have to ask themself: "How can I motivate businesses (i.e., managers and/or respondents) to cooperate?" This question is addressed in this section. Motivation strategies include both external (authority) and internal motivation, and can be applied at both the management level, addressing managers, and the employee level, addressing respondents (see Figures 9.2; Torres van Grinsven et al. 2011).

Our basic assumption is that managers and respondents are not motivated to cooperate with a business survey. This influences how they process information and reach a decision. In persuasion theory, two information processing approaches are identified: the systematic approach and the heuristic approach (Groves and Cialdini 1991; Groves and Couper 1998; Hoeken et al. 2009; Perloff 2010).

[1] In Chapter 2 we discussed *capacity* following the business organizational sciences terminology; in this chapter we use *ability* following the social psychological literature; however, both terms relate to the same factors.

If managers and respondents were motivated, for instance, because of the survey topic, they would apply a *systematic approach*, which means that they would carefully read the provided information. This holds for (among others) accountants, who have professional standards and are committed to accuracy (they would even prefer to leave an item blank than provide an estimate). The systematic approach is based on thorough processing of information and arguments. For the provided information and arguments to be persuasive, they need to be relevant to businesses. Generally relevant and persuasive information are derived from analyzing the business context, as discussed in Section 9.3. In Figure 9.7 we listed questions businesses can ask about the survey in each of the three stages. These questions should be used as a guide in formulating a motivation strategy.

Once the questions are known, a motivation strategy that can be applied is to list as many arguments as possible, to catch the respondents' attention. This approach is referred to as "*length is strength*" (Hoeken et al. 2009). Research has shown that (also for casual readers) more arguments are more persuasive than fewer but strong arguments. This approach can be applied by listing the arguments in a bulleted style, with one argument per line, and listing the strongest arguments first. A general rule of thumb to make salient what is relevant.

This approach is helpful when addressing the manager of a business to get commitment in the prefield or field stage. It may be useful when businesses are contacted for the first time. A study sponsored in the UK by Her Majesty's Revenues and Customs (Ipsos MORI 2010) discusses the attitude of large businesses toward research. The reasons identified for participating in government sponsored surveys were

- Influencing policy and making a difference, survey participation is a way to have your voice heard by the government and influence decisions.
- Improving or building relationships with government departments.
- Keeping current: surveys are a way to be informed about topics that are of interest to the government.
- Finding out what other businesses think; surveys provide a source of information about the business world and a source for benchmarking.

In addition, the businesses in this study felt that the purposes and uses of the survey should be clarified.

However, since we assume that managers and respondents are not motivated, the chances that they will carefully read letters and other communication materials are very small. Instead of applying the systematic approach, they more likely apply a *heuristic approach*, the second approach in information processing. Heuristics are shortcuts that help people make decisions, without much effort in information processing. In the social-psychological literature on persuasion, a number of general heuristic cues or compliance principles are described that people use as shortcuts. Cialdini (2001, 1990) [Groves and Couper (1998)] identified six general compliance principles that guide heuristic decisionmaking: reciprocation, commitment and consistency, social validation, authority, scarcity, and "liking" (these principles were mentioned in Section 2.1.2). In addition, "helping" is also described by social psychologists as a heuristic norm (Vonk 2009; Groves et al. 1992; De Leeuw et al. 2007; ABS 2002). We will briefly discuss these principles here. [For a more detailed discussion, we refer the reader to Cialdini (2001); and Snijkers et al. (2007a), who have applied these principles to business survey communication; Dale and Haraldsen (2007) have applied them with regard to perceived response burden.]

Reciprocation This term refers to a willingness to comply with a request in expectation of receiving something in return, such as a gift, favor, or concession. This principle implies that by receiving a gift, businesses would be more likely to comply with a survey request. Survey participation is seen as a repayment of the gift. This principle is closely related to the social exchange theory [discussed in Section 2.2.1.2; see also Weisberg (2005) and Dillman et al. (2009)].

Examples of applying this heuristic are providing an incentive, providing feedback (e.g., benchmark information), thanking respondents, offering assistance and information, and doing a concession (e.g., informing respondents that the questionnaire could be much longer but the number of questions is reduced as much as possible). Incentives, feedback, and gratitude work best if provided with receipt of the questionnaire or prior to that (in prefield communication) (Singer 2002; Groves et al. 2004). However, the actual number of studies on improving compliance by incentives in business surveys is limited (see Section 9.6.2.2). A meta-analysis carried out by De Leeuw et al. (2007) on the influence of advance letters regarding response in telephone household surveys shows that reciprocation is effective. They conclude that using reciprocity arguments in the advance letter, like thanking the respondent in advance and offering assistance and information, improves the response rate.

Commitment and Consistency These terms refer to the tendency to behave in similar ways in similar situations, and then commitment to that position. This principle states that once a business (e.g., because of a business policy), has agreed to a survey request, the business will probably retain this position. Then one has committed oneself to a position, and since we want to be consistent in our views and behavior, we stick to this position.

Getting commitment is especially important in business surveys, since many business surveys are recurrent in nature. This principle states that it is important to try to get commitment from newly sampled businesses, and explains the significant effect of past behavior on survey participation as found by Davis and Pihama (2009). This principle also states that it is important to make an effort to get commitment in the prefield stage and build on that in later stages. Keywords are "building a relationship" and "maintaining interaction".

Grounds or rationale for commitment should be established for businesses to initially comply with a survey. A study in The Netherlands (Vendrig 2005; Snijkers 2008) showed that as soon as businesses see the value of participating in a survey and there is a connection with their business activities, their irritation about participation decreases. This can be achieved by better explaining the reason for conducting surveys, what will happen with the data (Giesen 2007), and tailoring the design to the businesses' processes. Also, businesses could be involved in the survey design process. Experiences around the world provide evidence that this is, indeed, the case. In Austria (Pfeiffer and Walter 2007), Finland (Jeskanen-Sundström 2007) Spain (Ortega 2007), the United States (Gruber 2007), and New Zealand (Barrow 2007), open and active contact strategies have been implemented to improve survey compliance. According to these experiences, improvement of the relationship with business respondents has a positive effect on survey participation. The way to build commitment is by starting small, and to build a relationship from there. Commitment is not just one measure, but a series of consistent measures, such as involving businesses (or business organizations e.g., trade associations) in the survey design process (this also relates to social validation); when getting into contact, trying to get the name of a contact person (this is a first step); and whenever there is personal contact, trying to get a concrete agreement as to when the completed questionnaire will be returned.

Social Validation This is the tendency to behave according to the norms and values of the social group to which one belongs or others that one identifies with. This principle states that if other businesses that one feels associated with comply with a survey request, one is more likely to do so as well.

Social validation is effective on two levels: on the external environment level, which cannot be affected by a surveyor, and on the individual level. This external environmental level includes the survey climate in general, such as when politicians discuss the burden of surveys, and that life for businesses should be made easier, with fewer regulations (Snijkers et al. 2007), or when opinion leaders in the economy address the burden and usefulness of surveys. On the individual level this principle becomes increasingly important with the use of social media (e.g., Twitter, facebook). Managers and respondents use these media to utter their views (very often frustrations) about surveys (Torres van Grinsven and Snijkers 2013). They feel strengthened in their opinions and attitudes when they are supported by other businesses.

According to Cialdini (2007), the persuasive impact of social validation is underestimated, and as a result is underemployed. Examples of applying this principle are having ambassadors or spokespeople tell the story of the relevance of the survey [(e.g., by using endorsement letters; an ambassador could be someone from a sampled business (Fisher et al. 2003)], cooperation with trade organizations, referring to general norms that are agreed in the business world (i.e., being a responsible business), and informing sampled businesses that most of the businesses participated in previous surveys.

Authority This refers to willingingness to comply if the request comes from a person or an organiation that one perceives as a legitimate authority. If a survey request comes from an organization that is in the legal position to conduct a survey, businesses are more likely to participate. National statistical agencies and national banks are such authorities. This principle is evoked by using the name of the survey organization in the communication with businesses. If a survey organization does not have an official status related to conducting surveys, a reference could be made to the sponsors or stakeholders who have commissioned the survey (e.g., ministries, trade organizations, or research bodies).

Execution of a survey by a national statistical agency evokes this principle by itself, even if the survey is voluntary. The mandatory status of a survey is a separate application of the authority principle. This has a positive effect on survey participation. Willimack et al. (2002) report that mandatory surveys achieve higher response rates than do voluntary ones. A study among respondents for the 2005 Dutch Annual Structural Business Survey showed that about 90% of businesses reported that the major reason for participation was the legal obligation (Snijkers et al. 2007a).

Using the mandatory status is, however, not a very positive stimulus for achieving compliance (Snijkers et al. 2007a). This is confirmed by experiences in, e.g., Austria. Pfeiffer and Walter (2007) state that in a communication strategy the survey organization should stress the service orientation toward businesses, and utilize the fact that it is mandatory as little as possible. On the other hand, business respondents do not always have negative attitudes about the mandatory status of surveys. Respondents use this argument to convince their manager that it was, indeed, necessary for them to spend time on the questionnaire (Giesen 2007).

Apart from the official status, authority has a number of dimensions that are relevant to survey communication. Related to this principle are concepts such as expertise, experience, reliability, honesty, and trust (Hoeken et al. 2009). *Trust* relates this principle to the social exchange theory (Dillman et al. 2009; Weisberg 2005), showing that if one is a trusted party, this may be effective in getting survey participation. On the flip side of this, surveys by unreliable parties will have low response rates. Expertise (exposed, e.g., by having an understanding of the basic aspects of the industry), experience, reliability, and honesty show that one can be trusted.

The authority principle can be applied in many ways, such as showing that one is a professional, and has a professional image. Be credible, and make trustworthy communication items that can be distinguished from junk mail. Refer to informed consent and confidentiality of the data. Show that the survey organization can be trusted and that the data are safe in their hands. Show the logo of the survey organization in letters and other communication items. Refer to official organizations that are involved in the survey (as stakeholders or sponsors) or make use of the survey data. List their logos in communication items. Have an authority (a senior manager of the survey organization, or an expert on the topic) explain the relevance of the survey, and the importance of the business's data for the survey. As with social validation, have someone else tell the story of the survey. Involve senior managers of the survey organization in the communication strategy. They have authority, and can more easily access middle or senior managers in a business to obtain cooperation (Fisher et al. 2003). Refer to official policies of authorities with regard to survey participation. An example is, of course, the mandatory status of official business surveys. Relevant policies of other authorities may also be referred to, such as those of trade organizations.

Liking This is the tendency to comply with requests that come from others whom one "likes." A business is more willing to comply if the survey request comes from a likeable person or organization. Liking is very basic in the heuristic approach; if the organization, a second dimension of liking is the personal level. Having personal contact by professional staff is very important to ensure compliance. They are the face or facade of the organization to the outside world. Interviewers, field agents, help desk employees, and others do their utmost to be respondent-friendly, understand the respondents' viewpoints, and try to help them. Although respondents may not feel that the survey organization is sympathetic, they may respond because they liked the "friendly" person at the other end of the (phone) line (Snijkers et al. 2007a).

Liking can be established in the following ways. Treat businesses with respect, and not as numbers. Personalize the communication (Dillman et al. 2009). Show that you are genuinely interested in their story, and not just their data. Don't evoke negative feelings. Use the language of the business, and be friendly in your communication. Adapt as much as possible to their conditions. Be flexible. Be open, sincere, and service-oriented. In case of mistakes, admit them. Train survey staff involved in the communication process (see Figure 9.8) in human interaction, persuasion techniques, and negotiating survey participation with businesses (Ipsos MORI 2010). At the survey organizational level, have a proactive service-oriented marketing strategy directed at targeted groups in their role as data providers, like small and medium-sized businesses and large businesses.

Scarcity This principle refers to the tendency to comply because one has the feeling of being in a scarce or unique position. According to this principle, the possibility that is offered with inclusion in a survey is considered to be a rare opportunity. Also, the business itself can be in a unique position, and have a unique contribution to the survey (e.g., very large businesses).

The scarcity principle has almost no effect on survey participation for smaller businesses. These businesses don't participate either because they feel that the survey does not offer a unique opportunity or they do not consider themselves unique or feel that their input is important for the survey. Stressing this fact may even backfire. On the other hand, this principle is applicable to larger businesses, since they are relevant in the sample to get reliable estimates. Those establishments are aware of the fact that they are unique, and that their contributions are crucial to achieving reliable statistics. This fact can be used in a tailored communication strategy.

Helping A final principle is the helping norm (Groves et al. 1992). "Helping" refers to social responsibility; it is a general norm to do something good for others or for society as a whole when asked to help (Vonk 2009). "A helping norm . . . motivates individuals to help others who are in need and who are dependent upon them for aid. . . . Thus, even a simple request to participate in a survey will be significantly more successful when it includes an appeal to the helping norm" (Groves and Ciadini 1991; Cialdini 1990; Luppes 1995). Helping is also related to altruism (De Leeuw et al. 2007): doing something for others without getting something in return. Groves et al. (1992) report that a study conducted by Mowen and Cialdini (1980) resulted in a 19% increase in response rate by adding the phrase "it would really help us out" to the end of the letter. As such, businesses can be asked to help out by complying with a survey request.

9.5.3.2 Facilitation Strategies

The second step in getting survey cooperation is by facilitation. Now that we have tried to motivate managers and respondents to cooperate, we have to facilitate them. Facilitative measures are designed to facilitate the response process, making it easy to respond and reducing as much as possible (perceived) response burden. Surveyors must ask themselves "How can I make it easy for respondents, and help them to respond?" This should be apparent to the respondent in all elements of a survey, but especially in communication items; the first things they experience when being contacted are communication items (envelopes, letters, emails, or personal contacts). Being aware of response burden and adopting a service-oriented approach are important when designing a survey communication strategy. Note that a first impression counts.

Examples of facilitation measures are addressing the right contact person (so that letters don't have to be passed around); using language that businesses understand (talking the language of businesses); the timing of sending the questionnaires (when businesses have the data ready and time to comply); providing a survey calendar, and informing them about the next survey so they can be prepared; making the process of completing and returning the questionnaire as easy as possible; and coordinating or bundling communication for multisurveyed businesses. Specific facilitation as well as motivation strategies depend on the business context (as in Section 9.3), and will be discussed in the next section.

9.6 BUSINESS SURVEY COMMUNICATION MEASURES, PRACTICES, AND GUIDELINES

So far we have discussed survey communication in a more general way. Now, it is time to discuss practices and guidelines for designing a business survey communication strategy. We will focus on prefield and field strategies aimed at achieving the predefined objectives, which in general is to increae the likelihood of getting response. The practices and guidelines are related to the process map (Section 9.2) and the communication components (Figure 9.10). Here, we will not discuss postfield measures; enforcement has already been discussed in Section 9.4.2.2, and postfield communication related to data validation will be discussed in Chapter 11.

To increase the likelihood of getting response, we will discuss a number of prefield and field measures:

- Prefield measures to
  - Increase the chances of establishing contact (e.g., dealing with gatekeepers)
  - Introduce the survey and gaining cooperation (e.g., prenotifications)
  - Reduce response burden in the survey design

- Field measures to
  - Make contact, dispatch the questionnaire, and inform the business about the survey
  - Gain survey participation: motivating and facilitating the response process, timing, incentives, and feedback measures
  - Support and assist respondents in the response process: help desk, websites, ombudsman (mediator)
  - Increase response rates: nonresponse follow-ups and refusal conversion.

Prefield measures can also be applied in the fieldstage. In addition, we will discuss measures to maintain a relationship, to deal with large and multisurveyed businesses, and how to increase web takeup rates in a mixed-mode design.

9.6.1 Prefield Communication Measures

9.6.1.1 Measures for Increasing the Chances of Establishing Contact

The first step in obtaining response is to prevent noncontacts from the beginning. Increasing the chances of establishing contact includes three aspects: checking eligibility, getting up-to-date contact information, and identifying and getting through to the appropriate contact person to introduce the survey.

As we discussed before, the business world is dynamic. Since there are time delays between drawing the sample and the actual data collection, the frame information needs to be updated. To check contact information, an Internet search can be undertaken [if time and resources are available (Ipsos MORI 2010; Fisher et al. 2003)]. If no detailed information about the business is available, there is no other option than to contact the business and ask for the required information. Businesses can be contacted by paper mail (letters and leaflets), email (with attachments), telephone, or in fact by any communication mode for which contact information is available—although mode of contact might be governed by survey organizational policies and procedures. Eligibility can also be determined during the fieldwork stage, such as by using a questionnaire that starts with a number of screening questions.

Next, the appropriate person needs to be identified and contacted to introduce the survey. In the case of panel or recurring surveys, the contact person typically is known. In these cases the prenotification letter (e.g., informing the business about any changes to the survey) can be directed to this person. However, the contact person may change temporarily or structurally, because of personal or organizational reasons: a new job, retirement, reorganization, or temporarily absence. As a consequence, emails will be waiting in the inbox for replies and letters end up on a desk (or in a wastebasket). To prevent this, letters are preferred over email, and the envelope can show information about the survey request, including the topic of the survey. This also demonstrates the legitimacy of the source. Professional envelopes, company letterheads or logos, graphics, and high-quality paper can help distinguish it from junk mail.

If the (name of the) contact person is not known, the appropriate contact person will usually depend on the topic of the survey, and the kind of data requested. This is someone who can decide on the survey request. For large businesses this very likely is the senior-level manager of the most appropriate department; for small businesses the owner/manager can be contacted. Reaching the right person can usually be achieved by sending a prenotification letter to introduce the survey, or by making "refinement" calls. These calls include asking receptionists (in their role as boundary spanners) for a particular functional area or department (external reporting, accounting, human resources, etc.). Once this department is reached, the task is to get referrals to likely managers, coordinators, or respondents who will take responsibility for ensuring that the survey is completed.

Establishing contact with businesses is fairly easy since this is generally vital for them [see, e.g., Lynn and Sala (2004); also Section 9.7], which is in contrast to household surveys. However, getting through to the appropriate contact person can be a time-consuming and cumbersome task. For large enterprises, this is much more difficult than for small and medium-sized businesses, because of their greater complexity. For instance, for the Economic Census, the US Census Bureau sends prenotifications to large enterprises requesting them to identify response coordinators (Petroni et al. 2004a). Where prenotification letters are sent, the address information on the envelope can be specified as much as possible to indicate the reporting unit, and type of data requested, or other details. This prevents letters being circulated unnecessarily. Where the telephone is used, interviewers have to deal with gatekeepers, voicemail, agenda issues, automated call systems, and other factors (Petroni et al. 2004a).

In dealing with gatekeepers, in a research report published by HM Revenue & Customs (HMRC) in the UK (Ipsos MORI 2010, p. 31) recommends being "flexible and understanding in order to build rapport with gatekeepers or personal assistants (PA), e.g., by offering to resend the advance letter, or to call back at a more convenient time." This report also recommends "to get PA's and gatekeepers on your side, give them as much information as possible, and be as accommodating and as flexible as you can." If you don't get through to the right person, return to the PA, who will know whom to contact next and their PA's name. Willimack and Dalzell (2006) suggest that identifying an alternative contact person does not necessarily result in nonresponse. They recommend having telephone staff conducting contact and follow-up calls to implement a flexible calling procedure. Lynn and Sala (2004) found that training telephone staff in negotiation skills and providing them with detailed knowledge about the survey (e.g., the background information that would be in letters and web sites) was very useful in negotiating with gatekeepers.

To handle voicemail, Willimack and Dalzell (2006) found that "voice mail may be less likely to be encountered during early morning and late afternoon."

As for the timing of contact, Houlihan (2007) suggests sending letters near the end of the week, since there may be a backlog in the paperwork on Friday. When businesses are contacted by phone, Willimack and Dalzell (2006) showed that the contact rate does not differ significantly by day of the week (Tuesday to Friday;

Mondays are the worst days for businesses to receive a call). As for the time during the day, they did not find a statistical significance, but the likelihood of reaching someone peaked early in the afternoon.

9.6.1.2 Measures for Introducing the Survey and Seeking Cooperation

Various prefield measures can be taken to introduce the survey and to gain cooperation. A number of measures are described by the US Interagency Group of Establishment Nonresponse (IGEN). This group was established in the late 1990s to research the various practices applied by statistical agencies in the United States with regard to nonresponse reduction (IGEN 1998; Ramirez and McCarthy 2007).

The IGEN reports a variety of marketing and public relation measures to seek cooperation, which are all associated with one or more motivational strategy as discussed in Section 9.5.3.1. A combination of mailings, telephone calls, and/or personal visits is used. The message stresses the importance, the purpose, and the legitimacy of a survey. Mailings usually consist of specially developed respondent materials, such as brochures and fact sheets. These materials are sent to respondents with a prenotification letter. Assurance of data confidentiality is also stressed in these mailings. When requesting sensitive data, the assurance of confidentiality is one of the main concerns of respondents.

Some agencies use a (personalized) survey calendar that can be accessed online to inform businesses about upcoming surveys. The survey calendar indicates the number of surveys they will receive, when surveys will be sent out, and when the data are due. As such, survey calendars can help businesses to plan their resources. Survey calendars also increase the transparency of the number of survey requests. Marske and Stempowski [(2008); see also Marske et al. (2007)] report that when a total overview of data requests is given, businesses are often surprised that the total burden is less than they had anticipated.

In case of a mandatory survey, statistical agencies mainly rely on mailings in the prefield stage. The UK Office for National Statistics, Bureau of Labor Statistics (BLS), and the US Census Bureau send prenotification letters to new respondents to inform them that they have been selected for a survey (Petroni et al. 2004b). In the UK, existing respondents in recurring surveys are not sent a prenotification letter; this is done only in case of changes in the design (e.g., introduction of new modes) (Jones et al. 2008b). Communicating changes in the questionnaire prior to the fieldwork helps businesses to plan the response process in advance; a side effect is that it reduces perceived response burden.

Sometimes prefield mailings are followed by phone calls or personal visits (IGEN 1998). Especially in the case of voluntary surveys, a more personal approach will result in higher response rates. Petroni et al. (2004b) report that because of the voluntary nature of its surveys, the US BLS uses interviewers in the prefield stage to followup on prenotification letters. Only after having survey commitment, BLS turns to one-way self-administered modes like paper. Telephone staff are trained in skills to gain cooperation and refusal conversion. Also, the UK HMRC research report (Ipsos MORI 2010) states that managers in large businesses prefer that an initial contact be followedup with a phone call, "to provide them with further information about the research and to enable them to ask questions." They appreciate a "personal" approach that helps make them feel valued.

Businesses may also get back to the survey organization on their own initiative (by phone, sending an email or using a web form), to ask questions, make a complaint, get additional information on the survey, or report changes. To facilitate businesses, survey organizations have a help desk and online help site (e.g., with FAQs; we will discuss these help facilities in more detail in Section 9.6.2.3). The phone number and the web address are listed in the prenotification letter. The typical questions that businesses pose are listed in Figure 9.7; these questions need to guide all communication measures. When businesses call the help desk, this personal contact provides the survey organization with an opportunity to get commitment. The contact information collected during this stage can be used in the field stage to personalize the survey communication, which helps in gaining cooperation (Dillman et al. 2009).

Some US agencies seek cooperation with business communities (IGEN 1998). Industry organizations and associations, like the Chambers of Commerce, are involved in efforts to legitimize the survey or promote survey participation. These measures are also used by other survey organizations such as Statistics Canada (Sear 2011) and Statistics Finland (Jeskanen-Sundström 2007). Business organizations may also be involved in the design of surveys, as we have seen in Section 9.5.3.1.

Now let's illustrate these measures with an example (Tuttle et al. 2010b; Mulrow 2008). In 2005 the US National Science Foundation and the US Census Bureau started a redesign of the Annual Survey of Industrial Research and Development (SIRD). In a 3-year period this survey was redesigned, resulting in a new Business R&D and Innovation Survey (BRDIS). Studies on the response process (as in step 2 of our seven-step approach; see Section 9.5.1) revealed that a number of business departments were involved in completion of the questionnaire. As a consequence, the identification of a response coordinator who is capable of getting assistance from competent respondents/data providers was critical, since obtaining accurate and complete data is directly linked to the competence of these employees. The communication strategy and the questionnaire were redesigned to facilitate identification of the most appropriate respondents for each section of the questionnaire.

Several months before the survey was mailed out, a prenotification was sent to inform businesses of the new survey. They were also asked to confirm or update the contact information, using an included paper form or the Census Bureau's website. This prefield contact strategy served two objectives: (1) businesses that responded well to the SIRD were informed about the new survey design by providing them with a draft of the new questionnaire; and (2) businesses that had poor responses in the past were contacted to obtain cooperation.

Strategies to secure cooperation included providing information about the survey, a request for help, notification that the survey was mandatory, getting commitment by asking for updated contact information, and including an endorsement letter by a former chief executive officer (CEO) of the Lockheed Martin Corporation. An experiment with reluctant businesses was carried out with a letter addressed to the CEO/president of the company, requesting contact information for an employee who would be capable of acting as a response coordinator for the BRDIS. The results showed the survey response rate of the CEO group (72%) to be only slightly higher than that of the control group (68%). However, a more striking result was that businesses who replied with updated contact information in the prefield stage showed a much higher response rate in the field stage (79%) than businesses who did not (62%).

Prefield measures may be costly and time-consuming; however, the BRDIS experiment shows that prefield contacts increase the chances of getting response in the field stage substantially. Since less respondent follow-up is needed during fieldwork, these measures reduce costs in the field stage, as well as perceived response burden. The commitment principle seems to be especially applicable; businesses who reacted in the prefield stage felt committed to participate in the survey. This requires a proactive approach. As Tuttle et al. (2010b, p. 5254) write: "the National Science Foundation and the Census Bureau implemented a proactive communication strategy to assist the survey respondents in understanding their new role and responsibilities." Thus, guiding them to be good respondents.

9.6.1.3 Reducing Response Burden

A final set of measures in the prefield stage, to increase the probability of response, is to design the survey in such a way as to ensure a response burden that is as low as possible. Response burden is associated with survey compliance costs (as discussed in Chapter 6). Throughout this book we have discussed measures to reduce both actual and perceived response burden. We have discussed measures associated with the sample, mode of data collection, and the questionnaire [see also Hedlin et al. (2005), Dale and Haraldsen (2007), and Willeboordse (1998)]. In this chapter we have discussed how to reduce perceived response burden by tailoring the communication strategy to the business context and by motivating and facilitating businesses (Giesen and Snijkers 2010; IGEN 1998).

As we have seen in Chapter 2, perceived costs and benefits are important factors that affect the response process (keeping the social exchange theory in mind). Following two principles from information processing theory [the least-effort principle, and the sufficiency principle (Hoeken et al. 2009; Gass and Seiter 2011)], we can state that when people are about to make a decision, they process just enough information (not more and not less), at a minimum level of effort, to ascertain what they need to know to make that decision. This implies that costs for businesses should be kept low (i.e., both perceived and actual response burden) to obtain survey participation instead of increasing benefits (e.g., by using incentives). A general guideline is: Make it easy and simple, and try to get respondents in the right mood.

9.6.2 Field Communication Measures

9.6.2.1 Making Contact, Dispatching the Questionnaire, and Informing the Business about the Survey

Fieldwork activities start with initiating contact by dispatching the questionnaire. In Section 9.6.1.1 we discussed measures to get updated contact information, to establish contact with the right person, and to deal with gatekeepers and voicemail.

If no separate prefield measures are carried out, these measures are also applicable in the field stage.

A common practice is to send an advance letter, independent of the mode of data collection, to ensure that the questionnaire ends up in the hands of someone in the business who will either be responsible for ensuring that it is completed on time (a manager or response coordinator) or will complete it personally (a respondent). The advance letter informs selected businesses about the survey. Typically, this is a one-page letter, containing information on the survey organization, the topic of the survey, what the data are being used for, why and how the business has been selected, the mandatory status of the survey (if applicable), when the data are due (the response deadline), and assurance of confidentiality. The letter also mentions contact information (phone number and email address of the help desk, a web address with FAQs) if assistance is needed. The objective of the letter is to introduc the questionnaire and gain survey cooperation. [For guidelines on letters and envelopes see Dillman et al. (2009, pp. 247, 261).]

Although emails in lieu of mailed advance letters may be used for notification, we recommend using emails only once a contact has been established. Emails can be used for recontacting respondents in a recurring web survey, but then only when the relationship is built and with small time intervals between each survey, as in a monthly survey. With longer time intervals, the chances of change, increase (e.g., new respondents or email address change), thus decreasing the probability of establishing contact.

Harrell et al. (2007) mention the ubiquity, low cast, and easy transmission of email. A disadvantage is the possibility of spam blocking the email. This is the survey organization's perspective. As to the respondent perspective, the UK HMRC research report (Ipsos MORI 2010) indicates that some managers would prefer emails, while others felt that because of the large number of emails they receive, the survey-related ones might go unread. An advantage of using emails in web surveys is that respondents don't need to retype the web address of the questionnaire (which is error-prone). [For guidelines on emails, see Dillman et al. (2009, p. 298).]

The paper questionnaire is sent in an envelope that also includes a toll-free return envelope and the advance letter. Sometimes the information that is in the advance letter is on the first page of the questionnaire as a cover letter (Jones et al. 2008b). In case of a web questionnaire, a web address and login information are listed (sometimes a paper questionnaire is also included). For CATI or CAPI surveys, the letter indicates that the business will be contacted by phone soon to conduct the interview or schedule an appointment.

If there is insufficient space on the single-page letter for all the required information, additional information may be presented online (see Section 9.6.2.3) or in a brochure that is enclosed in the envelope with the advance letter. Enclosures in envelopes can include brochures (with more information on the survey, the purpose of the data, etc.), questionnaire instruction booklets, and so on. With emails, PDFs (portable document files) can be enclosed, or hyperlinks where these documents can be downloaded. Not much research has been done on the effectiveness of brochures. It is our experience that advance letters and other peripheral mail package materials are often ignored, discarded, or lost as the questionnaire changes hands in a company. Groves et al. (1997) describe an experiment to test the effect of providing varied amounts of information about a US BLS business survey, and found that sending additional advance material had no effect on response rates.

9.6.2.2 Measures for Increasing Survey Participation

In Section 9.6.1.2 we discussed prefield communication measures for seeking cooperation. These measures can also be used in field communication to gain survey participation. In addition measures with regard to the mode of the data collection and questionnaire design can be taken: Questionnaire design was discussed in Chapter 8; in Section 9.6.3.2 we will discuss mixed-mode designs. Field measures for obtaining survey participation are discussed in the following paragraphs.

In Section 9.5.3 we discussed many measures associated with motivation and facilitation strategies. In addition to these measures, examples of measures used by the US Census Bureau and the US BLS to encourage cooperation include (Paxson et al. 1995; IGEN 1998; Fisher et al. 2003; Petroni et al. 2004a, 2004b; Dillman et al. 2009): training staff in gaining and maintaining cooperation (Section 9.6.1.2); marketing the public image of the survey organization (to increase BLS awareness on the site of sampled businesses); tailoring and personalization of the communication to business characteristics (e.g., by size, adapting to the complex structure of large businesses; see Section 9.3.2); enclosure of promotional materials with survey mailings, toll-free telephone help lines, and online help facilities to provide assistance to respondents completing questionnaires; ambassadors for surveys—employees of the sampled business acting as an intermediate (this is someone who respondents can easily turn to); and emphasizing the relevance of the data and asking for help.

In 1994 the US Census Bureau conducted a nonresponse follow-up for the SIRD. In this study businesses were asked about the following measures and which one(s) would "make it easier for them to report" (Petroni et al. 2004a):

- Clarifying instructions
- Providing toll-free assistance
- Mailing the form at a different time of the year
- Allowing more time to report
- Using an electronic reporting format

A discouraging result was that "most of the study participants said that none of these methods would likely encourage their response to the R&D Survey" (ibid., p. 43) Some participants, however, anticipated that toll-free assistance, allowing more time to report, and mailing at a different time of year might improve response. The most frequent responses to a closed question about strategies to encourage response were (1) making the survey mandatory and (2) limiting the amount of data requested or simplifying the questionnaire. The latter (response 2) involves response burden associated with the questionnaire (see Chapters 6 and 8). A mandatory status of a survey is a major reason for businesses to respond, as we have seen in Section 9.5.3.1 (authority); options for handling this are discussed in Section 9.4.2.2. In Section 9.6.1.2 we saw how to deal with voluntary surveys.

Mailing the questionnaire at a different time of year and allowing more time to respond is related to scheduling of the survey. The timing of the survey request is considered essential to obtain response. Busy periods may vary according to the business' size and sector; they include (Ipsos MORI 2010) the end of the month and the end of the financial year (depending on the business' accounting practice, this will be either a calendar or fiscal year, e.g., April to March). Monday is usually the busiest day of the week. Lynn and Sala (2004) found that the lunch break and afternoon (12:00–4:00 P.M.) were the most productive times for completing telephone interviews. For annual, quarterly, and ad hoc surveys, the ideal months for scheduling the survey may conflict with the delivery date of the survey data. The fieldwork period can be prolonged, provided the delivery date of the survey data is not jeopardized. A general recommendation is to tailor the timing to the business context, and send the questionnaire once the data are available and the respondents are able to complete the questionnaire.

Other measures to gain survey cooperation include incentives (accompanying the mailed questionnaire) and promised feedback (on receipt of the data). There are several issues with incentives in business surveys: who should receive the incentive—the manager or the respondent, what incentive should be used, when many people might be involved in the response process, and whether the incentive should be monetary or non-monetary. These issues are discussed by Cook et al. (2009) for business surveys. Incentives are not employed much; feedback is.

Research on improving survey participation by incentives for business surveys is limited [in contrast to social surveys (Singer 2002)], and the research does not give definite answers on its effectiveness (Cook et al. 2009). However, most of the studies examining monetary incentives show positive effects on response rates (Moore and Ollinger 2007; Barton and DesRoches 2007; Beckler and Ott 2007; Newby et al. 2003; Kaplan and White 2002; Jobber and O'Reilly 1998), regardless of

- The amount of the monetary incentive—amounts ranged from $2, $5, or $10 to $50, or indirect monetary incentives in the form of $20 ATM (automatic teller machine) cards.
- The size of the businesses—the studies included small businesses, like new businesses or farms, as well as large businesses in meat manufacturing, for example.
- The contact—the contact person for a business varied from the owner to a manager or a respondent.
- The moment of offering the incentive—both enclosed and promised monetary incentives raised response rates, but the former was more effective.

In contrast to these results, a large-scale experiment conducted by the US Research Triangle Institute showed no evidence of response rate improvements when using a $20 incentive (Biemer et al. 2007). Also, two studies by Cycyota and Harrison (2002, 2006) showed minimal effects of monetary incentives (the 2006 paper reports a meta-analysis of 231 published papers over a 10-year period).

Studies examining the effect of nonmonetary incentives do not give straightforward results. Beckler and Ott (2007) investigated incentive stimuli in the form of a wall clock and the promise of a report that compared the business's data to an overall group average. The results did not show an effect on the response rates. Jobber and O'Reilly (1998) reviewed different studies on the effect of various nonmonetary incentives. An enclosure of an academic or trade article actually resulted in lower response rates (Kalafatis and Tsogas 1994), as did the offer to send the respondent a copy of the survey results (Dommeyer 1989). By contrast, a study by London and Dommeyer (1990) had a positive effect on response; mail survey respondents who completed a questionnaire were told that they would be included in a raffle for three Hewlett-Packard (HP) calculators. Jobber et al. (1991) analyzed the effect of incentives on response rates of business surveys implemented in different countries. When the incentive was given in the form of a bookmark, response rates increased. However, a promise to send a free copy of the results to respondents had no effect on response rates.

On the other hand, experiences in a number of countries with improved communication strategies show that reciprocation is a key element of these strategies. For example, in Spain, feedback to information providers is an essential pillar in a communication strategy of a voluntary survey conducted by the Central Bank (Ortega 2007). This is also the case for the communication strategy used by the Federal Reserve Bank of New York (Gruber 2007). Statistics Spain (Revilla 2008) and Statistics Finland (Jeskanen-Sundström 2002) provide feedback for the structural businesses survey (SBS). With positive effects on response rates, they offer tailored benchmark information (e.g., the market share of the business, turnover related to businesses in the same sector) in exchange for the requested data. Statistics Netherlands offers a quarterly mailing with benchmark information for the monthly survey (Snijkers et al. 2007a; Vennix 2009). And, a study at the Australian Bureau of Statistics (Burnside et al. 2005), which offered a secure download of the survey results publication, showed a positive effect on response rates. The issue is that businesses receive something in return, which is key in building and maintaining a relationship. Willeboordse (1998) warns that the information should be of interest to the business; otherwise the effect might be negative.

9.6.2.3 Providing Assistance with Questions and Complaints: Help Desk, Website, Web Care, and Ombudsman

Many NSIs provide assistance to businesses, via a help desk and/or a website. These are important components of the communication strategy, especially in self-administered survey designs, since no one is present to help. Both the help desk and the website should provide respondents with a satisfactory reply. It is important that information communicated with all modes (also provided in advance and reminder letters, brochures, etc.) be consistent.

Businesses may contact the help desk to ask questions or to complain about the survey request (see Figure 9.4). They may send a letter, an email, or a web form; or call on the phone. If this concerns refusals, we recommend phone calls to the business (also when the business has sent a letter or an email). Personal contact provides the survey organization with a valuable opportunity to talk to respondents: to explore their problems, encourage survey participation, and for reluctant response, apply refusal conversion (see Section 9.6.2.4). We recommend training help desk staff in these measures as well.

Help desk staff may also be involved in web care. In addition to contacting the help desk directly, the business may also circulate questions and complaints about the survey on social media (Facebook, Twitter, etc.). The main objective of web care is to identify these messages (using alerts) and act if necessary. This involves answering questions, providing assistance, and providing relevant information on those media. Generally this is highly appreciated, and this appreciation will likely also be communicated on the Internet; this will have a positive effect on the image of the survey organization. However, we recommend reacting only to concrete questions and complaints that can be solved and not getting involved in discussions about, for instance, the relevance of surveys; leave this work to company spokespeople or ambassadors.

Help desks are an important but costly measure in business survey communication. A less expensive option is to set up an online help facility. Websites can be used to provide additional information on the survey and how the data are used, and to provide FAQs. Websites can also be used to facilitate businesses by showing a survey calendar. This can be personalized by offering businesses to log in to their own site, and providing a list of surveys that are applicable only to them. In addition, personalized information can be presented. However, businesses need to be made aware of the existence of such a website.

Almost all national statistical agencies nowadays have websites that provide additional information and assistance for businesses in their role as data providers. The following are some good examples

- US Census Bureau: "Business Help Site" (Marske and Stempowksi, 2008) (`http://bhs.econ.census.gov/bhs/index.html`). This site provides an overview of general services (e.g., requesting a new questionnaire via mail, requesting deadline extensions, correcting mailing information) and FAQs, background information for the various economic censuses and surveys conducted by the US Census Bureau, and survey-specific FAQs. Links to the main products for each survey are provided. A "survey schedule" shows a survey calendar. Questionnaires for all surveys can be downloaded from the "form archive." Business can log in to access customized information.
- Statistics Finland: "Data collection, enterprises" (`www.stat.fi/keruu/yritys_en.html`) (in English). Statistics Finland presents background information for each individual survey, including the questionnaire and advance letter, and results of the survey. General FAQs are available.
- Statistics Netherlands: "CBS for your business" (`www.cbsvooruwbedrijf.nl`) (only in Dutch). This site combines general information for businesses as data providers and statistical output. For surveyed businesses, general and survey-specific FAQs are available. In the self-service section businesses can ask for

postponement, a paper questionnaire, login codes, or permission to have their accountant act as the respondent. They can also report changes in the contact information, such as a new address or a new contact person.

Additional examples are

- Statistics Canada: "Information for survey participants" (www.statcan.gc.ca/survey-enquete/participant06-eng.htm)
- Statistics New Zealand: "Surveys and methods" (www.stats.govt.nz/surveys_and_methods/completing-a-survey.aspx)
- UK Office for National Statistics: "Release calendar" (www.ons.gov.uk/ons/release-calendar/index.html)
- US BLS: "Survey respondents" (www.bls.gov/respondents/home.htm),
- Statistics Norway (in Norwegian) (www.ssb.no/omssb/erapp/)

As to the design of websites, a vast body of literature is available. A comprehensive overview (including a checklist of issues to consider) is given by Van der Geest (2001). Keywords in website design are easy to find, accessibility, visual design/look-and-feel, and usability (see also Section 9.6.3.2 on web mixed-mode designs). According to Wroblewski (2008), good guidelines on accessibility are the Web Content Accessibility Guidelines (WCAG), produced by the W3C organisation (www.w3.org/TR/WCAG10). For websites as part of a survey communication strategy, the same look-and-feel as in other items such as letters and questionnaires should be applied. We recommend consulting these sources during the website design stage, and conducting a usability test before using it in the field.

For complaints about response burden, Statistics Canada has established an ombudsman (mediator; intermediary) (Sear 2011). The ombudsman has three tasks: resolution of complaints, measurement of response burden, and building a relationship with business associations. The latter has an indirect effect on survey cooperation by promoting a clearer understanding of the relevance and importance of surveys to the business community.

9.6.2.4 Nonresponse Follow-ups and Refusal Conversion

The disadvantage of initial mail contacts using prenotifications and advance letters is that it is easy for businesses to drop out, even with the application of motivational and facilitation strategies. Therefore, a follow-up strategy is needed to get response from reluctant businesses. Before starting with follow-up contacts, the background and contact information may be checked again.

A common follow-up strategy is to send one or more reminder letters. These letters contain more or less the same information as the advance letter, but with stronger appeals in successive mailings, and new response deadlines. The number of reminder letters depends on the amount of time remaining in the fieldwork period. The follow-up procedures at the UK ONS (Jones et al. 2008b) includes one paper reminder for monthly surveys, two for quarterly surveys, and up to three reminders for annual surveys. For annual surveys, the chief executives of very large nonresponding businesses will receive a paper letter and a phone call as well. Failure to respond results in legal enforcement actions, which occurs in the postfield period, as discussed above.

Similar follow-up strategies are applied by other NSIs, with variation in the number of reminder letters for a survey. For the Dutch annual SBS, an advance letter is sent, followed by a maximum of three reminder letters, and a telephone follow-up for large businesses. For annual surveys, Statistics Netherlands sends "remember" letters prior to the first deadline to encourage businesses to respond, rather than a first reminder letter. Erikson (2010) reports for Sweden one advance and two reminder letters for the monthly survey, and three reminders letters for quarterly, annual, and biannual surveys, with or without a duplicate of the questionnaire.

In the United States, almost every federal agency uses some sort of follow-up strategy (IGEN 1998). This includes the usage of letters, postcards, personal visits, telephone, fax, and email. The mode of the follow-ups depends on survey characteristics (e.g., size of the survey, recurring frequency, number of data collection days available), size of the business, ability to receive a follow-up, and costs. The US Census Bureau relies on self-administration modes for nearly all of its surveys or censuses; one to four mailings are used, with or without a duplicate questionnaire, and the telephone is used for nonresponse follow-ups for selected nonrespondents (Paxson et al. 1995; Petroni et al. 2004a). In addition, the following measures can be applied, as used by BLS and the US Census Bureau (Petroni et al. 2004a): stronger appeals in successive follow-up mailings, involving senior managers of the survey organisation (since they have more authority) by contacting senior business managers to gain cooperation by explaining the importance of their data for the statistics, and searching for a specific or alternative informant to report the desired data. The message and mode can be varied in successive follow-ups. The goal for all follow-up measures basically is the same: encouraging the respondent to respond, using the same motivational and facilitation strategies as in the initial contacts.

For large businesses, the final reminder can be one or a combination of these measures, usually involving some kind of personal contact, such as telephone follow-ups (more on large businesses in Section 9.6.3.1). Telephone follow-ups are also important in voluntary surveys, to allow for personalized refusal conversion. According to the IGEN (1998) most US agencies attach "great importance to the refusal conversion attempt, because it is usually the last chance to gain the respondent's cooperation." O'Neill (2007) describes a refusal avoidance training set-up at the US Census Bureau [which is based on theories and practices as used in social surveys; see Groves and Couper (1998), Groves and McGonagle (2001), and Morton-Williams (1993); for calling strategies, see Gwartney (2007)]. Telephone listening and motivation skills are taught; experiences are shared among interviewers. The key is that they learn to identify, analyze, and deal with reluctance and refusal. Interviewers are instructed to listen to reluctant respondents with a sympathetic ear and address the specific reason given; this enables a personalized motivational approach to encourage response.

Some US agencies [e.g., National Agricultural Statistics Service (NASS) and BLS (IGEN 1998)] developed respondent profiles from comments and other information collected during previous contacts with the business, which helps the second interviewer in the refusal conversion process. Unlike the initial interviewer, the second interviewer will be more familiar with the profile of the respondent. The interviewer may also try to seek out a higher-level company official if cooperation from the initial contact is impossible.

The timing of nonresponse follow-ups is also critical (IGEN 1998). It should not be too soon after a previous mailing, so respondents may get the feeling that are being pursued or closely monitored. It should not be too late, to ensure receipt of the data before the survey deadline. The BLS has a tailored follow-up strategy based on a number of factors, including size of the business, number of collection days available, data collection mode, and past responding history. This strategy allows the BLS to estimate the best moment in time to send mailings, namely, when the internal business reports are ready (Rosen et al. 1991). As a result, the mailings occur at various moments in time, thus reducing the number of follow-ups and collection costs. This approach can also be used to tailor the mailing of advance letters for businesses in the sample of recurring surveys. Paradata (on response patterns, and call patterns in telephone follow-ups) can be used to optimize this approach (this is discussed further in Chapter 10).

Statistics Canada has implemented a selective approach for follow-ups: the *score function*, a collection management tool that manages response rates by prioritizing collection units relative to their impact on estimates (Berthelot and Latouche 1993; Evra and DeBlois 2007; Sear et al. 2007). This function identifies businesses that are most important for follow-up because of their impact on estimates by both province and industry. In this way set levels of quality are reached with effective use of limited follow-up resources. An effective tailored approach can be achieved by combining the score function and the BLS approach. The score function can also be applied in the postfield stage to manage recontacts for data editing (Berthelot and Latouche 1993).

9.6.2.5 Measures for Maintaining a Relationship

Even when businesses have responded, the communication process is not complete. Maintaining a relationship is important to ensure survey participation in the future. A relationship can be maintained by simply thanking business for their cooperation and if in a recurring survey, informing businesses when the next questionnaire can be expected, and providing feedback. When using electronic questionnaires, this can easily be done with a return message on a "next" screen, thanking the respondent and providing feedback. Providing feedback can be as easy as informing them that the data have been received. A more sophisticated technique is to report back the provided data and asking if they are correct, since we want the business to correctly contribute to the survey results. This may result in businesses contacting the survey organization if they discover errors in the reported data. Examples of benchmark feedback were discussed in Section 9.6.2.2.

The Australian Bureau of Statistics has developed the "gold star provider" initiative (Gates 2008) to maintain a relationship. Businesses with a good response record, that is, those that respond prior to the duedate and don't need telephone follow-ups, are "rewarded" with a gold star rating. For future surveys, they are allowed more time to return their data without irritating telephone reminders.

9.6.3 Special Cases: Large Businesses and Mixed-Mode Designs

At the end of our journey in business survey communication, we discuss two cases that need special attention: dealing with large businesses and introducing mixed-mode designs.

9.6.3.1 Dealing with Large and Multisurveyed Businesses

For very large businesses in recurring surveys, many NSIs have established integrated communication procedures {e.g., Statistics Sweden [Erikson (A.), 2007], Statistics Canada (Sear et al. 2007), US Census Bureau (Marske et al. 2007), Statistics Netherlands (Pustjens and Wieser 2011)}. These procedures are aimed at building and maintaining a good relationship with these key businesses, and obtaining coherent data. Special procedures for dealing with large businesses become more important in the context of globalization (UNECE 2011), with businesses organized on a global rather than national level.

This task is assigned to a specially trained group of managers, often called *customer relations managers, large enterprise managers*, or *account managers*. They are the personal contact in the NSI for large businesses, and have special knowledge of these businesses. For Sweden, Erikson (A. 2007) reports 11 large enterprise managers (in 2007) as being responsible for 51 enterprise groups and 65 enterprises.

As with the regular communication strategy, the objective is to get timely, accurate, and complete data; in addition, the goal is to obtain coherent data over all surveys for each business. Customer relations managers are involved in updating the business register for these businesses, profiling [i.e., determining and updating the structure of businesses (Pietsch 1995)], and the data collection and editing processes. Since these are multisurveyed businesses, their actual response burden is high, so procedures are put in place to reduce their actual and perceived response burden. With this integrated approach, the response burden for these businesses is reduced as much as possible, by assisting these businesses in the completion process, and providing information about the surveys (background, output, and changes). The customer relations managers' knowledge of the business is essential in this business-oriented approach. They know their situation, know what surveys they receive, know the questionnaires, and understand the burden of being surveyed multiple times.

Statistics Canada (Sear et al. 2007; Huges 2008) has expanded this approach and implemented a tailored "holistic response management system" for business surveys. This system consists of four so-called "tiers":

- Tier 1 represents the largest businesses. The contact with these businesses is managed by enterprise portfolio managers; one manager has the responsibility for maintaining the relationship with the enterprises in his/her portfolio. This tier contained about 250 enterprises in 2005 (and 300 in 2006), covering 1065 questionnaires, and resulting in a weighted response rate of 85%.

- Tier 2 represents businesses that are smaller or less complex than the tier 1 enterprises, but still are significant to statistics. In this tier, the personal tier 1 approach is applied to address critical nonresponding businesses by a team of relationship managers. Their job is to investigate and resolve the problems in the data collection for these businesses. Actually, the general problem for these multisurveyed businesses was not the data providers themselves, but the way they were approached in the multisurvey context (i.e., there was no coordinated communication). In 2005, 145 businesses were in this tier (in 2006, about 200), covering 300 questionnaires, with a resolution rate of 93%.
- Tier 3 includes medium-sized businesses. These businesses represent the bulk of the samples. For these businesses, the regular communication approach is followed. Response rates are managed by use of the score function (as discussed above).
- Tier 4 represents the smallest businesses. These businesses are excluded from survey data collection, since Statistics Canada relies on administrative data for these businesses.

Sear et al. (2007) conclude that this approach has been very successful in improving the response rates and data quality for tier 1 enterprises and for tier 2 businesses by converting key delinquent respondents.

Marske et al. (2007) report for the US Census Bureau the same reasons for reluctance as found in Canada for tier 2, namely, inconsistencies in inbound and outbound communication. The Census Bureau was previously inconsistent in acting on business information updates and sharing it across surveys; the outbound communication was not streamlined. A newly installed survey calendar helped in coordinating outbound communication, both within the Census Bureau and toward businesses. These findings point in the direction of a proactive and business–oriented communication strategy with large businesses in order to avoid nonresponse. Marske and Stempowski (2008) explain how centralizing information to improve communication with large businesses was also helpful in coordinating internal information. Coordinating information in reporting calendars and through customer relations management made clear who had contact with a business, helped in preparing for visits and calls, and showed that various survey registers were out of sync. This, in itself, can help reduce response burden, for example, by circumventing the need for businesses to report changes several times, and may reduce the need for recontacting respondents.

9.6.3.2 Introducing Web Mixed-Mode Designs

An issue that has received much attention is the introduction of a mixed-mode survey design. Next to paper self-completion questionnaires, there is a recent trend toward alternative modes of data collection [web, telephone (touchtone) data entry (TDE)], primarily to enhance efficiency (Snijkers et al. 2011a; Brodeur and Ravindra 2010; Jones 2011; Thomas 2007). As a consequence, the survey design becomes a mixed-mode design. In this section we will discuss maximization of web takeup rates. The surveyor must be aware, however, that mixed-mode designs may introduce mode effects and thus increase measurement bias. On the other hand, mixed-mode designs are applied to reduce nonresponse rates and nonresponse bias (Sala and Lynn 2005).

There are two basic ways to introduce a mixed-mode design: a *simultaneous approach*, in which the selection of mode options is offered to the respondent; and a *sequential approach*, in which a primary mode is offered first, followed by alternative modes to facilitate respondents, and to increase response rates. Typically the sequential approach will start with the least expensive mode and switch to more expensive modes through the stages, with the most expensive mode retained to the final stage. As well as effectively using survey finances, this approach diminishes the need for respondents to choose a mode.

Traditionally, survey organizations have offered a primary mode of data collection, and any alternative modes are simultaneously offered as an option. The result of this strategy is higher takeup of the primary mode (i.e., the mode presented) in comparison to the alternative modes; and no or little improvement to the response rate. For example, TDE is used for several UK ONS business surveys that collect nine or fewer data items. When respondents were sent a paper questionnaire offering TDE as a simultaneous data collection mode, generally response via TDE averaged around 20–30%. However, when some surveys were moved to TDE as the primary mode of data collection and paper offered as a simultaneous data collection mode (indicated on the TDE letter; respondents had to request the paper questionnaire), response via TDE rose to 80–90% (Jones 2011; Thomas 2007).

A small-scale study by Hak et al. (2003) on factors that affect business web uptake explains this process. They conclude that the respondents continued to use the existing mode because it was convenient for them. This decision was based on comparing the burdens of the existing mode with the perceived burden of the new mode. This implies a skeptical attitude toward new developments (Haraldsen 2009).

Respondents also tend to use the mode to which they have immediate access. Holmberg et al. (2010) call this the "mode-in-the-hand principle." They conducted a large-scale experiment (embedded in a social survey of Statistics Sweden) with a number of sequential approaches compared to the simultaneous approach. The results indicated that the designs in which the paper option was presented at a later stage showed the highest web takeup rates, with comparable overall response rates. The results also suggested that the later in the fieldwork the alternative mail option was introduced, the higher the web takeup rate. This strategy was implemented in a Statistics Sweden survey, which confirmed the experimental results: 51% of respondents used the web, compared to 15% in the previous round of the survey. The mode-in-hand principle also seems applicable to business surveys as a guiding principle for increasing web takeup rates, and points toward the sequential approach.

A sequential approach was adopted to introduce the web mode for the annual Structural Business Survey (SBS2006) conducted by Statistics Netherlands in 2007. In the first two contacts, the 63,644 respondents were offered only the web; with the second or third reminder (depending on sector of the economy) a paper questionnaire was enclosed. In all letters the web option was clearly promoted; login codes were included in the middle of the letter. The paper questionnaire was not mentioned in the letter, but was available on request. Before implementation, it was tested in a field pilot in 2006 with positive results (Giesen and Vis 2006; Snijkers et al. 2007b).

In total, the takeup rate of the web questionnaire was 80.4%, while 19.6% of the responding businesses used the paper questionnaire (Morren 2008). The overall response rate of the SBS2006 was comparable to that of previous years: approximately 80%. The percentage of web reporting increased with size of the business: rising from 78% for very small businesses to 89% for large businesses; and 24% of businesses changed their mode of data collection with the second or third reminder (from web to paper). From these businesses, 56.5% responded. Although these businesses received a paper questionnaire, about one-third used the web questionnaire. The mode could also be changed on request by the business. An interesting result is the high response rate for businesses that requested a mode change: 91.2% (out of 4084). This seems a strong predictor for response. The implication for a communication strategy is that making an alternative mode obtainable, but only on initiative of the respondent, would be a way to obtain commitment.

Other studies in the United States and Europe show the same results. The US BLS, for example, introduced web data collection as a means of lowering postage, processing, and printing costs. In 2005 and 2006 web data collection experiments were carried out for the federally mandated Survey of Occupational Injuries and Illnesses: a survey of approximately 230,000 businesses. Downey et al. (2007) compared several ways of communicating the web reporting option. The control group received a paper questionnaire with additional information on how to report via the web. In this group 23% of respondents used the web option. In the experimental group, which received information only on how to provide the data via the web, 50% responded using the web mode (paper questionnaires were provided when requested).

Between 2005 and 2007, Statistics Sweden conducted a number of follow-up studies for web data collection designs in business surveys [Erikson (J.) 2007, 2010]. Haraldsen (2009) studied web takeup for business surveys conducted by Statistics Norway. These studies showed results in the same direction:

- Spontaneous takeup in the simultaneous approach is low (5–25%).
- Takeup rates can be increased significantly by eliminating paper in the first contacts.
- To obtain a sufficiently high response rate, alternative modes (e.g., paper questionnaires) can be provided with the reminders.
- Takeup rates seem to be higher for respondents who already have non-paper-questionnaire experience (e.g., using TDE).
- The lowest takeup rates for web (when offered as an option) were seen in small businesses.
- Web takeup rates have increased over the years.

Web takeup rates are determined not only by respondent characteristics and an effective communication strategy but also by questionnaire and web design factors. An overview of these factors is presented by Dowling and Stettler (2007) as based on qualitative research in the United Kingdom and the United States. Snijkers et al. (2007b) describe results based on cognitive pretests with the Dutch electronic SBS questionnaire. These guidelines were discussed in Chapter 8, but with regard to the communication strategy, focused at motivating and facilitating businesses to respond, we would like to stress a number of simple communication principles that affect web takeup rates. These include accessibility, consistency, providing feedback, and making it easy, simple, clear, straightforward, and logical for the respondent (Wroblewski 2008; Jarrett and Gaffney 2009; Haraldsen 2009). Remember that people don't read, which is even worse when reading from the screen. Consequently, people apply a click-and-rush behavior (Snijkers et al. 2007b); they click on the first button that seems appropriate. They may also click their way to other links and get lost. At all times the communication design should provide instant answers to questions like (Dillman et al. 2005):

- What am I supposed to do (next)?
- What will happen when I press this button?
- Where am I? Is this right?
- Where can I get help?

If something goes wrong, or when they are confused, they call the help desk, which is extra work for you.

Now, the following communication strategy could be considered when the objective is to increase businesses use of a web data collection mode [Erikson (J.) 2010; Haraldsen 2009]:

- Restrict access to paper, and use a sequential approach; don't provide paper in the first contacts, but as an alternative delivered with a reminder. Indicate, however, that a paper questionnaire is available on request. A mode switch can be performed sooner or later. When to offer the paper questionnaire depends on the number of reminders, and the response rate in a responsive design (if response is falling behind, changing modes is recommended). This can be tailored to size and industry.
- Make access to web questionnaires easy from the start. Ensure that the website is trusted, and not blocked by firewalls, and that the questionnaire works properly for all (major) web browsers. The process of finding the questionnaire, logging in, opening and completing the questionnaire, and sending the data should be simple and straightforward. At the end of the process, provide feedback; confirm the receipt of the data, say "thank you," and consider providing benchmark information.
- In all letters, including the reminders with which a paper questionnaire is presented, clearly promote the web option; make it visible. Apply motivation

strategies to encourage web takeup. The helping principle could be used, for instance, by indicating that it would help you if the respondent would use the web option; social validity could be applied by saying that most businesses use the web questionnaire; and reciprocation could be applied by offering relevant feedback information when the data have been submitted, which is not offered for paper.

- Inform existing businesses (in recurring surveys) about changes in mode in the prefield stage. Don't ask what mode they would prefer, but inform them about the new primary mode. If asked, people stick to their habits as well as to established procedures. In the advance letter the existing mode can be presented as an option. Be service-oriented: assist respondents in using the new mode.
- Finally but not least, make the communication items consistent, easy, simple, clear, straightforward, and logical to the respondent. Adopt the perspective of the respondent, and pretest all items individually and in the course of the process, such as, by usability testing (see Chapter 7).

9.7 SUMMARY AND EXAMPLE OF A COMMUNICATION STRATEGY

In this chapter we have discussed the role and objectives of business survey communication, a design methodology to ensure an effective strategy including the components of a strategy, quality aspects and paradata, mode characteristics, and motivation and facilitation strategies. We have discussed tailoring to the business context, and survey organization policies and procedures. We have discussed measures, practices, and guidelines. The final question we will deal with in this chapter is how effective these strategies are in getting response. Unfortunately, there is little descriptive research in this area. Paxson et al. (1995) compared the procedures of obtaining response for a number of business surveys to those for social surveys. One of their conclusions was (ibid., p. 314) "low response rates result more from not using available knowledge about obtaining high response than from inherent features of business surveys. The main suggestions for organizations without governmental mandatory authority are to send business surveys to named individuals and to use telephone follow-up methods to encourage response."

Now, you might ask "Is it that simple?" The answer is "no." Although these suggestions are consistent with the practices described above, we only partly agree with the conclusion. Both available methodological knowledge and the special features of business surveys (as discussed in Chapter 1) need to be considered! To quote Snijkers and Luppes (2000, p. 369), this comes down to "asking the *right person* (the person who has access to the requested data and is authorised to provide that information) the *right information* (the data that are really needed, nothing less and nothing more) at the *right moment in time* (when the data are available and the contact person has the time to complete the questionnaire) with the *right communication mode* (the mode that is most effective)."

In addition to Paxson et al. (1995), our recommendations are [see also Snijkers and Luppes (2000) and Willimack et al. (2002)]

- Prefield measures:
  - Make an effort to check eligibility, identify the appropriate contact persons, and get commitment; prevention is better than repair.
  - Be flexible in dealing with gatekeepers, and try to get them on your side.
  - Have a proactive service-oriented marketing strategy to assist survey respondents in understanding their role as data providers and their responsibilities, and facilitate them in preparing the internal response process. Keep the business context in mind. Train businesses to be respondents, instead of nonrespondents.
  - Build a strong relationship.
  - Make an effort to get a commitment for businesses entering recurring surveys; once they are committed, the chances of responding are high.
- Field measures:
  - Be coherent with prefield communication and over all field communication measures.
  - Make an effort to get as much response as soon as possible, by tailoring the strategy to business characteristics (e.g., size and industry), past behavior, and timing (availability of the data and resources).
  - Be proactive and service-oriented; don't wait until the response deadline has passed, but establish contact prior to that. Facilitate businesses in their role as data providers; make it easy for them to respond (this is expected, specially if compliance is mandatory). Build a relationship.
  - Use motivational strategies, such as informing businesses about the "why" of the survey (relevance and usage of the data). For mandatory surveys, although the authority principle is a major reason for response, this should be communicated clearly but not stressed too much.
  - Have a help desk and an online help facility for businesses. Follow up on questions, complaints, and requests by businesses.
  - For businesses, surveyed multiple times, ensure that all communication strategies are coherent. For large businesses, establish business-oriented customer relations management procedures.
  - In mixed-mode designs, apply the sequential approach.
  - If reminder follow-up is needed, be flexible and vary procedures; change the communication and the data collection mode, change the message (use another message or motivational strategy), possibly change to an alternative contact person, and put more effort in for relevant businesses. Two-way communication modes, like telephone follow-ups, are more effective for establishing contact and for refusal conversion, but also more expensive.
  - On receipt of the response, thank the respondent and provide feedback. Maintain a good relationship.

- Postfield measures:
  - Be coherent with prefield and field communication and over all field communication measures.
  - Try to avoid repeated contacts because of data validation by providing better questionnaires (with built-in error checks) and an effective communication strategy. If recontacts are needed, ask for help. Recontacting businesses also causes response burden. Timely checks on possible inconsistencies are less burdensome than checks that occur a long time after the data have been collected.
  - Try to avoid enforcement as much as possible. If enforcement is needed, ask for help.
  - Maintain a good relationship.
  - In case of a panel or recurring survey, apply the PDCA cycle, and monitor the communication process using paradata in order to improve future designs.

In general, we recommend applying the seven-step design approach as discussed in Section 9.5.1. Make all communication measures and actions coherent over all stages and over surveys. Also, tailor to the business context (keep their questions in mind; see Figure 9.7); getting to know businesses helps to understand their internal processes and prevents burdensome data requests.

However, this approach is costly and time-consuming. It also requires survey organization systems and procedures that support such a strategy, like a good case–contact management system, well-established internal communication sharing information among departments, and integrated sampling. If one chooses an initial less costly approach (e.g., by using one-way and publication modes), the tradeoff will be later costs from more follow-up and higher perceived response burden for the business as a (undesirable) result. In general, we can conclude that much effort is needed to obtain an acceptable level of response, even for mandatory surveys (Morren 2008; Gates 2008).

We will end this chapter with an example that illustrates the effectiveness of these recommendations. However, these recommendations are only partly based on research findings. Therefore, we would like to encourage business survey methodologists around the world to do more research into the effectiveness of communication measures and strategies. A detailed description of a strategy with a detailed analysis of the results (also based on paradata) would be a good start. We would like to invite you to send research documents to WileyBookBusinessSurveys@gmail.com.

In 2003, Lynn and Sala [(2004); see also Sala and Lynn (2005)] conducted an academic voluntary employer survey in the UK. This survey had a mixed-mode design, with a postal stage followed by nonresponse telephone follow-ups. Their goal was to maximize response rates for minimum costs. In July 2003 a questionnaire was mailed to employers, followed by a reminder letter 2 weeks later, and a second reminder with a questionnaire in August. Employers who had not responded by then were contacted by telephone (October 2003–January 2004). (This strategy closely resembles the strategies discussed above.)

In the prefield stage, contact information of employers was collected; employees participating in an employee survey were asked for permission and to provide contact details of their employer. Of a survey of 434 employed respondents, 254 consented to the employer survey. For one employer, no contact information was provided, yielding a net sample of 253 employers. To get commitment, the permission form signed by the employees in the prefield stage was vital to overcome the employer's concerns with regard to confidentiality and legitimacy of the survey. [This again indicates commitment and authority (with regard to trust) as important heuristics in encouraging businesses to participate.]

The advance letter explained the purpose of the survey; contact information in case of questions was listed, and a prepaid return envelope was included. The paper questionnaire was used for phone follow-ups. For the interviewers, a coversheet was added to the questionnaire, including information about the previous mailings and the business. On the coversheet the interviewers could also report details and results of contact attempts.

The results of this strategy are listed in Table 9.1. A first conclusion is that many actions are needed to obtain an acceptable final response rate of 71.5% (181 businesses). According to Lynn and Sala (2004), this response rate is comparable with (American) academic surveys of establishments (Dillman 2000, p. 331; Paxson et al. 1995, pp. 307–308). (Response rate for government mandates varies between 50–95% (IGEN 1998). Phone follow-ups in stage 2 contributed almost 30%, making this stage vital in the communication strategy. In total 686 actions were needed to obtain this result, with 595 letters in stage 1 and 91 businesses contacted by phone in

Table 9.1 Response Results of Employer Survey 2003

		Result of Actions (*n*)		Result per Action (%)		
Action	Number of Actions (*n*)	Response (*n*)	Non-response (*n*)	Response (%)	Non-response (%)	Result per Action (% Total Response)
			First Stage			
Advance letter	253	60	—	23.7	—	33.2
1st reminder	193	44	—	22.8	—	24.3
2nd reminder	149	25	—	16.8	—	13.8
Refusal	—	—	33	—	13.0	—
No reply	—	—	91	—	36.0	—
Total	595	129	124	51.0	49.0	71.3
			Second Stage			
Telephone follow-up	91	52	—	57.1	—	28.7
Refusal	—	—	34	—	37.4	—
Noncontact	—	—	5	—	5.5	—
Total	91	52	39	57.1	42.9	28.7
			First/Second Stages Combined			
Total	686	181	72	71.5	28.5	100

Source: Lynn and Sala (2004).

stage 2. On average, this yields 4.6 mail contact attempts per responding business (129 businesses), at a cost per respondent of €5.56 on average (with a total cost of 717 euros).

In stage 2, more telephone contacts were needed to obtain a result. A total of 553 calls were made to 91 employers, yielding an average number of attempts of 10.6 per responding business. The distribution of call attempts was skewed (Lynn and Sala 2004, p. 11): "Two thirds of the sample (64%) required at least four telephone calls and one third (33%) required at least seven calls." The maximum number of attempts was 30. The average cost per respondent in the telephone stage was € 60.62 (total 3152 euros). Although the phone stage was vital, it was at a high cost; about 2.3 times more attempts per respondent were needed at 11 times higher cost rate than for stage 1.

To conclude with the words of Lynn and Sala (2004, p. 29):

> To achieve a good response rate, considerable effort and a flexible approach (to the telephone stage) were required. It is necessary to make contact with the organisation, to overcome gatekeepers (usually receptionists or secretaries), to make contact with respondents, to persuade them to cooperate, and to encourage and allow them to retrieve information that may be held by other persons within the organisation.

ACKNOWLEDGMENT

In writing this chapter we would like to thank Martin Luppes (Statistics Netherlands), Dave Tuttle (US Census Bureau), and Lonne Snijkers for their reviews, comments, and suggestions.

CHAPTER 10

Managing the Data Collection

Ger Snijkers and Gustav Haraldsen

10.1 INTRODUCTION

In Chapters 1–9 we have been working towards collection of data from businesses. In Chapters 1–3 we set the scene with an overview of the survey production process, the business context, and quality challenges in business surveys; in Chapter 4 we discussed survey planning, and in Chapter 6, compliance cost issues (i.e., response burden). Chapters 5, 7, 8 and 9 focused on designing the survey components: the sample, the questionnaire, and the communication strategy. We have now approached the point at which all survey components are ready for implementation and the survey is launched. In this chapter, we will briefly discuss implementation of the survey components. The main focus of the chapter, however, is on managing the fieldwork: monitoring and controlling the data collection process and the quality of survey results. Now, we will see whether the planned survey design is effective, particularly whether the quality challenges as discussed in Chapter 3 have been dealt with appropriately. We will discuss an active fieldwork management approach, which may entail changing the prepared design during the fieldwork.

When we discussed the planning of a survey (in Chapter 4), we mentioned the need to apply the principles of project management, such as planning a survey in advance and working accordingly, to ensure that we reach the survey outputs within time and budget, and according to preagreed levels of quality. We also discussed risk management, in dealing with risks and planning mitigating actions (using quality gates). We also noted that an important part of the production process is beyond the control of the survey organization: the response process in

Designing and Conducting Business Surveys, First Edition.
Ger Snijkers, Gustav Haraldsen, Jacqui Jones, and Diane K. Willimack.

businesses (as discussed in Chapter 2). In Chapters 8 and 9 we discussed how to take account for this response process in the design. But still, even with careful preparation, the effectiveness of the design to get quality responses is uncertain. In the design stage, pretesting and piloting (discussed in Chapter 7) can help estimate these uncertainties and improve the initial design. These uncertainties threaten the ability of survey organizations to meet targeted survey results within time and budget (Groves and Heeringa 2006; Wagner 2010; Kreuter et al. 2010).

As a consequence, in addition to careful planning and designing, we need to take an active fieldwork management approach to manage these uncertainties during the data collection. Instead of relying on achieving the targeted results, we need to monitor the fieldwork continuously in real time. This relies heavily on the real-time availability and statistical analysis of suitable process data or paradata. Paradata-based indicators allow for identification of problems in the data collection process and appropriate intervention to increase the likelihood of achieving the survey targets (Groves and Heeringa 2006; Kreuter et al. 2010). This, however, also needs to be planned carefully, as we have seen.

In this chapter we will discuss this approach. Active fieldwork management and responsive designs are discussed in Section 10.3, following discussion of survey implementation in Section 10.2. Paradata are discussed in Section 10.4, where we relate paradata and indicators to the errors in surveys. This provides us with a framework for selecting suitable indicators to monitor the fieldwork. These indicators can be related to (1) fieldwork output (i.e., the survey data) and (2) the fieldwork process, which relates to the quality and costs of the survey design (Renssen et al. 1998). Key indicators are discussed in Sections 10.5 and 10.6: Section 10.5 discusses quality indicators of the survey data, including the response rate; Section 10.6 discusses indicators of the effectiveness and efficiency of the survey design (with regard to costs and resources). A brief section on evaluation and quality reports (10.7), and a summary (10.8) conclude this chapter.

10.2 IMPLEMENTING THE SURVEY

Before we can move from the design–build–test (DBT) stage to conducting data collection, we need to implement all components of the survey. At this point in time, the sample, the questionnaire, communication materials and procedures, and all systems, tools, procedures, workflows, and operational issues pertaining to the data collection, and data capture, coding, and editing methods and procedures will have been designed, built and, ideally, tested. Now, they can be implemented.

Implementing the survey involves getting everything ready for the fieldwork, getting everything working together. All components of the data collection process are put in place, and everyone is ready for takeoff. It is like launching a rocket to the moon; the rocket is loaded onto the launch platform; everyone in the control room is ready; and all roles, tasks, and procedures are clear. Everyone is waiting for the start signal. The engine of the rocket is not yet running, but the signal to start the engine can be given any moment now. Implementing a survey is the transition from deciding how the survey will be and designing it, to conducting it. This includes [see, e.g., US Census Bureau (2011) and Aitken et al. 2003):

- The sample is drawn, resulting in a list of selected businesses; in a prefield stage sample refinements may have been completed, to provide up-to-date information on the selected businesses.
- The finalized questionnaire is printed on paper, electronic questionnaires are uploaded on the internet, CATI and CAPI questionnaires are uploaded on interviewers' computers, and questionnaires for other data collection modes [e.g., touchtone (telephone) data entry (TDE)] are implemented.
- The finalized communication strategy results in a data collection schedule (a detailed process map of the data collection with regard to survey communication, including a timetable of communication measures and actions, procedures, and staff involved); all communication items are printed, uploaded on the Internet, or otherwise implemented; and respondents are allocated to interviewers.
- The finalized data capture, coding, and data editing methods and procedures are implemented in tools and instructions (see Chapter 11).
- All production systems, tools, and procedures are finalized, including data, metadata, and paradata systems.
- Staff is instructed with regard to their roles and are trained in skills, procedures, systems, and tools.
- The data collection process is set up; logistics, workflows, and operations are laid out. The sequence of activities aimed at successfully conducting a survey and meeting the survey objectives is ready to go.

All this requires advance planning, as we discussed in detail in Chapter 4. During the DBT stage the planning of the survey may have changed; initiation of the fieldwork may have been delayed, perhaps due to extra time needed to design and test the questionnaire. This does not necessarily mean that the completion of the fieldwork is at a later stage. A later start of the fieldwork does however affect the survey components; these need to be updated. Contact information may need to be updated, a new timetable might be needed, and staff involved in the fieldwork needs to be rescheduled.

For the fieldwork to be successful, all components must work properly and synchronously. In the design stage, the individual components, procedures, systems, and tools have been tested, but they have not yet been tested as a joint system in real-life conditions. This can be done by conducting a pilot test before starting the actual survey (see Chapter 7). We recommend conducting a pilot at least in case of a major redesign of a survey, such as a fully redesigned questionnaire or a completely new data collection system. In a recurring survey, this can be done for example, with a representative part of the sample. Rather than risk a problem arising with the whole survey, risk is reduced to a small part of the sample, thus ensuring the delivery of results. Again, conducting a pilot, needs advance planning.

10.3 ACTIVE FIELDWORK MANAGEMENT

Finally, after a long period of hard work, the sampled businesses are contacted with the questionnaire (or even in the prefield stage are contacted for survey cooperation, as we have seen in Chapter 9), and the response comes in. Now, everything is in operation: the data collection is up and running; the rocket is launched—and, as NASA would for a flight to the moon, we have to make sure that the survey hits the right goal. Everyone in the control room is monitoring the flight to reach the target, to get the rocket to the moon in time safely: Do we have enough fuel, is the engine working properly, are the astronauts doing well, are there any disturbances from outside—in short, are we on track as planned? In a business survey, to ensure that we achieve the targeted results (as described in the project plan; see Chapter 4), we need to do the same; we need to monitor the data collection process: Are the response rates okay, is the response distribution representative, do the data show measurement errors, do we have enough time, resources, and money to achieve the survey objectives? The need for monitoring the fieldwork in business surveys is crucial since we are dealing with uncertainties that threaten the ability of the survey organization to meet the targeted survey results within time and budget. In business survey data collection, we cannot rely on autopilot!

Monitoring is not a single action; it is part of a number of steps. To effectively monitor the fieldwork, we must adopt an active fieldwork management approach. According to Hunter and Carbonneau [(2005); see also Laflamme et al. (2008)], this includes

- Planning
- Monitoring
- Indentifying problems and finding solutions
- Proposing appropriate corrective actions
- Communicating and taking action
- Evaluation and documentation.

Planning involves formulation of the data collection process, which includes planning of fieldwork monitoring (as we have discussed in Chapter 4). Planning fieldwork monitoring includes two aspects: (1) identify relevant indicators to monitor the process in general; and (2) in addition, identify critical design features and plan for appropriate mitigating interventions (planning quality gates in risk management, as discussed in Chapter 4). In active fieldwork management, both have to be factored in.

In practice, survey organizations have to deal with many uncertainties that affect the survey costs and errors—uncertainties that may not have been preidentified during pretesting. In active fieldwork management, the general objective of monitoring is to identify collection problems during the data collection as early in the process as possible, and not at the end (when it is too late to act). At this point the relevant question is: "Are we still on track?" In addition, we recommend monitoring critical design features more closely. This relates to the responsive design approach as proposed by Groves and Heeringa (2006). In this approach the monitoring is targeted at preidentified potentially problematic design features, thus reducing the risk of not meeting the survey objectives; this approach includes planning of appropriate interventions in the design.

The responsive design approach, as presented by Groves and Heeringa (2006; see also Kreuter 2013) in social surveys (see also Section 4.3.4.4), involves the planning of four steps (ibid., p. 440):

1. Preidentification of design features that potentially affect survey errors and costs, e.g. based on pretesting. These design features may likely jeopardize the achievement of predetermined survey outputs.
2. Identification of indicators showing the cost and error properties of those design features, and monitor those indicators during data collection. To calculate these indicators, we need to define appropriate paradata.
3. Preparing the real-time collection and analysis of the paradata; during data collection the relevant indicators need to be calculated in real time.
4. In subsequent phases of data collection, the features of the survey can be altered according to "cost-error tradeoff decision rules." Decision rules must be finalized prior to the fieldwork, and should ensure that the predetermined quality levels of the survey results will be achieved. On the basis of these decision rules, alternative designs and appropriate actions must be prepared.

In responsive designs, the survey design is changed during the course of data collection. The challenge is to design the data collection in such a way that the design actually can be changed. Therefore, in responsive designs, the data collection process is divided into phases. Each phase is monitored by computing the identified set of indicators. The results are matched with the decision rules. At the end of each phase, if the indicators manifest unacceptable levels according to the decision rules, the appropriate actions are taken. For business surveys this includes for instance, a change in data collection mode for targeted subpopulations (e.g., from web to paper); in a subsequent phase for nonresponse follow-up, a change from paper to telephone reminders may be implemented; and in a final phase the questionnaire might be shortened with only key questions: the basic key question approach (Bethlehem 2009). This requires advance planning and a flexible data collection process.

Examples of decision rules can be found in the US Census Bureau Statistical Quality Standards (2011a). Sub-Requirement D3-3.6 (ibid., p. 58) states that "non-response bias analyses must be conducted when unit, item, or quantity (i.e., weighted) response rates for the total sample or important subpopulations fall below the following thresholds:"

1. The threshold for unit response rates is 80%.
2. The threshold for item response rates of key items is 70%.
3. The threshold for total quantity response rates is 70%. (Thresholds 1 and 2 do not apply for surveys that use total quantity response rates.)

If the precision of the resulting statistics is unacceptable, an additional sample may be drawn in subpopulations with low response rates to compensate for this (if time and money allow). If low response rates are expected, the additional sample can be prepared in advance (by oversampling).

For simplicity, in this chapter we will use the term *active fieldwork management*, to include the responsive design approach. Both approaches require real-time collection and analysis of paradata. The results of this analysis are presented to managers in status reports. On a continuous basis, they are informed about the status of the fieldwork through a number of process indicators and control charts. Key indicators such as response rate will be discussed in Sections 10.5 and 10.6.

If collection problems are detected, (prepared) appropriate corrective actions are proposed to the responsible managers. They are in the position to make decisions. When they do, the process of tackling these problems and handling the staff involved are identified. They are informed about the decisions, and consequently take action. When the process ends, it is evaluated and documented, for later improvements if needed (according to the PDCA cycle; see Section 4.2).

10.4 PARADATA

Paradata, or process data, are essential for managing the data collection process. With the active fieldwork management approach, they allow statistically informed monitoring of and decisionmaking for this process. The paradata concept was introduced by Mick Couper (1998), and since then this term has been used for survey process data [for an overview, see Couper and Lyberg (2005), Kreuter et al. (2010), Lynn and Nicolaas (2010), and West (2011)]. Currently, paradata are attracting attention. Studies using paradata have been presented at various methodology conferences (Kreuter et al. 2010). Many of these studies focus on social surveys [see, e.g., Kreuter (2013), Laflamme (2008), Laflamme and Karaganis (2010), and O'Reilly (2009)]. Here, we will discuss this concept for business surveys.

Paradata include a wide variety of data. A number of classifications of paradata have been proposed, e.g. depending on the type of data [e.g., call records, interviewers observations, vocal characteristics, keystrokes (Kreuter et al. 2010)] or macro/microparadata [overall process indicators or indicators at the unit/questionnaire/records level (Scheuren 2000)]. We feel that a useful classification should be objective-driven (Lynn and Nicolaas 2010). A number of objectives for using paradata can be identified (Lyberg 2011; Couper and Lyberg 2005; Aitken et al. 2003):

- *Monitoring and evaluation of an individual survey:*
  - Process management (of an individual survey): monitoring the data collection process with continuous updates of progress and process stability checks, within active fieldwork management
  - Quality management (of an individual survey): monitoring the quality of the survey process output (i.e., the survey data) within active fieldwork management
  - Client communication: providing customers with quality assurance

- *Monitoring and evaluation of the survey production environment (PDCA cycle):*
  - Short-term process improvements: continuous feedback to the survey production process
  - Long-term process improvements: monitoring process efficiency and quality over time as input for future process improvements projects, and perhaps even organizational change
  - Methodological improvements: research aimed at improving the survey components (e.g., the questionnaire or the communication strategy), or identifying root causes of problems and finding methodological solutions
- *Corporate image:* providing the general public with quality reports, to achieve confidentiality and trust in the statistical outputs

The use of paradata in the context of active fieldwork management to monitor an individual survey requires a joint focus on process and quality management. From a *process management perspective*, paradata are related to the various processes in the data collection process and their costs:

- *Paradata associated with processes in the survey organization:*
  - Logistics: when specific actions take place, such as sending out the questionnaire and later reminders, beginning of telephone reminders (per respondent, when contacted the first time, second time, etc.)
  - Performance of the staff involved (e.g., interviewers, help desk staff): number of contacts, timing of contact and results (call records), recordings of telephone contacts and keystroke data of interviews, number of questions processed (help desk)
  - Data capture process: number of questionnaires processed
  - Course and status of the response: return rate, response rate, timing of response
  - Editing process: records and (within records) variables that have been changed; number of businesses recontacted
- *Paradata associated with process at the respondent side:*
  - Performance during questionnaire completion (keystroke data or audit trails): date completion started, completion profile; time needed to complete, number of questions changed, edit checks triggered
  - Result of completion process: status of the response (responded: empty or completed questionnaire returned, not yet responded), item nonresponse, response burden
  - Help desk: number of questions and complaints (and the topic)
  - Web log data: when have websites been visited and how many times
- *Project management paradata:* consumption of money, human and material resources, and time

When adopting a *quality management perspective*, we focus on survey errors. From this perspective, paradata need to be related to sources of error in surveys as presented by Groves et al. (2004) in their total survey error model; this model was also central in our discussion of business survey errors in Chapter 3. The resulting framework is shown in Figure 10.1; it provides an overview of paradata and paradata-based indicators for assessment of survey errors. This framework has been presented by Kreuter and Casas-Cordero [(2010) see also Kreuter et al. (2010)], which we have adapted to business surveys [considering discussion and examples from, e.g., Biemer and Cantor (2007), Gates (2008), Green et al. (2008), Haraldsen et al. (2010), Giesen and Kruiskamp (2011), Giesen et al. (2009a), Regan and Green (2010), Snijkers and Lammers (2007), Snijkers and Morren (2010), and US Census Bureau (2011a)].

These paradata and indicators can be used to assess the quality of the survey components (during the DBT stage; see Chapter 4) and the survey data. They can be used after completion of a process, and most of them also during the collection process. In the next two Sections, we will discuss a number of key indicators.

With regard to the use of paradata, it should be noted that paradata are like survey data, as Groves and Heeringa (2006, p. 455) point out: "they need conceptual development, measurement development, and pre-testing." The same quality requirements hold for paradata: they should be valid and reliable, and their measurement and processing are also subject to measurement, nonresponse, and processing errors (Kreuter and Casas-Cordero 2010). Paradata are inexpensive to collect when the systems are adopted to

Sample-related errors	Measurement-related errors
Coverage errors: Overcoverage: number of ineligible units, number of duplications (related to total number of units) Completeness: relevant contact and background information missing Actuality: date of last update, number of changes made (records, variables) in, for example, the last month	Construct validity errors: Prefield methods in conceptualization stage to assess construct validity (see Chapter 7), such as recordkeeping studies Completeness of questionnaire: questionnaire is incomplete (questions are missing) or overcomplete (number of questions that can be deleted from it)
Initial sample errors: Precision: sample size not too large or too small to obtain precise estimates Efficiency of sample procedure: sample procedure to reduce sample size for the same level of precision Representativeness of sample: sample in comparison to population distribution (if known) Rotating samples: check sample procedure Sample procedures with regard to response burden (coordinated samples, survey holiday): check sample procedure	

Figure 10.1 Business survey errors and paradata-based indicators for their assessment.

Nonresponse errors:
- Prefield communication: number of noncontacts and reasons for noncontacts, number of ineligible units
- Field communication, in addition to prefield communication: number of advance letters, reminder letters, phone calls associated with timing of communication (including costs); as for phone calls—time and outcome of call (e.g., noncontact, gatekeeper, voicemail, refusal, appointment, response); return rate associated with timing of communication (split up for relevant strata); response rate associated with timing of communication (for relevant strata) and reasons for nonresponse; number of refusals (for relevant strata) and reasons for refusing; number of postponements (for relevant strata) and reasons for postponement; (for electronic questionnaires) audit trails and keystroke data related to timing of survey request (date of starting to work on questionnaire, peaks in completion activities, and date of response)
- Contacts by businesses: number and classification of questions by businesses; number and classification of complaints
- Data editing—postfield communication: recontact rate (for relevant strata)
- Postfield communication: enforcement rate (for relevant strata) and reasons for nonresponse

Measurement errors:
- Questionnaire: prefield methods in the operationalization stage to assess validity and usability, such as pretesting and eye-tracking data (see Chapter 7)
- Questionnaire—item nonresponse: questions with high item nonresponse rate
- Mode of data collection: response rate for each mode, changes in modes
- Respondent—field paradata (for electronic questionnaires); keystroke data and audit trails: completion profile and time, questions that have been corrected many times (identify difficult questions and sections), activated help functions, activated error checks
- Respondent—response burden: time needed to complete questionnaire, number of people involved, time needed to retrieve data
- Interviewer: interviewer performance information (audio recordings and keystroke information of interviews)
- Data editing—postfield communication: recontact rate (for relevant strata) and reasons for recontacting, number of records changed, number of times a variable has been changed, magnitude of value change

Unit processing errors:
- Data capture: number of records processed, number of records missing (related to response rates), error rates
- Data editing: number of records edited (manually, automatically), error rates

Response processing errors:
- Data coding: number of questions coded
- Data editing: number of edit failures

Figure 10.1 *(Continued)*

register them. However, they still need to be analyzed, and analysis may not be straightforward. For some processes, huge numbers of paradata may be collected. Audit trails or keystroke data, for instance, may easily result in millions of records (depending on the length of the questionnaire and the sample size), with each record containing information on one action in the completion process (Snijkers and Morren 2010).

10.5 MONITORING THE QUALITY OF THE RESPONSE

The survey data that are collected during the fieldwork are the results of the survey. In the project specification stage, the quality levels of the survey output have been specified in agreement with the stakeholders (as discussed in Chapter 4). During data collection, we need to monitor the response to ensure that these quality levels are reached. We need to monitor the product quality; product quality indicators for business survey response include

- At the unit level: response rate, nonresponse rate, and return rate
- With regard to the relative importance of businesses: weighted response rate
- With regard to distribution of the response: representativity (*R*-) indicator (defined in Section 10.5.4)
- As to the quality of the measurements: item nonresponse, and the process efficiency graph

In this section, we will discuss these survey data quality indicators. They indicate how close we have come to the original sample that was drawn to obtain accurate, complete, and timely measurements of variables. In survey reports, often only the (non)response rate is reported, simply because this indicator is easy to calculate and interpret. However, we need to consider all four aspects of the response to get a clear idea of the quality of the response.

Because we have a sample, we have sampling variance [sampling error (Bethlehem 2009)]. This variance is controlled by the sample procedure. The unit nonresponse that occurs during data collection introduces additional variance. Unit nonresponse means that the data for one unit are completely missing. This additional variance is beyond the surveyor's control, and therefore needs monitoring. Nonresponse also may lead to bias if the nonresponse distribution is skewed for auxiliary and target variables. Since businesses may have unequal weights with regard to target variables, the bias may be quite large when important businesses are not accounted for. The response is selective, and no longer representative. These are nonresponse errors.

A second type of nonresponse is at the level of individual questions; for some questions the data are missing, resulting in item nonresponse. This introduces additional variance and bias at the level of the measurements. Additional bias may also be introduced by measurement errors in the response. (Nonresponse and measurement errors have been discussed in Chapter 3.)

During the data collection process, product quality indicators serve as process indicators in active fieldwork management, indicating the effectiveness of the process up to a given moment in time. At the end of the data collection process, they become quality indicators; the quality levels are fixed (although the effect of errors can be reduced by statistical techniques in the postfield stage; see Chapters 11 and 12). To calculate these indicators, we first, need to look into data collection outcome codes. On the basis of these codes, the indicators can be calculated.

10.5.1 A Classification of Data Collection Outcome Codes

Calculation of the response quality indicators is not straightforward; it requires a detailed analysis of units that are counted as response and those that are counted as nonresponse. This analysis involves breakdown of the initial sample into all kinds of nonresponse and response codes. An example of a detailed overview of data collection outcome codes is provided by the American Association for Public Opinion Research in their response rate calculator for social surveys (AAPOR 2011b). Response outcome code schemes for business surveys have been discussed by Thomas (2002), Petroni et al. (2004a), and Morren (2008). On basis of these schemes, Figure 10.2 presents a detailed overview of response outcome codes for business surveys [see also Snijkers (1992) and Bethlehem (2009)]. This classification scheme provides us with an instrument that can be used to calculate indicators of the data collection process.

The scheme in Figure 10.2 is associated with the business survey communication process as described in Chapter 9 (Section 9.2.3). It can be adapted to the process in your survey. If, for example, the prefield stage is combined with the field stage, the prefield section is skipped. Or in case of a mixed-mode design, each mode can be overviewed seperately.

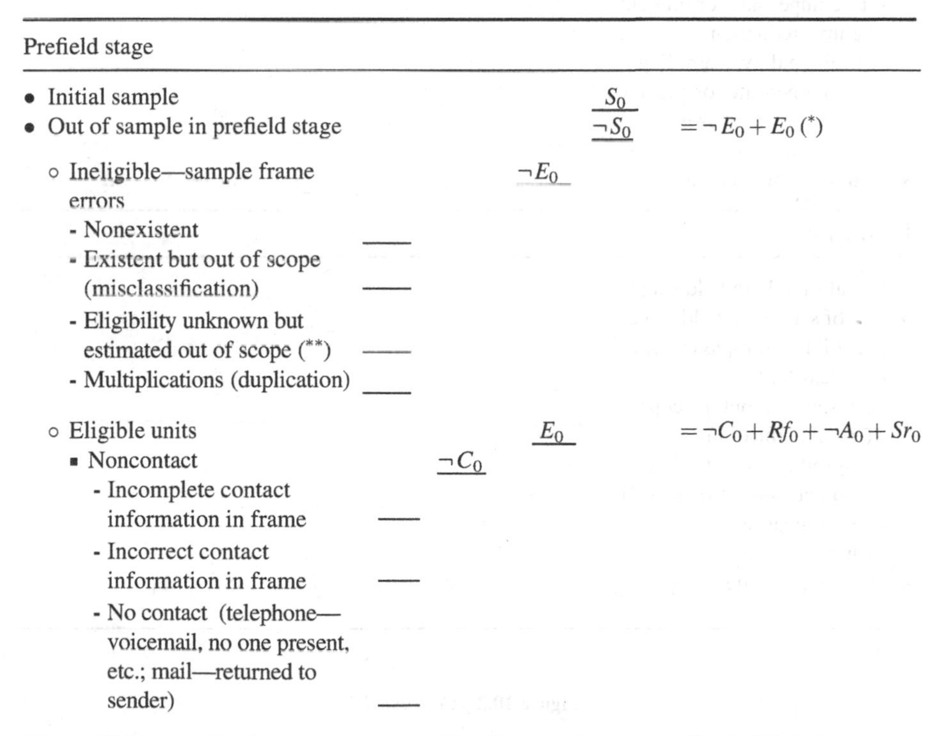

Figure 10.2 Data collection outcome codes. *Key*: (*) where '¬' denotes "not"; (**) businesses are considered to be "out of scope" as described by Thomas (2002), Ritzen (2007), and Struijs and Willeboordse (1995): no activity in country of interest, dead or dormant at end of period, not yet trading at end of period, or structure change.

- Refusal $\underline{Rf_0}$
  - Unwilling to cooperate ___
  - Reason for noncooperation unknown ___
- Inability to cooperate $\underline{\neg A_0}$
  - Unable to cooperate (e.g., data not yet available, no time, contact person not present) ___
  - Unable to communicate survey request (e.g., language problem) ___
- Other reasons for sample reduction (e.g., because of cost reasons, or interviewing capacity) $\underline{Sr_0}$

- Additional sample in prefield stage $\underline{Sa}$
  - To compensate for prefield sample reduction ___
  - Additional oversampling (to compensate for expected low response rates in subpopulations) ___
- Initial sample in field stage $\underline{S_1}$ $= S_0 - \neg S_0 + Sa$

Field stage

- Initial sample in field stage $\underline{S_1}$
- Out of sample in field stage $\underline{\neg E_1}$
  Ineligible—sample frame error:
  - Nonexistent ___
  - Existent but out of scope (misclassification) ___
  - Eligibility unknown but estimated out of scope (**) ___
  - Multiplications (duplication) ___
- Net sample in field stage $\underline{S_2}$ $= S_1 - \neg E_1$

Figure 10.2 (*Continued*)

- Net sample in field stage: eligible units and eligibility unknown but estimated in scope ____ ____ $\underline{S_2}$
- Number of questionnaires not dispatched and no follow-up (e.g., because of human or technical error, hardware malfunction, inadequate capacity) ____ ____ $\underline{\neg Qs}$
- Number of questionnaires dispatched: of which ____ ____ $\underline{Qs}$ $= S_2 - \neg Qs$
  - o Number of units followed up (no postponement) of which ____
    - Number of units: reminder ____
    - Number of units: reminder ____
    - Number of units: reminder ____
    - Number of units: telephone follow-up ____
  - o Number of units with postponement of which ____
    - Number of units reminded after postponement ____

Returns

- Number of questionnaires dispatched ____ ____ $\underline{Qs}$
- Number of questionnaires not returned or not received ____ ____ $\underline{\neg Rt}$ $= \neg C_1 + Rf_1 + \neg A_1 + Op$
  - o Noncontact ____ $\underline{\neg C_1}$
    - Incomplete contact information in frame ____
    - Incorrect contact information in frame ____
    - No contact (telephone—voicemail, no one present, etc.; mail—returned to sender) ____

Figure 10.2 *(Continued)*

- Refusal and breakoff — Rf_1
  - Unwilling to cooperate ___
  - Questionnaire (partially) completed but not returned ___
  - Reason for noncooperation unknown ___
- Inability to cooperate — $\neg A_1$
  - Unable to cooperate [e.g., data not yet available, no time, contact person not present (vacation, sick leave), etc.] ___
  - Unable to communicate survey request (e.g., language problem) ___
  - Technical/usability problems (e.g., computer problems, questionnaire doesn't work in browser) ___
- Operational errors during fieldwork — Op
  - Wrong questionnaire dispatched and no follow-up ___
  - Questionnaire dispatched to wrong unit address ___
  - Questionnaire sent back by business but not received ___
  - Other reasons ___
- Number of questionnaires returned and received — Qr $= Qs - \neg Rt$

Response

- Number of questionnaires returned and received — Qr
- Number of questionnaires returned but not processed — $\neg Qp$
  - Not processed for data quality reasons ___
    - Returned too late to process ___
    - Too suspect to process ___

Figure 10.2 *(Continued)*

- Too incomplete to process	___		
- Data processing errors (e.g., data lost)	___		
• Number of questionnaires returned and used		R	$= Qr - \neg Qp$
- Returned with complete data	F		
- Returned with partial data (item nonresponse, incomplete combined data)	P		
Edited data file			
• Number of questionnaires returned and used		R	
• Number of records in edited data file of which		Ed	$= R$
o Number of questionnaires edited	___		
- Automatic editing	___		
- Manual editing	___		
o Number of questionnaires for which key variables are edited	___		
o Number of businesses recontacted	___		

Figure 10.2 (*Continued*)

10.5.2 Response and Return Rates at Unit Level

On the basis of the response outcome codes in Figure 10.2, we can calculate the response, nonresponse, and return rates. Unfortunately, there is no standard definition of response or return rate. A definition that is generally followed is the one proposed by AAPOR (2011a, p. 44): "the response rate is the number of complete interviews with reporting units divided by the number of eligible reporting units in the sample." The following formula gives a definition for an unweighted response rate (*RR*) in businesses surveys, a response rate at the unit level:

$$RR = \frac{\text{number of responding reporting units}}{(\text{number of eligible units}) + (\text{number of units with eligibility unknown but estimated in scope})}$$

This definition is also proposed by the OECD (Martelli 2005), the UK Office for National Statistics (Thomas 2002), Statistics Norway (Haraldsen et al. 2010) and in the US by the Bureau of Labor Statistics and the Census Bureau (Petroni et al. 2004a; US Census Bureau 2011a). To estimate the number of units with unknown eligibility but estimated in scope, Thomas (2002) proposes using the same proportion as for the proportion of units with known eligibility in the sample: $S_2/(S_2 + \text{nonexistent} + \text{misclassification})$. With regard to this estimate, AAPOR (2011a, p. 45) states that it must be "guided by the best available scientific information . . . and one must not select a proportion in order to boost the response rate. . . . It may consist of separate estimates for subcomponents of unknowns. . . . In each case, the basis of all estimates must be indicated."

This definition can be operationalized in various ways, however, depending on what is counted in the denominator and the nominator. The denominator may be counted as the number of units in the initial sample (S_0), the initial sample in the field stage (S_1), the net sample in the field stage (S_2), or the number of questionnaires sent out (Qs). This relates to the question of whether out-of-sample eligible units in the prefield and ineligible units in the field stage should be counted. As for the nominator, the questions is whether partly completed questionnaires should be counted. For a design in which the prefield stage is combined with the field stage, we suggest the following formula [comparable to AAPOR (2011a, p. 45): RR4]:

$$RR = \frac{\text{number of questionnaires returned and used}}{(\text{eligible units}) + (\text{eligibility unknown but estimated in scope})} = \frac{F + P}{S_2} = \frac{R}{S_2}$$

This is the number of returned and used questionnaires relative to the net sample size. In this formula, both fully and partially completed questionnaires are counted. AAPOR also proposed publishing the response rate of fully completed questionnaires: $RRf = F/S_2$. In these formulas, the number of questionnaires not dispatched, and without follow-up ($\neg Qs$) is counted in the denominator. The response rate may be higher if this number is excluded. In practice, we hope that $\neg Qs = 0$; however, as we know, at all stages in the process errors may occur, so it may turn out that $\neg Qs > 0$.

The nonresponse rate is the complement of the response rate: $NR = 1 - RR = (S_2 - R)/S_2$. It indicates the proportion of questionnaires not returned, not received, and not processed. For an example, see Table 10.1.

In addition to the response rate, the return rate can be calculated:

$$RtR = \frac{\text{number of reporting units that returned a questionnaire}}{(\text{eligible units}) + (\text{eligibility unknown but estimated in scope})}$$

$$= \frac{(F + P) + \neg Qp}{S_2} = \frac{R + \neg Qp}{S_2}$$

This is the total number of returned questionnaires relative to the net sample. In the return rate the questionnaires that are returned but not processed ($\neg Qp$) are counted as well. The return rate is often used in the data collection process, since this indicator does not need the identification of returned questionnaires as qualified response. Businesses that return questionnaires that cannot be processed very likely will be recontacted with data editing.

These indicators can be specified for various designs. For instance, if the survey has a mixed-mode design, a response rate for each data collection mode can be calculated, like the web takeup rate. The indicators can also be specified for relevant population subgroups. For monitoring of the data collection process, they can be specified as a function of time and shown in a control chart, as in Figure 10.3.

10.5.3 Weights of Businesses

The (non)response rate and the return rate as defined above are at the unit level. A typical feature of business surveys, however, is that some businesses are more relevant than others. Businesses may have different weights with regard to the estimates of target variables; a small number of businesses contribute to a major proportion of the estimated population total. If this applies to your survey, these differences in relative importance need to be considered when monitoring the fieldwork. A number of procedures can be applied to do so:

- Weighted or quantity response rate
- Stratification
- Score function

In a weighted or quantity response rate, the relative importance of business units is factored in (Thomas 2002):

$$RRw = \frac{\text{total weighted quantity for responding reporting units}}{\text{total estimated quantity for all in-scope reporting units}}$$

For the weights, various variables may be taken, such as, turnover/revenue or size (number of employees). For turnover/revenue, for example, the weighted response rate indicates the contribution of each unit to the total turnover/revenue. As such, businesses relevant to the estimates have a higher weight. Also sample weights can be used, which are the inversion of the selection probabilities for each unit in the sample (see Section 5.4.1 IGEN 1998; Bethlehem 2009). A detailed procedure for calculating weighted (or quantity) response rates is presented by the US Census Bureau (2011a).

The relevance of weighted response rates is illustrated by Smith (2004) for the Canadian Unified Enterprise Survey. Table 10.1 shows both unit and weighted response rates relative to annual gross establishment revenues for the reference year 2002. From this table we can conclude that although the unit response rate is quite low (it would be 50% if the unmailed questionnaires were excluded), the weighted response rate indicates a level of 67% (or 73% of the number of questionnaires dispatched). This discrepancy is quite large because of the importance of businesses. The table also shows that the partial response is very low in counts, but they add 2.5 times as much to the weighted counts.

Table 10.1 Response Distribution for Canadian Unified Enterprise Survey for 2002

	Sample		Sample Weight	
Response Code	Count	%	Value[a]	%
Not mailed (data from tax records, excused from survey)	7,711	14.3	139,922	8.3
Ineligible units (frame errors)	5,247	9.7	55,535	3.3
Nonreturn (refusal) or late return (after survey deadline)	17,652	32.7	360,607	21.4
Response	23,322	43.2	1,128,545	67.0
Complete	21,462	39.8	992,082	58.9
Partial	1,860	3.4	136,463	8.1
Total original sample	53,932	100.0	1,684,609	100.0

[a]In million Canadian dollars.
Source: Smith (2004).

This indicator requires that the quantities for all units be known or estimable (e.g., using data from a previous survey), which makes it more complicated to use for monitoring the data collection in real time. An alternative to monitor the response of relevant businesses is stratification, where one or more strata for important businesses are predefined, and for each stratum the unweighted response rate *RR* is calculated. The response for relevant strata should be a 100% to obtain unbiased estimates. This procedure also provides information to efficiently allocate resources for nonresponse follow-up; relevant strata for which the response is falling behind require priority in nonresponse follow-up.

At Statistics Canada the score function is implemented for efficiently allocating nonresponse follow-up resources. The score function identifies businesses that are most important for follow-up because of their impact on estimates (see also Section 9.6.2.4) (Berthelot and Latouche 1993; Evra and DeBlois 2007; Sear et al. 2007). The goal of this procedure is equivalent to the weighted response rate: to obtain a high weighted coverage rate. The score function is a mathematical formula that assigns a score to each business. The scores are based on the variables of interest (e.g., size, turnover, or province), the size and number of potential errors, and practical considerations. With these scores, business can be ordered in decreasing order of importance. [See Evra and DeBlois (2007) for a detailed description of this procedure.]

10.5.4 Distribution of the Response: Representativeness

After monitoring the weighted response rate, we need to monitor the representativeness of the response distribution for auxiliary variables. With the weighted response rates, we do not factor in this distribution. We can do this by applying the *R* indicator, and response–sample/population distribution comparisons.

In a survey, a sample is drawn so as to provide a representative selection of businesses with regard to a number of relevant auxiliary variables, including size, industry, location, and perhaps even target variables such as turnover and revenue. All subpopulations are represented in the sample with the same proportion as in the population. In the sample procedure, the weights of businesses may be included, to give important businesses a higher chance of being selected (see Chapter 5). The representative sample results in unbiased estimates of target variables. The distribution of the actual response, however, may not mirror the distribution in the initial sample, because of various nonresponse reasons (see Figure 10.2). The response distribution is not representative; it is skewed compared to the sample. Some businesses are overrepresented, and others are underrepresented. We also say that the response is selective (Bethlehem 2009), resulting in biased estimates.

When samples are drawn from a register, we know the population distribution of auxiliary variables in the register. We can use this information to estimate any possible correlation between response distributions and population distributions. A group of European researchers developed a *representativity* (*R*) *indicator*: a metric that gives a combined measurement of this type of comparison between response and known population characteristics (Schouten et al. 2009; Ouwehand and Schouten 2011). The calculation is based on the formula $R(X) = 1 - 2S(\rho)$, where X denotes an auxiliary variable or a combination of auxiliary variables; ρ is the response probability for X, and $S(p)$ the standard deviation to the response probability; $R(X)$ is a measure of the distance between the response distribution and the known population distribution for X. If $R(X) = 1$, the response is perfectly representative, which implies unbiased estimates; a value of 0 indicates that the response distribution is far from being representative, which would result in biased estimates. Essential in this procedure is the choice of X. The auxiliary variables ideally need to be chosen such that they relate to the main survey variables (business size, business type, etc.). This procedure can be compared to a standardized weighting procedure for nonresponse adjustment. On the basis of known auxiliary variables, we can calculate weights. The magnitude of the weights is an indication of the distance between the response and the population distribution. If the weights are large, the distributions differ considerably, and the response is selective. Adjustment weighting is applied in the postfield stage to adjust for bias caused by nonresponse [see Chapter 12; also Bethlehem (2009)]; the R indicator, on the other hand, can be used during data collection to monitor the representativeness of the response distribution. It shows how representativeness changes over time. Partial R indicators can also be calculated for subgroups. [See Ouwehand and Schouten (2011) for a detailed discussion of the use of R indicators in business surveys.]

An alternative way to monitor the selectiveness of the response distribution for auxiliary variables is to simply compare this distribution with the population distribution for these variables, using, for instance, a stratification based on these variables (as discussed above in Section 10.5.3). This becomes complex, however, for multidimensional distributions. The R indicator is a statistical procedure involving comparison of multidimensional distributions.

10.5.5 Quality of the Measurements

The relation between a representative response distribution for auxiliary variables and the bias of estimates for target variables holds only if the auxiliary variables are correlated with the target variables. If this correlation is weak, the R indicator may show that the response distribution mirrors the population distribution, but we cannot draw conclusions as to the bias of estimates. An alternative and more direct way for bias assessment is to study how the response affects the target variables by item nonresponse and using the process efficiency graph.

Businesses may respond to surveys, which would result in high response rates and even representative response distributions, but their answers to specific variables may be missing (item nonresponse) or incorrect (measurement error). Item nonresponse occurs if the questionnaire is partially completed. The item nonresponse rate $NRi(q)$ for a question q can be defined as the number of questionnaires with no answer for q relative to the total number of questionnaires returned and used. The fraction of partially completed questionnaires (P/R; see Figure 10.2) denotes the fraction of the response with one or more answers missing from the questionnaire. Since questionnaires can be long, the number of item nonresponse rates can also be quite large. We, therefore suggest monitoring the item nonresponse rate for target variables (Regan and Green 2010). To monitor these variables, we can use the same procedures as discussed above to compute unit level rates and indicators of representativeness.

A final indicator that we will discuss for real-time monitoring of response quality is the process efficiency graph (Haraldsen et al. 2010). This graph monitors the joint effect of nonresponse and measurements errors on key estimates, instead of studying auxiliary variables.

The process efficiency graph is a control graph that charts the development of the estimates for a target variable when response is coming in. We can best illustrate this with an example, as is shown in Figure 10.3 for the Norwegian Business Tendency Survey. In 2003 the response rate was 85%, and has increased to over 90% since then. This is high, even if mandatory business surveys conducted by the same institution have response rates close to 100% (Wang 2004). The key estimates calculated from this

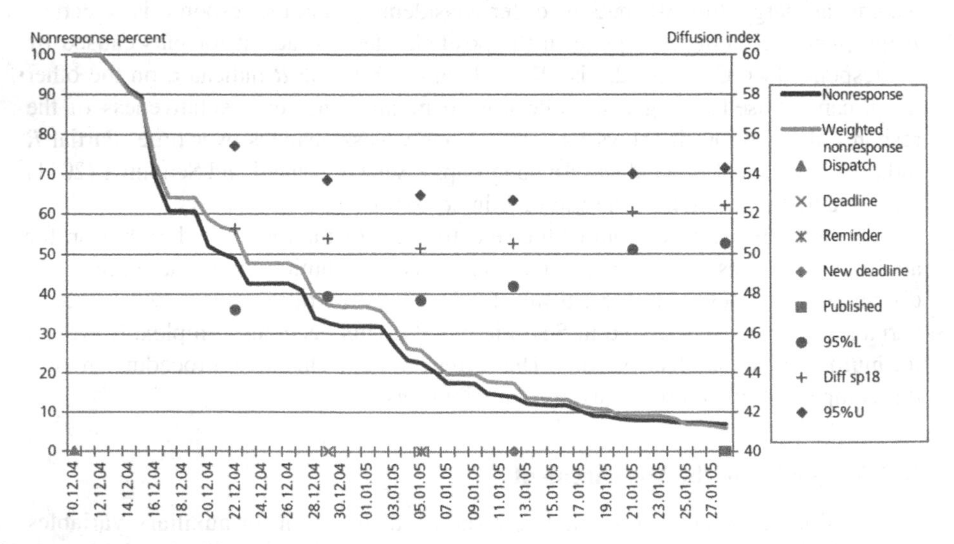

Figure 10.3 Process efficiency graph: nonresponse and general diffusion index in the Norwegian Business Tendency Survey, 4th quarter 2004.

survey are called *diffusion indices*. These indices measure how businesses judge, for instance, their production or financial result in the next quarter compared to the current quarter. In Figure 10.3 we have included the question as to how they generally judge the outlook for their enterprise in the forthcoming quarter. The response alternatives are "better," "unchanged," or "worse." The index gives the proportion that judges the prospects as better. The graph gives the nonresponse rate, the diffusion index, and the confidence interval for the index value as the data collection proceeds.

Figure 10.3 shows that even for response rates barely above 50%, the estimates indicate that a majority judge the prospects for the future to be better than in the past quarter, but it is not before the response rate exceeds 90% that we can be sufficiently confident about this conclusion. Note also that it took close to 1.5 month before this conclusion could be drawn. The confidence interval decreases as the net sample increases, and in this case, where the result is close to 50/50, we need a quite small confidence interval in order to draw a significant conclusion—or, to phrase it more generally, the need for a high response rate depends on the distribution of the true answers for the questions that we are posing.

10.6 MONITORING THE SURVEY PRODUCTION PROCESS

So far, we have discussed indicators that can be used to monitor the quality of the response; that is, the product of the survey production process. We also need to monitor the field and postfield processes in order to study the effectiveness and cost-efficiency of the survey design (Regan and Green 2010). The survey design is effective if it produces accurate, complete, and timely survey data. Monitoring the process therefore includes monitoring of the sample, the questionnaire, the data collection mode, the communication strategy, and the data processing, in relation to their error sources (using adequate paradata, as listed in Figure 10.1). The design is efficient if it is cost-efficient with regard to both internal survey organization costs and survey compliance costs for businesses. This relates to paradata on costs, resources, and response burden. Next to process management of the production process, this serves short-term and long-term improvements of the process, as well as methodological improvements of the design (as we will discuss in the next section).

To monitor the various processes in the production process, we need a variety of indicators. An overview of indicators is presented in Figures 10.4 and 10.5 [based on considerations and examples from, e.g., Aitken et al. (2003), Green (2010), Giesen et al. (2009a), Haraldsen et al. (2010), Morren (2008), Morren and Snijkers (2011), Regan and Green (2010), Renssen et al. (1998), and US Census Bureau 2011a)]. A number of these indicators can be calculated from the data collection outcome scores in Figure 10.2; others require collection of other paradata [e.g., the editing process; see, e.g., Giesen and Kruiskamp (2011)], the question completion process within businesses using audit trails or keystroke data (Snijkers and Morren 2010; Chan et al. 2011), or the project administration (cost and resource data). They can be used as process indicators during fieldwork; after fieldwork they are quality indicators with regard to the effectiveness and efficiency of the survey design. In Figures 10.4 and 10.5, key indicators are highlighted.

Sample	• **Noncontact rate** (in prefield and field stages) = (number of noncontacts due to incorrect or incomplete information) / (sample size) • **Ineligibility rate** (in prefield and field stages) = (number of ineligible units in sample) / (sample size) • **Duplicate rate** (in prefield and field stages) = (number of duplicates) / (sample size)
Questionnaire and response quality	• **Recontact rate** (during editing) = (number of units recontacted) / (number of returned and processed questionnaires) • **Edit rate for key variables** (if question is applicable) = (number of units for which variable is edited) / (number of units for which question is applicable) • Average absolute magnitude of edits for key variable = (total |edited value − original value|) / (number of units for which question is edited) • Relative magnitude of edits for key variable = (total edited value − total original value) / (total original value) • Activation rate of edit checks (per question) = number of times edit checks have been activated during questionnaire completion (audit trail data at respondent level) • Change rate (per question) = number of times an answer has been changed during questionnaire completion (audit trail data)
Dispatching of questionnaire	• **Number of questionnaires dispatched** (*Qs*) • Number of questionnaires not dispatched and no follow-up (including reasons) ($\neg Qs$)
Data collection mode	• **Mode response and return rate** (takeup rate) = distribution of response and returns for data collection modes • **Mode change rate** = number of units that received or requested another mode
Communication strategy	• **Response and return rates before first reminder** • **Response and return rates after each reminder and before the next one** • **Reminding rate** = number of units that needed one or more reminders, relative to either sample size (including nonresponse) or number of questionnaires returned • **Contact attempt rate** = average number of contact attempts per responding unit (e.g., average number of letters needed to obtain one response) • **Noncontact rate** = (number of units not contacted) / (total sample size), in telephone contacts due to no contact (voicemail, agenda problems, etc.), in mail/web survey envelopes returned to sender, or undeliverable emails • **Refusal rate** (in prefield and field stages) = (number of refusing units) / (total sample size) • Postponement rate = number of units asked for postponement relative to sample size

Figure 10.4 Effectiveness indicators of the survey production process.

Helpdesk and web site	• **Number of telephone calls, emails, web forms, relative to sample size** • Percentage of questions/complaints answered/solved • Percentage of questions/complaints for different categories: - Postponement - Assistance - Change in business information (out of scope, out of business, inactive, contact information) - Refusal - New questionnaire (in another mode) - Questions about the survey - Complaints about response burden • Problems with completion: login problems, software problems, problems with returning the data • Website visit rate = number of times websites visited (related to timing of dispatching of questionnaire and reminding)
Data capture	• Number of questionnaires received but lost in process • **Estimated scanning error rate** = rate at which answers have been misinterpreted by scanning software • **Estimated keying error rate** = rate at which answers have been keyed incorrectly into the computer
Data editing	• **Edit failure rate** = rate at which units are rejected by edit checks • Rate of manual imputation = rate at which units are manually imputed

Figure 10.4 (*Continued*)

Internal costs	• **Fieldwork costs per unit response**: contact attempt cost rate = average costs for number of contact attempts per responding unit • **Fieldwork cost per unit response, for the data collection modes** • Costs for reminding nonrespondents • Total costs for producing each survey component • Costs per unit response, based on total costs • **Control charts showing the consumption of money**
Resources	• **Control charts showing consumption of resources** (based on registration of working hours)
Actual response burden	• **Time needed to complete questionnaire** (see Chapter 6)
Perceived response burden	• **Questionnaire to measuring perceived response burden** (see Chapter 6)

Figure 10.5 Cost-efficiency indicators for the survey production process.

10.7 SURVEY EVALUATION AND QUALITY REPORT

After the fieldwork is finished, the survey process is evaluated. This includes

- An evaluation of the fieldwork process
- A customer evaluation
- An evaluation of the management of the fieldwork as a project
- An evaluation with regard to response burden

These evaluations relate to improvement goals of the use of paradata, following the PDCA cycle (as discussed in Chapter 4). As a result of these evaluations, measures for improvement are proposed: short-term process, long-term process, and/or methodological improvement. The results of these evaluations are documented in a quality report. These evaluations need to be planned in the project plan of the fieldwork (as we have discussed in Chapter 4).

The fieldwork process is evaluated on the basis of the indicators as discussed above (Sections 10.5 and 10.6). This evaluation shows whether (1) the results have been achieved according to specifications (as predefined in the project plan)—the achieved levels of the quality indicators are compared with the predefined quality levels, and (2) the fieldwork process was effective in achieving the results, and where the process can be improved (see Figure 10.4). This evaluation can be done either as a quality self-assessment, or by independent quality audits, and may be based on survey organisation quality guidelines (as we discussed in Chapter 4; see Section 4.3.6 for an overview of relevant references).

An evaluation of the survey also includes a customer evaluation. Users of the data are asked about a number of quality aspects of the deliverables. Eurostat (2011b; see Chapter 3), for instance, has defined a number of output-related quality dimensions from the user's perspective: relevance, accuracy and reliability, timeliness and punctuality, coherence and comparability, accessibility, and clarity (Ehling and Körner 2007; van Nederpelt 2009). Statistics Norway (Haraldsen et al. 2010) has developed a short customer evaluation questionnaire that is focused at the production process (see Figure 10.6). This form (1) asks for verification of whether the job was carried out according to the intended specification, (2) checks the extent of customer satisfaction with the work done, and (3) checks the extent of customer satisfaction with the contact and information given by the data collection unit.

The customer evaluation form leaves enough space to elaborate (in free text) on the different aspects of the process and the outputs. For negative evaluations, the project manager needs to ask for a follow-up meeting with the customer, to clarify the customer's experiences and expectations, and the reasons for the customer's discontent.

Next, the fieldwork project is evaluated with regard to its management (see Chapter 4). This involves an evaluation of the consumption of money, resources, and time. This evaluation focuses on the cost-efficiency of the fieldwork (see

Customer Evaluation

Project name

Main objectives

1. To what extent did the delivery from the Department of Data Collection fulfill the main objectives of the project?

- Fulfilled the objectives in all respects
- Fulfilled the objectives in most respects
- Had several shortcomings
- Did not fulfill the main objectives of the project

Please elaborate on your answer to the previous question:

2. To what extent were you satisfied with the communication you had with the Data Collection Department during the project period?

Very satisfied	Quite satisfied	Neither satisfied nor dissatisfied	Quite dissatisfied	Very dissatisfied
O	O	O	O	O

Please elaborate on your answer to the previous question:

3. To what extent were you satisfied with how we carried out our job tasks?

Very satisfied	Quite satisfied	Neither satisfied nor dissatisfied	Quite dissatisfied	Very dissatisfied
O	O	O	O	O

Please elaborate on your answer to the previous question:

Figure 10.6 Customer evaluation form.

Figure 10.5): whether the results were achieved within time and budget. The anchor for this evaluation is the project plan. As to external costs, the response burden is also monitored (see Chapter 6).

All relevant indicators (as discussed in Sections 10.5 and 10.6) are documented in a quality report on the survey production process (US Census Bureau 2011b). We suggest publishing the overall unweighted response rate (*RR*), full response rate (*RRf*), nonresponse (*NR*) rate, and return rate (*RtR*). The response rate as a function of time can be shown in control charts. Distribution of the response rates for relevant subgroups or strata can be tabulated, as can the response distribution for the various data collection modes. To show the quality of the data, the weighted response rates, and for key variables, the process efficiency graph and item response rates, can be published. In follow-up with AAPOR (2011a), we recommend not only stating these rates but also showing the formula, and including a table showing the final response outcome codes. In addition, the report should include a section on costs and response burden, indicating the total costs of the survey, the average costs per responding unit, and actual and perceived response burdens.

The quality report should also include a methodology section describing the design of the survey. This section discusses the survey design and what measures have been taken during the design stage to ensure its quality, including minimalization of survey errors, costs, and response burden (see Figure 10.1). This includes update procedures for the sample frame, an efficient sample procedure, pretesting of the questionnaire in the conceptualization–operationalization stage (see Chapter 7), tailoring of the questionnaire, applying the seven-step design approach for the communication strategy (as discussed in Chapter 9), and conducting a field pilot.

Within a total process quality management approach, based on the PDCA cycle, this information is used to improve the survey design and the survey production environment (see Figure 4.4). These measures conclude the quality report.

10.8 SUMMARY

In this chapter, we have discussed a systematic approach to managing the fieldwork of individual surveys: the active fieldwork management approach. In this approach we incorporated the responsive design approach to reduce total survey errors for critical design features. Key to active fieldwork management is the real-time collection and analysis of process data, that is, paradata. Like systematically monitoring the flight of a rocket to the moon, we proposed systematically monitoring the fieldwork process using statistically analyzed paradata. Various paradata-based indicators have been listed, and associated with survey errors and costs.

In business surveys, uncertainties threaten the ability of survey organisations to meet the predefined survey outputs, within time and budget. As a consequence, monitoring the fieldwork process with continuous updates and stability checks is of utmost importance as we also indicated in Chapter 4, when discussing risk management and quality gates (Section 4.7). In a total process quality approach (see Section 1.3.2.1), based on the PDCA cycle, monitoring and evaluation of the survey process are also aimed at identifying areas for improvement of the survey production environment (see Section 4.3.2.2).

The implementation of a process quality management system based on active fieldwork management provides the data collection unit with a number of challenges (Haraldsen et al. 2010). To monitor a single survey, the paradata need to be collected and analyzed in real time, indicators need to be easily available for project and process managers at all times during the data collection process, and the data collection process needs to be flexible in order to implement required interventions. For improvements of the survey production system, indicators from all processes need to be connected and linked. The process quality management system should be deeply rooted in the survey organization and management.

ACKNOWLEDGMENT

In writing this chapter, we would like to thank Barry Schouten (from Statistics Netherlands) for his reviews, comments, and suggestions.

CHAPTER 11

Capturing, Coding, and Cleaning Survey Data

Jacqui Jones and Mike Hidiroglou

11.1 INTRODUCTION

In comparison to other components of the survey process, the literature on capturing and coding data is sparse. In contrast, there is a whole body of literature on cleaning data. All three components are essential when designing and conducting a survey. The capture of survey data takes place in both the collection and processing phases of the survey. Data capture occurs initially during the survey field period, as survey data are entered into the data collection instrument. The phase(s) of the survey process where survey data coding and cleaning occurs (are) largely dependent on the mode of data collection. If electronic modes of data collection are used, some coding, such as respondents assigning their businesses to a standard industrial classification, may take place during the survey field period. Some data cleaning may also take place in the survey field period if the electronic data collection instrument has inbuilt validation. Data capture, coding, and cleaning then proceeds after the survey field period as data are returned to the survey organization for processing.

The connections between what is typically regarded as the *process* phase of the survey and the *collection* phase have implications in terms of the mode, questions, questionnaire, and respondent communication. These need to be considered, as these parts of the survey process are designed and built. This chapter provides an overview of capturing, coding, and cleaning survey data (see Figure 11.1) highlighting aspects that can influence other parts of the survey process.

Designing and Conducting Business Surveys, First Edition.
Ger Snijkers, Gustav Haraldsen, Jacqui Jones, and Diane K. Willimack.

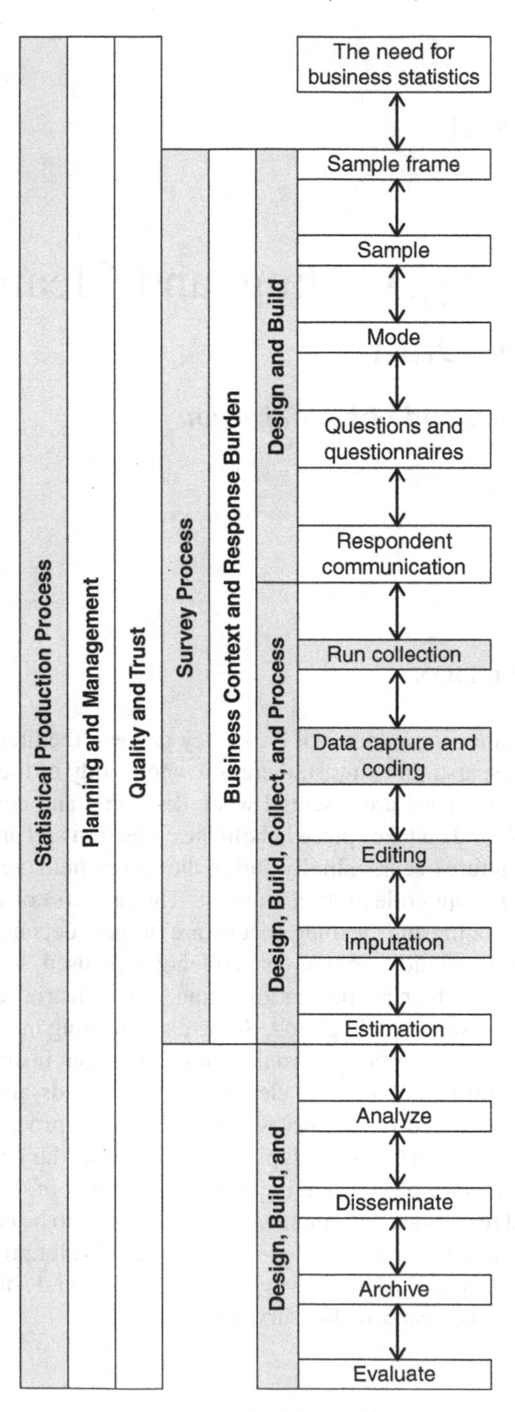

Figure 11.1 Statistical production process using surveys: data capture, coding, and cleaning.

For further information on the survey process, phases, subprocesses and components, the reader is referred to discussions of the generic statistical business process model (GSBPM) and survey process in Chapter 1.

11.2 DATA CAPTURE

The objective of this function is to "capture" responses to survey questions and transfer them into a single electronic datafile. There are various ways of capturing and transferring data. These depend on the mode of data collection and the available resources (time, money, expertise, and equipment). Traditionally, data capture focused on how survey organizations would capture the data into a datafile that was machine readable (Lyberg and Kasprzyk 1997). However, data capture starts prior to this as responses to questions are captured in the data collection instrument. The data capture process actually takes place in three steps; the first step begins with the respondent, interviewer, or recognition software, and the third step ends with data capture processing by the survey organization (see Figure 11.2); these steps impact on the design and functionality of the questionnaire, with the risk of creating different types of errors; this is explained in the text below.

11.2.1 Step One: Initial Data Capture

Initial data capture includes the physical or verbal action of recording a response to a question. Although not previously identified as a component of data capture, it has been identified as part of the response process (Tourangeau et al. 1984; Edward and Cantor 1991; Willimack and Nichols 2010). The response process models focus on the cognitive processes that lead up to and include the reporting (i.e., communication) of the response. There is therefore some overlap between the response process (collection) and initial data capture (process).

The mode of data collection determines how data are initially captured. From an initial data capture perspective, electronic modes of data collection can be categorized into those that require the data to be keyed by respondents (e.g., web questionnaires, spreadsheets, telephone data entry, or computer-assisted self-interviewing) and those that require verbal responses from respondents (e.g.,

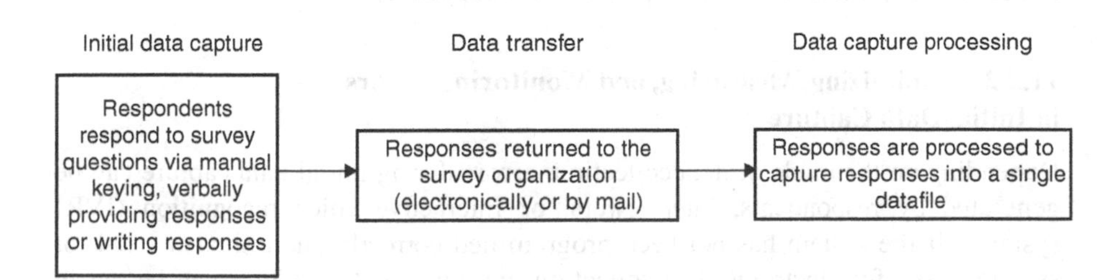

Figure 11.2 The three step data capture process.

Table 11.1 Initial Data Capture Using Different Modes of Data Collection

Data Collection Mode	Initial Data Capture by	Initial Data Capture Potential Sources of Error
Web questionnaires	Respondents	Keying
Spreadsheets	Respondents	Keying
TDE	Respondents	Keying
IVR	IVR recognition software	Software recognition
CATI	Interviewers	Mishearing and keying
CAPI	Interviewers	Mishearing and keying
Paper self-completion	Respondents	Writing

interactive voice recognition (IVR), computer-assisted telephone interviewing (CATI), or computer-assisted personal interviewing (CAPI)). The verbal category can be further subdivided into manual keying by interviewers and electronic recognition. If respondents use nonelectronic modes (i.e., paper), they will initially capture their data by writing their responses onto the questionnaire. Important features in the success of initial data capture are the design and functionality of the data collection instrument, and the instructions for responding to each question. Similarly to other components of the survey process, initial data capture is subject to potential errors; these errors may arise from keying, hearing, writing or electronic recognition. Table 11.1 summarizes how data are initially captured using different modes of data collection and potential sources of error.

For electronic self-completion modes of data collection, initial data capture also requires respondents to carry out authentication to ensure that the correct business is responding to the correct survey, and for web surveys to ensure a secure connection and check the integrity of software. In web surveys digital certificates are generally used and are regarded as solutions for securing communications and transactions (Zissis et al. 2011). Increasingly survey organizations are establishing web portals for respondents to carry out initial data capture and data transfer; these processes include registration, activation, enrolment, data entry, submission, and logout activities. For example, in 2004 Statistics Norway launched its own web portal for all its web business survey questionnaires. Despite the development of web portals Haraldsen et al. (2011) note that little usability testing has been carried out to improve web portals from the respondent's perspective.

11.2.2 Minimizing, Measuring, and Monitoring Errors in Initial Data Capture

Depending on the mode of data collection, errors during initial data capture can be generated by respondents, interviewers, or interactive voice recognition (IVR) systems. If the system has not been programmed correctly, there is also a risk of system errors for electronic data collection instruments. Procedures can be put in place to minimize the potential for errors, such as in questionnaire completion instructions, interviewer training or electronic systems reading back-keyed data and asking respondents to confirm that the data are correct.

When manual keying is used in initial data capture (e.g., web, TDE, spreadsheets, CATI and CAPI), validation can be built in to assist in minimizing errors. For example, consistency checks can be performed between responses such as an aggregate and its associated breakdown. In some electronic questionnaires (e.g., web and spreadsheets), even though aggregates can be automatically generated from the breakdowns, this does not necessarily provide a validation of the breakdowns unless respondents are asked to confirm the generated aggregate. If the business responded to the survey in a previous reporting period, current data initially captured into some types of electronic questionnaire (e.g., web, TDE, CATI and CAPI) might be concurrently checked against what was previously reported. This serves two purposes: (1) to check for potential keying errors and (2) to perform validation for cleaning the data (Peck and Maclean 2008). For example, if a response to a survey question is more than a certain percentage difference from the response in the previous reporting period, then the respondent may be prompted to confirm that the entered value is correct.

For the TDE and IVR data collection instruments, it is also possible for the electronic system to read back-captured data so that respondents can determine whether the captured data are correct. Although this is useful for minimizing keying or system recognition errors, this approach increases the length of time for initial data capture, and this can potentially lead to respondents breaking out of the process (i.e., ending the call).

For electronic respondent self-completion modes, with inbuilt validation (e.g., reading back entered data, consistency checks between questions), a measure that can be put in place for measuring and monitoring initial data capture errors is the number of data items changed, following validation. For example, in TDE and IVR systems, if responses are read back by the electronic system for respondents to confirm, then the number of instances where changes are made to responses to individual questions can be recorded. For questions with inbuilt validation, the initial data capture error rate can be formulated as follows

$$\frac{\text{Total number of responses changed for each question}}{\text{Total number of responses to each question}} \times 100$$

Monitoring the percentage of changes to specific questions can ultimately identify problems that respondents are having in using the electronic system, or where there is a potential system error. Errors generated and ignored during initial data capture will be identified and dealt with in the editing and imputation process (see Section 11.4).

For paper self-completion questionnaires there are no options for implementing interactive checks to reduce the risk of initial data capture errors. Survey organizations can only rely on the questionnaire completion instructions included on (or with) the questionnaire (see Section 11.2.8). For surveys collecting numerical values, aggregates as well as breakdowns may be requested in attempts to both minimize the risk of initial data capture errors and validate the data (i.e., that the breakdowns equal the aggregate).

An advantage of using electronic questionnaires is that it is possible to enable quality control to minimize errors at the initial data capture stage via the design and functionality of the questionnaire. In addition, for CAPI and CATI, interviewer training and monitoring can focus on initial data capture. For paper self-completion questionnaires, quality control is more difficult and generally relies on questionnaire instructions and design.

11.2.3 Step Two: Data Transfer

How data are transferred from the respondent to the survey organization depends on the mode of data collection. Data collected using electronic modes will be electronically transferred, and the focus of this process is on security generally using advanced encryption technology. In contrast, data collected using paper questionnaires will be transferred to the survey organization either using the postal service or by respondents faxing their questionnaires. These transfer methods also pose security issues, but the general perception is that the postal service and faxing are more secure than electronic transfer; whether this is actually the case is an interesting research question. The focus for postal data transfer is usually on the timeliness issue.

On receipt in the survey organization, regardless of mode, the transferred data must be identified in terms of which survey they relate to and which business they are associated with; the former is more relevant for survey organizations that undertake multiple business surveys during the same time period. The key to the identification of the survey and business is the use of unique survey-specific identifiers, namely, barcodes and unique reference numbers. Barcodes are generally used for paper self-completion questionnaires that are either scanned or manually keyed, and unique reference numbers used for electronic modes. The unique survey-specific identifiers ensure that the data are captured and saved to the relevant survey, with reference to the correct business. They are also required to update central file(s) of which businesses have responded so that reminders are sent only to those businesses that have not responded. There is, however, often a time overlap between reminders being sent out and questionnaires still being returned; this is especially the case for paper questionnaires. In many national statistical institutes (NSIs), business surveys attempt to overcome this time overlap by including in the reminder letters a note to ignore the reminder if the business has already responded to the survey.

If a questionnaire is scanned, the barcode is read during the scanning process and the captured data are saved according to the survey and business identification held within the barcode, and the central response file is updated. If a questionnaire is manually keyed, the barcode is generally read separately by a barcode recognition tool prior to manual keying. At UK ONS, approximately 4 in every 100 questionnaires cannot be identified from the questionnaire barcode; these are sent for manual identification of the survey, which is taken from the reference codes on the questionnaire.

Unique survey-specific reference numbers are sent to businesses electronically or via the postal system. Respondents enter their unique reference number when logging into the electronic system (web, TDE, IVR); or an interviewer who is collecting the data will enter the reference number at the start of the interview. A point to note is that these unique reference numbers are often long and ideally should be presented to respondents or interviewers with spaces between the digits to reduce the risk of keying the incorrect reference number (e.g., 234 123 345 456, not 234123345456).

During the data capture process it is essential that the survey organization manage the process, including safeguarding survey responses and associated returned materials. Smith et al. (2007) refer to this as *inventory control.*

11.2.4 Step Three: Data Capture Processing

Data capture processing takes place when the data are returned to the survey organization and involves transferring the returned data into a datafile that can be interpreted by a computer (Lyberg and Kasprzyk 1997). The returned data, depending on the mode of data collection, will be in either paper or electronic format. Data capture processing methods depend on the data collection mode (see Table 11.2). With the exception of data collected using IVR, data collected via electronic modes are already in the form of alpha characters or numerical values, and each returned case is electronically transferred to a server to be added to the survey datafile or data warehouse. In contrast, data collected using IVR will be returned to the survey organization as verbal recordings, transferred to a server, and then translated into alpha characters and numerical values. The translated data are then added to the survey datafile or data warehouse. Data returned on paper questionnaires need to be captured electronically either by manual keying, electronic scanning, or a combination of the two. The electronically captured data are then added to the survey datafile or data warehouse. Data capture from surveys using multimode data collection pose additional challenges for survey organizations such as the ability to concurrently process, integrate, and store data from multiple modes.

The different data capture processes can potentially create different types of errors. Table 11.3 displays potential sources of error in initial data capture and data

Table 11.2 Mode of Data Collection and Data Capture Processing Methods

Data Collection Mode	Data Capture Processing Method
Web Spreadsheets TDE CATI CAPI	Data are transferred to a server and ideally are loaded into an integrated data management system (e.g., a data warehouse); raw data are electronically combined into a datafile
IVR	Data are transferred to a server; verbal recordings are translated to alpha characters and numerical values using speech recognition; translated data are electronically combined into a datafile
Paper questionnaires	Manual keying, electronic scanning, or combination thereof

Table 11.3 Initial Data Capture and Data Capturing Processing Potential Sources of Error

Data Collection Mode	Initial Data Capture by	Initial Data Capture Potential Source(s) of Error	Data Capture Processing by	Data Capture Processing Potential Source(s) of Error
Web, spreadsheets, TDE	Respondents	Keying	Electronic integration	System specification
CATI, CAPI	Interviewers	Mishearing and keying	Electronic integration	System specification
IVR	Recognition software	Software recognition	Translating verbal responses to characters and values; and electronic integration	Translation and system specification
Paper self-completion	Respondents	Writing	Keying, scanning, or a combination of keying and scanning Electronic integration	Keying, scanning, system specification

capture processing. Note that there are more potential sources of error for data collected using paper self-completion and IVR than for any other modes. The modes with the less potential sources of errors are the electronic respondent self-completion (i.e., web, spreadsheets, and TDE).

The following sections focus on capturing data from paper questionnaires as this mode requires "actual" data capture processing to capture the data from the paper questionnaire into an electronic format.

11.2.5 Capturing Data from Paper Questionnaires

Although many NSIs nowadays use electronic modes of data collection [e.g., Statistics Norway, Netherlands Central Bureau of Statistics, Statistics Canada, and US Census Bureau use web questionnaires; UK Office for National Statistics (ONS) uses TDE], there remains a heavy reliance on paper self-completion questionnaires in business surveys. Traditionally nearly all business surveys used paper self-completion questionnaires, and data capture processing was highly resource-intensive, expensive, and prone to errors as data were manually sorted, counted, and tallied. For some of these reasons there was a move to implement less manual data capture processing systems such as the 1916 introduction of punchcards in the Dutch Central Bureau of Statistics (Van Den Ende 1994).

Until the mid-1970s, at the ONS, all business survey data were manually keyed from the paper questionnaires and captured by punching holes onto paper tape. A keying operator would take a batch of questionnaires and data-punched (via keying) onto the paper tape. As keying operators were trained to read the punched paper tape to check for errors, a second operator would then validate the paper tapes and note any errors. Any errors would be dealt with by the supervisor. As each operator had a unique operator number, their work could be identified to them, so there was a process of feedback to assist in continuous quality improvement. When paper tapes had been punched and checked, they were wound onto spools and taken to the magnetic tape library (old mainframe).

The mid-1970s then saw the introduction of visual display units (VDUs) as a replacement for paper tapes. This development used the strokes from the keyboard to input the data to disk; the data were then loaded onto the mainframe computer system. Capturing the volume of data returned involved approximately 80 keying operators who had to average 11,000 key depressions per hour; if this was exceeded and 16,000 key depressions/hour rate reached and maintained, the operators would receive a bonus. All data items were double-keyed: when the data were keyed the first time by a keying operator, the questionnaire was marked with a P (punched); when independently keyed a second time, by another keying operator, the questionnaire was marked with a V (verified). During the second keying, P – V differences would be flagged by the system for manual confirmation inspection; this is known as *independent data confirmation*. When successfully passed through the double-keying process, the questionnaire would be marked with PV (punched and verified) along with the operator's names.

In the late 1990s, ONS introduced electronic scanning of data; this was a major advance and reduced the time it took to capture data and the number of people involved in the process. Nowadays five Kodak scanners are used with `OCR_for_Forms` software. The actual process involves the software taking an image of each page of the questionnaire, which then uses the questionnaire identifier number, to identify which specific survey the questionnaire is intended. Once the survey is identified, the returned data are located on each page and the recognition engine captures data that it can correctly interpret, within prespecified confidence levels. At ONS, the confidence level is determined by testing but is generally set at around a 96% confidence level (i.e., the recognition engine will capture data that it is 96% confident that it recognizes). The confidence level has to be carefully balanced between the need for minimizing the amount of manual keying and minimizing errors. Data that cannot be correctly interpreted are flagged and sent with the relevant image to one of the 13 keying operators for manual verification. The advances in data capture processing has reduced the time and expenditure on data capture processing and increased the capability to process more surveys; currently at ONS approximately 100 different business surveys are carried out each year, and this translates into sending out approximately 2 million questionnaires annually.

The advance to electronic scanning did, however, raise an issue for the highly efficient keying operators at ONS when new desktop computers were purchased. Although these desktop computers had standard keyboards, in the additional numerical keypad the keys were in a position reverse to that used for years; the numerals 7, 8, and 9 appeared at the top of the new numerical keypads, in contrast to

1, 2, and 3 appearing at the top of the traditional numerical keypads. To overcome this issue and maintain keying efficiency, a software package was introduced that enabled use of the keypad in either the traditional or new numerical format. Each operator could specify a particular orientation; this method is still used today as many of the keying operators were trained with the original keypads.

The data capture process includes opening the envelopes containing the returned questionnaires and preparing the questionnaires for processing. Over the years this process has been streamlined, and the physical movement of questionnaires minimized. In the days of data capture onto paper tapes, returned questionnaires would first be taken to the survey results section, where people would open the envelopes and separate the questionnaire into survey-specific batches. These batches would then be taken to the processing center for the data to be punched onto paper tapes and loaded onto the mainframe. The questionnaires would then be returned to the survey results section for archiving. Nowadays, this all takes place in the survey processing center, where data capture also takes place. This development not only expedites the process but also safeguards against loss of questionnaires.

11.2.6 Electronic Scanning

Electronic scanning of paper questionnaires has vastly improved with the move from optical mark recognition (OMR) to optical character recognition (OCR) and intelligent character recognition (ICR) (see Table 11.4). The key improvements were the ability to not only recognize when the questionnaire had been marked (OMR) but with OCR, to also recognize individual numbers and characters, and with ICR recognize and convert handwritten text and numbers. These improvements have led to many survey organizations implementing electronic scanning to reduce the costs and time required for data capture processing.

A number of steps need to be undertaken to electronically scan and capture data from paper questionnaires (Mudryk et al. 2004; Rosen et al. 2005; Smith et al. 2007; Nguyen et al. 2008). They are divided into two main categories: (1) development and testing, and (2) scanning and capturing the data (see Table 11.5).

Nowadays, many NSIs use electronic scanning combined with manual keying to collect and capture business survey data from paper questionnaires. For example, for the 2002 and 2007 US economic censuses data returned on paper questionnaires were captured using a combination of OMR, ICR, and manual keying (Studds 2008).

11.2.7 Design of Paper Self-Completion Questionnaires for Electronic Scanning

Dillman (2000) provides an overview of some of the historical constraints placed on the design of paper self-completion questionnaires that were electronically scanned to capture data (e.g., questions printed on only one side of the paper; response categories on only one part of the page, e.g., the right-hand column; minimizing the number of questionnaire pages, leaving fewer pages to scan; including questionnaire markers and identification for scanning purposes). Although some of these historical

Table 11.4 Differences between OMR, OCR, and ICR

Method	Description
Optical mark recognition (OMR)	OMR looks for marks in a predefined location. The earliest use of OMR was for paper tape (in 1857) and punchcards (created in 1890). Nowadays OMR uses an optical scanner or mark reader to look for the marks in a predefined location on the questionnaire. OMR can detect the mark but cannot identify what the mark is, as there is no recognition engine. OMR is best used for discrete data, with predefined response categories. Accuracy rates are typically high (~99.8%), and more accurate than manual keying (Smith et al. 2007). OMR employs special paper, special ink, and/or a special input reader.
Optical character recognition (OCR)	Through the recognition engine, OCR translates scanned printed character images into machine-readable characters (ASCII). It looks at individual characters rather than whole words or numerical amounts. Accuracy rates are ~80% (Smith et al. 2007). It therefore reduces the cost of data capture processing by reducing the amount of manual keying. OCR uses a pattern recognition engine, requiring clear contrasts between completed responses and the paper background.
Intelligent character recognition (ICR)	More recently the term ICR has been used to describe the process of interpreting image data, in particular alphanumeric text. ICR recognizes and converts handwritten characters to machine-readable characters. ICR is generally contained as a module of OCR and can provide real-time recognition accuracy reports.

Source: United Nations (2001).

constraints are no longer required, with the development of OCR and ICR, some still remain. For example, many survey organizations face the challenge of publishing statistical results as quickly, efficiently, and cheaply as possible; this continues to place constraints on the number of pages allowed for a questionnaire. Markers and identifiers used for scanning purposes are often still included on questionnaires. Both of these points remain valid at ONS. In relation to the inclusion of markers and identifiers on questionnaires, an ONS initiative was put in place to try to minimize their effects from the respondent's perspective. Figure 11.3 illustrates how the design of response box unique identifiers has changed with the objective of deemphasizing information on the questionnaire that is needed for survey processing but not by respondents. The older ONS design of response identification codes printed the code between the question and the response category box, thus theoretically breaking the link between the question and the response box. The newer design places response

Table 11.5 Required Steps for Electronically Scanning and Capturing Data from Questionnaires

Development and Testing	Scanning and Capturing the Data
Designing the questionnaires to optimize scanning recognition (e.g., boxes for individual alpha and numeric characters, barcodes to identify the survey, ink color to enhance scanning, markers for image alignment)	Preparing the questionnaires (e.g., questionnaires unfolded, staples removed, respondent-attached notes removed, e.g., attached notes with comments and compliment slips
Creating a template of the questionnaire that individually defines each data item that will be captured (e.g., whether fields are alpha or numeric, number of characters per field, page location of each data field, order in which to capture the fields)	Sorting questionnaires into batches Scanning questionnaires and implementing quality control parameters Flagging and manually dealing with suspect data items Electronically exporting the scanned data Implementing procedures to detect paper jams (which can stop the scanners) and dust (which can cause data capture errors)
Defining confidence levels for individual fields; or developing logical edits	
Testing the questionnaire template for electronic scanning and data capture, including quality control parameters	

identification codes to the right-hand side of response boxes, and they are deemphasized as much as possible using a smaller font size. Interestingly, however, data validation staff often use these response identification codes, instead of question numbers, when talking to respondents; this was one of reason for keeping the identification codes on the questionnaires.

The use of specific ink color (dropout color) is another constraint that is often placed on the design of questionnaires that are electronically scanned. At ONS,

Old ONS design—question—response id code – response box

Catagory		Total Value of Acquisitions		Total Value of Proceeds from Disposals	
Textiles and soft furnishings (see note 1)	1700	000	28	000	29

New ONS design—question—response box – response id code (deemphasized)

11. How many GB sites, in total, does this business operate from?
For example, a workshop, factory, warehouse, office, mine, depot, or constructoin site where one or more persons carry out economic activity for a minimum of 20 hours per week <u>and</u> the planned activity for your business is for more than one year.. 90

Figure 11.3 Response box unique identification codes on ONS paper questionnaires. © Crown copyright (2012).

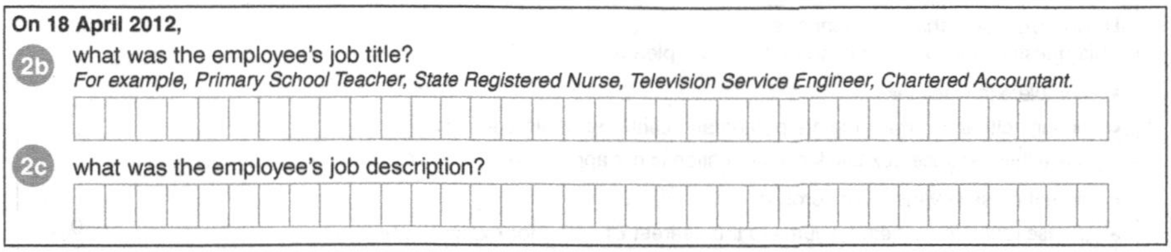

On 18 April 2012,

2b what was the employee's job title?
For example, Primary School Teacher, State Registered Nurse, Television Service Engineer, Chartered Accountant.

2c what was the employee's job description?

Figure 11.4 An example of individual alpha character boxes. © Crown copyright (2012).

dropout color technology improves the scanning recognition of data entered by respondents, reduces the amount of data referred for manual keying, and reduces the amount of electronic storage space. Dropout color technology results in only required discrete fields left on the scanned questionnaire image (e.g., questions and responses), as any text or shape in the dropout color will be dropped out of the image. This reduces the rate of scanning rejections. For example, if dropout color technology is not used and a returned data item touches the line of the response box, the response will be unidentifiable and will be flagged for manual intervention. As dropout color technology drops out the response box lines, a data item touching the response box can still be identifiable. The constraint placed on questionnaires using dropout color technology is that the dropout color be red, green, or blue.

Individual alpha and numerical character boxes are another historical constraint that remains (see Figure 11.4 for an example), whether using OCR or ICR, as the electronic scanning system cannot recognize freehand text, although it can take an image of it. For alpha characters in particular, this can cause problems during initial data capture as respondents find it unnatural to enter each individual character of a word into a separate box. Ensuring that there are enough individual character boxes for the maximum likely response is also challenging from a design perspective. If the text is likely to be longer than one row, moving from one row to the next can result in empty character boxes, which may be captured as spurious spaces.

11.2.8 Instructions for Completing a Paper Self-Completion Questionnaire to be Scanned

Dillman (2000) also provides some historical information on constraints placed on respondents completing scannable paper self-completion questionnaires (e.g., instructing respondents to complete the questionnaire using a soft-lead pencil, coloring in circles or ovals instead of ticking or crossing). These types of questionnaire completion constraints are still evident today, although they have improved as scanning has become more sophisticated. In 2002, ONS business survey questionnaires included a half to a whole page of instructions just on how to complete the questionnaire as it was to be scanned. During 2003, internal research was undertaken with ONS staff in the Survey Processing Centre to identify common problems associated with scanning the questionnaires and use this information to improve and

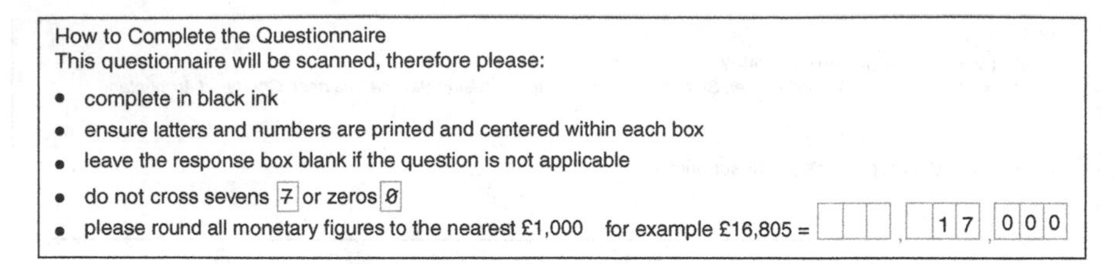
How to Complete the Questionnaire
This questionnaire will be scanned, therefore please:

- complete in black ink
- ensure latters and numbers are printed and centered within each box
- leave the response box blank if the question is not applicable
- do not cross sevens 7 or zeros 0
- please round all monetary figures to the nearest £1,000 for example £16,805 = , 1 7 , 0 0 0

Figure 11.5 Example of the shortened questionnaire instructions. © Crown copyright (2012).

shorten the instructions. The research revealed that the most prevalent problems, when scanning the questionnaires, were caused by respondents:

- Ticking response categories when they were required to insert crosses (crosses improved the ability to capture the correct marked response box)
- Putting a line through the whole page if the questions were not relevant to them; this often meant that part of the line would go through a response box and be captured as a data item.
- Using red ink, when completion in black ink was required as it improved the ability to capture the correct marked response box.
- Crossing sevens and zeros, which made it difficult to identify the number.
- Entering £ thousands values when thousands (000) were already preprinted.

Further research with respondents was undertaken to identify the key information that respondents needed to avoid these problems when completing the questionnaire. The outcome of the research resulted in the questionnaire completion instructions reducing (see Figure 11.5).

The new instructions successfully minimized the occurrence of most of the common problems listed above. However, the issue of crossing and not ticking boxes remained, and increased the risk of scanning errors and ultimately data accuracy. The real issue here was that it was more natural for respondents to tick rather than cross their chosen response categories. In 2003, an experiment was carried out on different treatments for instructing respondents to cross and not tick response boxes. The experiment was carried out using the e-commerce survey questionnaire, which included many questions with yes/no response options. Two treatment groups and a control group were designed:

Control group — Standard instructions, asking respondents to cross response boxes at the beginning of the questionnaire, with no shaded cross boxes

Treatment group 1 — Instruction to cross response boxes at the top of each page, with no shaded cross boxes

Treatment group 2 — Instructions to cross response boxes at the beginning of the questionnaire, with shaded cross boxes

The results of the experiment showed that

- Control group respondents with instructions at the beginning of the questionnaire with no shaded cross boxes would remember to cross response categories for the first page. Yet when they turned the page, they would generally completely forget the instructions and start ticking.
- Treatment 1 respondents with instructions to cross response boxes at the top of each page, with no shaded cross boxes, would remember to cross response boxes for the first few questions on the page. They would then as they moved down the page either completely forget that they had to use crosses or would remember partway down the page, and then go back up the page and cross their ticks.
- Treatment 2 respondents with instructions to cross response boxes at the beginning of the questionnaire, with shaded cross boxes, had no issue in remembering to cross their selected response boxes. So, eventually we had found a solution that worked—shaded cross boxes: X

11.2.9 Minimizing, Measuring, and Monitoring Errors in Data Capture Processing

It is important to measure and monitor process quality during the data capture process, as substitution errors can be generated from either electronic or manual capture. A substitution error is generated when a returned alpha character or numerical value is incorrectly interpreted, and consequently an incorrect character or value is captured; this is known as a *false positive*. Substitution errors therefore introduce errors into the data. To reduce the risk of substitution errors, it is imperative that measures for monitoring and controlling the data capture process be designed and implemented (Lyberg and Kasprzyk 1997; Rosen et al. 2005).

Minimizing errors in the data capture processing stage centers around the ability to capture data with some degree of confidence. For paper self-completion questionnaires, this starts at the document preparation stage, regardless of whether the data are to be captured manually or with a combination of electronic and manual methods. The document preparation stage is largely a manual process with the objective of minimizing substitution errors. It generally involves opening the envelopes of returned questionnaires, receipting them, and sorting them (e.g., if automatically capturing the data, removing illegible questionnaires for manual capture or for transcription onto new questionnaires; batching the questionnaires into a logical order for processing; removing staples or questionnaire booklet spines; and checking that barcodes are intact) (Mudryk et al. 2004; Rosen et al. 2005).

Quality controls for the document preparation stage usually consists of checks on, for example

- The percentage of questionnaires containing the correct number of pages:

$$\frac{\text{Number of questionnaires containing the correct number of pages}}{\text{Total number of questionnaires}} \times 100$$

- The percentage of questionnaires with the correct identifiers (e.g., barcodes):

$$\frac{\text{Number of questionnaires containing the correct identifiers}}{\text{Total number of questionnaires}} \times 100$$

Ideally, to promote continuous quality improvement, the outcomes from the quality controls should be recorded and fed back to the person or people undertaking the document preparation, or it may be an issue for the design of the questionnaire.

To monitor and control substitution errors, confidence levels are set in automatic capture systems; these are the levels of confidence that the system has in interpreting the data. Each data item might have an associated confidence level set. This depends largely on the fact that different types of data (e.g., alpha, numerical; tick or cross boxes) will have different rates of recognition. For example, it is easier to automatically recognize tick or cross boxes, and typed responses compared to written responses. Different calibration checks, with predefined tolerance limits, should therefore be used to cover the different types of reported data that the survey or surveys include. Data items with recognition confidence levels below a predefined confidence level should be sent for manual verification and keying (Mudryk et al. 2004). The United Nations (2001, p. 71) defines the *confidence score* as a "number (usually between 0 and 999) assigned by the recognition engine as to the confidence of the recognized character".

As well as substitution errors, rejection errors can be introduced during data capture processing; these occur when the system cannot recognize the returned data and it is passed for manual intervention. Whereas substitution errors introduce errors, rejection errors theoretically should not introduce errors unless the manual intervention does not correctly key the data (Lyberg and Kasprzyk 1997; Rosen et al. 2005). The United Nations (2001) defines the *rejection rate* as the proportion of data that are rejected (due to low confidence or excessive length or basic edits). Rejection rates increase as the required level of confidence increases.

Scanning calibration checks are carried out to minimize the risk of substitution errors. The scanning of questionnaires creates an image of the questionnaire, which is then either processed by the system (automated data entry) or manually keyed from the image; the latter approach usually applies for poorer-quality images. The scanning system needs to be regularly calibrated and maintained to optimize the scanning process. This largely involves spotchecks during the course of the scanning process. The spotchecks focus on calibration settings and the quality of the images. Calibration settings will typically be checked 2–4 times during the day. The settings should be checked by comparing a batch of scanned test questionnaires with a batch of test questionnaires that have been scanned and manually verified. Differences between the actual and expected results are then calculated; any discrepancies represent the percentage of substitution error. High rates of substitution error should result in scanning ceasing until recalibration is successfully undertaken and shown to reduce, within predefined tolerance limits, the substitution error rate (Mudryk et al. 2004).

Another way to minimize substitution errors is to identify key questions where no substitution error can occur. These may be questions that are particularly important in the survey or written responses to open-ended questions, as these may be subsequently coded and any errors will make coding more difficult. To minimize the risk of substitution errors from manual keying, responses to key questions should be 100% independently verified. Ideally, this would be carried out by independent double manual keying by two different keying operators; any discrepancies between the double keying would then be passed for manual review (Mudryk et al. 2004; Nguyen et al. 2008).

Additionally, statistical acceptance sampling might be used for all data items in the survey. This involves agreeing a sample design for either dependent (reviewing processed data items) or independent (repeating the processing of the data items) verification of processed data items (Minton 1972).

Quality control is an essential activity when using electronic data capture. Experiences at Statistics Canada have led to the identification of three key stages where quality control is required: (1) *document preparation*—at this stage the objective is to ensure that only readable documents enter the scanning process; (2) *scanning calibration setup*—to maximize and maintain the scanning and recognition process; and (3) *capture of survey documents*—to maintain output quality. There is also an acknowledgment of dependences between these three stages, thus increasing the risk of errors (Mudryk et al. 2004).

If undertaking manual data capture, either of selected data items or the whole questionnaire, it is still essential that the accuracy of the keyed data be checked. This can be carried out either independently by rekeying some of the questionnaires and identifying substitution errors or by rechecking keyed data against some of the completed questionnaires. Any errors, whether systematic or random, need to be investigated and corrected.

Having put so much effort into minimizing errors when designing, building, and running the survey, the last thing you want to happen at the data capture stage is to capture incorrect data. This is why it is so important to have a well-planned data capture process with quality controls and checks.

Further Reading

Useful references for further information on data capture are Nguyen et al. (2008), Studds (2008), and United Nations (2001).

11.3 DATA CODING

Responses to different types of business survey questions collect different types of data; for example:

- Responses to financial questions provide numerical currency responses:

2.	**What were your holdings of the following liquid assets:**	**Balance at the End of the Quarter, Millions of £**	
	(a) Cash? Include notes and coins of all denominations held on the premises. . . .	£ ☐☐☐☐☐☐ . ☐	1000

- Responses to whether a business has or has not undertaken a particular type of activity or has access to a particular system or process can provide qualitative data (if asked as an open-ended question) or provide yes/no responses (if asked as a closed question); for example:

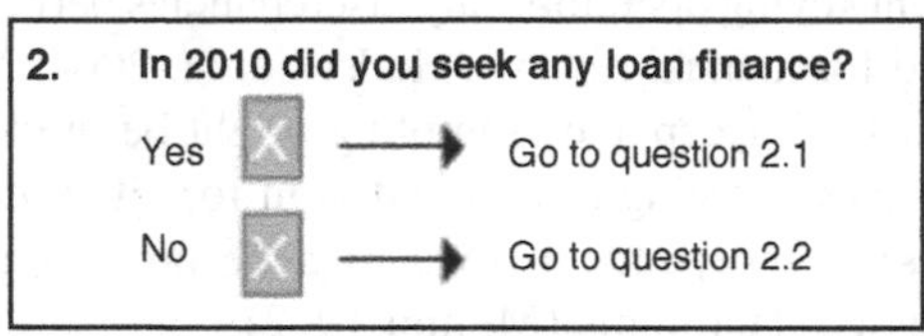
2. In 2010 did you seek any loan finance?
Yes ☒ → Go to question 2.1
No ☒ → Go to question 2.2

- Responses to open-ended questions, for example, to classify the business to an industrial classification, can provide quantitative responses if respondents are self-coding in an electronic questionnaire, or provide qualitative descriptive responses; for example:

5. If the description above is incorrect, please describe the business activity of this site in the boxes below.
For example:
- Manufacture of UPVC windows
- Licensed hotel and restaurant

Ensure letters are printed and centred within each box. Leave a space between each word. 110

The objective of data coding is to code collected data so that they can be summarized and analyzed. The coding approach will vary with the type of data collected and the objective of your analysis. For *numerical data*, you may wish to recode into more manageable groups, such as recoding into bands, responses to the number of employees in the business (e.g., from 0 to 5, from 6 to 10, from 11 to 20, from 21 to 30); for *categorical data*, you will need to recode categories into numerical values (e.g., coding a "yes" response to 1 and a "no" response to 0); or for *qualitative data* (e.g., descriptions), you will need to recode into numerical codes (e.g., responses to industrial classification questions into classification codes). In summarizing and analyzing the data, each numerical code will then represent a particular response or group of responses. To carry out data coding you will need to either develop a coding list (nomenclature) or use an existing one (de Vaus 1991).

For business surveys there is often a need to code, to existing nomenclature, businesses to industries, employees to occupations, and the products produced, exported, or imported; examples of existing nomenclature are the:

- International Standard Industrial Classification (ISIC)
- Nomenclature génerale des Activités économiques dans les Communautés Européennes (NACE)

- Central Product Classification (CPC)
- Classification of Products by Activity (CPA)
- Products of the European Community (PRODCOM)
- Standard International Trade Classification (SITC)
- Standard Occupation Classification (SOC)

To classify the activities of a business there are both international and national industrial classifications. At the international level there is the United Nations system for classifying economic data, known as the International Standard Industrial Classification of All Economic Activities (ISCI); at the European level there is the Classification of Economic Activities in the European Community (NACE); and for North America, there is the North American Industry Classification System (NAICS), which in 1997 replaced the Standard Industrial Classification (SIC). In addition, there are also national classifications, for example, the UK Standard Industrial Classification (UK SIC). All these classifications are hierarchical systems. At the aggregate level there are no differences between the classifications, but there are at the detailed level. NAICS groups together economic units that use the same or similar processes to produce goods or services. It is therefore a production–oriented system, unlike SIC, which groups both demand-based and production-based economic units together. NAICS is a six-digit system in contrast to SIC, which is largely a four-digit system. Table 11.6 summarizes the different classification systems. Periodically the industrial classifications, and other classifications, are updated to reflect changes in industries, products etc. (see Section 11.3.4). This is a time-consuming activity, requiring considerable coordination across countries and input from coding experts, statisticians, and subject matter experts (Lyberg and Kasprzyk 1997). For example, in 2002 the NACE system was updated. While retaining a mapping to ISCI at the first and second levels, the third and fourth levels were subdivided to reflect European needs.

A business does not necessarily map to only one industry. For example, supermarkets have moved from exclusively selling food to also selling clothes, insurance, and fuel, providing banking and credit cards, and having restaurants inside their stores; this all impacts on how the business is classified to an industry. The industry classification of a business can be affected by how the business classification is coded (i.e., respondents, automatic coding, expert coders) and whether a "sales" or "value-added" approach is followed (see Chapter 5, Box 5.1).

Table 11.6 Comparison of Different Industrial Classifications

Code	ISIC	NACE	UK SIC	NAICS
Letter	Section	Section	Section	None
2-digit	Division	Division	Division	Industry sector
3-digit	Group	Group	Group	Industry subsector
4-digit	Class	Class	Class	Industry group
5-digit	None	None (national versions only)	Subclass	Industry
6-digit	None	None	None	National-specific

Nowadays, data coding of qualitative text responses can be undertaken manually, automatically, or by a combination of the two. Traditionally all data coding was undertaken manually, generally by classification experts. However, manual data coding has three key disadvantages: (1) it is expensive, (2) it is time consuming, and (3) it can introduce systematic errors. Manual data coding also requires the expertise and knowledge of the coders to be maintained, so they make the correct judgments during coding. This requires coding training and often years of experience (Lyberg and Kasprzyk 1997). In many NSIs the majority of text coding is now undertaken automatically; manual intervention occurs only if a direct match is not found.

Some NSIs have a dedicated classification unit. For example, ONS has a classification unit that deals with industrial and occupational classifications. The people who work in this unit have expert knowledge of these classifications, and when updates are being carried out, they are generally involved in the groups undertaking the work. For example, they were involved in the development of the 2007 UK Standard Industrial Classification (SIC07). They are also responsible for answering external classification queries.

11.3.1 Coding Descriptions to Classifications

Responses to classification questions such as industry, occupation, and product can be returned as alpha fields with varying lengths. These responses, which are generally descriptions, are then coded to the nomenclature (coding list). For industrial classification coding, the coding can be at a one-, two-, three-, four-, or five-digit level (if using NAICS can be coded to a six-digit level). Table 11.7 provides an example, down to the four-digit level of the manufacture of textiles in the manufacturing industry as classified by the UK Standard Industrial Classification (SIC07). To code description responses, keywords need to be contained within the alpha fields, and it is these keywords that are used to allocate the description to a classification code. There are, however, problems with this approach as the keywords can relate to more than one category in the classification; for example, the manufacture of carpets and rugs (13.93) might also include the preparation and spinning of textile fibers (13.10). The challenge is to get respondents to provide detailed descriptions and designing questionnaires, questions, and response boxes that will facilitate this.

11.3.2 Questionnaire Design Implications

The key to collecting relevant and accurate responses to classification questions is the wording and design of the questionnaire (Perry 2007). Ultimately the question(s) should be worded to elicit a description that will provide all the relevant information so that a code can be assigned. Combined with this, sufficient space must be provided for the respondent to enter a detailed description. This is a particular issue for self-completion paper questionnaires. If combined with electronic data capture, respondents will be expected to enter each alpha character in its individual character box. For some types of capture systems, respondents will also be required to wrap words

Table 11.7 2007 UK SIC for Manufacture of Textiles in the Manufacturing Industry

1-Digit Level	2-Digit Level		3-Digit Level		4-Digit Level	
	13	Manufacture of textiles	13.1	Preparation and spinning of textile fibers	13.10	Preparation and spinning of textile fibers
			13.2	Weaving of textiles	13.20	Weaving of textiles
			13.3	Finishing of textiles	13.30	Finishing of textiles
			13.9	Manufacture of other textiles	13.91	Manufacture of knitted and crocheted fabrics
					13.92[a]	Manufacture of madeup textile articles, except apparel
					13.93[a]	Manufacture of carpets and rugs
					13.94	Manufacture of cordage, rope, twine, and netting
					13.95	Manufacture of nonwovens and articles made from nonwovens, except apparel
					13.96	Manufacture of other technical and industrial textiles
					13.99	Manufacture of other textiles n.e.c.[b]

[a]These categories have a fifth level of classification.
[b]Not elsewhere classified.

from one line to the next. Neither of these provides a natural way for respondents to write (see Section 11.2.7).

The use of web questionnaires can shift the process of data coding from the survey organization to the respondent, as respondents can self-select their codes rather than providing a description. This provides a new twist to quality control, as survey organizations will not have the detailed descriptions and will be unable to carry out verification activities unless respondents are recontacted for further clarification; this potentially increases respondent burden. An example of respondent self-classification is the UK system for company registration, where respondents select their own industrial classification from a copy of the standard industrial classification. The results from this self-classification show a higher incidence of coding to property and business services compared to internal coding from business descriptions (Perry 2007).

In Canada, a partnership between Statistics Canada and the Canada Revenue Agency (CRA) has helped to overcome the potential issue of respondents not providing enough detail to code industry descriptions. New businesses have to apply for a business number from the CRA; this can be done in person, over the phone, or on the Internet. Part of the business number application process involves providing a business activity description, which is then sent to Statistics Canada for industrial classification coding. To improve the collection of descriptions, a decision tree tool has been developed. The reported description is entered into the decision tree tool either by the respondent (over the Internet) or by someone in the CRA if the respondent is responding in person, or by telephone. An algorithm then identifies keywords and prompts the respondent or CRA person for further details before an industry code can be assigned, for example, details of the activities that are undertaken (Beaucage 2007).

11.3.3 Two Examples of Data Coding Tools

As mentioned earlier, data coding can be undertaken manually, automatically, or using a combination of the two methods. If a combination approach is used, the manual coding is usually carried out for cases in which automatic coding has failed or where the automatic coding is below a predefined confidence threshold. This section provides two examples of automatic coding tools that can be used for coding responses to industry classification questions.

At ONS, automatic coding tools have been used since the early 1990s, progressing from text searching to precision data coder software (which relies on linguistic engineering techniques) onto automatic coding by text recognition (ACTR) (Perry 2007). The original philosophy for ACTR was developed at the US Census Bureau (Hellerman 1982), and the matching algorithms were then developed at Statistics Canada (Wenzowski 1988). The principles of ACTR are based on text recognition and mapping to a single word. This can be problematic, as text responses can also include "trivial words, inconsistent use of double letters, and multiple word groupings which may be hyphenated, joined, or entered as separate, multiple words" (Wenzowski 1995, p. 3). ACTR overcomes these issues by parsing to standardize words, thereby removing grammatical and syntactical differences, and superfluous characters. ACTR also requires the development, implementation, and maintenance of a dictionary (a knowledge base) that includes standard text descriptions and associated codes. ACTR uses the dictionary to match the parsing text. Maintenance (updating of expected text) of the dictionary is paramount, as this improves the quality of automatic coding. Each match is allocated a score based on a word weighting algorithm, and if the score is above a specified threshold, the code is allocated (a direct match); if the score is outside the specified threshold, the description is referred for manual coding (an indirect match). The ultimate objective is to automatically match as much of the data as possible but within specified quality constraints (Wenzowski 1995; Perry 2007; Mohl 2007; Ferrillo et al. 2008).

The ACTR technique can also be used for interactive coding for text that could not be coded automatically. This requires intervention by coders who make judgments on the returned responses. Experience with ACTR at the ONS revealed that different rates of coding were achieved depending on who was doing the coding. For expert coders in the Industrial Classification Branch, fewer than 40% of residual descriptions were coded; in contrast, over 80% were coded by nonexpert coders. This difference was explained by the fact that expert coders would assign codes only if they believed they had all the information they required—this is an issue of quality versus quantity. This last point needs to be considered when planning who will carry out any manual or interactive coding (Williams 2006).

The Statistics Canada ACTR system was originally developed in the mid-1980s. The ACTR interface and database were updated to keep pace with modern technology (e.g., 64-bit processing). The new system, known as Jasper, has an improved user interface to assist with interactive coding when descriptions cannot be coded automatically (Mohl 2007).

In the US Census Bureau, new businesses have to apply to the Internal Revenue Service (IRS) for an employer identification number (EIN); this is undertaken on a standard administrative form. The information requested includes business name and a description of the business. Autocoder, an automatic coding tool, was developed and implemented (in 2004 and 2006) to reduce the internal burden and cost of coding business names and descriptions to an industry code. As for any automatic coding tool, a coding dictionary was compiled as industry codes were assigned, based on a combination of business name and business descriptions. In fact, five coding dictionaries were developed on the basis of how the code was assigned, for example, one- and two-word business names up to full business description. The development of these dictionaries involved clerical coding of over 4 million records. The use of more than a single coding dictionary results in the assignment of more than one industry code; the choice of the final code to assign is based on a scoring method. Autocoder is able to assign approximately 80% of new businesses to an industry code, but there is variation in the quality of the assigned codes. It was found that setting the coding rate to 60% produced comparable coding quality to clerical coding; for Autocoder version 2, this was increased to 70% (Kornbau et al. 2007).

11.3.4 Minimizing, Measuring, and Monitoring Errors in Data Coding

A key quality concern for automatic and manual coding is the ability to maintain coding accuracy over time. To alleviate this concern, organizations often develop quality control processes that include regular measurement of precision rates. These measurements typically involve either taking a subsample of the derived codes, repeating the coding manually (ideally by expert coders), and then comparing the original codes with the repeat codes to produce the precision rate (independent verification); or taking a subsample of the derived codes and reviewing them (dependent verification). If the precision rates are outside of agreed tolerance levels, then further work is generally undertaken, such as modifying coding dictionaries and training staff. For the process to be effective, there has to be a continuous quality improvement loop back to either the automatic coding system or the original coders so that identified errors can be prevented in the future (Campanelli et al. 1997; Kornbau et al. 2007; Perry 2007):

- Precision rate (for automatic coding):

$$\frac{\text{Number of correct codes automatically assigned}}{\text{Total number of codes automatically assigned}} \times 100$$

- Precision rate (for manual coding):

$$\frac{\text{Number of correct codes manually assigned}}{\text{Total number of codes manually assigned}} \times 100$$

To monitor automatic coding, it is also useful to monitor over time the percentage of codes automatically assigned (i.e., direct matches):

- Percentage of codes automatically assigned

$$\frac{\text{Number of codes automatically assigned}}{\text{Total number of codes assigned}} \times 100$$

A change in data coding methods or a classification change will undoubtedly produce different coding results, although these are generally improvements. From the survey organization's perspective, a classification change will require updates to coding lists and dictionaries, as well as training for coders. For example, in the Italian Statistical Agency (ISTAT), the industrial classification revision from NACE rev. 1 to NACE rev. 2 increased the number of industrial sections by 4, divisions by 26, classification classes by just over 100, and categories by around 35. In conjunction with these changes, approximately 45% of four-digit codes and 35% of five-digit codes were split between two or more new codes. All these changes had to be implemented into the ACTR dictionary, which involved removing irrelevant descriptions under NACE rev. 2 and adding new descriptions. In total, updating the dictionary required the addition of 3000 new descriptions, 800 revisions, and 281 deletions (Ferrillo et al. 2008). This updating process also included quality assurance procedures.

There will also be a requirement to inform users of the changes and produce quality measures on the impact of the changes so they are fully informed of the differences between methods and classifications and the resulting quality of the coded data. Where possible, to assist time series analysis, mappings should be made between the different versions of the classification. For example, US trade product codes are frequently updated so that new products can be included and obsolete products removed; this poses problems for time series trade analysis. Pierce and Schott (2012) describe the development of an algorithm for mapping the periodic revisions to product codes, which also enables characterization of the extent of code changes over time and procedures for controlling revisions so that growth in US trade can be analyzed over time.

We would like to finish the data coding section by summarizing the key aspects for improving the quality of industry coding (Yu 2007, p. 23):

- Developing a good coding index
- Capturing, at the front of the collection process, adequate descriptions of activities from businesses
- Improving, and balancing, accuracy and efficiency of Autocoders
- Continuously improving the manual coding process through proper training and feedback processes
- Minimizing the effect of backlogs, such as through prioritizing of coding effort
- Evaluating quality to ensure the quality of Autocoders and to make visible the quality of coding outcomes
- Collaborating with agencies that supply information
- Updating industry classifications to improve their relevance to measuring the economy

These key aspects all need to be considered when designing, building, and implementing the coding method and associated process. Many of these aspects are especially important when undertaking coding where there are short time periods between the dispatch of the questionnaire, requested "reply by" date for the data, and the closedown date for the datafile; so there is time to produce the statistical outputs.

Further Reading

Useful references for further information on data coding are Beaucage (2007), Conrad (1997), Groves et al. (2004), Lyberg and Kasprzyk (1997), Wenzowski (1995), and Yu (2007).

11.4 EDITING AND IMPUTATION

This section provides an overview of editing and imputation in business surveys and, where applicable, describes the potential impact of editing on questionnaire design and survey communication. Further detail is then provided on types of errors, methods for detecting errors, treatment of errors, macroediting, imputation, and finally editing strategies.

11.4.1 An Overview of Editing and Imputation in Business Surveys

If using electronic modes of data collection (e.g., web, TDE, CATI and CAPI), data editing can start in the collection phase of the survey process [e.g., see Barlow et al. (2009)]. The collection phase edits generally have to be limited to reduce the potential of respondents becoming annoyed with error warnings and consequently breaking out and providing only a partial response. On receipt of the data in the survey organization, the data will be captured and coded and then it must be cleaned; this is usually done by editing and imputation. Editing is generally carried out first at the microlevel, as data are initially verified, and then at the macro aggregate level as you get nearer to producing the final statistical outputs. Editing involves two key processes: (1) identification of errors and (2) treatment (cleaning) of errors. The overall challenge when editing data is to determine how much is sufficient. Do you attempt to clean all data identified as being in error, or do you prioritize data cleaning in some way? When the editing has been completed, you will inevitably be left with some missing data; this is when imputation is carried out—to replace missing values with imputed values.

Despite survey design attempts to collect from respondents the right data the first time, errors will occur. These errors will be generated from a multitude of factors. Examples include, respondents: misunderstanding questions, not reading definitions and/or includes and excludes, entering data incorrectly during initial data capture, or, for recurring or panel surveys, different respondents responding and selecting different data in the business records. Ultimately, though, data errors introduce measurement error into the survey data (de Waal et al. 2011), and if left in the data, they may distort the final estimates, making it difficult to carry out further steps in the survey and statistical production process, and reducing trust in the statistical outputs produced (Granquist 1995).

From the survey organizations perspective, the traditional approach to data editing has been that it is an essential survey subprocess to ensuring high-quality data. However, data editing uses a considerable amount of the survey budget (estimates are between 20–40% of the survey budget) and more recently, concerns regarding respondent burden and overediting have become more prominent. The focus now is more on editing only data that will have an impact on the final data outputs, instead of the traditional approach of editing as much as you can in the time that you have (Thomas 2002a). In more recent years there has also been shift toward the value of editing for collecting information on errors and sources of error, and then using this information to improve the survey process (Luzi et al. 2007).

Errors can be categorized by their source into systematic and deterministic errors. *Systematic errors* occur as a result of a feature of the survey design, such as unclear definitions, poorly worded questions, requesting the data from the wrong person in the business, and/or data capture problems. In contrast, *deterministic errors* (otherwise known as *random errors*) have probability distributions with their own variance and expectation (Luzi et al. 2007).

At this stage it is important to highlight some of the key characteristics of business surveys that can potentially impact on editing:

1. Business survey samples are often skewed toward large businesses that will contribute the most to the variable of interest.
2. Businesses are generally in a sample for a specified period of time or for large businesses for life; so there is previous historical survey information available that is generally strongly correlated between survey periods.
3. There may be definitional and reporting period differences between what is being asked for in the survey and what is available from the business records.

Also, some of the quality positives and negatives that can arise from editing are (Granquist et al. 2006)

1. Time is needed to carry out editing, which will impact on the timeliness of the published statistics.
2. Statistical accuracy can be affected both positively (by correcting errors) and negatively (by introducing new systematic errors).
3. Coherence can be improved between the different time periods in the survey.
4. As long as the editing methods are documented for users, it can assist users in interpreting the data.

Editing can be applied either at the record level or across records. Microediting detects errors within a record, while macroediting detects errors in weighted aggregates of the records (cells), these aggregates often correspond to the level at which the data will be disseminated. Cells are declared as suspicious by the macroediting process because their values are either too high or low in comparison to other cells, or in comparison to other variables. The records that belong to cells that have been declared suspicious by the macroediting process are once more examined with microediting. Figure 11.6 provides an overview of the micro/macroediting process. Data in error can be "fixed" either manually or automatically. A manual fix takes place if the coder (data editor) deems that there is enough information to do so; this may involve recontacting the respondent. An automatic fix is done by imputing the data (see Section 11.4.7).

11.4.2 Microediting

Microediting starts during the actual survey field period (collection) as the data are returned to the survey organization and continues until the survey datafile is closed for analysis and then the production of results. Further microediting may take place if errors are identified during macroediting. The time available to carry out editing is thereby linked to the frequency of the survey, with monthly surveys allotting the shortest period in which to collect and edit the data and annual surveys, the longest. Table 11.8 provides examples of how length of the survey field period varies with survey frequency. The monthly retail sales survey collects two data item, but on the quarter-months, five employment questions are added to the questionnaire. The additional employment questions increase the number of editing queries and potentially impact on the ability to clean the data in time for the closedown of the survey datafile. So, from an editing perspective, for surveys with short data collection and processing timetables, one must carefully consider the number of data items collected, or it may be necessary to include errors in the final datafile or increase the use of methods such as imputation.

11.4.3 Types of Microdata Error

Record-level data errors can be classified into five types: missing values, systematic errors, random errors, influential errors, and outliers (Luzi et al. 2007; de Waal et al.

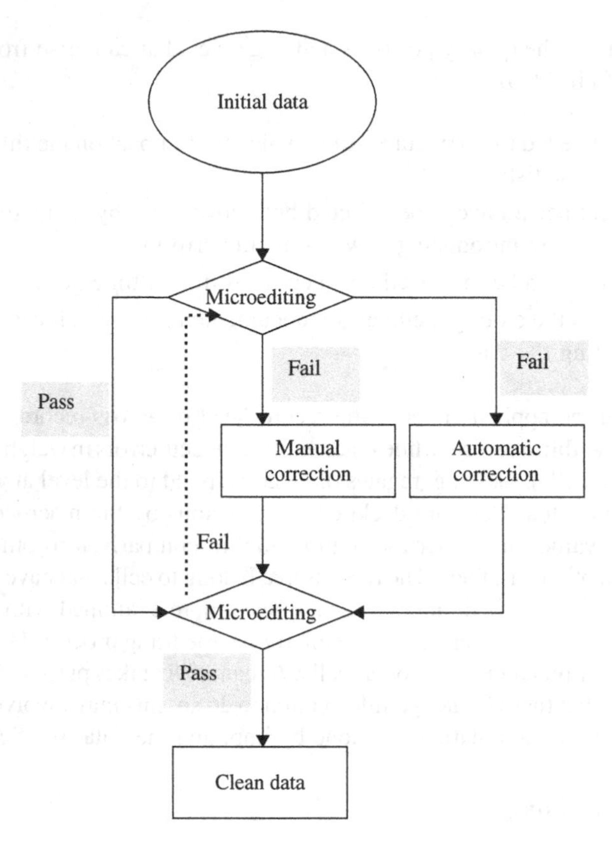

Figure 11.6 General flow of an editing process.

2011). The first three types are part of data validation and are detected via edits with fixed rules. The last two types use statistical editing procedures for detection. In practice, data validation is carried out before statistical editing. These five error types are outlined below.

1. *Missing values*, as the term infers, means that fields within a questionnaire are not completed. The editing process should be able to differentiate between true and false missing fields. True missing fields occur when a response was expected but not given by the respondent. False missing fields occur when a response was not expected, but given by the respondent. For example, if a questionnaire has skip-level questions, the respondent is not expected to complete fields that are skipped. If the respondent does complete these fields, the fields are in error. It is therefore important that the editing process factors in the sequencing of questions and associated patterns, so as to differentiate between true and false missing values. Even declaring fields with true missing

Table 11.8 Examples of ONS Survey Collection and Processing Timetables

Survey Sample size Number of Data Items	Number of Days from Questionnaire Dispatch to Reply by Date	Number of Days from Reply by Date to Closing the Survey Datafile	Total Number of Days for Collecting and Editing the Data
Monthly retail sales (5000 businesses) Two data items (turnover) but on the quarter-month, five additional data items are collected (employment)	11	6	17
Quarterly capital expenditure (month 1) (27,000 businesses) 10–11 data items	16	11	27
Annual business survey (62,000 businesses)	There is a staggered dispatch of questionnaires		
The survey has different questionnaire versions:			
Nonproduction (40,000 businesses)	66–71	241	307–312
Production (11,500 businesses)	72–73	—	—
International trade in services (nonproduction and production) (10,500 businesses)	94	—	—
The number of data items varies from 8 to 111	—	—	—

values in error might not be totally correct. There could be legitimate reasons why the respondent has not completed these fields. Consequently, fields declared as having true missing values may need to be examined further using a probabilistic framework. The probability that the field has a true missing value can be estimated using auxiliary information (e.g., historical data). Also in some instances the field is clearly missing. For example, a sequence of fields is expected to obey exact conditions set by accounting principles, but does not.

2. *Systematic errors* are errors that are systematically reported by respondents for some questions. Missing such errors may lead to serious biases in the estimates if

there are a sufficient number of them in the dataset. These errors may occur when the respondent provides answers on a scale different from what is expected. An example would be a change in the measure of the metric that is requested. The respondents would continue to provide their answers in the old scale (e.g., in pounds) when in fact they should respond in the new scale (e.g., in kilograms). These errors may be quite difficult to detect, especially if the change in scale is not very significant such as a change from pounds to kilograms. In some instances, however, they are easier to detect, such as reporting in scales that are 1000 times larger or smaller than the one being requested. Other reasons for systematic errors include misunderstanding of the skip rules related to filter questions in the questionnaire; errors due to the respondent's inability to provide information based on the required classification/definition; sign errors that occur when respondents systematically omit a minus sign for variables, such as profit, which can be negative; and systematic missing values such as omission of the total value for a number of component items. The editing to detect systematic errors requires prior knowledge that they may occur. Ratio edits can detect such errors by declaring them as such if their comparison (via ratios) with other variables in the questionnaire is outside acceptable bounds set for these ratios. However, one should not rely entirely on edits to detect such errors in surveys, and "tightening" the edits is not a solution. Instead, one should consider using other methods, such as traditional quality control methods, careful analysis, review of concepts and definitions, postinterview studies, data validation, and data confrontation with other data sources that might be available for some records, to detect such systematic errors.

3. *Random errors* are errors that occur accidentally. They may be generated by the respondent or interviewers during initial data capture or during data capture processing by the survey organization (see Section 11.2). They are detected via edit rules that determine whether the data are in error. Once this detection has occurred, the next step is to determine which field(s) is (are) in error, and this may not be straightforward. For example, an edit rule states that the sum of a number of fields will be exactly equal to another field (their total). Suppose that the data do not satisfy this edit, any of the fields or combinations thereof could be in error: which one is it? This decision is called *error localization*. The fields in error may be declared as such either using deterministic rules or via the use of general guiding principles. The deterministic rules will point to fields that are more likely to be in error than others using some predefined criteria. For example, if the sum of the fields does not add up to a field that represents their total, then the field in error will be the total. If all fields are considered to be in error, then the guiding principle is to change the fewest of them so as to ensure that the edit rule is respected. This idea, first developed by Fellegi and Holt (1976), has three principles: (a) changes are made to the minimum number of fields to ensure that the completed record passes all of the edits, (b) distribution of the datafile should be maintained as much as possible, and (c) the associated imputation rules have to correspond to the edit rules. It is

better to use Fellegi–Holt principles as opposed to deterministic rules for "fixing" records with random errors. Although each deterministic rule may be simple to specify, it will be tedious to specify a large number of them if they are to account for as many failure scenarios as possible. Furthermore, because they are deterministic, they may introduce a bias. This is not the case for edits based on the Fellegi–Holt principles, as the resulting imputed data preserve as much as possible of the structure of the original dataset.

4. *Influential errors* are associated with records whose values have a significant impact on statistics (usually totals). The values associated with such records may or may not be correct. They occur for the following reasons: (a) they may arise from records that are unusually large in comparison to records in the same stratum; (b) the record may be associated with a record with a large sample weight, and their contribution to the target estimates is significant; or (c) their weight may be unusually high (relative to the weights of other records in the specified domain) due to problems with stratum jumping, sampling of birth records or highly seasonal records, large nonresponse adjustment factors arising from unusually low response rates within a given adjustment cell, unusual calibration weighting effects, or other factors. These records are detected via statistical procedures; selective editing is the most common one (Farwell and Raine 2000; Hedlin, 2003; Hidiroglou and Berthelot 1986).
5. *Outliers* are observations that are numerically distant from the rest of the data. Grubbs (1969, p. 1) defined an outlier as follows: "An outlying observation, or outlier, is one that appears to deviate markedly from other members of the sample in which it occurs." Outliers can occur by chance in any distribution, but they are often indicative of whether the population has a heavy-tailed distribution; such is the case for business surveys. Chambers (1986) differentiates between representative and nonrepresentative outliers. Representative outliers are correct observations that may have similar units in the population (for further information on representative outliers see Section 5.5). Nonrepresentative outliers are either incorrect observations or are unique. If they are incorrect, they are much too large in comparison to their true value, and their value should consequently be reduced. If they are unique and correct, their impact on the estimates needs to be dampened by either trimming their values (winzorizing) or reducing their weights. Outliers can be detected via univariate methods, or multivariate methods depending on the dimensionality of the variables. These detection methods should be based on robust estimates of the centrality and dispersion parameters (i.e., variance).

11.4.4 Methods for Identifying Microerrors

Typically, implausible data are identified by applying editing rules to the microsurvey data. If using electronic modes of data collection some of these editing rules can be built into the data collection instrument, but a careful balance needs to be struck between taking the opportunity to edit data as the respondent or interviewer completes the questionnaire and having too much editing so as to annoy respondents, thereby increasing the risk of partial nonresponse.

When the survey data have been returned, captured, and coded, any data that fail edit rules are categorized either as fatal (or hard edits) or query edits (or soft edits). Hard edits will identify fields that cannot possibly be correct, whereas soft edits will identify fields in error that may in fact be correct. Fields declared in error by a hard edit must be dealt with by the reviewer. Examples of hard edits include economic activities that cannot exist within a list of acceptable economic activities and balance edits. An example of a soft edit is a field declared in error because its current value may seem quite different as, compared to its previous value. It should be noted that both types of edit rules may be incorrect; that is, they will falsely declare records in error when in fact they are not. A signal that queries if edits are erroneous is triggered when respondent follow-ups indicate that a significant proportion of records are correct. It is therefore important to analyze edit failures and revise the edit rules if appropriate. Erroneous or questionable data are corrected, identified for follow-up, or flagged as missing to be imputed later. Microediting includes edits that determine data validation and statistical consistency. Data validation is determined using validation edits, consistency edits, range edits, and logical edits. Edits associated with statistical consistency detect data values that are likely to be wrong. They are based on the estimated distribution of the dataset. These distributions are estimated using historical and current data. The next section describes these edits.

11.4.4.1 Data Validation Edits

Validity edits verify the syntax within a questionnaire. For example, characters in numeric fields are checked, or the number of observed digits is ensured to be smaller than or equal to the maximum number of positions allowed within the questionnaire. They also check for missing values.

Consistency edits compare different answers from the same record to check their logical consistency. The rules are determined by subject matter experts, accounting principles, and by other means. An example of such an edit is that totals are equal to the sum of their parts.

Range edits determine whether values reported are outside of bounds (i.e., $0 <$ sales $<$ upper limit).

Logical edits are specified by linear equalities or inequalities such as

$$\begin{aligned} a_{11}x_1 + \cdots + a_{1n}x_n &\leq b_1 \\ \vdots \qquad\qquad & \vdots \quad \vdots \\ a_{m1}x_1 + \cdots + a_{mn}x_n &\leq b_m \end{aligned}$$

where $x_1, \ldots, x_n$ are survey responses, and $a_{11}, \ldots, a_{mn}$ and $b_1, \ldots, b_m$ are user-specified constants. These inequalities specify an *acceptance* region in which acceptable records fall. For example, suppose that the edit set is defined as $x \geq 0, y \leq x + 3$, $x \leq 6, y \geq 0$. The acceptance region is represented by the following diagram.

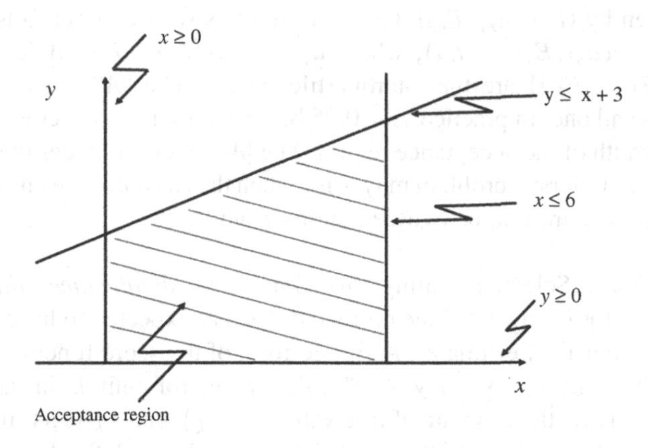

11.4.4.2 Statistical Edits

Statistical edits detect data values that are likely to be wrong. They are based on the estimated distribution of the dataset. These distributions are estimated using historical and current data. An example of this type of edit is testing whether the ratio of two variables lie between bounds. These methods should use robust estimators of location and scale. We next provide a brief description of each method.

Quartile Method The quartile method is a simple procedure measure for detecting outliers. Given that a dataset consists of n observations, define $d_i = (y_i - m)/s$, where m is a location estimate and s is a scale estimate. Robust estimates for m and s are the median and the interquartile distance. Tolerance intervals based on these robust statistics are $(m - c_L s, m + c_U s)$, where c_L and c_U (chosen by examining past data) are not necessarily equal. The lower bound c_L is often set to zero for positive data. Data points falling outside the interval are declared as outliers. The procedure needs to be modified for detecting outliers in periodic surveys. There is more emphasis on small changes of large values than on large changes in small values, and the size masking effect needs to be removed.

Hidiroglou–Berthelot Method Hidiroglou and Berthelot (1986) used a transformation to identify outliers. Their procedure uses a ratio procedure that factors in the size of the observations. Although the procedure was originally developed to detect errors in repeated surveys, it can also be used to determine outliers within a questionnaire. Let x_i and y_i (where $i = 1, \ldots, n$) denote measures on variables x and y for unit i: we assume that their realized values are positive. The ratio of these two variables is given by $r_i = x_i/y_i$. The data are transformed to ensure outliers can be identified at both tails of the distribution. This transformation is given by $s_i = 1 - (\text{med}\, r_i)/r_i$ if $0 < r_i < \text{med}\, r_i$ and $s_i = r_i/(\text{med}\, r_i) - 1$ otherwise. The magnitude of data is incorporated via the following transformation $E_i = s_i \max(x_i, y_i)^U$, where the $0 < U < 1$; E_i terms are called *effects* and the parameter U controls the shape of the curve defining the upper and lower boundaries. The first quartile, median, and thirds quartile of the E_i

values is given by (E_{q1}, E_m, E_{q3}). Outliers are units whose effect falls outside the interval $(E_M - cd_{q1}, E_M + cd_{q3})$, where $d_{q1} = \max(E_M - E_{q1}, |AE_M|)$ and $d_{q3} = \max(E_{q3} - E_M, |AE_M|)$ are the interquartile ranges. The "A" value is a constant between zero and one: in practice, $A = 0.05$ has been found satisfactory. The c value controls the width of the acceptance region. The $|AE_M|$ term reduces the tendency of declaring false outliers; a problem may arise when the E_i values are clustered around a single value and are one or two deviations from it.

Selective Editing Selective editing (also known as *significance editing*) selects records in error for follow-up if the corrected data are expected to have a significant impact on the statistical estimates. A simple form of the score function for unit k is $score_k = w_k|y_k - E(y_k)|/\sum_s w_k|y_k - E(y_k)|$, where, for unit k in sample s, the observed value is y_k, the estimate of true value is $E(y_k)$, and w_k is its survey weight. Suspicious records are ranked by their associated score. Records that have scores above a given threshold are followed up. Methods based on this idea have been developed by Latouche and Berthelot (1992), Lawrence and McDavitt (1994), Lawrence and McKenzie (2000), and Hedlin (2003). Records in error that are not followed up are imputed. Latouche and Berthelot's (1992) score function was based on the magnitude of historical change for that record, the number of variables in error for that record, and the importance of each variable in error. One of the score functions that they suggested is given by

$$score_k(t) = \sum_{q=1}^{Q} \frac{w_k(t)E_{k,q}(t)I_q\left(x_{k,q}(t) - x_{k,q}(t-1)\right)}{\sum_s w_k(t)E_{k,q}(t)x_{k,q}(t-1)}$$

where $E_{k,q}(t)$ is equal to 1 if there is an edit failure or partial nonresponse, and 0 otherwise; $w_k(t)$ is the weight of unit k at time t; and I_q reflects the relative importance for variable q. An estimate of the true value is given by its previous value $x_{k,q}(t-1)$.

Hedlin (2003) minimizes the bias incurred by accepting records in error. For a sample s, let the raw data be denoted as $y_1, y_2, \ldots, y_n$ and the clean data as $z_1, z_2, \ldots, z_n$. The score for the kth record is computed as $score_k = w_k \times |y_k - E(y_k)|/\hat{Z}$, where $\hat{Z} = \sum_{k \in s_d} w_k z_k$, and s_d is part of the current sample for a specified domain d of interest (say, the dth industrial sector). The $E(y_k)$ term is usually the previous "clean" value of that record, or the median of the corresponding domain. A record k will be rejected if $score_k$ exceeds a prespecified threshold based on historical data. The threshold is set so that the coverage probability of the estimate is nearly unchanged.

11.4.5 Treatment of Microdata Errors

The treatment of identified microdata errors can be undertaken using estimation (see Chapter 5), imputation (see Section 11.4.7), or automatic or interactive methods. As the name suggests automatic treatment of errors is undertaken automatically via an editing tool to deal with known, generally systematic type errors. The automatic treatment of errors reduces respondent recontacts and costs. For example, for £ thousands errors, an automatic method can be implemented that compares the accepted value from the previous period with the latest value. If the ratio of these two values is within a 1000 centered range, then the latest value is divided by 1000 and rounded to give the edited value. When totaling values do not equal the breakdowns, again the latest reported value is compared with the previous accepted value; if the difference lies within an acceptable range, then the total is changed.

In contrast, interactive methods involve human intervention to review flagged errors and make judgments on how or if they should be corrected; this can lead to nonuniform treatment of errors and can potentially generate errors from the editing process (Nordbotten 1963).

Business survey interactive editing treatments pose problems for both survey organizations and respondents. From the survey organization perspective, the editing treatments can be expensive, time-consuming, and affect the quality of the final data. From the respondent perspective, they have already returned their data and may then be contacted again to check if the data are correct; this is often irritating to respondents. Now think about the process of recontacting respondents from the business survey response process perspective (for further information, see Chapter 2). We know that especially in medium-sized and large businesses, a number of people in the business maybe involved in the survey response process, for example, collating the data, and completing and returning the questionnaire. So, when businesses are recontacted, the issues are (1) who we are actually making contact with, and whether they will have easy access to the business data that we are querying, (2) whether there are any confidentiality issues in giving the reported data to someone in the business who was not the actual respondent(s), and (3) whether the person whom we are contacting will have access to a copy of the questionnaire. We do know that many businesses responding to statutory business surveys keep a copy of their completed questionnaire as proof that they have returned it (Jones et al. 2004). Potentially, then, the very act of recontacting businesses to verify data may introduce measurement errors.

When an error has been flagged for interactive treatment, editing staff will have different visual views of the returned data based on the mode of data collection. Errors found in data collected using paper questionnaires and captured using electronic images of the questionnaire will generally be presented to editing staff as an electronic image of the questionnaire. What is presented therefore includes the questions, instructions, and the responses. In contrast, for example, for data collected using TDE or IVR there is no questionnaire to present to the editing staff; so generally only the captured data are presented, often in a spreadsheet format. Figure 11.7 gives an example of how TDE data are presented to ONS editing staff. The last two columns shown in the data window show the question numbers and the associated returned data. This fictitious example shows that the respondent has not responded to five employment questions (50–54) and there are differences between what was reported for the previous period (last six rows) and what has been reported for the current period. When presented with this information, the editor then has to decide what to do (i.e., pass the data or recontact the respondent). The different mode-specific presentations could potentially impact the editor's judgment on how to treat the error and if businesses are recontacted the wording of the questions asked.

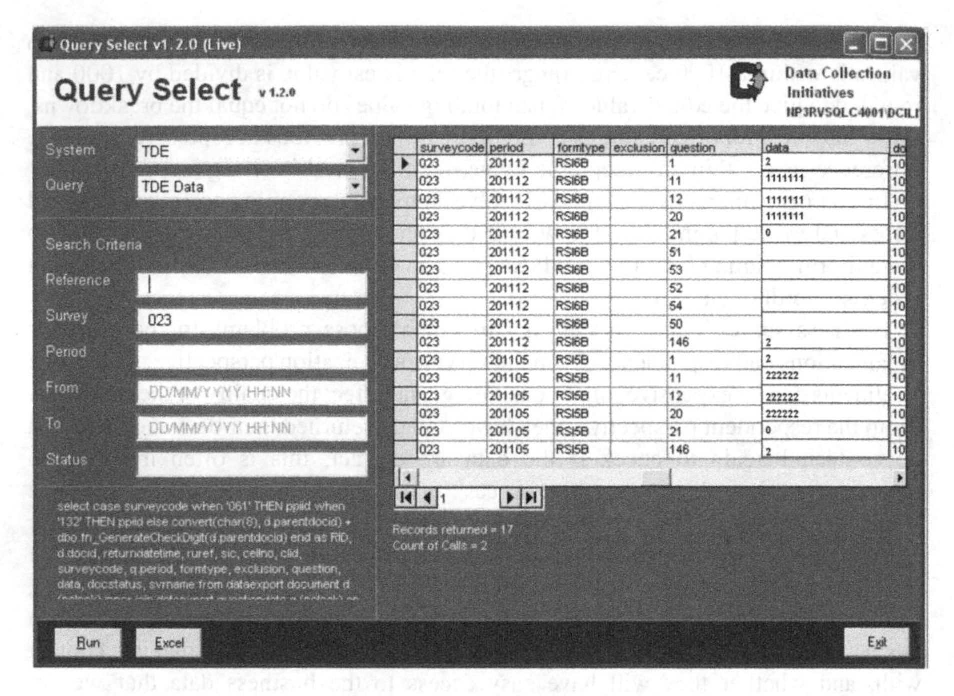

Figure 11.7 Presentation of TDE data to data editors. © Crown copyright (2012).

Because of mode-related differences in the visual presentation of data to editing staff, the number of questions asked using particular modes (see Chapter 8) needs to be considered from not only the respondent's perspective but also a data editing perspective. For example, at ONS TDE is used only for business surveys collecting nine of fewer data items, primarily because of concerns of respondent burden and the possibility of breakouts during initial data capture. However, if more than nine data items were collected, it might also pose issues for editing staff when visually presented with the returned data and no corresponding questions and instructions. This is an issue that really needs further research.

Before recontacting respondents, editors use their judgment (generally based on previous responses from the business or industry knowledge) to decide whether the data items are in error or can be passed. A possible method for improving the judgment decision is to request, on the questionnaire, explanatory comments on any changes to data from the last reporting period. This approach is used in many ONS business surveys (see Figure 11.8); alternatively, a specific question can be included at the end of the questionnaire asking about any changes to the data in comparison to the last reporting period. This second approach is used in some Statistics Canada business surveys. As far as we are aware, no research has been carried out, in recurring and panel business surveys, on the utility to the interactive treatment of microerrors, of requesting information to explain changes in the data.

The treatment of microerrors depends on the type of error and the detection method (see Table 11.9). Figure 11.9 provides a high-level overview of the

Comments

- **Please use the space below to comment on significant movements in data between the current and previous quarterly returns.**
- **Please include details of any significant impacts upon your data resulting from changes in the accounting approaches you have taken.**

Figure 11.8 A comments box example and associated text. © Crown copyright (2012).

automatic and interactive treatment of microerrors; for the latter a number of decisions then must take place, which all have the potential to introduce further errors. At the end of the process when the "data are loaded into the survey datafile" there may still be errors that need to be treated using estimation or imputation, such as outliers, influential errors, and missing data.

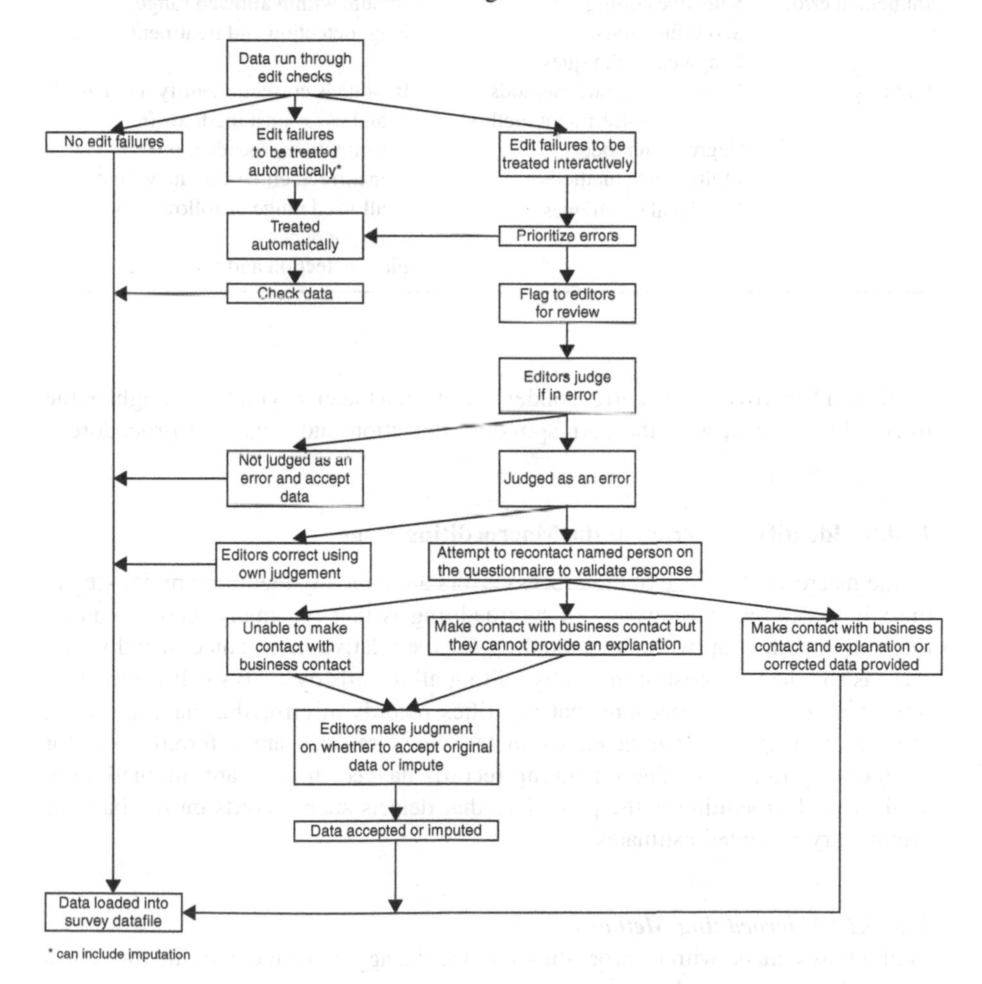

Figure 11.9 A high-level overview of the treatment of microerrors.

Table 11.9 Summary of Error Types and Associated Actions for Microediting

Error Types	Detection	Treatment
Missing values	Logical edits	If missing: impute
	Validity and range edits	If not missing: use special character to identify it
	Consistency edits	Flag: detection and treatment
	Fellegi–Holt edits	
Systematic errors	Expert subject matter knowledge	Flag: detection and treatment
	Statistical edits: ratio edits	
Random errors	Logical edits	Use deterministic imputation
	Validity and range edits	Use Fellegi–Holt principles to impute correct values
	Consistency edits	Flag: detection and treatment
Influential errors	Selective editing	Impute within allowed range
	Score functions	Flag: detection and treatment
	Graphical techniques	
Outliers	Robust univariate methods Hidiroglou–Berthelot methods Regression models Multivariate methods Graphical techniques	If value is correct, identify as such, and use robust methods in estimation procedures; if incorrect, either impute within allowed range or follow up with respondent
		Flag: detection and treatment

Table 11.9 provides the correspondence between the errors that are sought at the microediting stage, with the corresponding detection and treatment procedures.

11.4.6 Identifying Errors at the Macroediting Stage

At the macroediting stage of the process errors are typically identified by looking at the whole dataset. A drawback of microediting is that too many records can be flagged for follow-up without accounting for the relative importance of individual records and the high cost of manually editing all records by analysts. It is therefore important to have a procedure that identifies records in error that have the most impact on weighted aggregated estimates. Such records are referred back for microediting treatment. The remaining records flagged in error are imputed automatically. Macroediting is the procedure that detects such records on the basis of preliminary weighted estimates.

11.4.6.1 Macroediting Methods

Methods associated with macroediting include the aggregate method, the top–down approach, and graphical methods.

Aggregate Method The weighted data are aggregated to a level that corresponds to publication levels. These aggregates are compared to previous values. If the change is not deemed reasonable, the individual records of the suspicious aggregates are scrutinized further. Records within the suspicious aggregates are ordered from high to low, and their impact on the aggregate is analyzed. Records that have values falling outside preset tolerance bounds are sent back into the microediting process, which might include recontacting respondents to ensure that their reported values are valid, data in error are corrected and flags are set to record these steps.

Top–Down Method This simple procedure sorts the largest entered values (top 10 or 20) and starts the manual review from the top or the bottom of the list. Units that have an abnormally large contribution to an estimator of interest such as the sample total are flagged and followed up. Let $y_{(1)} \leq \ldots \leq y_{(n)}$ denote the ordered values of the observed y values in the sample S. The cumulative percent distribution of the top j units to the y total of all the sampled units is computed. Unweighted or weighted versions of the cumulative percent distribution are computed. The unweighted version identifies units that may be in error. Once the unweighted top–down method is performed, the weighted version provides us with an idea of units that will be influential on account of their very large $w_k y_k$ product. We illustrate how the cumulative percent distribution is computed. The cumulative percent contribution $P_{(j)}$ to the total for each of the j top units is given by $P_{(j)} = 100 \times \sum_{k=j}^{n} w_k y_{(k)} / \hat{Y}_s$, where $\hat{Y}_s = \sum_{k=1}^{n} w_k y_k$, and w_k is the associated survey weight. For the record with the highest observed data, $j=(n)$, $P_{(n)} = 100 \times w_n y_{(n)} / \hat{Y}_s$. For the next highest record, $j=n-1$, $P_{(n-1)} = 100 \times (w_{n-1} y_{(n-1)} + w_n y_{(n)}) / \hat{Y}_s$, and so on. If the percentages exceed preset values, the associated are flagged for review.

Graphical Methods Graphical editing is deemed to be most useful at the macro-editing stage and can be used for the setting of editing bounds and identifying outliers (Hughes et al. 1990). Graphical representations such as boxplots, scatter-plots, and histograms are used for graphical editing. Graphical editing has been considered by Statistics Sweden (Granquist 1991) and the Australian Bureau of Statistics (Hughes et al. 1990); and has been put into practice at the US Bureau of Labor Statistics (Esposito et al. 1994) and at Statistics New Zealand (Houston and Bruce 1993). Graphical editing is an efficient, powerful, online output editing method that can significantly reduce clerical effort without a measurable loss in data quality (Hughes et al. 1990). Graphical methods are user-friendly, parameter-free, easy to understand, and flexible. The idea is to start with the worst values and stop when further editing has no notable impact on estimates. The reviewer selects the worst cases by looking at the edit values visualized in scatterplots and then interactively retrieves the corresponding record. To guide the reviewer in choosing records for manual review, acceptance regions of any shape can be provided and overlaid on the scatterplot. A change is entered interactively, and statistics on the impact of the change are displayed. The method is also useful as tool for providing query edits with efficient acceptance limits (Hughes et al. 1990) and, consequently, for developing and evaluating edits.

11.4.7 Imputation

Records that fail edits and cannot be treated automatically or interactively will need to be treated. A subset of these records may have been declared as outliers as well, regardless of their edit status. Records also considered as having failed edits are those that have not responded to the survey (or unit nonresponse), or that have provided incomplete data (partial response). Furthermore, some data items may have been manually deleted if they have been considered in error as a result of the editing process. The overall impact of edit failure is that it results in *missing data*.

There are several options available for dealing with missing data. The simplest one is to do nothing; that is, missing values are flagged on the output datafile, leaving it up to the data user or analyst to deal with them. This "solution" to missing data is usually adopted when it is revealed to be too difficult to impute values with sufficient accuracy. For example, this occurs for variables that have no direct relationship with any other collected variable. In farm surveys, for example, when livestock and crops cannot be used to impute each other, it is preferable not to impute the missing values.

Another option is to adjust the survey weights for nonresponse. Although this procedure is intended mainly for unit nonresponse, it can be used for partial non-response. However, the drawback is that there will be as many weight adjustments as there are missing fields across the records. Methods such as calibration, or mass imputation, are used to ensure consistency between the resulting tabulations. These approaches have been considered by Statistics Netherlands for the construction of a consistent set of estimates based on data from different sources (Kroese et al. 2000).

The preferred option for survey users is to impute missing data within individual records. The imputation procedures should be based on the Fellegi–Holt principles (Fellegi and Holt 1976): (1) data within individual records must satisfy all specified edits, (2) data in each record should satisfy all edits by changing the fewest possible variables (fields), (3) imputation rules should be derived automatically from edit rules, (4) imputation should maintain the joint distribution of variables.

In business surveys, because there are usually strong accounting relationships between collected variables, manual imputation is often considered, especially for small surveys where the resources devoted to imputation systems are minimal. This approach is not reasonable as it can lead to different tabulations, thereby yielding inconsistent results. The resulting manual imputation may not be the best method to use if the imputation is based on incomplete knowledge.

Imputation methods can be classified as deterministic or stochastic. A *deterministic* imputation results in unique imputed data. Examples of deterministic imputation methods often used for business surveys include: logical, mean, historical, sequential (ordered) hot-deck, ratio and regression, and nearest-neighbor imputation. A *stochastic* imputation results in data that are not unique, as some random noise has been added to each imputed value. Stochastic imputation is usually composed of a deterministic component with random error added to it. Stochastic imputation, unlike deterministic imputation, attempts to preserve the distribution of the data. Examples of stochastic imputation include random hot-deck, regression with random residuals, and any deterministic method with random residuals added.

Imputation of plausible values in lieu of missing values results in internally consistent records. Imputation is also the most feasible option for partial nonresponse. Good imputation techniques can preserve known relationships between variables, which is an important issue in business surveys. Imputation also addresses systematic biases, and reduces nonresponse bias. However, imputation may introduce false relationships between the reported data by creating "consistent" records that fit preconceived models. For example, assume that $x < y$ for two response variables x and y. If this constraint is false, imputing the missing variable x (or y) will result in incorrect imputed data. Imputation will normally use reported data grouped within subsets of the sample or population. Such subsets are known as *imputation groups* or *imputation classes*.

11.4.7.1 Deterministic Imputation Methods

Logical (or Deductive) Imputation Missing values are obtained by deduction, using logical constraints and reported values within a record. Typical examples include deriving a missing subcomponent of a total. This type of imputation is often used in business surveys if there is a strong relationship between variables (especially financial ones).

Mean-Value Imputation Missing data are assigned the mean of the reported values for that imputation class. Mean-value imputation should be used only for quantitative variables. Respondent means are preserved, but distributions and multivariate relationships are distorted by creating an artificial spike at the class mean value. The method also performs poorly when nonresponse is not random, even within imputation classes. It is often used only as a last resort. Note that the effect on the estimates of mean value imputation corresponds exactly to a weight adjustment for nonresponse within imputation groups. Weighting is useful for unit nonresponse as relationships between variables are preserved.

Historical Imputation This is the most useful method in repeated business surveys. It is effective when variables are stable over time, and when there is a good correlation between occasions for given variables within a record. The procedure imputes current missing values on the basis of the reported values for the same unit on a previous occasion. Historical trend imputation is a variant of the procedure; previous values are adjusted by a measure of trend, based on other variables in the record. Historical imputation can be seen as a special case of regression imputation (see text below).

Sequential Hot-Deck Method This method assumes that the order of the data items is fixed, even though they might have been sorted according to some criterion (measure of size or geography). Missing data are replaced by the corresponding value from the preceding responding unit in the datafile. The datafile is processed sequentially, storing the values of clean records for later use, or using previously stored values to impute missing variables. Care is needed to ensure that no systematic bias is introduced by forcing *donors* (reported data items) to always be smaller or larger than *recipients* (missing data items). The sequential hot-deck method uses actual observed data for imputation. The distribution between variables tends to be preserved, and no invalid values are imputed, unlike with mean, ratio, or regression imputation. This is ensured if the imputed variables are not correlated with other variables within the record. If the variables are correlated (as is the case, e.g., with financial variables), then imputing the missing variables by those from the preceding unit will not preserve the distribution. This problem is resolved, in practice, by imputing complete blocks of variables at a time; a block being a set of variables with no relationship with variables outside the block.

Nearest-Neighbor Imputation Nearest-neighbor imputation uses data from clean records to impute missing values of recipients. It uses actual observed data from recipients. Donors are chosen such that some measure of distance between the donor and recipient is minimized. This distance is calculated as a multivariate measure based on reported data. Nearest-neighbor imputation may use donors repeatedly when the nonresponse rate is high within the class.

Ratio and Regression Imputation Methods These imputation methods use auxiliary variables to replace missing values with a predicted value that is based on a ratio or regression. This method is useful for business surveys when auxiliary information is well correlated with the imputed variable. However, accounting relationships between variables need to be respected, and this leads to the need for some adjustment of the predicted values. For example, suppose that the relationship $x + y = z$ holds for three variables, x, y, and z. If x and y are imputed, we might need to adjust the imputed values (e.g., by prorating) to ensure that this relationship still holds.

The response variable needs to be continuous for these methods to be effective. For regression, the independent regression variables may be continuous or dummy variables if they are discrete. A disadvantage of this method is that the distributions of the overall dataset may have spikes.

11.4.7.2 Stochastic Imputation Methods

Deterministic methods tend to reduce the variability of the data, and to distort the data distributions. Stochastic imputation techniques counter these negative effects by adding a residual error to deterministic imputations that include regression imputation. Another approach for stochastic methods is to use some form of sampling in the imputation process, as is the case of hot-deck imputation.

Variance estimation of summary statistics of survey data needs to account for imputation. Haziza (2009) provides a good overview of the different procedures that are currently available for this purpose.

11.4.8 Minimizing, Measuring, and Monitoring Errors in Editing and Imputation

Survey data editing can be expensive, and attempts to identify and deal with both systematic and deterministic errors can be an additional burden for businesses, and directly impact on the final data that will be used to produce statistical outputs.

Therefore it is essential that quality process measures be put in place to ensure that the editing process is effectively measured and monitored.

Ideally, especially for surveys producing monthly statistical outputs when there is a very short time period between finalization of the datafile and dissemination of the statistical outputs; there is a need to have editing process quality measures monitored in real time, so that any corrective action can be undertaken immediately. For example, imagine that you are responsible for producing monthly statistical outputs from a monthly survey, and a preliminary run of the statistical outputs shows unusual current price data (turnover/sales) for the time of year and the economic conditions; further investigation might reveal that the editing staff have not been recontacting as many businesses as usual to validate data that have failed edit checks, as they are short of staff and are desperately trying to clear the backlog of errors. Instead, they are using their knowledge and judging a large percentage of the errors as correct data, and this is impacting on the accuracy of the data. Real-time measures of the number of errors and the number of businesses recontacted would identify this issue immediately, and give you time to raise this as an issue before the survey datafile is finalized.

Essentially, the very act of detecting data errors provides useful data quality information. Detection of systematic errors can also identify problems with the questionnaire that ideally can then be resolved (Pannekoek 2009).

11.4.8.1 Editing Quality Process Measures

In data editing, several different perspectives need to be considered in relation to measurement and monitoring. Some of them are described below.

Overall Perspective The formula for determining the overall edit failure rate is (Regan and Green 2010):

$$\frac{\text{Number of questionnaires failing edit checks}}{\text{Total number of questionnaires}} \times 100$$

The following formula can be used to determine whether there are issues with particular questions:

$$\frac{\text{Number of data items failing edit checks per question}}{\text{Total number of data items per question}} \times 100$$

How Errors Are Investigated The formula for calculating the business recontact rate is:

$$\frac{\text{Number of questionnaires flagged for manual editing and businesses recontacted}}{\text{Total number of questionnaires flagged for manual editing}} \times 100$$

(*Note*: This needs to be considered in relation to how the editing works, i.e., if an error is identified, will the entire questionnaire be flagged for investigation, or just the particular data items that fail the edit checks?)

By default, the business noncontact rate (where editing staff apply their own knowledge to clear edits, or can see that there has been a substitution error in data capture processing) is the remaining proportion. From a quality improvement perspective, this is useful to measure and monitor data capture substitution errors.

The Outcome of Error Investigations Three rates are relevant here:

- The editing rate (Regan and Green 2010):

$$\frac{\text{Number of questionnaires changed by editing for the item}}{\text{Total number of questionnaires in scope for the item}} \times 100$$

- The rate of business recontacts that result in changes to data:

$$\frac{\text{Number of business recontacts where changes are made to data}}{\text{Total number of business recontacts}} \times 100$$

- The rate of business noncontacts that result in changes to data:

$$\frac{\text{Number of business noncontacts where changes are made to data}}{\text{Total number of business noncontacts}} \times 100$$

It is also useful, from a quality assurance perspective, and if wishing to collect anecdotal evidence to provide context to the statistical outputs, to collect the reasons for changes to data items.

11.4.8.2 Imputation Quality Process Measures

In relation to imputation, useful quality process measures are

- The item imputation rate (Regan and Green 2010):

$$\frac{\text{Number of units where the item is imputed}}{\text{Total number of units in scope for the item}} \times 100$$

- The unit imputation rate (Regan and Green 2010):

$$\frac{\text{Number of units imputed}}{\text{Total number of units}} \times 100$$

11.4.9 Editing Strategies

Some survey organizations (e.g., ABS, Statistics Canada, ONS) have started to take an integrated approach to editing by developing and implementing editing strategies; the key focus of the strategies is a holistic selective editing approach to both micro- and macroediting. The objectives of these strategies are to minimize respondent recontacts, maintain data quality, and reduce costs (Skentelbery et al. 2011; Saint-Pierre and Bricault 2011). End users, whether they are internal users (e.g., national accounts) or external users, are the focus of the strategy as they are required to assess and agree on each survey's priority variables for editing (Brinkley et al. 2007).

At ONS, the microediting aspects of the editing strategy have already been implemented into two monthly business surveys (retail sales survey and the monthly business survey). Work in 2011 was carried out on implementing a similar approach to the annual business survey (the ONS equivalent to the structural business survey), which is more complicated as it includes a larger number of variables. To assist the approach in the annual business survey, SELEKT, a tool developed by Statistics Sweden, is being used. The macroediting approach is still being developed but will provide an end-to-end approach to editing (Skentelbery et al. 2011).

At Statistics Canada investigative work identified that for many records in the unified enterprise survey, the same records would be studied several times as they pass through the editing process, without any knowledge of the impact to the final estimates. To optimize this approach across the survey processing procedures, comprehensive indicators and decision rules are being developed to target records that need to be treated. The process will be undertaken in parallel rather than with the traditional sequential approach. For example, early estimates will be produced as soon as enough data are available, to identify from the estimates suspicious microdata for follow-up treatment (Saint-Pierre and Bricault 2011).

In the US National Agriculture Statistics Service, the focus has been on developing a comprehensive editing and imputation structure that can be used across its range of surveys. The approach has included reviewing all existing editing and imputation systems with the objective of implementing an enterprise level system that can deal with a variety of types of surveys and the agricultural census, and have a common look with a graphical user interface (Manning and Atkinson 2009).

The examples presented in this section generally indicate that many survey organizations are critically assessing how they carry out editing and imputation in an attempt to reduce costs and maximize the benefits to data quality.

Further Reading

A useful reference for further information on editing and imputation is de Waal et al. (2011).

11.5 CONCLUDING COMMENTS

This chapter has provided an overview of capturing, coding, and cleaning survey responses and how these processes can take place in both the collection and process phases of the survey process. Traditionally, in the survey methodology literature, there was very little focus on capturing and coding business survey data; yet they are intrinsically linked to the design of the questionnaire, respondent communication, and ultimately data quality.

In the interactive treatment of microerrors by editing staff, there are several areas where further research is needed, for example, how decisions are made on whether to pass the data or recontact a business, the impact of the visual presentation of data to editing staff, and the costs and benefits of requesting information from businesses on why there have been changes to the data from the last reporting period (see Section 11.4.5).

From a quality perspective, survey errors can be generated from initial data capture, data capture processing, data coding, editing, and imputation. These potential sources of error should be measured and monitored in an attempt to ensure the quality of the survey data, implement quality improvements, and provide relevant process measures to both internal and external survey data users.

ACKNOWLEDGMENTS

We would like to thank the following people for their comments, suggestions, help, and reviews while we were writing this chapter: Simon Compton, Julie Curzon, Linda Scott, Nick Barford, Martin Brand, Maria Morris, John Crabb, and the people in the ONS Survey Processing Centre, in particular Jayne Evans, Gaynor Morgan, and Katherine Jones.

CHAPTER 12

From Survey Data to Statistics

Jacqui Jones

12.1 INTRODUCTION

Once the survey data have been captured, coded, and cleaned (see Chapter 11), and, where relevant, estimated to the total population (see Chapter 5), analysis needs to be carried out to bring together the key messages contained within the data; then the derived outputs must be prepared for dissemination and disseminated. This chapter provides an introductory overview of the phases involved in moving from survey data to statistics: analysis and dissemination (see Figure 12.1). The chapter also reviews the archive phase in the generic statistical business process model (GSBPM) (see Chapter 1), which is another dissemination process. The final phase evaluation was covered in Section 10.7. This chapter presents an introductory overview that merely scratches the surface of some specialized areas; key sources of further information are cited throughout this chapter.

12.2 ANALYSIS

The ultimate aim of the analysis phase is to produce statistics that meet the users' (or the researchers') statistical needs in relation to the output quality dimensions (see Chapter 3), and in doing so, build or maintain trust in the produced statistics. Part of the trust dimension will include maintaining data security and confidentiality during analysis; this is especially important for economic data that are used to compile market-sensitive statistics such as gross domestic product (GDP), balance of payments, and retail trade. The output of statistical analysis can include descriptions, relationships, comparisons, and predictions, which are then used to produce statistics.

Designing and Conducting Business Surveys, First Edition.
Ger Snijkers, Gustav Haraldsen, Jacqui Jones, and Diane K. Willimack.

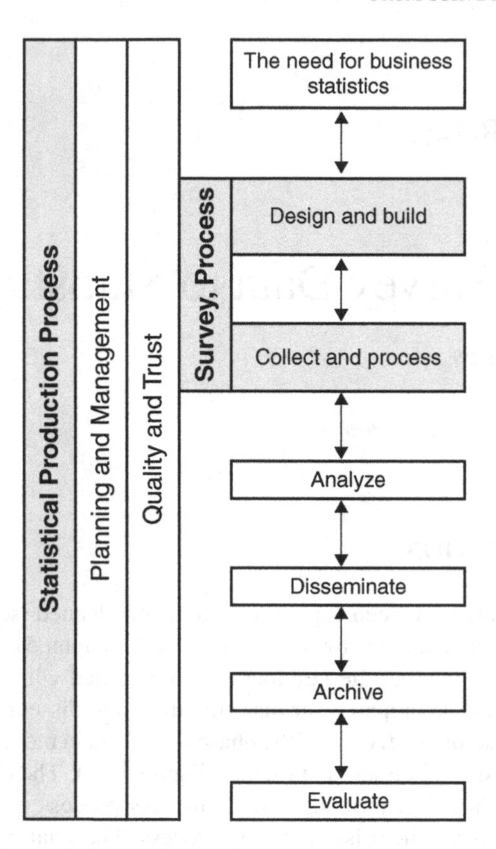

Figure 12.1 Statistical production process using surveys.

Analysis can be categorized into primary or secondary analysis. *Primary analysis* involves analysis of the data for the first time, so analysis of the data from the business survey you have just conducted would be classified as primary analysis. In contrast, *secondary analysis* uses available data and often combines different datasets to undertake *value-added* analysis (Glass 1976). Depending on the type of data collected and the objectives of the analysis, the same types of analytical techniques might be used for primary or secondary analysis. These analytical techniques might focus on summarizing and describing the patterns in the data (descriptive analysis) or studying relationships between variables, and explanations of relationships (inferential analysis). For economic data, analysis can also include modeling techniques to forecast (predict) future movements; examples of economic forecasting can be seen in the economic reports prepared by Central Banks and the International Monetary Fund.

There are many books on analyzing survey data [see, e.g., Heeringa et al. (2010), Chambers and Skinner (2003), Dale et al. (1988), Fink (1995), and Hyman (1972)]. This analysis section provides an introduction to some different types of analytical techniques that can be used when analyzing business survey data and begins by discussing three different types of statistical outputs.

12.2.1 Analysis and Different Types of Statistical Outputs

In national statistical institutes (NSIs), primary analysis of economic data from business surveys often includes analytical techniques to derive volumes from collected monetary values (see Section 12.2.3.3), and, where appropriate, seasonally adjust the data (see Section 12.2.3.1). Descriptive analysis (univariate analysis) at the weighted aggregate data level is then used to summarize and describe patterns in the data over time, including comparisons. Some caution is needed in the choice of measures for making comparisons over time. For example, month-to-month movements in monthly time series can often be dominated by random (stochastic) variation, which may be behavioral, or may simply reflect period-to-period sampling and nonsampling error. This can be easily overinterpreted by data users. Quarter-to-quarter comparisons or annual change will be less prone to large sampling and nonsampling errors, although the longer the span of data being compared, the less timely the comparison is. This can be an important consideration for users of economic data, particularly for users of short-term economic indicators.

The analysis might also include reviewing anecdotal evidence given by business respondents either on the questionnaire, or collected during editing checks (see Section 11.4.5) or nonresponse follow-ups. The anecdotal evidence provides the "why" to the data, that is, why there has or has not been a change in the data. Or it may provide an indication of external factors that took place at the time that should be considered when assessing the statistics. An example of primary descriptive analysis from a business survey and how anecdotal evidence can indicate the external factors that existed at the time is provided in Figure 12.2; an extract from the August 2011 ONS retail sales statistical release.

In August 2011, all retailing sales volumes showed no growth compared with August 2010; however, the sectors that make up retail sales showed a mixed picture. Year on year, August 2010 compared with August 2011:

- Nonstore retailing increased by 13.7% and provided the greatest upward pressure to the all-retailing figure contributing 0.6 index points
- Predominantly automotive fuel increased by 1.6% and contributed 0.1 index points to the all-retailing figure
- Predominantly nonfood stores decreased by 1.2% and provided the greatest downward pressure, pulling the index down by 0.4 index points
- Predominantly food stores decreased by 0.8%, pulling the index down 0.3 index points

There was some feedback from retailers that the August 2011 riots, seen across parts of England, resulted in some stores closing early, thereby reducing their opening hours; and some stores were directly affected by the riots. The economic impact of these closures cannot be measured in relation to the all-retailing index, as there was also some feedback from some retailers that they saw increases in sales due to changes in shopping behavior.

Figure 12.2 Example of primary descriptive analysis and anecdotal evidence. © Crown copyright (2011).

Secondary analysis of data can combine data from different business surveys, as well as administrative data, to produce statistical outputs. A good example of this is the compilation of economic data to produce the annual national accounts. The annual national accounts process takes microdata from a variety of sources, including business surveys, and aggregates them to an intermediate level, such as industry and institutional sectors; the intermediate aggregations are then adjusted for items such as undercoverage, and timing; then the data are reconciled to produce consistent data. The annual national accounts produce statistics broken down by nonfinancial corporations, financial corporations, general government, households, nonprofit institutions serving households, and the rest of the world. These breakdowns are produced at a total economy level and for the following accounts (United Nations 1999):

- Production account (GDP)
- Generation of income account (operating surplus)
- Allocation of primary income account (national income)
- Secondary distribution of income accounts (national disposable income)
- Use of disposable income account (national saving)
- Capital account (changes in net worth due to saving and capital transfers: net lending)
- Financial account (net lending)
- Other volume changes account (changes in net worth due to other changes in volume of assets)
- Revaluation account (changes in net worth due to holding gains/losses)
- Opening and closing balance sheet (net worth)

If the objective of the analysis is to study relationships between variables, and explanations of relationships (inferential analysis using bivariate or multivariate analysis), then various techniques and tests can be used depending on the type of data you have. You might start by looking at frequency distributions to identify patterns in the data and then, depending on the type of data and the objective of the analysis, progress to analysis using cross-tabulations, correlation matrixes, or regression analysis; and testing relationships between variables. In the following paragraphs an example from the 2009 European Company Survey is presented, a survey that was carried out with just over 27,000 businesses in 30 European countries. The results were weighted by the total number of businesses and employees in each country. The analysis focused on both employers and employees proportional analysis, with different employer and employee results emerging when there was strong correlation between size of business and the issue being investigated (European Foundation for the Improvement of Living and Working Conditions 2010, pp. 89–91).

The 2009 European Company Survey included questions on overtime, part-time working, flexitime, and nonstandard working arrangements. Looking explicitly at the flexitime analysis, descriptive analysis of the collected survey data identified the following (European Foundation for the Improvement of Living and Working Conditions 2010, pp. 5–10):

1. Two types of flexitime arrangements: (a) basic, allowing employees to vary their start and finish times, but with a set number of hours that have to be worked each day, and (b) advanced, which allow employees to work a set number of hours over a month or year.
2. From the employer's perspective, flexitime arrangements provided the ability to manage changes in demand.
3. From the employee's perspective, flexitime arrangements enabled them to improve their work–life balance.
4. Just over half of sampled businesses with 10 or more employees offered flexitime arrangements, but not always available to everyone in the business, with 45% of sampled businesses having flexitime arrangements for all employees.

Further bivariate and multivariate analysis then identified relationships between variables and explanations of relationships, for example:

- Countries with high rates of businesses offering flexitime arrangements had a higher proportion of employees permitted to use the flexitime.
- Businesses that did not operate "normal" working hours (e.g., shiftwork) tended to limit flexitime to particular positions in the business, with the service sector permitting more positions to use flexitime than the industry sector.
- Flexitime was most used in Finland and least used in Bulgaria, Greece, and European Union candidate countries.
- The proportion of businesses using flexitime arrangements has increased since the survey four years ago, but there was little difference in the proportion of employees in businesses that were permitted to use flexitime.
- Businesses employing large numbers of employees were more likely to have flexitime arrangements than smaller businesses.
- Businesses with variable demand cycles were more likely to use flexitime arrangements, but the differences were only five percentage points.

It is important to remember when analyzing survey data that the data may contain sampling and nonsampling errors. To provide an indication of the precision of the statistics, you can test the reliability and accuracy of the data by producing confidence intervals, coefficients of variation, or standard errors (Bethlehem 2009). From both analyst and user perspectives, these are useful output quality indicators.

Further Reading

Useful references for further information on analysing survey data are Chambers and Skinner (2003), Proctor (1994), and United Nations (1999).

12.2.2 Analytical Technique Considerations

When compiling statistical or research outputs, various analytical techniques can be used. It is important to ensure that appropriate techniques are used to meet the statistical output needs, for the type of data being analyzed, for the design of the survey that has collected the data, and for the impact of sampling and nonsampling errors.

The decision on which analytical technique to use generally depends on

- The type of data collected (numerical, categorical, etc.)
- Whether the analysis will be undertaken at the aggregate or microdata level
- Whether the data are part of a time series
- Whether there is more than one component series
- The design of the survey sample

12.2.2.1 Analyzing Different Types of Data

The type of survey question asked and the type of response categories provided determine the type of data collected. Business surveys can include questions that collect qualitative data (from open-ended responses) and quantitative data (e.g., attitudes, perceptions, values).

Survey data can also be classified by the scale of measurement: nominal, ordinal, or continuous (interval and ratio). A useful overview of scales of measurement and appropriate statistics can be found in Stevens (1946).

The analysis you perform must account for the type of data you have. The range of analytical techniques is vast, and relevant texts [e.g., Koop (2009)] should be studied to ensure that appropriate analysis is carried out for the type of data you have.

12.2.2.2 Aggregate and Microlevel Data Analysis

Many business surveys, especially those conducted by NSIs, are repeated at regular intervals (e.g., monthly, quarterly, annually). If the objective is to analyze changes over time, then you need to consider if the analysis will be carried out on aggregate- or microlevel data. *Aggregate analysis* involves analysis of aggregated weighted microdata (estimates). For example, you may have survey data on the number of employees, broken down by full-time and part-time, from businesses in the service industry. If you wanted to undertake longitudinal analysis of individual businesses to see how their employment changes over time, then you would use microlevel data for your analysis. In contrast, if you wanted to investigate how, for example, employment in the service industry as a whole changes over time, then you would use weighted aggregate-level data.

Typically in producing economic statistics from business survey data, NSIs carry out analysis on weighted aggregate-level data focusing on industries, sectors, and/or the whole economy; for researchers, analysis is generally carried out on microlevel data, focusing on the business level (McGuckin 1995). Microlevel data can be grouped into different levels of aggregation, including full aggregation. In contrast, aggregate-level data cannot be decomposed into microlevel data. Accessing micro-level data is often an issue for researchers not employed by the survey organization that originally collected the data, but many organizations do provide access to microdata via controlled research microdata laboratories or public user files (see Section 12.5).

12.2.2.3 Analysis of Time Series Data

Time series data are data that are repeatedly collected over time. There are two different types of time series: (1) data that are collected as a stocktake at a given point in time, such as the number of people employed on a certain date—these are often referred to as *discrete values*; and (2) data that are collected as a flow over a given period, such as retail sales for the month—these are often referred to as *continuous* but may actually collect discrete values. Unlike stocktake data, flow data can contain trading-day effects (Chatfield 2004; Foldesi et al. 2007).

As in other types of analysis, the first analytical step should involve getting acquainted with the data using descriptive methods; an example would be to plot the time series over a time period to determine whether there are any seasonal factors and outliers. Analysis of time series data will often include removing the effects of any seasonal factors and calendar effects so that short term and long term movements can be interpreted more readily (e.g., identification of turning points in the time series). Adjusting data for seasonal factors and calendar effects also enables analysts to see if any one-off special events (e.g., unusual weather, strikes, earthquakes) have had an effect on the series, as they will remain in the series following seasonal adjustment (see Section 12.2.3.1).

12.2.2.4 Analyzing Single- or Multicomponent Series

Just as the goal of cross-sectional data analysis is often to discern relationship between variables, analysts of business survey data are often interested in the relationships between different variables over time. This can involve either different variables, or components, of a time series drawn from the same survey, or time series derived from different sources.

This type of analysis has to be conducted with great care. Many time series exhibit an upward or downward trend over time, and many monthly and quarterly time series have seasonal patterns. Simple comparisons of time series, or summary statistics such as correlation coefficients, often indicate a strong relationship. However, these are seldom meaningful. (A frequently cited example is based on an analysis showing a positive correlation between church attendance and crime!) There is a large literature on time series econometrics, which deals with methods for modeling the relationship between time series (Hamilton 1994). Many of these techniques are necessarily quite complex in an attempt to overcome the problems outlined above. Techniques such as cointegration take account of trend and seasonal features of time series in the modeling process (Hatanaka 1996). The literature covers a wide range of different issues that can arise, such as modeling potential time lags between cause and effect in time series, different periodicity or time periods for the observations in time series (e.g., modeling a calendar year series with a financial year series), dealing with different types of data (e.g., continuous variables, ordinal variables), and dealing with sample design effects.

Table 12.1 Types of Sample Design and Analytical Considerations

Sample Design	Considerations
Single cross-sectional	Does not allow for analysis of changes over time at aggregate and micro levels; is only a snapshot
Repeated cross-sectional	Does allow for analysis of changes over time at the aggregate but not micro level—as the sampled units (businesses) will vary between surveys as each sample is independent; analysis of changes at aggregate level need to account for any potential impact from changes in survey design and conduct between surveys (e.g., question wording changes, changes in imputation methods)
Panel	Allows for analysis of changes over time at aggregate and micro levels; need to account for sample attrition (i.e., nonresponse and business deaths)
Combination of repeated and panel—respondents are gradually rotated out of the sample, possibly in overlapping waves	Allows for analysis of changes over time at aggregate level. Allows for some microlevel analysis of changes over time but only for the period that the sampled unit remains in the sample

Note: Further information on the analysis of repeated surveys can be found in Smith (1978).

12.2.2.5 Analyzing Data with Different Sample Designs

From an analysis perspective, the sample design of the survey that collected the data must also be considered, as it impacts on the level (aggregate or micro) and type of analysis. The design effects of the sampling also need to be considered, such as ratio of variance, for x sampling units, between a hypothetical simple random sample and the variance obtained (Lohr 1999). Table 12.1 summarizes different types of sample designs and some of the analytical considerations that need to be considered.

The sample design may also affect the analytical outputs, and Skinner and others have described analytical modification procedures for taking this factor into account (Skinner et al. 1986).

12.2.2.6 Pitfalls to Avoid in Statistical Analysis

Finally it is worth listing here some of the main pitfalls that analysts of business survey data should avoid; these are:

- Having insufficient understanding of the provenance and quality of the data being used, particularly with regard to potential sources of error
- Not having a sufficiently clear definition of what you are trying to achieve with your analysis
- Lacking sufficient understanding of the patterns in the data (e.g., distributions for each variable being used, potential for zeros, missing values, outliers, spurious values)

- Inappropriate use of techniques
- Overinterpretation or inappropriate presentation of results (e.g., reporting the results of a model without properly considering the appropriateness of the model, including how well the data fit the model)

12.2.3 Analytical Techniques for Improving Statistical Interpretation

This sections discusses some analytical techniques often used to improve statistical interpretation when analyzing business survey data: seasonal adjustment, creating an index, and deriving volumes.

12.2.3.1 Seasonal Adjustment

Why Data Are Seasonally Adjusted. Business surveys can collect monthly, quarterly, or annual data from businesses, such as turnover data, number of full-time employees, capital expenditure, and stocks. Generally in NSIs, the monthly and quarterly data are used to produce short-term economic statistics and the annual data, to produce structural business statistics. Data collected from these surveys are used to estimate the same economic phenomena in different periods or at different points in time, and are combined together to create time series that are used to measure and analyze changes over time. These changes can be repeatable or nonrepeatable; repeatable changes are regarded as seasonal variation as they occur at regular times each year. Removing seasonal variation therefore allows improved interpretation of the time series, that is, comparison of two time periods that have different seasonal variation (Atuk and Ural 2002).

Most economic data series are affected by seasonal variation. For example, there may be seasonal variation in the retail sector at Christmas and Easter, and seasonal variation in the consumption of electricity and gas sector between winter and summer. In addition to seasonal variation, there may be calendar effects, such as changes in the number of trading days, in months that have public holidays, and between months with different numbers of days. The seasonal variations are assumed to occur with similar intensity at the same time each year; this is assumed on the basis of previous movements in the data series. From an economic monitoring perspective, carrying out seasonal adjustment helps identify underlying movement in the data by estimating and removing seasonal and calendar related variation. This assists users in interpreting the time series and in making comparisons with other seasonally adjusted time series.

A good example of the effect of seasonal adjustment on a data series is the retail sales data series. Figure 12.3 shows the seasonally adjusted and nonseasonally adjusted UK retail sales volume data series from October 2009 to January 2012. The nonseasonally adjusted series shows the annual peaks in December from the Christmas spending effect. Seasonal adjustment removes these seasonal variations as it occurs at a regular time each year, namely, Christmas, and leaves a series that allows interpretation of the underlying movement in retail sales volumes, which includes any irregular movements.

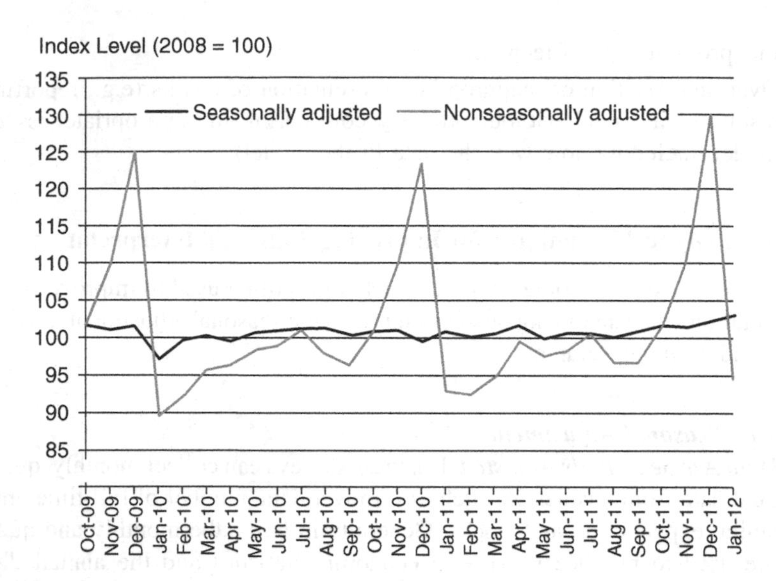

Figure 12.3 All retail sales volumes, October 2009 to January 2012. *Source*: UK Office for National Statistics (2012). © Crown copyright (2012).

How Seasonal Adjustment Works Time series data can be decomposed into four unobserved components: trend, cyclical, seasonal, and irregular (Foldesi et al. 2007). Where:

Component	Symbol	Description
Trend cycle	*T*	Medium- and long-term movements in the series; assumed to include the cyclical component
Cyclical	*C*	Medium-term fluctuation; examined in long time series
Seasonal component	*S*	Movements due to repeatable seasonal factors (e.g., routine public holidays)
Irregular component	*I*	Includes components such as sampling error and nonsampling error, as well as planned and unplanned special events (e.g., additional public holidays and/or natural disasters)

Additional components are calendar effects (e.g., varying trading days from period to period) and outliers (Foldesi et al. 2007).

As seasonal adjustment is based on observed seasonal variations in the data series, data for several years are needed before you can seasonally adjust the series. Seasonal adjustment may be possible for series with as few as 3 years of data, but recommended practice is to have at least 5 years of data before undertaking seasonal adjustment (Eurostat 2009a).

The most commonly used seasonal adjustment methods work by decomposing the time series into the four unobserved components: trend, cyclical, seasonal, and irregular (noise). For data series where there is an assumption that the seasonal fluctuations are not affected by the level of the series, this is represented by the additive model: $Y = T + C + S + I$.

If however, it is assumed that the seasonal fluctuations move in proportion to changes in the level of the data series, this is represented by the multiplicative model: $Y = T \times C \times S \times I$.

Seasonal Adjustment Methods Seasonal adjustment can be carried out at different levels of a time series hierarchy; direct seasonal adjustment adjusts the data (removes the usual seasonal fluctuations) at an aggregate level, in contrast to indirect seasonal adjustment that adjusts the lower-level data, and then the data are aggregated. For example, an indirect seasonal adjustment of total employees would be derived by seasonally adjusting the time series of part-time and full-time employees and adding the two resulting series. Direct seasonal adjustment would be obtained by seasonally adjusting the total employee series. Indirect seasonal adjustment is seen as more appropriate for data series that contain components that have very different seasonal patterns, but when aggregated may not sum to the aggregate series (Hood and Findley 2003).

For decades time series data have been seasonally adjusted. During the 1920s Frederick R. Macaulay developed "classical decomposition" (ratio-to-moving-average method), which is regarded as the first overall nonparametric seasonal adjustment methodology, and has been the basis for many seasonal adjustment approaches. The most widely used has been the X-11 family of methods, originally developed by Julius Shiskin in the US Census Bureau in the 1960s (Shiskin et al. 1967). Since then there have been various developments in seasonal adjustment, all building on Macaulay's approach. One development that has had the greatest impact has been the introduction of computers, which resulted in the ability to perform decomposition calculations in seconds, rather than days, when manually calculated (Foldesi et al. 2007).

Seasonal adjustment methods can be categorized into nonparametric (non-model-based), semiparametric (partially model-based), and parametric (model-based). Nonparametric methods are based on ratio to moving average methods. These rely on the repeated application of symmetric weighted averages of combinations of observations in the series. One issue with this is that at the ends of the series not all observations are available to apply these symmetric averages, and asymmetric weighting structures have to be applied instead. As new data points are added to the time series, the asymmetric weights can be replaced with symmetric ones, thus generating revisions in the seasonally adjusted time series (Dagum 1987; Foldesi et al. 2007).

The semiparametric methods extend the moving-average algorithm by reducing the use of forecasts and backcasts using autoregressive integrated moving-average (ARIMA) models. Examples of semiparametric seasonal adjustment software are X11-ARIMA, X12-ARIMA, and X13-ARIMA-SEATS. In the X12-ARIMA software a preprogram for detecting and correcting outliers and calander components was introduced (Atuk and Ural 2002; Foldesi et al. 2007). Table 12.2 provides an

Table 12.2 Timeline of the X11 Family of Methods

Year	Method	Description
1954	Method I	A manual method that extended Macaulay's ratio-to-moving-average method to forecast and backcast data
1955	Method II	An electronic version of method I
1960	X3	Improved replacement of extreme values and method of computing seasonal factors for recent years
1961	X9	Used as a standard program
1961	X10	Used for erratic data series
1965	X11	Replaced X9 and X10 as the standard program and included regression techniques to make trading day adjustments; also included additional time series analysis tools (e.g., analysts could choose additive or multiplicative versions)
1980	X11-ARIMA	Statistics Canada incorporated into X11, autoregressive integrated moving average (ARIMA), which allowed forecasting and backcasting of the series
1998	X12-ARIMA	Replaced X11 as the standard program with four key enhancements: alternative adjustment capabilities, new diagnostics, improved time series modeling, and a new user interface
2006	X13-ARIMA-SEATS	Included an improved user interface, new user diagnostics to help overcome any inadequate adjustments, and the SEATS algorithm for regARIMA-model-based seasonal adjustment

Sources: Shiskin, et al. (1967), Dagum (1980), Findley et al. (1998), and Foldesi et al. (2007).

overview of the development of nonparametric and semiparametric seasonal adjustment software for the X11 family of methods.

Parametric (model-based) seasonal adjustment methods can be subdivided into (1) deterministic, and (2) stochastic methods. Deterministic methods use models based on regression analysis, and trends and seasonality are considered against a predetermined behavior curve. The essence of deterministic methods is that future values can be predicted from past values; therefore, seasonality is assumed to be constant throughout the time series. Examples of deterministic software models are DAINTIES and BV4. In contrast stochastic methods are modeled based either using an autoregressive integrated moving average (ARIMA) model or structural model. Examples of stochastic software models are SEATS (for ARIMA models) and STAMP (for structural models). The ARIMA model first models the time series, from which it derives components. In the structural model the starting point is with estimation of the components. In stochastic time series analysis seasonality is assumed to move or change throughout the time series. Both methods require some manual intervention, by studying diagnostics (quality information) generated from the seasonal adjustment tools and decisions on where to make adjustments (Foldesi et al. 2007).

In the European statistical system, the most widely used are TRAMO-SEATS (a parametric approach, supported by Banco de España) and X12-ARIMA (a semi-parametric approach, supported by the US Census Bureau) (Eurostat 2009a).

Impact on Revisions from Seasonal Adjustment There are three predominant sources of revisions to time series data: (1) changes to past data (e.g., respondents returning late data), (2) new data points, and (3) changes to seasonal adjustment parameters (Elliott et al. 2008). The latter source of revisions is necessary to ensure that appropriate adjustments are made as usual seasonal fluctuations change; this is achieved by regular seasonal adjustment reviews. For a short data series (3–7 years old), the seasonal adjustment review should take place every 6 months; for a long data series (over 7 years old) the review should take place once a year (Eurostat 2009a). Following a seasonal adjustment review, previously used adjustments may change and produce revisions to the data series as they are applied across the time series.

At the UK Office for National Statistics (ONS), each seasonally adjusted data series undergoes an annual seasonal adjustment review to ensure that the most appropriate parameters are used for the next 12 months. Once reviewed, any parameter changes can affect the data series and cause revisions to the back series. At ONS users are informed of when the annual season adjustment review will take place, as this helps explain any revisions following the review.

Advantages and Disadvantages of Seasonal Adjustment As with any other analytical technique, there are advantages and disadvantages to carrying out seasonal adjustment. The Eurostat (2009a) seasonal adjustment guidelines provide a good overview of the advantages and disadvantages:

- *Advantages*

  Helps interpret short-term (subannual) changes in times series.

  Improves the ability to compare long-term and short-term movements between sectors.

  Assists business cycle analysis and detection of turning points in the data series.
- *Disadvantages*

  Reliant on manual intervention for setting priors.

  Dependent on quality of the collected data.

  Can be time-consuming and is best dealt with by experts.

  Ideally requires the use of specialist tools.

  Seasonally adjusted series are subject to revisions when subsequent data become known.

  If not carried out correctly, can give misleading results, including residual seasonality.

  Ideally requires a series of 7 years (long series) to provide stable parameters. Seasonal adjustment can be carried out on series that are 3–7 years in length (short series) but should ideally have more frequent seasonal adjustment reviews (e.g., every 6 months).

Further Reading

Useful references for further information on seasonally adjustment are Atuk and Ural (2002), Bell and Hillmer (1984), Eurostat (2009a), Findley et al. (1998), Foldesi et al.

(2007), and US Census Bureau website (`http://www.census.gov/srd/www/x12a/`).

12.2.3.2 Index Numbers

Index numbers are measures that represent the change of aggregated data in a convenient manner; they enable disaggregated data with differing trends over time to be brought together and combined to show an overall trend. An example is the *consumer price index* (CPI), which combines the different price movements of many goods and services into one convenient, summary measure showing how overall (or average) price has changed between two time periods.

Index numbers can provide a measure of real growth and inflation and are another way of assisting interpretation of economic changes over time. This is done by decomposing the data into price change and quantity change components (Diewert 2007). To do this, microdata are summarized for the consumption of individual commodities and production of individual products and services, and based on Keynesian economic policy, assume that changes in the price of individual products and services are linked to the wider economy (Diewert 1987). The assumption is, from the economic approach that consumers engage in cost minimizing behavior.

If detailed price and expenditure (or value) data are collected periodically in a survey, then you may wish to use a statistical measure to identify the overall price change for the aggregated data—the measure you need is a *price index*. Price index numbers can be used as indicators in their own right [e.g., CPI and retail price index (RPI)] and can also be used to remove the effect of price change from expenditure data (*deflation*—see Section 12.2.3.3). To provide a measure of change, index numbers use a price reference period from which price change is referred; this is also sometimes called the *base period*. It is convenient to set the value of an index to 100 at a defined period—the index reference period; this is commonly the same period as the price reference period. Then, if the index number at a subsequent period (after the reference period) exceeds 100, the series is showing positive growth; if the index number is less than 100, the series is showing negative growth or *contraction*.

The most common approach to estimating an overall price change is to construct a fixed basket of representative goods and services where the items are priced for a set of consecutive time periods. A price index is calculated as the change in prices from the current period to the price reference period, where the quantities of each commodity are fixed. However, quantities are seldom available, and so the price changes are more usually weighted by the expenditure shares for commodities, which are typically derived from a survey. The basket is updated on a regular basis (usually once a year) and adjustments are made for some goods disappearing during a year and new goods appearing. Weights are updated frequently; either yearly, or every few years.

The simplest price index number is the ratio of prices for one good between two time periods [current price (P_t) by the base period price (P_0)], scaled to 100 in the price reference period:

$$P_{0,t} = 100\left(\frac{p_t}{p_0}\right)$$

The ratio (p_t/p_0) is called a *price relative*.

This is easily extended to create a simple price index for multiple goods—an aggregate price index:

$$P_{0,t} = 100\frac{1}{n}\sum_{i}^{n}\left(\frac{p_{ti}}{p_{0i}}\right)$$

Just combining the price relatives for very different items rarely produces a fair measure of overall price change; consumers typically purchase very different quantities of goods, so a fairer measure can be achieved by weighting the prices (p_t) by the quantities bought (q_t). When both price and quantity information is available, the total expenditure (or value) in period t, can be calculated as

$$V_t = p_t\,q_t$$

Comparing the aggregate expenditure between two time periods does not provide a measure of price change, as it includes contributions to change from both price and quantity. A number of economists studied this problem in the 19th and 20th centuries and a number of solutions for separating these effects were proposed.

Aggregate Laspeyres Price Index The solution proposed by Étienne Laspeyres weights the current period prices (p_t) and the base period prices (p_0) with the base period quantities (q_0). A price index is formed by multiplying current prices by the base period quantities, all divided by the base period prices multiplied by the base period quantities:

$$P^L_{0,t} = 100\frac{\sum_i^n p_{ti}\,q_{0i}}{\sum_i^n p_{0i}\,q_{0i}}$$

The Laspeyres price index formula attempts to measure only the effect of changing prices by removing the effect of the change in quantities by fixing the quantities as those from the base period. In circumstances where consumers reduce the relative quantities of goods bought, relative prices increase; the Laspeyres formula will overstate the price change, and it is therefore considered to be an upper bound.

In practice, it is rarely possible to capture quantities; instead, expenditures on goods are captured along with prices. This is easily accommodated into the Laspeyres formula by simple algebra, which allows us to express the formula in terms of price relatives and expenditure shares:

$$P^L_{0,t} = 100\sum_{i}^{n}\frac{p_{ti}}{p_{0i}}w_{0i} \qquad \text{where} \quad w_{0i} = \frac{p_{0i}\,q_{0i}}{\sum_i^n p_{0i}\,q_{0i}}$$

The Laspeyres price index formula is a base period expenditure weighted arithmetic mean of price relatives.

The issue of quantities (or expenditure shares) not reflecting changing consumer habits is partially addressed by updating both the basket of goods and services and the weights on a regular basis, usually annually but no less frequently than every 5 years. In practice, the expenditure weighting information is usually collected from a different period to the price reference period; the weight reference period may be a year earlier than the price reference period. This leads to a formula that is a variant of the Laspeyres formula, known as the *Lowe index*.

Paasche Price Index An alternative formula was proposed by Hermann Paasche. Unlike the Laspeyres price index, which uses fixed weights from a single base year for each subsequent time period, the Paasche price index uses current period weights for each subsequent period. While this may seem to provide a significant advantage, it has a serious practical disadvantage in requiring continually updated weighting information. The Paasche price index formula is

$$P^P_{0,t} = 100\frac{\sum_i^n p_{ti}\, q_{ti}}{\sum_i^n p_{0i}\, q_{ti}}$$

Written in price-relative, expenditure form, this is

$$P^P_{0,t} = 100\left\{\sum_i^n \left(\frac{p_{ti}}{p_{0i}}\right)^{-1} w_{ti}\right\}^{-1} \quad \text{where} \quad w_{ti} = \frac{p_{ti}\, q_{ti}}{\sum_i^n p_{ti}\, q_{ti}}$$

The Paasche index is a current expenditure weighted harmonic mean of price relatives.

Fisher Price Index The economist Irving Fisher studied a wide variety of alternative formulas for price indices using axiomatic properties to identify the "best" formula to use. This work identified formulas that use weighting information from both the base period and the current period as possessing the largest set of desirable properties. This included Fisher's own index, which is the geometric average of the Laspeyres and Paasche indices; it uses weights from both the base and current periods and is sometimes called a *symmetric index*. The later development of the economic approach to index numbers, using models of consumer preference and utility, also led to identification of the *Fisher index* (and similar index formulas) as "superlative" measures of price change. The Fisher price index formula is

$$P^F_{0,t} = 100\sqrt{P^L_{0,t} P^P_{0,t}} = 100\sqrt{\frac{\sum_i^n p_{ti}\, q_{0i} \sum_i^n p_{ti}\, q_{ti}}{\sum_i^n p_{0i}\, q_{0i} \sum_i^n p_{0i}\, q_{ti}}}$$

This shares the key disadvantage of the Paasche index of requiring current period weights, which makes it impractical for NSIs to use in their regular CPI production. However, in many countries it is considered the "theoretical target" to which NSIs aspire. In practical, operational usage, a Laspeyres-type formula, with regular updates of weights and basket contents is judged to be a sufficiently close approximation.

Quantity (or volume) indices can also be created by simply interchanging the p and q values in the equations above. They provide a measure of the flow of quantities of goods and services consumed.

The change in expenditure (or value) of a set of commodities between two time periods can be split into a change in price and a change in quantity in the following way:

$$\frac{V_t}{V_0} = P^L_{0,t}\, Q^P_{0,t} = P^P_{0,t}\, Q^L_{0,t} \left(= P^F_{0,t}\, Q^F_{0,t} \right)$$

This is an important result that is used widely to calculate consumption expenditures at constant base period prices.

Direct and Chained Indices There are two distinct approaches to producing an index number series: the direct and the chained approaches. The *direct* index compares the current period directly to the base period; it takes no account of what happened in the periods between the current and base periods. For example, if an index series is required for time periods 0,1,2, . . . , T — then a direct index will measure change between each time period t and time period 0 using only data from these two time periods. In contrast, a *chained* index does include the effects of intermediate time periods. So, for a chained index, the price index between time periods t and 0 will consist of an index between time periods 0 and 1, multiplied by an index between time periods 1 and 2 all the way up to an index between time periods $t-1$ and t. This uses data from all time periods.

For example, in the UK, the basket of goods is updated using chain linking each January (the "link month"). In this month, the basket of goods and weighting structures are collected on a consistent basis with the previous months; prices are also collected for a second updated basket of goods, including some new goods and a new index constructed using updated expenditure weights. This new index is scaled to the index value of the "old" basket of goods. The new basket is then used to measure inflation for the subsequent 12 months until the next chaining takes place. In practical terms, the chaining method requires new expenditure weights each year and also involves additional price collection and index number construction in the link month. From a data perspective, chained indices are costly and time-consuming to produce as they require period-to-period information on prices and quantities, but they are generally regarded as an improvement to direct indices as they reduce index number spread (Johnson 1996; Balk 2010) and keep weighting structures current.

Further Reading

Useful references for further information on index numbers are Allen (1975), Balk (2008, 2010), Diewert (1987), Hill (1988), and Johnson (1996).

12.2.3.3 Values and Volumes

Monetary values are often collected in business surveys collecting economic data. For example, the financial amount generated from turnover/receipts/sales for the month/quarter/year. Although useful in their own right, the problem with these monetary values is that you cannot see the effects of changes in prices and volumes. For example, if the results of a retail sales survey show the sales to have increased by 2% over the last year, but prices are known to have increased by 5% over the same period, this suggests that the volume of sales (quantity bought) has decreased. This section provides an overview of why value data are collected and volume data are derived and how to remove the effects of price changes.

Why Value Data Are Collected and Volume Data Are Derived Volume data (sometimes described as *constant price data*) enable analysts and policymakers to monitor economic growth of a particular sector, a number of sectors, or the country as a whole. So, how is this done? Taking retail sales as an example, we first need to ask what data businesses can provide. Imagine that you are an owner of a store that sells food, household goods, and some clothes, and we need to know, from an economic perspective, how your store is performing. As a business, you can provide the monetary value of your sales for a specified period, but to collect the volume of products sold could lead to burden and measurement issues; for instance, would you expect respondents to weigh or count the number of items that they sold? To make this easier from the respondent perspective and hopefully to improve the quality of reported data, the value of sales (amount spent or turnover) is collected from the business, as this allows you to aggregate different goods and services and obtain meaningful totals. However, from an economic analysis perspective, the reasons for changes between different periods are important; that is how much of the change is driven by price and how much by quantity? Volumes are subsequently derived, from the value of sales, by removing the effect of prices.

Value data can be decomposed into price and quantity, which is of more interest to analysts:

$$\text{Value} = \text{price} \times \text{quantity}$$

For example, imagine that you have a market stall selling apples. In month 1 the price of 1 kg of apples on your market stall is £1.00, and during that month you sell 50 kg of apples. The total value for month one is

$$\text{Value} = £1.00 \times 50 = £50$$

In month 2, due to increased transportation costs, you increase the price of 1 kg of apples to £1.20, but your regular customers still buy the same quantity of apples. The total value for month 2 is

$$\text{Value} = £1.20 \times 50 = £60$$

However, in month 3, although the price stays the same as in month 2, some of your regular customers decide that after the price rise last month they will go elsewhere to purchase their apples. So, the quantity of apples you sell falls to 40 kg. The total value for month 3 is

$$\text{Value} = £1.20 \times 40 = £48$$

These three hypothetical months are represented in Table 12.3.

If the value is not decomposed into price and quantity, you would know only that the value had increased in month 2 and then decreased in month 3, but not why.

Now let's consider a real example: UK retail sales for July 2011. In this statistical monthly release seasonally adjusted estimates of all retailing (including automotive

Table 12.3 Three Months of Selling Apples

Month	Price of 1 kg of Apples (£)	Quantity of Apples Sold (kg)	Value
1	1.00	50	50
2	1.20	50	60
3	1.20	40	48

fuel), value, volume, and implied price deflators (estimated prices) were released (see Table 12.4).

The percentage change between each of the latest three months compared with the same three months a year ago show us that in terms of value, year on year, there were increases in all retail sales for the last 3 months. However, in volume terms (quantity), the percentage change shows a different picture with May 2011 showing negative volume percentage change and July 2011 showing no volume change compared with the same month a year ago. So why is the value increasing? If you look at the implied price deflator, which is an estimate of price change that can be used as an indicator of price movements (see Section 12.2.3.9), you will see that some of the changes can be attributed to estimated price changes.

This is why it is important to consider not only values but also volumes and prices; as value data include the effects of price and quantity changes, they do not reflect real growth.

Removing the Effects of Price Changes (Deflation) The effects of price changes are removed by deflating the data. This is common practice in the production of economic statistics by NSIs and is used to assess the real rate of change. Deflation is carried out by dividing current price value data by a price index; this is more practical, given the difficulties in measuring quantities (D'Aurizio and Tartaglia-Polcini 2008). To deflate current price value data, you need price indices for the goods and services for which you have collected value data. Deflation is then carried out by

1. Taking the value data collected in current prices
2. Dividing the value data by the price index in decimal form (i.e., dividing the price index by 100 prior to dividing the value data)

Table 12.4 Retail Sales, Seasonally Adjusted, Year-on-Year Percentage Change

All Retailing, Including Automotive Fuel	Value (Year-on-Year Percentage Change)	Volume (Year-on-Year Percentage Change)	Implied Price Deflator[a] (Year-on-Year Percentage Change)
May 2011	3.4	−0.1	3.8
June 2011	4.0	0.4	3.7
July 2011	4.3	0.0	4.4

[a]Note that the implied price deflator is calculated using nonseasonally adjusted data.

Source: UK Office for National Statistics, *Retail Sales Statistical Bulletin*, July 2011. © Crown copyright (2011).

For a hypothetical example, let's presume that

- The value of all retail sales for the latest period was £10 million.
- The price index for commodities in all retail sales was 110 (with a base year of 2010).
- The price index in decimal form would be

$$\frac{110}{100} = 1.10$$

- The deflated value data (i.e., volume) would be

$$\frac{10{,}000{,}000}{1.10} = £9{,}090{,}909$$

So, inflation adjusted to 2010 prices, the volume of retail sales in the latest period was just over £9 million; this can then be compared with volume estimates in previous periods.

Therefore, the key component in deflating collected value data is to have an appropriate price index. There are well-known price indexes available, such as the CPI, RPI, PPI, SPPIs (see Abbreviations list preceding Chapter 1), and import/export indices. The sections below provide an overview of these price indices.

12.2.3.4 Consumer Price Index (CPI)

Consumer prices index theory was developed by the English economist Joseph Lowe in 1823 and is based on a fixed basket approach measuring price change between two periods (see Box 12.1). The CPI collects consumer expenditure for the product, including taxes, and estimates price change between two periods. The CPI is based on prices for a basket of predefined consumption goods, and represents the expenditure of the "average" household. As well as average household expenditure, a CPI can be used to estimate the effect of price changes for a cost-of-living index (COLI), thereby measuring the extent that household incomes and expenditures will need to change for them to continue to have the same standard of living. In the UK the CPI is produced by collecting approximately 180,000 price quotes each month from a list of almost 700 representative goods and services (Office for National Statistics 2011c). Excluded from the CPI basket are investment items, life insurance, changes in interest rates, illegal goods and services, and home-produced items. Included in the CPI basket are consumer items such as motoring expenses and food costs. The CPI and the closely related RPI (see Section 12.2.3.5) are the main measures of inflation. The CPI is therefore an economic indicator in its own right, is used for making adjustments to state provided welfare benefits, and is also used as a deflator (International Labour Organization 2003; Bureau of Labor Statistics 2007).

Quality concerns come from using a "basket of goods" approach. These concerns center on the fact that the basket is set for a period and therefore does not reflect new goods, changes in the quality of goods, and/or changes in prices as the weights are also

Box 12.1 BASKET OF GOODS APPROACH

Calculated by

$$\frac{\text{Value of the basket of goods in period of interest}}{\text{Value of the basket of goods in base year}}$$

Generally, for convenience and referencing the base year will be set to 100.

fixed for a specified period (International Labor Organization 2003); these issues are partially addressed by regularly updating the basket (Office for National Statistics 2011c).

12.2.3.5 Retail Prices Index (RPI)

The RPI has a longer history in comparison to the CPI (Office for National Statistics 2011c). In terms of construction, the RPI is similar to the CPI as both use a basket of good approach. The differences lie in the population base, the goods included in the basket, and the formulas used to combine price quotes in the first stage of aggregation (see Table 12.5 for examples of RPI and CPI differences).

12.2.3.6 Producer Price Index (PPI)

In the United States from 1902 to 1978 there was only one price index, the *wholesale price index* (WPI), which was a commodity index representing the price of a basket of wholesale goods. In 1978 the WPI was replaced by the *producer price index* (PPI), which reflected measurement of finished-goods price changes from both the

Table 12.5 Differences between the UK RPI and CPI

Factor	RPI	CPI
Population base	Excludes highest-earning households, pensioner households, and institutional households	Includes all private and institutional households
	Excludes spending by foreign visitors	Includes spending by foreign visitors
	Includes spending by UK households abroad	Excludes spending by UK households abroad
Formulas	Uses arithmetic means to combine price quotes in the first stage of aggregation	Mainly uses the geometric mean to combine price quotes in the first stage of aggregation
Commodities	Commodities included are informed by evidence of household spending by the living costs and food survey	Follows national account concepts on commodities consumed by households
	Includes owner-occupied housing costs	Excludes owner-occupied housing costs (e.g., mortgage interest payments, housing depreciation, buildings insurance)

consumer price and the business perspectives; industry based indices were produced (Archibald 1977; Swick and Bathgate 2007).

The PPI measures price changes for a fixed basket of goods and services intended to represent the output of all producers. The prices collected are those received by the producer; PPI includes goods and services purchased by producers as inputs to the production of their goods or services or capital investment, including imports. Data for PPIs are primarily collected using surveys, and the indices are based on the Laspeyres formula (Swick and Bathgate 2007).

In the United Kingdom the data source for PPI weights is the annual survey of products (ProdCom), which is required under European regulation. ProdCom includes both product- and industry-based manufacturing classes; these businesses may also produce services, but these are excluded (Wood et al. 2007).

12.2.3.7 Services Producer Price Index (SPPI)

As the service industry has expanded in its contribution to the output of nations, there has been a growing need to develop price indices for the service industry so that estimates of price inflation can be produced and service industry outputs deflated. The UK *services producer price indices* (SPPIs) were first published in 1996, but prior to 2006 they were published as *corporate services price indices* (CSPIs) and did not align with international standard terminology. SPPIs measure the average price movements for interbusiness services and in the UK were developed on a product by product basis starting with products that had the largest contribution to the total service industry. They are the service industry equivalent of PPIs and like PPIs are calculated as Laspeyres indices. Coverage of service industry products is not as good as PPIs, with the UK data source based on a fixed panel of products collected using periodic surveys; and internationally, the classification of service products are not well defined. There are also conceptual difficulties in collecting SPPIs as price movements in intermediate consumption need to be distinguished from final consumption to satisfy national accounts requirements. Because of all these issues, methodological SPPI developments continue (Wood et al. 2007).

12.2.3.8 Import and Export Indices

Import and export prices are often collected from customs declarations. For example, in the United States export declarations and import entries are collected by the US Customs Service and in the United Kingdom, by Her Majesty's Revenue and Customs. In the United States, a sample of commodities based on those that each month have significant value of trade are included in the import and export indices. Commodities only traded occasionally, and those that do not have a quantity associated with them, are excluded from the indices. The value of imports and exports used excludes cost of transportation, insurance and tax. The disadvantage with this approach is that it does not reflect changes in quality, for example, changes in importing or exporting cheap cars to expensive cars or vice versa. To overcome this issue and the fact that the customs approach excludes certain commodities, the US Bureau of Labor Statistics collects alternative import and export price data from the International Price Program. The program includes both a quarterly survey and a monthly survey. The quarterly survey includes approximately 9000 businesses who supply prices for approximately 23,000 items. The monthly survey includes approximately 2000 businesses supplying prices for around 4000 items. The prices collected are product-specific, so if the product specification is changed, this will be reflected. The indices are published according to three classifications including the Standard International Trade Classification (SITC), which enables international comparisons to be made (Alterman 1991). The indices are used to deflate foreign trade data, and contribute to inflation measurement.

12.2.3.9 Implied Deflators

When you have selected an appropriate price index and deflated the value (current price) data, you will have both value and volume (constant price) data. You can then calculate implied deflators (estimated price changes) for aggregates, and individual industries and sectors. This is carried out by dividing value by volume and multiplying by 100:

$$\text{Implied deflator} = \frac{\text{value}}{\text{volume}} \times 100$$

Remember that

$$\text{Volume} = \frac{\text{value}}{\text{price deflator}}$$

$$\text{Value} = \text{price} \times \text{quantity}$$

Further Reading

Useful references for further information on price indices and deflation are D'Aurizio and Tartaglia-Polcini (2008), Eurostat (2001), International Labour Organization (2003), International Monetary Fund (2004), and Office for National Statistics (2011c).

12.3 PREPARING FOR DISSEMINATION

When the analysis has been completed, the statistics need to be prepared for dissemination; and all too often this phase is carried out under tight time constraints as the dissemination date looms nearer. This phase involves preparing the statistical tables and datasets, ensuring the confidentiality of the data through disclosure control, creating appropriate presentation of the statistics, writing statistical commentary to describe and explain the statistics, if appropriate writing the headline statistical message, and preparing information that will be disseminated to accompany the statistics (e.g., metadata and quality reports) (see Figure 12.4). This section provides an introductory overview of preparing to disseminate activities.

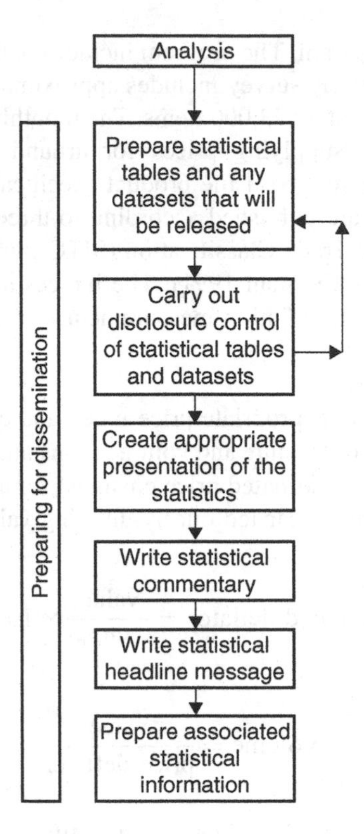

Figure 12.4 Steps involved in preparing to disseminate activities.

12.3.1 Preparing Statistical Tables and Datasets

Tables are a grid approach to the presentation of numbers (Miller 2004) and can include count and magnitude data. Most economic data are presented in tables using magnitude data (e.g., total value of retail sales) but with accompanying counts of businesses (Zayatz 2007). The design of statistical tables needs careful consideration in terms of not only the content of the tables and which variables to include but also the title of the table, column and row labels, and disclosure control (see Section 12.3.3). The table title and column and row labels need to be meaningful to enable users to easily identify the table content. Notations in statistical tables, such as for revised estimates and cells that cannot be released, need to be clearly identified in the tables and the footnotes. The key therefore to designing a good statistical table is to ensure that it is as self-explanatory as possible.

To improve the presentation of summary statistics, Table 12.6 was introduced into the ONS retail sales statistical bulletins in 2011 (similar tables were also introduced in the Index of services and Index of production statistical bulletins). From the column headings you can see that the table includes both comparison of time periods and contributions by weight that the different sectors provide for the latest period.

Table 12.6 Year-on-year Growth Rates and Contribution to All Retailing by Sector

	% of All Retailing	Volume (SA) Year-on-Year Growth (%)	Contribution to all Retailing (% Points)	Value (SA) Year-on-Year Growth (%)	Contribution to all Retailing (% Points)
All retailing	**100.0**	0.6		5.4	
Predominantly food stores	**41.7**	−0.3	−0.1	5.7	2.4
Predominantly nonfood stores					
Total	**43.2**	−0.7	−0.3	0.4	0.2
Nonspecialized stores	7.8	4.0	0.3	4.4	0.3
Textile, clothing and footwear stores	12.2	−2.1	−0.3	0.4	0.0
Household goods stores	9.7	−1.6	−0.2	−2.0	−0.2
Other stores	13.5	−1.6	−0.2	−0.1	0.0
Nonstore retailing	**4.9**	15.5	0.8	15.6	0.8
Automotive fuel	**10.2**	2.8	0.3	20.0	2.0

Source: UK Office for National Statistics, *Retail Sales Statistical Bulletin*, September 2011. © Crown copyright (2011).

Very positive user feedback has been received with the introduction of this table, predominantly as it summarizes the key statistics in the release. From Table 12.6 you can see that

- The predominantly nonfood stores total makes up the largest percentage of all retailing (43.2%)
- The largest year-on-year, seasonally adjusted growth in volume was in the nonstore retailing sector (15.5%), which contributed 0.8 percentage points to all retailing.
- Year-on-year, seasonally adjusted growth in value was also high in the nonstore retailing sector (15.6%), and in terms of weighted contribution this added 0.8 percentage points to all retailing. The largest weighted value contribution came from predominantly food stores.

When statistics are released, they are generally accompanied by the corresponding datasets; for business statistics produced from recurring surveys, these will be time series datasets containing back series data for all the periods. For example, for statistics produced from a monthly business survey, the time series datasets will contain annual, quarterly, and monthly data for the most recent period and the back series. Increasingly there is a move to providing online interactive facilities for users to access more than one dataset at a time. For example, interactive web facilities for users to build their own statistical tables from datasets available on the web are provided by Statistics New Zealand (TableBuilder) and the US Census Bureau (DataFerrett). These are very useful facilities for users and provide a flexibility that static statistical tables do not.

In the UK the transparency and open-government initiative is gaining pace with the objective of increasing government accountability by publishing more government data online (Berners-Lee 2009). For many types of data, there are privacy and confidentiality issues with this that still need to be thought through, especially in relation to the potential impact on disclosure control and survey response rates.

Further Reading

A useful reference for further information on creating statistical tables is Miller (2004).

12.3.2 The Importance of Maintaining Confidentiality

When survey organizations request data from respondents, they enter into a legal contract with them to protect their anonymity and maintain the confidentiality of data received. This guarantee is related to handling data, analysis, and dissemination and is put in place for legal, ethical, and practical reasons. Tradeoffs will often have to be made between the demands for detailed data and ensuring anonymity and confidentiality (Forth and Webster 2008). There are three general aspects to confidentiality, which are that, individual data items will (1) be used only for statistical purposes, (2) not be released to anybody else without consent of the respondent, and (3) not be identifiable directly or indirectly in statistical outputs.

The confidentiality assurance is generally provided to respondents in the survey organization's confidentiality guarantee or pledge (see Chapter 9) and supported by legislation. These confidentiality guarantees or pledges also fulfill the obligation laid out in principle 6 of the *fundamental principles of official statistics*, which state that "Individual data collected by statistical agencies for statistical compilation, whether they refer to natural or legal persons, are to be strictly confidential and used exclusively for statistical purposes" (United Nations Statistical Commission 1994).

From a policy perspective, the survey confidentiality assurance has implications for the sharing of data. This has always been an issue but is becoming more of an issue in the United States and UK, where there is an increasing drive to make more government data available online, and to increase the sharing of regulatory data in an attempt to "open up government" and reduce the regulatory burden on businesses (Berners-Lee 2009). There is a push to include NSI business survey data within this. To provide unambiguous confidentiality assurance to safeguard businesses, a clear line needs to be drawn between data collected for regulatory administrative purposes (e.g., taxation) and data collected for statistical purposes from surveys (Anderson and Seltzer 2005). A good overview of disclosure control problems in releasing microdata is provided in Bethlehem et al. (1990).

Over the decades there have been repeated attempts by governments to share survey data. In the United States, following the outbreak of World War II, there was increasing pressure to share data for the benefit of conducting the war, and only for this purpose. This provision was then added to the Second War Powers bill, which was passed in March 1942. Later in the same year further legislation was passed that enabled the sharing of data between federal agencies that had the mandatory authority to collect the same specific data; one benefit of this was seen as removing duplication of data requests. However, this led to confusion with different agencies trying to access data where they had no legal authority to collect them. In 1961 statistical confidentiality standards were defeated in a Supreme Court case, but in 1962, after several cases of conflict between federal agencies, confidentiality standards were reconfirmed in legislation. In 1974 these were again confirmed in the Privacy Act (Anderson and Seltzer 2005).

From an operational perspective, the confidentiality assurance requires survey organizations to prevent the risk of disclosing identities and attributes, while holding the data and when disseminating the statistics (Hundepool et al. 2010). In business surveys assurances of confidentiality are also seen as a vehicle for encouraging "full and accurate responses by businesses" (Anderson and Seltzer 2005, p. 1). Some research has indicated, for household survey respondents, that confidentiality concerns can affect survey participation (Couper et al. 2008). Later research, however, found, that even if you are able to numerically estimate the likelihood of statistical disclosure and provide this information to respondents as part of informing their consent to participate in the survey, it has no effect on survey participation—the exception to this was when respondents were also given information on the possible harm arising from disclosure of their information. What did appear to affect survey participation was the survey topic, with indications that sensitive topics can reduce participation. The researchers believed that sensitive topics could be intrinsically linked to perceptions of harm (Couper et al. 2010). If you consider this in relation to business survey responses, when the requested data might be market-sensitive, and possibly not even disseminated to shareholders, you can see that maintaining the confidentiality of responses is paramount.

Research undertaken by Nichols and Willimack (2001) identified that in relation to data sensitivity businesses were concerned about (1) the type of data being requested, (2) how detailed the requested data were, and (3) the timescale between the data being business sensitive and when they were requested by the survey organization. Businesses may also be concerned about data sharing, and unsure of what the survey confidentiality assurance actually means and the penalties for breaching confidentiality. Businesses may also have a variety of perceptions in relation to how much they can trust the survey organization (Stettler and Willimack 2001).

To maintain operational data confidentiality, survey organizations should implement both data access control and disclosure control techniques (Willenbourg and de Waal 2001). Access control includes two elements: (1) access to data during the phases of running the collection, processing the data, and analyzing the data prior to dissemination; and (2) access to microdata following dissemination (see

Section 12.5). Disclosure control procedures are then used as a method of controlling the risk of identifying an individual or business in released statistics, including tables, time series datasets, and microdata. There are three common approaches to disclosure control of tables, time series datasets, and microdata:

1. *Anonymizing*—removal of unique identifiers
2. *Nonperturbative methods*—suppression of data that may indirectly reveal an individual or business
3. *Perturbative methods*—introduction of uncertainty to the data

Let us consider an example of disclosure risk for a tabular release. Imagine that you work for a large specialized business that is in a monthly business survey sample. As your business is specialized, the sampling cell that your business is in includes only your business and one other, and you know that there is only one other business that undertakes the same activities as you. So, if the data were published at this level, you could ascertain the reported value for the other business in your cell in the table simply by subtracting the value of the data that you responded with from the total of the cell. This is one aspect that disclosure control procedures safeguard against.

At ONS mandatory business surveys are governed by the Statistics of Trade Act (1947). Within this act, Section 9 states that no published tables should disclose an individual business unless that business has given prior consent. It goes onto say that published data should not contain any cells that can identify the exact number of respondents represented in the cell (Lowthian et al. 2006). As disclosure control procedures are aimed at reducing the risk of identifying individuals and businesses, the net effect is the reduction of data quality as data are modified or suppressed—the goal for survey organizations is balancing the release of as much quality data as possible with the confidentiality assurance provided to respondents (Zayatz 2007).

12.3.3 Disclosure Control Methods

The purpose of applying statistical disclosure control methods is to reduce to an acceptable level, the risk of disclosing confidential information while minimizing the impact on data utility. Disclosure control can be applied to microdata and/or tabular data. The disclosure control methods first need to identify the risk of disclosure and then have to remove the risk, using either perturbative (distortion) or nonperturbative (suppression or reduction of detail) methods.

12.3.3.1 Microdata Disclosure Control Methods

There is a growing demand for the release of microdata to external researchers that places a greater need for appropriate access controls (see Section 12.5) and disclosure control methods to reduce the risk of identity and attribute identification (Hundepool et al. 2010). To reduce this risk, perturbative and nonperturbative disclosure control methods are used for microdata depending on whether the data are continuous or categorical, the risk of identification, who will have access to the data (e.g., internal or external), and a desire for low information loss.

Table 12.7 Microdata Disclosure Control Methods for Different Data Types

Method	Continuous Data	Categorical Data
Perturbative		
Additive noise	Yes	
Data distortion by probability distribution	Yes	Yes
Microaggregation	Yes	
Resampling	Yes	
Lossy compression	Yes	
Multiple imputation	Yes	
Vector camouflage	Yes	
Postrandomization method (PRAM)		Yes
Rank swapping	Yes	Yes
Rounding	Yes	
Nonperturbative		
Sampling		Yes
Global recoding	Yes	Yes
Top and bottom coding	Yes	Yes
Local suppression		Yes

Source: Domingo-Ferrer and Torra (2001), pp. 95, 103).

Domingo-Ferrer and Torra (2001) provide a useful overview of different microdata disclosure control methods for different data types, which are summarized in the Table 12.7. For further information on individual methods, the reader is referred to this paper (Domingo-Ferrer and Torra 2001) or alternatively Skinner (2009) or Hundepool et al. (2010).

National statistical institutes use microdata nonperturbative global recoding methods and local suppression extensively; this is often carried out using μ-ARGUS software (a relative of τ-ARGUS used for tabular data). μ-ARGUS is used to identify the proportion of microdata records at risk of identification. The software focuses on combinations of identifying variables and the optimal set of recodings and suppressions to reduce the risk of disclosure (Hundepool 2000); other perturbative disclosure methods can then be selected for application (Hundepool et al. 2010, pp. 103–105).

12.3.3.2 Tabular Disclosure Control Methods

Statistical outputs from business surveys can include tables of counts, magnitudes, and frequencies of, for example, businesses, geography, industry, sector, employment size, and/or product code. The level of geography and industrial classification can vary, for example, with some industrial classification variables at a two-, three-, or even four-digit level (Lowthian et al. 2006).

One way to minimize the risk of disclosive tabular data is to carry out pretabular disclosure control methods on the microdata as outlined in Section 12.3.3.1. Hundepool et al. (2010) note that one of the advantages of this approach is that output tables will be "consistent and additive" (ibid., p. 173), but the disadvantages are a possible "high level of perturbation," distorted data distributions, and a lack of transparency for users (ibid., p. 174).

Other methods for protecting tabular data are table designs and posttabular methods such as cell suppression, cell perturbation, rounding, small-cell adjustment, and barnardization [Hundepool et al. (2010, pp. 174–183) provides an overview of these posttabular methods].

In the first instance, when preparing data tables for publication, you should design the tables to avoid the possibility of disclosive (unsafe) cells. Here you may have conflict between maintaining confidentiality through disclosure control and meeting user needs for statistical outputs. So, you need to carefully consider user needs and how far you can combine table cells.

Once the tables have been designed, they must be checked for disclosive cells. Remember that for businesses included in the survey sample, there may be economic gain to elicit from the survey outputs information about their business competitors. This poses two risks to the data: (1) from businesses contributing to the cell value, and (2) from businesses not contributing to the cell value. When considering these two possible risks, information from secondary sources and the other tables contained within the statistical release also need to be considered as they can be used to identify businesses. However, in statistical outputs from business surveys the tables generally include weighted responses, so disclosure by differencing should not be possible unless the weights are known (Lowthian et al. 2006).

Cells in a table maybe unsafe because of low counts; if there is only a single respondent or two respondents, the cell is clearly disclosive. For higher frequencies, the cell could still be considered disclosive, especially for sensitive variables. Cells may also be unsafe if one or more respondent dominates the cell. A published cell total in combination with the value of another member of the cell could lead to an accurate estimate of the largest respondent.

A two-stage process for identifying unsafe cells is a recommended approach:

1. Threshold rule—there must be a specified number of respondents in a cell.
2. The $p\%$ rule—unsafe cells are identified using the $p\%$ rule. A cell is most vulnerable to disclosure when the respondent with the second largest value attempts to estimate the largest value provided by a respondent in the cell. The rule ensures that the respondent with the second largest value in the cell is unable to estimate the largest value in the cell to within $p\%$ (Cox 1980).

Ramanayake and Zayatz (2010, p. 5) show that for a cell with value T:

$$T = \sum_{i=1}^{N} \chi_i$$

where N is the total number of businesses and $\chi_i \geq 0$ are the individual values from businesses ordered from largest to smallest. The cell will result in disclosure if

$$\frac{p}{100} * \chi_i - \sum_{i=3}^{N} \chi_i > 0$$

and should be suppressed. There are two types of cell suppression:

1. *Primary suppressions*—where the originally identified unsafe cell (a cell that has failed either the threshold rule or the *p*% rule) is suppressed
2. *Secondary suppressions*—suppressing other cells in the table that provide values that could be used to calculate the primary suppressed cells (e.g., subtracting the total value of the remaining cells in the row or column from the marginal total)

In the UK Government Statistical Service, disclosure control risks are checked by ensuring that published cells in a magnitude table (aggregates), from business surveys, meet the following criteria (Lowthian et al. 2006):

- There must be at least *n* businesses in a cell (threshold rule).
- The total of the cell minus the largest value *m* from a reporting business must be greater than or equal to *p*% of the largest value (*p*% rule).
- The values of *n*, *m*, and *p* must not be disclosed.

For tables of count data, the UK Government Statistical Service recommend that cells be rounded to 5 and percentages or rates be derived from the rounded values (Lowthian et al. 2006).

When disclosure control methods have been applied to statistical output tables, it is imperative that this be communicated to users, although no parameter values should be mentioned. This should be carried out by including statements to inform users of the method and the impact on data quality.

It must also be remembered that disclosure control methods can reduce the amount of information published from surveys. For example, in business survey outputs, one primary suppression can result in up to three secondary suppressions. But this needs to be balanced against the confidentiality assurance that has been given to respondents; the tradeoff is confidentiality versus data quality. Unlike cell suppression, data swapping of microdata prior to the generation of tabular data, should maintain the quality of the tabulated data as the overall content of the table is maintained (Ramanayake and Zayatz 2010).

Many NSIs use τ-ARGUS or other software to identify tabular cells at risk of disclosure and to apply modifications to remove the disclosure risk and limit information loss. The focus of τ-ARGUS is on secondary cell suppression and redesigning tables by combining columns and rows (Hundepool 2000).

Further Reading

Useful references for further information on confidentiality are Anderson and Seltzer (2005, 2007) and Nichols and Willimack (2001). References on disclosure control are Hundepool (2000), Hundepool et al. (2010), Massell (2004), Massell and Funk (2007), Ramanayake and Zayatz (2010), Willenbourg and de Waal (2001), and Zayatz (2007).

12.3.4 Presenting the Statistics

Presenting the statistics is just as important as any other phase of the statistical production process, and the last thing you want, after all the effort of collecting capturing, cleaning, and analyzing the data, is to produce statistics that are poorly or inappropriately presented. Statistics can be presented in tables, graphs, and words, and effective presentation will generally include all three approaches as this assists in ensuring that the wider user base is catered for. The presentation of statistics also requires clear data labels, for example, for economic statistics to indicate which estimates are seasonally adjusted, and to distinguish value, volume, and indices.

Despite the old adage "a picture is worth a thousand words," when presenting statistics, this will be the case only if the statistics are visualized appropriately. Too often inappropriate graphical scales, colors, and types of graphs are used. There is a whole body of literature on how to appropriately visualize data (see Further Reading Section below). The key points to note are that data visualization should illustrate trends and relationships between the data, it should be kept simple, be easy to see the key messages, and be accessible for the range of users (United Nations Economic Commission for Europe 2009b; Bethlehem 2009, pp. 322–339). As Julian Champkin says: "The mark of a good graphic is that the user reads the diagram, not the caption. Information should be in the picture, not in the words underneath" (Champkin 2011, p. 41).

12.3.4.1 Writing Statistical Commentary

A useful overview of how to improve statistics commentary is provided in the report by the United Nations Economic Commission for Europe (2009a). One of the key messages in this publication is that writing commentary to accompany statistics should be approached as "statistical story-telling" in that the commentary needs to be engaging to the reader as it provides an overview of the key messages that the statistics show. What is important to remember is that there are numerous types of statistical users ranging from experts to members of the public who might access the statistics only through media articles or even just headlines. The goal should be to pitch the statistical story so that it represents the statistics, is interesting, and can be understood by a variety of types of users. It must also be remembered that people are more likely to trust the statistics if they can understand them; this is where the statistical story should help but also includes disseminating relevant metadata (see Section 12.3.4.3).

Writing about numbers can be difficult, and often the writer will naturally migrate to just describing increases and decreases. Miller (2004, pp. 2–3) acknowledges that there is little training in how to write about numbers and that this exacerbates the prevailing issue of poor statistical commentary. In the UK this issue has been identified across the Government Statistical Service, and there are efforts to raise awareness of the benefits of good statistical commentary and develop guidelines.

The importance of statistical commentary is being increasingly linked to maximizing the utility of the statistics. In December 2007, Mr. Almunia, the then European Commissioner for Economic and Monetary Affairs, concluded a speech at the Shaping Statistics for the 21st Century Conference with a call for better communication of statistics: "Producing relevant and accurate numerical data is all very well. But if statistics are to fulfill their role as a public good and a tool for democracy, they must be clearly explained and set in context so that they are accessible and easy to use for the general public" (SIGMA 2008).

12.3.4.2 Identifying the Headline Message

It is not essential to produce a headline statistical message, and some organizations do not produce them, however, if you are going to prepare one imagine that you are writing a media headline that can stand alone without any further explanation. When the statistics are disseminated if the media or key users use the headline that you produced, this can be your measure of success. A successfully produced headline should succinctly and accurately cover the overarching message that the statistics are representing and put it into a context. Some examples of succinct and context-based statistical headline messages are given below. It must be noted, however, that when headline statistical messages were reviewed, during the writing of this chapter, the majority of headlines still only focused on increases and decreases and did not place the message in a context.

Some examples of succinct and context-based statistical headline messages are

- "CPI annual inflation stands at 5.2% in September 2011. CPI annual inflation has never been higher, but was also 5.2% in September 2008" (UK Office for National Statistics, October 18, 2011).
- "Flattening of per capita residential energy consumption reflects multiple factors" (US Energy Information Administration, October 21, 2011).
- "The main strategic focus of enterprises in Canada is orientated toward exploiting existing products or current activities rather than developing new ones" (Industry Canada, *Highlights of the Survey of Innovation and Business Strategy*, 2009).
- "Still healthy, wealthy, and wise—but dropping in productivity" (Australian Bureau of Statistics, October 6, 2011).
- "The total energy used by the services sector in New Zealand during 2010 was almost 64 petajoules (PJ), Statistics New Zealand said today. This is about the same as the total amount used inside New Zealand homes each year" (Statistics New Zealand, October 13, 2011).

12.3.4.3 Information to Support Disseminated Statistics

For users to fully understand the released statistics, associated metadata should be disseminated to accompany the statistics. Metadata are "data that define and describe other data," and statistical metadata are "data about statistical data, and comprise data and other documentation that describe objects in a formalized ways" (United Nations Economic Commission for Europe 2009c, p. 7). Included in statistical metadata are information on how the data were collected and processed (survey design), definitions and classifications used, and an assessment of the quality of the derived statistics. Metadata may be standalone or part of an integrated centralized standard metadata system; the latter is the vision for many survey organizations. All too often metadata are created as a late addition and only for internal use by survey and analyst practitioners (Gillman and Appel 1999). Instead of this late addition and internal approach, metadata should be central to the production and dissemination of statistics, thereby benefiting users and internal practitioners.

In more recent years there has been an increasing focus on compiling and disseminating quality output measures to accompany statistical outputs. In Europe this initiative has been centrally driven by Eurostat, who have produced guidelines for both output quality measures (Eurostat 2005; Office for National Statistics 2007) and producing quality reports (Eurostat 2009b).

At ONS standalone quality reports now accompany all statistical outputs and are annually updated. The reports contain information relating to output quality dimensions (see Chapter 3) and also contain associated metadata on aspects such as the source of the data and methods used to compile the statistical output.

ONS quality reports were first introduced in 2005 and piloted with the Bank of England and Her Majesty's Treasury. Positive feedback was received from these two key users, who reported that the quality reports reduced the time needed for data quality assessment. Following the pilot, quality reports were initiated for all ONS statistical outputs, but by 2009 the number of web hits for these reports was low and there were concerns relating to web accessibility and usability.

To evaluate the accessibility and usefulness of ONS quality reports, a web user survey was carried out in 2010. Respondents to the survey included known users, professional government groups, people who contacted the ONS contact center, and unknown users accessing the ONS website; in total 138 responses were received. Analysis of responses showed that only 30% of respondents had previously seen a quality report, so access was an issue (the quality reports were on a separate web page to the statistical output). There was mixed feedback on the usefulness of the reports, 25% of respondents stating that they would change nothing at all, and a another 25% stating that they would like the level of detail changed and the key points more readily available in the reports. The remaining respondents suggested changing access to the reports or commented on specific changes. Overall, though, respondents were positive to the quality reports, and work continues on improving access to and content of the reports (Green et al. 2010a). Noting that to maximize the utility of the information that accompanies statistics, the information must be accessible and relevant to users' needs.

Further Reading

Useful references for further information on presenting statistics are Cleveland and McGill (1984, 1987), Few (2004), Robbins (2005), Tufte (2006), United Nations Economic Commission for Europe (2009b), and Wainer (1984). References on statistical commentary are Miller (2004) and United Nations Economic Commission for Europe (2009a). References on metadata and quality reporting are Eurostat (2005, 2009b) and United Nations Economic Commission for Europe (2009c).

12.4 DISSEMINATION

In NSIs statistical release dates are preannounced, sometimes as much as 12 months in advance, and should align with users needs for the statistics. At ONS, provisional statistical release dates must be made available 12 months in advance, and under guidance issued by the National Statistician (2009), dates must be confirmed 6 months in advance for market-sensitive statistics, and one month in advance for non-market-sensitive statistics.

Most official statistics are disseminated and made available to everyone at a standard daily time; in the UK Government Statistical Service this is 0930 hours. However, in some NSIs predetermined statistics are prereleased to a small number of preagreed individuals in advance of general release. This prerelease access is provided only if there is clear justification for the need for early access for briefing purposes, namely, preparing briefings for government ministers; this is often important for market-sensitive statistical outputs such as GDP, CPI, RPI, and trade statistics. Within the UK prerelease arrangements are specified in the 2007 Statistics and Registration Service Act and in the UK Code of Practice for Official Statistics. The key aspects of UK prerelease access are that the statistics be made available 24 hours prior to general release to named individuals who have provided justification for the need for prerelease access; the list of individuals is published. While in receipt of prerelease statistics the statistics are under embargo and cannot be disclosed to anyone else, or even an indication given of whether a statistic has increased or decreased (Cabinet Office 2008).

For obvious security reasons, prerelease access cannot be granted over the Internet; so prerelease documents are sent via email in encrypted password-protected files. In the UK, for certain statistical releases, particularly market-sensitive statistics, 45 minutes of prerelease access, prior to the 0930 hours general release time, is given to predetermined members of the media during "lockins," where they are literally locked in a room and telephone and Internet access are confiscated. The lockin provides journalists with the opportunity to study the statistics, ask the statistician in the lockin questions, and generate their stories prior to general release. This means that almost simultaneously with the general statistical release on the Internet, the headline statistics and stories also appear on the television, radio, and Internet newscasts.

When the statistics are released onto the Internet, they are often presented in a variety of dissemination products such as, summary documents of the key points, bulletins providing a fuller picture of the statistics with accompanying data reference tables, datasets containing the actual data, and more recently podcast summaries. The range of statistical products helps to meet the needs of different types of users, such as analysts who want to analyse the datasets, and the media who want to know the key points.

12.4.1 Evaluating Statistical Dissemination

The very essence of conducting a survey is to meet some known user needs. As outlined in Chapter 1, these needs might come from economic, political, or sociological objectives in relation to monitoring, policymaking, and decisionmaking.

They might also be driven by national or international legislative requirements. In addition to this, they might be used by members of the general public to inform their everyday decisions or general interest. It is important, therefore, to evaluate the dissemination of statistics. Evaluation can be undertaken by secondary research such as collating media coverage, or primary research such as conducting surveys of users, or organizing user groups. An example of using a variety of methods to gain an understanding of the needs of users can be found in Jones et al. (2011). In this example a combination of user surveys, a user event, and a review of responses to a previous consultation provided indepth insight of users' needs in relation to the presentation of the statistics and use of the statistics. Ultimately, the objective of user research is to investigate whether the released statistics meet user needs, not just in terms of the actual statistics but also the way in which they are disseminated and presented (United Nations Economic Commission for Europe 2009b).

Further Reading

A useful reference for further information on the statistics release aspect of dissemination is the London Cabinet Office (2008).

12.5 ARCHIVING

Data archiving includes internal and external archiving of microdata. Internal archiving is necessary so that data can be used for future releases of time series data or held for a period of time in accordance with internal archiving policies; for example, images of completed business survey questionnaires are kept at ONS for 4 years for an annual survey, 12 quarters for a quarterly survey, and 27 months for a monthly survey before they are destroyed. Archiving of internal microdata (data captured and saved in a datafile) are kept for much longer. In contrast, external archiving is used to maximize the utility of microdata by making it available to researchers (e.g., academics, nongovernment agencies, and international agencies) for secondary analysis. For NSIs, this access has to be provided in compliance with the sixth United Nations Fundamental Principle of Official Statistics (see Section 12.3.2). To comply with this principle, the United Nations Economic Commission for Europe (2007, p. 6) advocates the use of the following principles:

"Principle 1. It is appropriate for microdata collected for official statistical purposes to be used for statistical analysis to support research as long as confidentiality is protected.

Principle 2. Microdata should be made available only for statistical purposes.

Principle 3. Provision of microdata should be consistent with legal and other necessary arrangements that ensure that confidentiality of the released microdata is protected.

Principle 4. The procedures for researcher access to microdata, as well as the uses and users of microdata, should be transparent and publicly available."

Microdata are usually held in data matrixes in which each row represents a survey record and each column, a response to a particular survey question. Each survey record will generally include qualitative unique identifiers such as name and address. Microdata anonymization includes the removal of these unique identifiers. In the broadest sense, these identifiers are not sensitive as external people will know or can look up the names and addresses of individuals or businesses, but when contained within statistical datasets, the identifiers make it possible to identify the responses of individuals or businesses (Bethlehem 2009). So the first step in anonymizing microdata is to "deidentify" (anonymize) the microdata by removing identifiers such as name of business and address (Willenbourg and de Waal 2001; Ramanayake and Zayatz 2010), but this can still leave the risk of identification through uniqueness. To reduce this risk, further perturbative and/or nonperturbative methods (see Section 12.3.3.1, Table 12.7) are applied, such as grouping sectors, adding noise to the data, or data swapping.

External access to archived microdata is controlled either through anonymized public user microdata files or through actual or virtual microdata laboratories and approved researcher procedures. The level of disclosure risk left in the microdata is governed by how external access is provided. Microdata contained within public user files must be anonymized to the extent that identification will not be possible even if the data are combined with other microdata files; whereas microdata available in actual or virtual laboratories may contain potentially identifiable data if combined with other datafiles. This is why controls such as approved researcher, the justification of the research, and the checking of analytical outputs for disclosure control are carried out in the laboratories. From the survey organization perspective, as methods for data matching are improving and more data are made publicly available, the risk of identification in survey microdata increases; this could possibly lead to more reliance on microdata laboratories and less use of public user files (United Nations Economic Commission for Europe 2007).

For business survey data, additional reasons for increased use of microdata laboratories, or not having access to any or very much business survey microdata, are (1) the skewed design of many business survey samples towards large businesses, with the largest businesses being selected for many surveys with certainty—which reduces the availability of business survey microdata containing large businesses; and (2) that in some countries there are many sources of publicly available business data (e.g., company records) which enhances the potential for data matching (United Nations Economic Commission for Europe 2007). These reasons should, however, be offset with the fact that some data collected in business surveys remain sensitive for only a limited period of time before the businesses publicly release the information themselves, for example, profits and losses, company accounts, and dividends reports.

An Example of Available Microdata

The US Census Bureau makes available individual business responses to the 1963, 1967, and 1972 to present Annual Survey of Manufacturers and the Census of Manufacturers. Period-to-period individual business responses are linked via a unique longitudinally consistent business identifier. More recently, microdata files of 1950s and 1960s survey and census responses were recovered prior to the decommissioning on an old mainframe at the US Census Bureau. The challenges now are whether the recovered 1950s and 1960s microdata can be converted from their original file format, restructured, and then, using data matching techniques, linked together with the already available microdata; if they can, this will further enhance an already rich source of data for analyzing changes over time in manufacturing businesses. In addition, it will provide more periods for aggregate-level analysis, from the days of the manufacturing boom (1950s and 1960s) to present, thus opening up options for new research (Becker and Grim 2011).

Externally archived microdata must have associated metadata to accompany them, or users will not know the research potential of the data, and will not have the basic knowledge to analyze the data (Wirth 2011). An example of a microdata metadata system is the German Microdata Laboratory (MISSY), which focuses on the German microcensus (1% of households in Germany). MISSY contains metadata for 1973–2008 microcensus scientific user files. It includes detailed variable-level information, comparisons of variables over time, questionnaires, concepts, classifications and definitions, and microdata tools (Bohr 2011).

In addition to disclosure control concerns, access to microdata can raise concerns in relation to data quality, for example, differences between microdata and published aggregate data; and concerns in relation to microdata access control costs. From a benefits perspective, access to microdata should reduce the need for researchers to collect their own data, thereby reducing response burden; and enable the same data to be repeatedly used in research, thus maximizing their utility (United Nations Economic Commission for Europe 2007).

Further Reading

A useful reference for further information on archiving is United Nations Economic Commission for Europe (2007).

12.6 CONCLUDING COMMENTS

This chapter has provided an introductory overview of the phases and procedures needed to move from survey data to statistics. What is evident from the chapter are the numerous activities that need to be undertaken during the analysis and dissemination statistical production phases, and specific analytical techniques that may also need to be carried out. Many of the phases and procedures covered are specialties in their own right (e.g., seasonal adjustment, index numbers, disclosure control) and have a whole body of literature associated with them. This chapter therefore only scratches the surface of these specialized areas. Many of the areas covered require specialists to be involved in the methodological design and review of procedures.

Until recently, statistical presentation and commentary were not regarded as specialisms and have been neglected fields. As survey and statistical practitioners, we have put considerable effort into the design and implementation of surveys and statistical methods and then often let ourselves down in the way we present, describe, and explain our statistical outputs. There now appears to be an increasing focus on correcting this and ensuring that statistical commentary and presentation are improved to meet a wide range of different user needs.

ACKNOWLEDGMENTS

I would like to particularly thank Simon Compton (ONS) for his reviews, comments, and suggestions—I am not sure that I could have finished this chapter without his input. I would also like to thank the following people for their reviews, comments, and suggestions while I was writing this chapter: Jeff Ralph, Philip Lowthian, Alan Smith, Darren Morgan, Paul Smith, and Martin Brand (UK Office for National Statistics); and Mike Hidiroglou (Statistics Canada).

References[1]

AAPOR (2011a), *Standard Definitions: Final Disposition of Case Codes and Outcome Rates for Surveys*, 7th ed., The American Association for Public Opinion Research, Deerfield, IL (available at www.aapor.org/Standard_Definitions2.htm).

AAPOR (2011b), *AAPOR Response Rate Calculator*, version 3.1, the American Association for Public Opinion Research, Deerfield, IL (available at www.aapor.org/Standard_Definitions2.htm).

ABS (2002), *Advance letters – an ABS Approach*, Research report, Australian Bureau of Statistics, Canberra.

ABS (2006), *Forms Design Standards Manual*, Australian Bureau of Statistics, Canberra.

ABS (2010), *Quality Management of Statistical Processes Using Quality Gates*, Information paper 1540.0, Australian Bureau of Statistics, Canberra: (available at www.abs.gov.au/websitedbs/D3310114.nsf/home/Quality:+Data+Quality+Management).

Aitken, A., Hörngren, J., Jones, N., Lewis, D., and Zilhão, M. J. (2003), *Handbook on Improving Quality by Analysis of Process Variables*, Eurostat, Luxembourg (available at http://epp.eurostat.ec.europa.eu/portal/page/portal/quality/documents/HANDBOOK%20ON%20IMPROVING%20QUALITY.pdf).

Alba-Juez, L. (2009), *Perspectives on Discourse Analysis: Theory and Practice*, Cambridge Scholars Publishing, Newcastle upon Tyne, UK.

Allen, R. G. D. (1975), *Index Numbers in Economic Theory and Practice*, Transaction Publishers, Pistacaway, NJ.

Alterman, W. (1991), Price trends in U.S. trade: New data, new insights, in Hooper, D., and Richardson, J. D., eds., *International Economic Transactions: Issues in Measurement and Empirical Research*, University of Chicago Press, Chicago, pp. 109–143.

Ambler, C. A., Hyman, S. M., and Mesenbourg, T. L. (1995), Electronic data interchange, in Cox, B. G., Binder, D. A., Chinnappa, B. N., Christianson, A., and Kott, P. S., eds., *Business Survey Methods*, Wiley, New York, pp. 339–352.

[1] Many of the references listed in this list are available on the internet, even if no internet address is provided.

Designing and Conducting Business Surveys, First Edition.
Ger Snijkers, Gustav Haraldsen, Jacqui Jones, and Diane K. Willimack.

Anderson, A. E., and Harley, M. D. (2007), Using software to collect data electronically for the economic census, *Proceedings of the 3rd International Conference on Establishment Surveys (ICES-III)*, Montreal, June 18–21, American Statistical Association, Alexandria, VA, pp. 435–442.

Anderson, A. E., and Morrison, R. L. (2005), Applying knowledge of business survey response processes to the design of data collection software at the U.S. Census Bureau, paper presented at the Annual Federal Committee on Statistical Methodology (FCSM) Research Conference, Arlington, VA, Nov. 14–16.

Anderson, A. E., Nichols, E., and Pressley, K. (2001), Usability testing and cognitive interviewing to support economic forms development for the 2002 economic census, paper presented at the 2001 International Statistics Canada Methodology Symposium: Achieving Data Quality in a Statistical Agency: A Methodological Perspective. Hull, Quebec, Oct. 18–19, CD-rom, Statistics Canada, Ottawa.

Anderson, A. E., and Tancreto, J. G. (2011), Using the web for data collection at the United States Census Bureau, paper presented at the 58th World Statistics Congress, Annual Meetings of the International Statistical Institute (ISI), Dublin, Aug. 21–26.

Andersson, C., and Nordberg, L. (1994), A method for variance estimation of non-linear functions of totals in surveys – theory and software implementation, *Journal of Official Statistics* **10**(4): 395–405 (available at www.jos.nu).

Anderson, M. (2011), Guesses, estimates and adjustments: Websters's 1755 "Census" of Scotland revisited again, *Journal of Scottish Historical Studies* **31**(1): 26–45.

Anderson, M., and Seltzer, W. (2005), Federal statistical confidentiality and business data: Twentieth century challenges and continuing issues, paper presented at the Annual Federal Committee on Statistical Methodology (FCSM) Research Conference, Arlington, VA, Nov. 14–16.

Anderson, M., and Seltzer, W. (2007), Challenges to the confidentiality of U.S. Federal Statistics, 1920–1965, *Journal of Official Statistics* **23**(1): 1–34 (available at www.jos.nu).

Andrusiak, G. (1993), Re-engineering of industry statistics: Maintaining relevance in trying times, *Proceedings of the 1st International Conference on Establishment Surveys (ICES-I)*, Buffalo, NY, June 27–30, American Statistical Association, Alexandria, VA, pp. 724–728.

Archibald, R. B. (1977), On the theory of industrial price measurement output price indexes, *Annals of Economic and Social Measurement* **6**(1): 64–81.

Atuk, O., and Ural, B. (2002), *Seasonal Adjustment in Economic Time Series*, Discussion paper: 2002/1, Central Bank of the Republic of Turkey, Statistics Department, Ankara.

Babbie, E. (1995), *The Practice of Social Research*, 7th ed., Wadsworth Publishing Company, Belmont, CA.

Babyak, C., Gower, A., Mulvihill, J., and Zaroski, R. A. (2000), Testing of the questionnaires for Statistics Canada's unified enterprise survey, *Proceedings of the 2nd International Conference on Establishment Surveys (ICES-II)*, Buffalo, NY, June 17–21, American Statistical Association, Alexandria, VA, pp. 317–326.

Baillargeon, S., and Rivest, L.-P. (2009), A general algorithm for univariate stratification, *International Statistical Review* **77**(3): 331–344.

Baillargeon, S., and Rivest, L.-P. (2011), The construction of stratified designs in R with the package stratification, *Survey Methodology* **37**(1): 53–65.

Bakker, B., and Van Rooijen, J., eds. (2012), Methodological challenges in register-based research, *Statistica Neerlandica* (special issue) **66**(1): 1–84 (available gratis at http://onlinelibrary.wiley.com).

Balk, B. M. (2008), *Price and Quantity Index Numbers*, Cambridge University Press, Cambridge, UK.

Balk, B. M. (2010), Direct and chained indices: A review of two paradigms, in Diewert, W. E., Balk, B. M., Fixler, D., Fox, K. J., and Nakamura, A. O. (eds.), *Price and Productivity Measurement*, Vol. 6, *Index Number Theory*, Trafford Press, pp. 217–234.

Ballou, J., DesRoches, D., Zhao, Z., and Potter, F. (2007), Meeting the challenges of designing the Kauffman firm survey: Sampling frame, definitions, questionnaire development and respondent burden, *Proceedings of the 3rd International Conference on Establishment Surveys (ICES-III)*, Montreal, June 18–21, American Statistical Association, Alexandria, VA, pp. 464–468.

Bankier, M. D. (1988), Power allocations: Determining sample sizes for subnational areas, *The American Statistician* **42**(3): 174–177.

Barlow, B., Freedman, S., and Mason, P. (2009), The data editing aspect of EIA's Internet data collection system, paper presented at the Conference of European Statisticians, Work Session on Statistical Data Editing, Neuchâtel, Switzerland, Oct. 5–7.

Barrow, R. (2007), Working together. How good relationships with providers can improve the quality of official statistics, paper presented at the 56th Session of the International Statistical Institute (ISI), Lisbon, Aug. 22–29.

Barry, C. A. (1998), Choosing qualitative data analysis software: Atlas/ti and Nudist compared, *Sociological Research Online* **3**(3).

Barton, T., and DesRoches, D. (2007), Minimizing non-response in a survey of new businesses, *Proceedings of the 3rd International Conference on Establishment Surveys (ICES-III)*, Montreal, June 18–21, American Statistical Association, Alexandria, VA, pp. 214–219.

Bates, N., Conrey, F., Zuwallack, R., Billia, D., Harris, V., Jacobsen, L, and White, T. (2009), *Messaging to America: Results from the Census Barriers, Attitudes, and Motivators Survey* (*CBAMS*), C2PO 2010 Census Integrated Communications Research Memoranda Series 10, US Census Bureau, Washington, DC, May 12.

Bavdaz, M. (2006), The response process in recurring business surveys, paper presented at the 3rd European Conference on Quality in Official Statistics, Cardiff, UK, April 24–26.

Bavdaz, M. (2007), *Measurement Errors and the Response Process in Business Surveys*, PhD thesis, University of Ljubljana, Slovenia.

Bavdaz, M. (2009), Conducting research on the response process in business surveys, *Statistical Journal of the International Association of Official Statistics* **26**(1–2): 1–14.

Bavdaz, M. (2010a), The multidimensional integral business survey response model, *Survey Methodology* **36**(1): 81–93.

Bavdaz, M. (2010b), Sources of measurement errors in business surveys, *Journal of Official Statistics* **26**(1): 25–42. (available at www.jos.nu).

Bavdaz, M., Giesen, D., and Biffignandi, S. (2011), Businesses as users of NSI statistics: Opportunities for creating greater value added, paper presented at the 2011 European Conference on New Techniques and Technologies for Statistics (NTTS), Brussels, Feb. 22–24.

Beaucage, Y. (2007), The many ways of improving the industrial coding for Statistics Canada's Business Register, *Proceedings of the 3rd International Conference on Establishment Surveys (ICES-III)*, Montreal, June 18–21, American Statistical Association, Alexandria, VA, pp. 14–22.

Beaucage, Y., Hunsberger, P., and Pursey, S. (2005), The redesign of the Statistics Canada's Business Register, *Proceedings of the Section on Government Statistics*, Joint Statistical Meetings, Minneapolis, Aug. 7–11, American Statistical Association, Alexandria, VA, pp. 989–994.

Becker, R. A., and Grim, C. (2011), *Newly Discovered Microdata on U.S. Manufacturing Plants from the 1950s and 1960s: Some Early Glimpses*, Centre for Economic Studies, US Census Bureau paper, CES, Sept. 11–29.

Beckler, D., and Ott, K. (2007), Incentives in surveys with farmers, *Proceedings of the 3rd International Conference on Establishment Surveys (ICES-III)*, Montreal, June 18–21, American Statistical Association, Alexandria, VA, pp. 501–508.

Bell, W. R., and Hillmer, S. C. (1984), Issues involved with the seasonal adjustment of economic time series, *Journal of Business and Economic Statistics* **2**(4): 291–320.

Benedetti, R., Bee, M., and Espa, G. (2010), A framework for cut-off sampling in business survey design, *Journal of Official Statistics* **26**(4): 651–671 (available at `www.jos.nu`).

Berners-Lee, T. (2009), *Putting Government Data Online, Notes after Talking with Various People in UK and US Governments Who Would Like to Put Data on the Web and Want to Know the Next Steps* (available at `www.w3.org/DesignIssues/GovData.html`).

Berthelot, J.-M., and Latouche, M. (1993), Improving the efficiency of data collection: A generic respondent follow-up strategy for economic surveys, *Journal of Business & Economic Statistics* **11**(4): 417–424.

Besley, T. (2007), *Inflation and the Service Sector: A Speech to the Cardiff Breakfast Club*, Jan. 18 (available at `www.bankofengland.co.uk/publications/speeches/2007/speech299.pdf`).

Bethel, J. (1989), Sample allocation in multivariate surveys, *Survey Methodology* **15**(1): 47–57.

Bethlehem, J. (2008), Surveys without questions, in de Leeuw, E. D., Hox, J. J., and Dillman, D. A., eds., *International Handbook of Survey Methodology*, Lawrence Erlbaum Associates, New York, pp. 500–540.

Bethlehem, J. (2009), *Applied Survey Methods: A Statistical Perspective*, Wiley, Hoboken, NJ.

Bethlehem, J., and Biffignandi, S. (2012), *Handbook of Web Surveys*, Wiley, Hoboken, NJ.

Bethlehem, J. G., Keller, W. J., and Pannekoek, J. (1990), Disclosure control of microdata, *Journal of the American Statistical Association* **85**(409): 38–45.

Beukenhorst, D., and Giesen, D. (2010), Internet use for data collection at Statistics Netherlands, paper presented at the 2nd International Workshop on Internet Survey Methods, Statistics Korea, Daejeon, Sep. 8–9, South Korea.

Biemer, P., and Cantor, D. (2007), Introduction to survey methods for businesses and organizations, short course presented at the 3rd International Conference on Establishment Surveys (ICES-III), Montreal, June 18, American Statistical Association, Alexandria, VA.

Biemer, P. P., Ellis, C., Pitts, A., and Aspinwall, K. (2007), Do monetary incentives increase business survey response rates? Results from a large scale experiment, *Proceedings of the 3rd International Conference on Establishment Surveys (ICES-III)*, Montreal, June 18–21, American Statistical Association, Alexandria, VA, pp. 509–516.

Biemer, P. P., and Fecso, R. S. (1995), Evaluating and controlling measurement error in business surveys, in Cox, B. G., Binder, D. A., Chinnappa, B. N., Christianson, A., Colledge, M. J., and Kott, P. S., eds., *Business Survey Methods*, Wiley, New York, pp. 257–281.

Biemer, P. P., and Lyberg, L. E. (2003), *Introduction to Survey Quality*, Wiley, Hoboken, NJ.

Biffignandi, S., Oehler, M., Bolko, I., and Bavdaz, M. (2011), Use of NSI statistics in textbooks, in Giesen, D., and Bavdaz, M., eds., *Proceedings of the BLUE-ETS Conference on Burden and Motivation in Official Business Surveys*, Statistics Netherlands, Heerlen, March 22–23, pp. 43–52.

Birch, S., Schwede, L., and Gallagher, C. (1998), *Juvenile Residential Facility Census Questionnaire Redesign Project: Results from Phase 2 Cognitive Interviewing Testing*, Report prepared by the U.S. Census Bureau for the Department of Justice, Washington DC.

Birch, S., Schwede, L., and Gallagher, C. (1999), *Juvenile Residential Facility Census Questionnaire Redesign Project: Results from Phase 3 Mail-out Test Analysis*, Report prepared by the U.S. Census Bureau for the Department of Justice, Washington DC.

Birch, S., Schwede, L., and Gallagher, C. (2001), *Juvenile Probation Survey Development Project: Results of Phase 1 Exploratory Interviews*, Report prepared by the US Census Bureau for the Department of Justice, Washington DC.

Blanke, K. (2011), Scrolling or paging – it depends, paper presented at the 5th International Workshop on Internet Survey Methodology, Statistics Netherlands, The Hague, Aug. 29–31.

Blau, P. (1964), *Exchange and Power in Social Life*, Wiley, New York.

Blumberg, B., Cooper, D. R., and Schindler, P. S. (2008), *Business Research Methods*, 2nd European ed., McGraw-Hill, Maidenhead.

Boeije, H. (2010), *Analysis in Qualitative Research*, Sage, Thousand Oaks, CA.

Bogardus, E. S. (1925), Measuring social distances, *Journal of Applied Sociology*, **9**: 299–308.

Bogen, K. (1996), The effect of questionnaire length on response rates. A review of the literature, *Proceedings of the Section on Survey Research Methods*, American Statistical Association, Alexandria, VA, pp. 1020–1025.

Bohr, J. (2011), Microdata information system MISSY: Metadata for scientific research, paper presented at the 4th European Survey Research Association (ESRA) Conference, Lausanne, Switzerland, July 18–22.

Bradburn, N. (1978), Respondent burden, *Health Survey Research Methods* **79**(3207): 49–53.

Bradburn, N., Sudman, S., and Wansink, B. (2004), *Asking Questions: The Definite Guide to Questionnaire Design for Market Research, Political Polls, and Social and Health Questionnaires*, Jossey-Bass, San Francisco.

Brancato, G., Macchia, S., Murgia, M., Siognore, M., Simeoni, G., Blanke, K., Körner, T., Nimmergut, A., Lima, P., Paulino, R., and Hoffmeyer-Zlotnik, J. (2006), *Handbook of Recommended Practices for Questionnaire Development and Testing in the European Statistical System*, Eurostat, Luxembourg (available at http://epp.eurostat.ec.europa.eu/portal/page/portal/quality/documents/RPSQDET27062006.pdf).

Bremner, C. (2011), An investigation into the use of mixed mode data collection methods for UK business surveys, in Giesen, D., and Bavdaz, M., eds., *Proceedings of the BLUE-ETS Conference on Burden and Motivation in Official Business Surveys*, Statistics Netherlands, Heerlen, March 22–23, pp. 217–220.

Brewer, K. R. W. (1999), Design-based or prediction-based inference? Stratified random vs stratified balanced sampling, *International Statistical Review* **67**(1): 35–47.

Briggs, J., and Duoba, V. (2000), *STRAT2D: Optimal Bi-Variate Stratification System*, Statistics New Zealand, Wellington (available at `http://www2.stats.govt.nz/domino/external/web/prod_serv.nsf/4aa2fc565310869fcc256bb4007a16d9/880f3e651a212666cc256b3b00766320/$FILE/strat2d.pdf`).

Brinkley, E., MacDonald, A., and Bismire, L. (2007), Integration of annual economic collections in the Australian Bureau of Statistics, *Proceedings of the 3rd International Conference on Establishment Surveys (ICES-III)*, Montreal, June 18–21, American Statistical Association, Alexandria, VA, pp. 122–131.

Brodeur, M., and Ravindra, D. (2007), Unified enterprise survey, *Proceedings of the 3rd International Conference on Establishment Surveys (ICES-III)*, Montreal, June 18–21, American Statistical Association, Alexandria, VA, pp. 132–142.

Brodeur, M., and Ravindra, D. (2010), Statistics Canada's new use of administrative data for survey replacement, paper presented at the European SIMPLY Conference on Administrative Simplification in Official Statistics, Ghent, Belgium, Dec. 2–3.

Brodeur, M., Koumanakos, P., Leduc, J., Rancourt, E., and Wilson, K. (2006), *The Integrated Approach to Economic Surveys in Canada*, Catalog 68-514-X, Statistics Canada, Ottawa (available at `www.statcan.gc.ca/pub/68-514-x/68-514-x2006001-eng.pdf`).

Bureau, M. (1991), Experience with the use of cognitive methods in designing business survey questionnaires, *Proceedings of the Survey Research Methods Section*, Joint Statistical Meetings, Atlanta, Aug. 18–22, American Statistical Association, Alexandria, VA, pp. 713–717.

Bureau of Labor Statistics (2007), The Consumer Price Index, in *BLS Handbook of Methods*, Bureau of Labor Statistics, Washington, DC, pp. 1–45.

Burns, E. M., Carlson, L. T., French, D. K., Goldberg, M. L., Latta, R. B., and Leach, N. L. (1993), Surveying an uncharted field, *Proceedings of the 1st International Conference on Establishment Surveys (ICES-I)*, Buffalo, NY, June 27–30, American Statistical Association, Alexandria, VA, pp. 37–44.

Burnside, R. (2000), Towards best practice for the design of electronic data capture instruments, paper presented at the Statistics Methodology Advisory Committee meeting, Nov., Australian Bureau of Statistics, Canberra.

Burnside, R., Bishop, G., and Guiver, T. (2005), The effect of an incentive on response rates and timing in an economic survey, paper presented at the 55th Session of the International Statistical Institute (ISI), Sydney, Australia, April 5–12.

Burnside, R., and Farrell, E. (2001), Recent developments at the ABS in electronic data reporting by businesses, paper presented at the 2001 International Statistics Canada Symposium: Achieving Data Quality in a Statistical Agency: A Methodological Perspective, Hull, Quebec, Oct. 18–19, CD-rom, Statistics Canada, Ottawa.

Burr, M. A., Levin, K. Y., and Becher, A. (2001), Examining web vs. paper mode effects in a federal government customer satisfaction study, paper presented at the 56th Annual Conference of the American Association for Public Opinion Research (AAPOR), Montreal, May 17–20.

Burt, C. W., and Schappert, S. M. (2002), Evaluation of respondent and interviewer debriefing techniques on questionnaire development methods for health provider-based surveys, paper presented at the International Conference on Questionnaire Development, Evaluation, and Testing Methods (QDET), Charleston, SC, Nov. 14–17.

Cabinet Office (2008), *Limiting Pre-release Access to Statistics: The Government's Response to the Consultation Exercise*, Cabinet Office, London.

Campanelli, P., Thomson, K., Moon, N., and Staples, T. (1997), The quality of occupational coding in the United Kingdom, in Lyberg, L., Biemer, P., Collins, M., de Leeuw, E., Dippo, C., Schwarz, N., and Trewin, D., eds., *Survey Measurement and Process Quality*, Wiley, New York, pp. 437–456.

Canty, A. J., and Davison, A. C. (1999), Resampling-based variance estimation for labour force surveys, *The Statistician* **48**(3): 379–391.

Carlson, L. T., Preston, J. L., and French, D. K. (1993), Using focus groups to identify user needs and data availability, *Proceedings of the 1st International Conference on Establishment Surveys (ICES-I)*, Buffalo, NY, June 27–30, American Statistical Association, Alexandria, VA, pp. 300–308.

Carmines, E. G., and Zeller, R. A. (1979), *Reliability and Validity Assessment*, SAGE Publications, Minnesota.

Chambers, R. L. (1986), Outlier robust finite population estimation, *Journal of the American Statistical Association* **81**(396): 1063–1069.

Chambers, R., Cruddas, M., and Smith, P. (1998), Sample design for the PRODCOM (product sales) inquiry, *Proceedings of the 1997 International Statistics Canada Methodology Symposium*, Statistics Canada, Ottawa, pp. 342–345.

Chambers, R., Kokic, P., Smith, P., and Cruddas, M. (2001), Winsorization for identifying and treating outliers in business surveys, *Proceedings of the 2nd International Conference on Establishment Surveys (ICES-II)*, Buffalo, NY, June 17–21, American Statistical Association, Alexandria, VA, pp. 717–726.

Chambers, R. L., and Skinner, C. J., eds. (2003), *Analysis of Survey Data*, Wiley, Hoboken, NJ.

Champkin, J. (2011), Making information beautiful—and clear, *Significance* **8**(1): 39–41.

Chan, G., Morren, M., and Snijkers, G. (2011), *Audit Trials: The Log of the 2006 Structural Business Survey* (in Dutch: Audit trails: Het logboek van de PS-2006), Research report, Statistics Netherlands, Heerlen.

Chandra, H., and Chambers, R. (2009), Multipurpose weighting for small area estimation, *Journal of Official Statistics* **25**(3): 379–395 (available at `www.jos.nu`).

Chandra, H., and Chambers, R. (2011), Small area estimation under transformation to linearity, *Survey Methodology* **37**(1): 39–51.

Chatfield, C. (2004), *The Analysis of Time Series, An Introduction*, 6th ed., Chapman & Hall/CRC, Boca Raton, FL.

Chestnut, J. (2008), *Effects of using a grid versus a sequential form on the ACS basic demographic data*, Memorandum Series, Chapter ASC-MP-09. Washington, DC.

Choiand, B., and Ward, D. (2004), Analysis of statistical units deliniated by OECD member countries, paper presented at 18th International Roundtable on Business Survey Frames, Beijing, Oct. 18–22.

Christianson, A., and Tortora, R. D. (1995), Issues in surveying businesses: An International survey, in Cox, B. G., Binder, D. A., Chinnappa, B. N., Christianson, A., Colledge, M. J., and Kott, P. S., eds., *Business Survey Methods*, Wiley, New York, pp. 237–256.

Church, A. H., and Waclawski, J. (1998), *Designing and Using Organizational Surveys: A Seven-Step Process*, Jossey-Bass, San Francisco.

Cialdini, R. B. (1990), Deriving psychological concepts relevant to survey participation from the literature on compliance, helping and persuasion, paper presented at the 1st International Workshop on Household Survey Nonresponse, Stockholm.

Cialdini, R. B. (2001), *Influence. Science and Practice*, Allyn & Bacon, Boston.

Cialdini, R. B. (2007), Descriptive social norms as underappreciated sources of social control, *Psychometrika*, **72**(2): 263–268.

Clark, H. C., and Schober, M. F. (1992), Asking questions and influencing answers, in Tanur, J. M., ed., *Questions about Questions. Inquires into the Cognitive Bases of Surveys*, Russel Sage Foundation, New York, pp. 15–48.

Clayton, R., Searson, M., and Manning, C., (2000), Electronic data collection in selected BLS establishment programs, *Proceedings of the 2nd International Conference on Establishment Surveys (ICES-II)*, Buffalo, NY, June 17–21, American Statistical Association, Alexandria, VA, pp. 439–448.

Cleveland, W. S., and McGill, R. (1984), Graphical perception: Theory, experimentation, and application to the development of graphical methods, *Journal of American Statistical Association* **79**(387): 531–554.

Cleveland, W. S., and McGill, R. (1987), Graphical perception: The visual decoding of quantitative information on statistical graphs (with discussion), *Journal of the Royal Statistical Society Series A*, **150**: 192–229.

Cochran, W. G. (1977), *Sampling Techniques*, Wiley, New York.

Cohen, B., Zukerberg, A., and Pugh, K.W. (1999), Improving respondent selection procedures in establishment surveys: Implications from the Schools and Staffing Survey (SASS), paper presented at the 54th Annual Conference of the American Association of Public Opinion Research (AAPOR), St. Petersburg Beach, FL, May 13–16.

Colledge, M., and March, M. (1997), Quality policies, standards, guidelines, and recommended practices at national statistical agencies, in Lyberg, L., Biemer, P., Collins, M., de Leeuw, E., Dippo, C., Schwarz, N., and Trewin, D., eds., *Survey Measurement and Process Quality*, Wiley, New York, pp. 501–522.

Conrad, F. (1997), Using expert systems to model and improve survey classification processes, in Lyberg, L., Biemer, P., Collins, M., de Leeuw, E., Dippo, C., Schwarz, N., and Trewin, D., eds., *Survey Measurement and Process Quality*, Wiley, New York, pp. 393–414.

Conrad, F. G., and Schober, M. F. (2010), New frontiers in standardized survey interviewing, in Hesse-Biber, S. N., and Leavy, P., eds., *Handbook of Emergent Methods*, Guilford Press, New York.

Conrad, F. G., Schober, M. F., and Coiner, T. (2007). Bringing Features of Human Dialogue to Web Surveys, *Applied Cognitive Psychology* (Special Issue: Cognitive Psychology and Survey Methodology: Nurturing the Continuing Dialogue between Disciplines) **21**(2): 165–187.

Cook, S., LeBaron, P., Flicker, L., and Flanigan, T. S. (2009), Applying incentives to establishment surveys: A review of the literature, paper presented at the 64th Annual Conference of the American Association of Public Opinion Research (AAPOR), Hollywood, FL, May 14–17.

Corby, C. (1984), *Content Evaluation of the 1977 Economic Censuses (DE-2)*, Research report 84/29, US Census Bureau, Washington, DC.

Corby, C. (1987), Content evaluation of the 1982 economic censuses: Petroleum distributors, in *1982 Economic Censuses and Census of Governments Evaluation Studies*, US Census Bureau, Washington, DC, pp. 27–50.

Costa, V., Krsinich, F., and Van der Mescht, R. (2008), Editing and Imputation of aministrative data used for producing business statistics, *Proceedings of the Conference of European Statisticians (CES)*, United Nations Economic Commission for Europe (UNECE), Work Session on Statistical Data Editing, Vienna, Austria, April 21–23, pp. 1–9.

Couper, M. (1998), Measuring survey quality in a CASIC environment, paper presented at Joint Statistical Meetings, Dallas, Aug. 9–13, *Proceedings of the Survey Research Methods Section*, American Statistical Association, Alexandria, VA.

Couper, M. P. (2008), *Designing Effective Web Surveys*, Cambridge University Press, Cambridge.

Couper, M. P., Baker, R. P., Bethlehem, J., Clark, C. Z. F., Martin, J., Nicholls II, W. L., and O'Reilly, J. M. (1998), *Computer Assisted Survey Information Collection*, Wiley, New York.

Couper, M., and Lyberg, L. (2005), The use of paradata in survey research, paper presented at the 55th Session of the International Statistical Institute (ISI), Sydney, Australia, April 5–12.

Couper, M. P., Singer, E., Conrad, F. G., and Groves, R. M. (2008), Risk of disclosure, perceptions of risk, and concerns about privacy and confidentiality as factors in survey participation, *Journal of Official Statistics* **24**(2): 255–275 (available at www.jos.nu).

Couper, M. P., Singer, E., Conrad, F. G., and Groves, R. M. (2010), Experimental studies of disclosure risk, disclosure harm, topic sensitivity, and survey participation, *Journal of Official Statistics* **26**(2): 287–300 (available at www.jos.nu).

Couper, M., Tourangeau, R., Conrad, F. G., and Zhang, C. (2011), Further research on the design of complex grids, paper presented at the 5th International Workshop on Internet Survey Methodology, Statistics Netherlands, The Hague, Aug. 28–31.

Cox, L. H. (1980), Suppression methodology and statistical disclosure control, *Journal of American Statistical Association* **75**(370): 377–385.

Cox, B. G., Binder, D. A., Chinnappa, B. N., Christianson, A., Colledge, M. J., and Kott, P. S., eds. (1995), *Business Survey Methods*, Wiley, New York.

Cox, B. G., and Chinnappa, B. N. (1995), Unique features of business surveys, in Cox, B. G., Binder, D. A., Chinnappa, B. N., Christianson, A., Colledge, M. J., and Kott, P. S., eds. (1995). *Business Survey Methods*, Wiley, New York, pp. 1–17.

Cox, B., Elliehausen, G., and Wolken, J. (1989), Surveying small businesses about their finances, *Proceedings of the Section on Survey Research Methods*, Joint Statistical Meetings, Washington DC, Aug. 6–10, American Statistical Association, Alexandria, VA, pp. 553–557.

Cycyota, C. S., and Harrison, D. A. (2002), Enhancing survey response rates at the executive level: Are employee or consumer-level techniques effective? *Journal of Management* **28**(2): 151–176.

Cycyota, C. S., and Harrison, D. A. (2006), What (not) to expect when surveying executives: A meta analysis of top manager response rates and techniques over time, *Organizational Research Methods* **9**(2): 133–160.

Czaja, R., and Blair, J. (2005), *Designing Surveys: A Guide to Decisions and Procedures*, Pine Forge Press, Thousand Oaks, CA.

Dagum, E. B. (1980), *The X11ARIMA Seasonal Adjustment Method*, Catalog 12-564E, Statistics Canada, Ottawa.

Dagum, E. B. (1987), Revisions of trend-cycle estimators of moving average seasonal adjustment methods, *Journal of Business and Economic Statistics* **5**(2): 177–189.

Dale, A., Arber, S., and Procter, M. (1988), *Doing Secondary Analysis*, Unwin Hyman, London.

Dale, T., and Haraldsen, G., eds. (2007a), *Handbook for Monitoring and Evaluating Business Response Burdens*, Eurostat, Luxembourg.

Dale, T., Hole, B., and Nøtnæs, T. (2007b), Towards a more user-friendly reporting system for KOSTRA. Municipality-state-reporting in Norway, paper presented at the International Workshop on Questionnaire Pre-testing Methods (QUEST), Statistics Canada, Ottawa, April 24–26.

Dalenius, T., and Hodges, J. L. (1959), Minimum variance stratification, *Journal of the American Statistical Association* **54**(285): 88–101.

D'Aurizio, L., and Tartaglia-Polcini, R. (2008), Use of deflators in business surveys: An analysis based on Italian micro data, *Journal of Official Statistics* **24**(2): 277–300 (available at www.jos.nu).

Davidsson, G. (2002), Cognitive testing of mail surveys at Statistics Sweden, paper presented at the International Conference on Questionnaire Development, Evaluation, and Testing Methods (QDET), Charleston, SC, Nov. 14–17.

Davies, P., and Smith, P., eds. (1999), *Model Quality Report in Business Statistics* (4 volumes), UK Office for National Statistics, Newport, UK.

Davis, W., DeMaio, T. J., and Zukerberg, A. (1995), *Can Cognitive Information Be Collected through the Mail? Comparing Cognitive Data Collected in Written versus Verbal Format*, Working Paper in Survey Methodology SM95/02, US Census Bureau, Statistical Research Division, Washington, DC.

Davis, W. R., and Pihama, N. (2009), Survey response as organisational behaviour: An analysis of the annual enterprise survey, 2003–2007, paper presented at New Zealand Association of Economists Conference, Wellington, July 1–3.

De Groot, A. D. (1994), *Methodology: Foundations of research and thought in the behavioral sciences* (in Dutch: Methodologie: Grondslagen van onderzoek en denken in de gedragswetenschappen), Van Gorcum, Assen, The Netherlands (available at http://www.dbnl.org/tekst/groo004meth01_01/).

De Heer, W., de Leeuw, E. D., and van der Zouwen, J. (1999). Methodological issues in survey research: A historical review, *Bulletin de Methodologie Sociologique* **64**(Oct.): 25–48.

De Leeuw, E. D., Callegaro, M., Hox, J., Korendijk, E., and Lensvelt-Mulders, G. (2007), The influence of advance letters on response in telephone surveys, *Public Opinion Quarterly* **71**(3): 413–443.

De Leeuw, E. D., Hox, J. J., and Dillman, D. A. (2008), The cornerstones of survey research, in de Leeuw, E. D., Hox, J. J., and Dillman, D. A., eds., *International Handbook of Survey Methodology*, Lawrence Erlbaum Associates, New York, pp. 1–17.

DeMaio, T. J., and Jenkins, C. R. (1991), Questionnaire research in the census of construction industries, *Proceedings of the Survey Research Methods Section*, American Statistical Association, Alexandria, VA, pp. 496–501.

Detlefsen, R. E., and Veum, C. S. (1991), Design issues for the retail trade sample surveys of the U.S. Bureau of the Census, *Proceedings of the Survey Research Methods Section*, Joint Statistical Meetings, Atlanta, Aug. 18–22, American Statistical Association, Alexandria, VA, pp. 214–219.

De Vaus, D. A. (1991), *Surveys in Social Research*, 3rd ed., UCL Press, University College London.

Deville, J.-C., and Särndal, C.-E. (1992), Calibration estimators in survey sampling, *Journal of the American Statistical Association* **87**(418): 376–382.

De Waal, T., Pannekoek, J., and Scholtus, S. (2011), *Handbook of Statistical Data Editing and Imputation*, Wiley, Hoboken, NJ.

Diewert, W. E. (1987), Index numbers, in Eatwell, J., Milgate, M., and Newman, P., eds., *A Dictionary of Economics*, Vol. 2, Macmillan Press, New York, pp. 767–780.

Diewert, W. E. (2007), *Index Numbers*, Discussion Paper 07-02, Department of Economics, University of British Columbia, Vancouver, Canada.

Dillman, D. A. (1978), *Mail and Telephone Surveys: The Total Design Method*, Wiley, New York.

Dillman, D. A. (2000), *Mail and Internet Surveys: The Tailored Design Method*, 2nd ed., Wiley, New York.

Dillman, D. A. (2007), *Attending to a Thousand Details: The Challenge of Applying Visual Design Principles to Establishment Surveys*, course presented at Statistics Norway, Oslo, Sept. 21.

Dillman, D. A., Gertseva, A., and Mahon-Haft, T. (2005), Achieving usability in establishment surveys through the application of visual design principles, *Journal of Official Statistics* **21**(2): 183–214 (available at www.jos.nu).

Dillman, D. A., Smyth, J. D., and Christian, L. M. (2009), *Internet, Mail, and Mixed-Mode Surveys: The Tailored Design Method*, 3rd ed., Wiley, Hoboken, NJ.

Dippo, C. S., Chun, Y. I., and Sander, J. (1995), Designing the data collection process, in Cox, B. G., Binder, D. A., Chinnappa, B. N., Christianson, A., Colledge, M. J., and Kott, P. S., eds., *Business Survey Methods*, Wiley, New York, pp. 283–301.

Dodds, J. (2001), Experiences with Internet reporting on E-business surveys, paper presentation at the 34th International Field Directors and Technologies Conference, Montreal, May 20–23.

Domingo-Ferrer, J., and Torra, V. (2001), Disclosure control methods and information loss for microdata, in Doyle, P., Lane, J. I., Theeuwes, J. J. M., and Zayatz, L., eds., *Confidentiality, Disclosure and Data Access: Theory and Practical Applications for Statistical Agencies*, North-Holland, Amsterdam, pp. 91–110.

Dommeyer, C. (1989), Offering mail survey results in a lift letter, *Journal of the Market Research Society* **31**(3): 399–408.

Dowling, Z. (2005), The challenges of implementing Web data collection for mandatory business surveys: The respondent's perspective, paper presented at the 1st European Association of Survey Research Conference (EASR), Barcelona, July 18–22.

Dowling, Z., and Stettler, K. (2007), Factors influencing business respondents' decision to adopt web returns, *Proceedings of the 3rd International Conference on Establishment Surveys (ICES-III)*, Montreal, June 18–21, American Statistical Association, Alexandria, VA, pp. 1028–1039.

Downey, K., McCarthy, D., and McCarthy, W. (2007), Encouraging the use of alternative modes of electronic data collection: Results of two field studies, *Proceedings of the 3rd International Conference on Establishment Surveys (ICES-III)*, Montreal, June 18–21, American Statistical Association, Alexandria, VA, pp. 517–524.

Dubois, M.-A. (2008), Electronic questionnaire: Accessibility standards and guidelines, paper presented at the 2nd International Workshop on Business Data Collection Methodology (BDCM), Ottawa, Oct. 22–24.

Duncan, G. J., and Kalton, G. (1987), Issues of design and analysis of surveys across time, *International Statistical Review* **55**(1): 97–117.

Dyrberg, B. (2003), Timeliness, response burden and cooperation with respondents at Statistics Denmark, paper presented at the OECD Short-Term Economic Statistics Working Group.

Dyrberg, B. (2006), The Reduction of Response Burden: Methods Used and Improvement of Cooperation with the Respondents in Denmark, *Statistika, Journal for Economy and Statistics* **2006**(2): 163–169 (available at http://panda.hyperlink.cz/cestapdf/pdf06c2/dyrberg.pdf).

ECFIN (2007), *The Joint Harmonised EU Programme of Business and Consumer Surveys*, European Commission Directorate-General for Economic and Financial Affairs, Brussels.

Edwards, R. W. (2007), Business surveys: Past, present and challenges for the future, *Proceedings of the 3rd International Conference on Establishment Surveys (ICES-III)*, Montreal, June 18–21, American Statistical Association, Alexandria, VA, pp. 1–13.

Edwards, W. S., and Cantor, D. (1991), Towards a response model in establishment surveys, in Biemer, P. P., Groves, R. M., Lyberg, L. E., Mathiowetz, N. A., and Sudman, S., eds., *Measurement Errors in Surveys*, Wiley, New York, pp. 211–233.

Edwards, W. S., Gaertner, G. H., and Nieva, V. F. (1993), The effect of organizational context on survey response, paper presented at the 1st International Conference on Establishment Surveys (ICES-I), Buffalo, NY, June 27–30, American Statistical Association, Alexandria, VA.

Ehling, M., and Körner, T., eds. (2007), *Handbook on Data Quality Assessment Methods and Tools*, Eurostat, Luxembourg (available at http://unstats.un.org/unsd/dnss/docs-nqaf/Eurostat-HANDBOOK%20ON%20DATA%20QUALITY%20ASSESSMENT%20METHODS%20AND%20TOOLS%20%20I.pdf).

Eldridge, J., Martin, J., and White, A. (2000), The use of cognitive methods to improve establishment surveys in Britain, *Proceedings of the 2nd International Conference on Establishment Surveys (ICES-II)*, Buffalo, NY, June 17–21, American Statistical Association, Alexandria, VA, pp. 307–316.

Elliott, D., McLaren, C. H., and Chow, J. (2008), A decomposition approach to revisions in aggregated time series estimates, paper presented at the Joint Statistical Meeting, Denver, Aug. 3–7, American Statistical Association, Alexandria, VA.

Erikson, A.-G. (2007), Large enterprise management – higher quality and lower burden through better relations, *Proceedings of the 3rd International Conference on*

Establishment Surveys (ICES-III), Montreal, June 18–21, American Statistical Association, Alexandria, VA, pp. 825–829.

Erikson, J. (2002), Coherence analysis as a tool for questionnaire evaluation in enterprise statistics, paper presented at the International Conference on Questionnaire Development, Evaluation, and Testing (QDET), Charleston, SC, Nov. 14–17.

Erikson, J. (2007), Effects of offering web questionnaires as an option in enterprise surveys: The Swedish experience, *Proceedings of the 3rd International Conference on Establishment Surveys (ICES-III)*, Montreal, June 18–21, American Statistical Association, Alexandria, VA, pp. 1431–1435.

Erikson, J. (2010), Communication strategies in business surveys – implications when web data collection becomes the main mode, paper presented at the 3rd International Workshop on Business Data Collection Methodology (BDCM), Wiesbaden, Germany, April 28–30.

Eskenazi, J. (2011), iPad vs newspaper: What the eyes tell us, paper presented at the Tobii Eye Tracking Conference on User Experience, London, June 9–10.

ESOMAR (2008), *ICC/ESOMAR International Code on Market and Social Research*, World Association for Social, Opinion and Market Research (ESOMAR), Amsterdam (available at www.esomar.org/knowledge-and-standards/codes-and-guidelines.php).

Esposito, R., Fox, J. K., Lin, D., and Tidemann, K. (1994), ARIES: A visual path in the investigation of statistical data, *Journal of Computational and Graphical Statistics* **3**(2): 113–125.

ESS (2011), *Quality Assurance Framework*, European Statistical System, Eurostat, Luxembourg.

Estevao, V., Hidiroglou, M. A., and Särndal, C. E. (1995), Methodological principles for a generalized estimation system at Statistics Canada, *Journal of Official Statistics* **11**(2): 181–204 (available at www.jos.nu).

Eurofound (2010), *European Company Survey 2009: Overview*, Office for Official Publications of the European Communities, Luxembourg.

European Commission (2007), *The Joint Harmonised EU Programme of Business and Consumer Surveys: User Guide*, Directorate-General for Economic and Financial Affairs, Brussels (available at http://ec.europa.eu/economy_finance/db_indicators/surveys/documents/userguide_en.pdf).

European Commission (2008), *Action Programme Cutting Red Tape for Europe*, Directorate-General for Enterprise and Industry, Unit B.5, Brussels.

European Commission et al. (2009), *System of National Accounts 2008*, Joint report of the European Commission, International Monetary Fund, Organisation for Economic Co-operation and Development, United Nations, and World Bank, New York.

European Communities (2003), *Business Register Recommendations Manual*, Office for Official Publications of the European Communities, Luxembourg.

European Foundation for the Improvement of Living and Working Conditions (2010), *European Company Survey 2009: Overview*, Office for Official Publications of the European Communities, Luxembourg.

Eurostat (2001), *Handbook on Price and Volume Measures in National Accounts*, Office for Official Publications of the European Communities, Luxembourg.

Eurostat (2005), *Quality Measures for Economic Indicators*, Office for Official Publications of the European Communities, Luxembourg.

Eurostat (2008), *Statistical Classification of Economic Activities in the European Community, NACE rev. 2*, Eurostat Methodologies and Working Papers, Office for Official Publications of the European Communities, Luxembourg (available at http://circa.europa.eu/irc/dsis/nacecpacon/info/data/en/NACE%20Rev%202%20structure%20+%20explanatory%20notes%20-%20EN.pdf).

Eurostat (2009a), *ESS Guidelines on Seasonal Adjustment*, Eurostat Methodologies and Working Papers, Office for Official Publications of the European Communities, Luxembourg (available at http://epp.eurostat.ec.europa.eu/cache/ITY_OFFPUB/KS-RA-09-006/EN/KS-RA-09-006-EN.PDF).

Eurostat (2009b), *ESS Handbook for Quality Reports*, Office for Official Publications of the European Communities, Luxembourg.

Eurostat (2009c), *Principal European Economic Indicators: A Statistical Guide*, Office for Official Publications of the European Communities, Luxembourg.

Eurostat (2010), *Business Registers. Recommendations Manual*, Eurostat, Luxembourg.

Eurostat (2011a), *ESS Handbook for Quality Reports*, Eurostat, Luxembourg.

Eurostat (2011b), *European Statistics Code of Practice*, Eurostat, Luxembourg (available at http://epp.eurostat.ec.europa.eu/portal/page/portal/quality/code_of_practice).

Evra, R.-C., and DeBlois, S. (2007), Using paradata to monitor and improve the collection process in annual business surveys, *Proceedings of the 3rd International Conference on Establishment Surveys (ICES-III)*, Montreal, June 18–21, American Statistical Association, Alexandria, VA, pp. 227–232.

Faber, G., and Gagné, P. (2001), Measuring business entries and exits, *Proceedings of the 2nd International Conference on Establishment Surveys (ICES-II)*, Buffalo, NY, June 17–21, American Statistical Association, Alexandria, VA, pp. 1093–1098.

Falorsi, P. D., and Righi, P. (2008), A balanced sampling approach for multi-way stratification designs for small area estimation, *Survey Methodology* **34**(2): 223–234.

Farrell, E., and Hewett, K. (2011), Keeping up appearances: Maintaining standards during strategic changes in electronic reporting, paper presented at the 58th World Statistics Congress, Annual Meetings of the International Statistical Institute (ISI), Dublin, Aug. 21–26.

Farwell, K., and Raine, M. (2000), Some current approaches to editing in the ABS, *Proceedings of the 2nd International Conference on Establishment Surveys (ICES-II)*, Buffalo, NY, June 17–21, American Statistical Association, Alexandria, VA, pp. 529–538.

Featherston, F., and Stettler, K. (2011), Outside the answer boxes: Messages from respondents, paper presented at the 66th Annual Conference of the American Association for Public Opinion Research (AAPOR), Phoenix, AZ, May 12–15.

Fellegi, I. P., and Holt, D. (1976), A systematic approach to automatic edit and imputation, *Journal of the American Statistical Association* **71**(353): 17–35.

Ferrillo, A., Macchia, S., and Vicari, P. (2008), Different quality tests on the automatic coding procedure for the economic activities descriptions, paper presented at the 4th European Conference on Quality in Official Statistics, Rome, July 8–11.

Few, A. (2004), *Show Me the Numbers: Designing Tables and Graphs to Enlighten*, Analytics Press, Oakland, CA.

Findley, D. F., Monsell, B. C., Bell, W. R., Otto, M. C., and Chung Chen, B. (1998), New capabilities and methods of the X-12-ARIMA seasonal-adjustment program, *Journal of Business & Economic Statistics* **16**(2): 127–152.

Fink, A. (1995), *How to Analyze Survey Data*, Sage, Thousand Oaks, CA.

Fisher, J., and Kydoniefs, L. (2001), Using a theoretical model of response burden to identify sources of burden in surveys, paper presented at 12th International Workshop on Household Survey Nonresponse, Oslo, Sept. 12–14.

Fisher, S., and Adler, R. (1999), Using iterative cognitive testing to identify response problems on an establishment survey of energy consumption, paper presented at the 54th Annual Conference of the American Association of Public Opinion Research (AAPOR), St. Petersburg Beach, FL, May 13–16.

Fisher, S., Bosley, J., Goldenberg, K., Mockovak, W., and Tucker, C. (2003), A qualitative study of nonresponse factors affecting BLS establishment surveys: Results, *Proceedings of the Survey Research Methods Section*, Joint Statistical Meetings, San Francisco, Aug. 3–7, American Statistical Association, Alexandria, VA, pp. 679–684.

Fisher, S., Frampton, K., and Tran, R. (2001a), Pretesting the survey of respirator uses and practices: Cognitive and field testing of a new establishment survey, paper presented at the 56th Annual Conference of the American Association for Public Opinion Research (AAPOR), Montreal, May 17–20, also in *Proceedings of the Joint Statistical Meetings*, American Statistical Association, Alexandria, VA, CD-rom.

Fisher, S., Goldenberg, K. L., O'Brien, E., Tucker, C., and Willimack, D. K. (2001b), Measuring employee hours in government surveys, paper presented to the Federal Economic Statistics Advisory Committee (FESAC), US Bureau of Labor Statistics, Washington, DC, June 7.

Foldesi, E., Bauer, P., Horvath, B., and Urr, B. (2007), *Seasonal Adjustment Methods and Practices*, European Commission Grant 10300. 2005. 021-2005. 709, Hungarian Central Statistical Office, Budapest (available at http://epp.eurostat.ec.europa.eu/portal/page/portal/ver-1/quality/documents/SEASONAL_ADJUSTMENT_METHODS_PRACTICES.pdf).

Forsyth, B. H., Levin, K., and Fisher, S. K. (1999), Test of an appraisal method for establishment survey questionnaires, *Proceedings of the Survey Research Methods Section*, American Statistical Association, Alexandria, VA, pp. 145–149.

Forth, J., and Webster, S. (2008), *Methodological Review of Research with Large Businesses*, Paper 4: *Confidentiality and Disclosure*, Research report 60, UK: Her Majesty's Revenue and Customs, London (available at www.hmrc.gov.uk/research/paper4-confidentiality.pdf).

Fosen, J., Haraldsen, G., and Olsson, U. (2008), Does perceived response burden (PRB) affect response error?, paper presented at the 2nd International Workshop on Business Data Collection Methodology (BDCM), Ottawa, Oct. 22–24.

Fowler, F. J., and Mangione, T. W. (1990), *Standardized Survey Interviewing: Minimizing Interviewer-Related Error*, Sage, Newbury Park, CA.

Fowler, F. J., Jr., and Cosenza, C. (2008), Writing effective questions, in de Leeuw, E. D., Hox, J. J., and Dillman, D. A., eds., *International Handbook on Survey Methodology*, Lawrence Erlbaum Associates, New York, pp. 136–161.

Fox, J. E. (2001), Usability methods for designing a computer-assisted data collection instrument for the CPI, *Proceedings of the Federal Committee on Statistical Methodology (FCSM) Research Conference*, Arlington, VA, Nov. 14–16, pp. 107–112.

Frame, J. D. (1995), *Managing Projects in Organizations: How to Make the Best Use of Time, Techniques, and People*, Jossey-Bass, San Francisco.

Frame, J. D. (2003a), *Managing Projects in Organizations: How to Make the Best Use of Time, Techniques, and People*, 3rd ed., Jossey-Bass, San Francisco.

Frame, J. D. (2003b), *Managing Risk in Organizations: A Guide for Managers*, Jossey-Bass, San Francisco.

Francoz, C. (2002), Review of the French industrial R&D survey, paper presented at the International Conference on Questionnaire Development, Evaluation, and Testing (QDET), Charleston, SC, Nov. 14–17.

Freedman, H., and Mitchell, B. (1993), The development of surveys of waste management: The canadian experience, *Proceedings of the 1st International Conference on Establishment Surveys (ICES-I)*, Buffalo, NY, June 27–30, American Statistical Association, Alexandria, VA, pp. 52–61.

Freedman, S. R., and Rutchik, R. H. (2002), Information collection challenges in electric power and natural gas, paper presented at the Joint Statistical Meetings, New York, Aug. 11–15, American Statistical Association, Alexandria, VA.

Frits, B. (1992), *The History of National Accounting*, MPRA Paper 5952, Statistics Netherlands, Voorburg (available at http://mpra.ub.uni-muenchen.de/5952).

Frost, J.-M., Green, S., Jones, J., and Williams, D. (2010), Measuring respondent burden to statistical surveys, paper presented at the European SIMPLY Conference on Administrative Simplification in Official Statistics, Ghent, Belgium, Dec. 2–3.

Gallagher, C., and Schwede, L. (1997), *Facility Questionnaire Redesign Project: Results from Phase 1 Unstructured Interviews and Recommendations for Facility-Level Questionnaire*, Report prepared by the US Census Bureau for the Department of Justice.

Gass, R. H., and Seiter, J. S. (2011), *Persuasion: Social Influence and Compliance Gaining*, Allyn & Bacon (Pearson), Boston.

Gates, L. (2008), Identification of optimal call patterns for intensive follow-up in business surveys using paradata, paper presented at the 2008 International Statistics Canada Methodology Symposium: Data Collection: Challenges, Achievements and New Directions, Catalog 11-522-X, Statistics Canada, Ottawa, Oct. 28–31 (available at www.statcan.gc.ca/pub/11-522-x/2008000/article/10998-eng.pdf).

Gerber, E., and DeMaio, T. J. (1999), Probing strategies for establishment surveys, paper presented at the 54th Annual Conference of the American Association of Public Opinion Research (AAPOR), St. Petersburg Beach, FL, May 13–16.

Gerber, E., Wellens, T., and Keeley, C. (1996), Who lives here? The use of vignettes in household roster research, *Proceedings of the Section on Survey Research Methods*, American Statistical Association, Alexandria, VA, pp. 962–967.

Gernsbacher, M. A. (1990), *Language Comprehension as Structure Building*, Lawrence Erlbaum Associates, Hillsdale, NJ.

Gevers, T., and Zijlstra, T. (1998), *Project Management in Practice* (in Dutch: Praktisch projectmanagement), Academic Service, Schoonhoven, The Netherlands.

Ghosh, M., and Meeden, G. (1997), *Bayesian Methods for Finite Population Sampling*, Chapman & Hall, London.

Giesen, D. (2007), The response process model as a tool for evaluating business surveys, *Proceedings of the 3rd International Conference on Establishment Surveys (ICES-III)*, Montreal, June 18–21, American Statistical Association, Alexandria, VA, pp. 871–880.

Giesen, D., ed. (2011), *Response Burden in Official Business Surveys: Measurement and Reduction Practices of National Statistical Institutes*, Deliverable 2.2, BLUE-Enterprise and Trade Statistics Project (BLUE-ETS), European Commission, European Research Area (available at `www.blue-ets.istat.it/fileadmin/deliverables/Deliverable2.2.pdf`).

Giesen, D., and Hak, T. (2005), Revising the structural business survey: From a multi-method evaluation to design, paper presented at the Annual Federal Committee on Statistical Methodology (FCSM) Research Conference, Arlington, VA, Nov. 14–16.

Giesen, D., and Kruiskamp, P. (2011), *Using Edit Process Information for Questionnaire Development: A First Study* (in Dutch: Gebruik van gaafmaakinformatie voor vragenlijstontwikkeling: Eerste verkenning van een feedback loop NOPS), Internal report, Statistics Netherlands, Heerlen.

Giesen, D., and Raymond-Blaess, V., eds. (2011), *Inventory of Published Research: Response Burden Measurement and Reduction in Official Business Statistics. A literature review of national statistical institutes' practices and experiences*, Deliverable 2.1, BLUE-Enterprise and Trade Statistics Project (BLUE-ETS), European Commission, European Research Area (available at `www.blue-ets.istat.it/fileadmin/deliverables/Deliverable2.1.pdf`).

Giesen, D., and Snijkers, G. (2010), Affecting response burden in business surveys using communication strategies, paper presented at the European SIMPLY Conference on Administrative Simplification in Official Statistics, Ghent, Belgium, Dec. 2–3.

Giesen, D., and Vis, R. (2006), *Pilot e-PS: Preliminary Results* (in Dutch: Tussentijdse evaluatie Pilot e-PS), Internal report, Statistics Netherlands, Heerlen.

Giesen, D., Morren, M., and Snijkers, G. (2009a), *Proposal to Continuously Monitoring Business Surveys* (in Dutch: Voorstel permanente monitoring bedrijfsenquêtes), Internal report, Statistics Netherlands, Heerlen.

Giesen, D., Morren, M., and Snijkers, G. (2009b), The effect of survey redesign on response burden: An evaluation of the redesign of the SBS questionnaires, paper presented at the 3rd European Survey Research Association Conference (ESRA), Warsaw, Poland, June 29 – July 3.

Giesen, D., Meertens, V., Vis-Visschers, R., and Beukenhorst, D. J. (2010), *Questionnaire Development* (in Dutch: Methodenreeks: Vragenlijstontwikkeling), Methodology Series, Statistics Netherlands, Heerlen/The Hague.

Gilbert, N. (2011), Quality gates framework for statistical risk management, paper presented at South-African Reserve Bank Statistics Seminar, Bela-Bela, South Africa, March 28–31.

Gillman, D. W., and Appel, M. V. (1999), Statistical metadata research at the census bureau, paper presented at the Annual Federal Committee on Statistical Methodology (FCSM) Research Conference, Washington DC, Nov. 15–17.

Giovannini, E. (2008). *Understanding Economic Statistics: An OECD Perspective*, OECD, Paris (available at `www.oecd.org/std/41746710.pdf`).

Glass, G. V. (1976), Primary, secondary and meta-analysis of research, *Educational Researcher* **6**(28): 3–8.

Goddard, G. A. M. (1993), How not to collect fire statistics from fire brigades, *Proceedings of the 1st International Conference on Establishment Surveys (ICES-I)*, Buffalo, NY, June 27–30, American Statistical Association, Alexandria, VA, pp. 107–110.

Goldenberg, K. L. (1994), Answering questions, questioning answers: Evaluating data quality in an establishment survey, *Proceedings of the Section on Survey Research Methods*, American Statistical Association, Alexandria, VA, pp. 1357–1362.

Goldenberg, K. L. (1996), Using cognitive testing in the design of a business survey questionnaire, *Proceedings of the Section on Survey Research Methods*, American Statistical Association, Alexandria, VA, pp. 944–949.

Goldenberg, K. L., and Phillips, M. A. (2000), Now that the study is over, what did you really tell us? Identifying and correcting measurement error in the job openings and labor turnover survey pilot test, *Proceedings of the 2nd International Conference on Establishment Surveys (ICES-II)*, Buffalo, NY, June 17–21, American Statistical Association, Alexandria, VA, pp. 1482–1487.

Goldenberg, K. L., and Stewart, J. (1999), Earnings concepts and data availability for the current employment statistics survey: Findings from cognitive interviews, *Proceedings of the Section on Survey Research Methods*, American Statistical Association, Alexandria, VA, pp. 139–144.

Goldenberg, K. L., Butani, S., and Phipps, P. A. (1993), Response analysis surveys for assessing response errors in establishment surveys, *Proceedings of the 1st International Conference on Establishment Surveys (ICES-I)*, Buffalo, NY, June 27–30, American Statistical Association, Alexandria, VA, pp. 290–299.

Goldenberg, K. L., Levin, K., Hagerty, T., Shen, T., and Cantor, D. (1997), Procedures for reducing measurement errors in establishment surveys, *Proceedings of the Section on Survey Research Methods*, American Statistical Association, Alexandria, VA, pp. 994–999.

Goldenberg, K. L., Gomes, A., Manser, M., and Stewart, J. (2000), Collecting all-employee earnings data in the current employment statistics survey, paper presented at the Joint Statistical Meetings, Indianapolis, Aug. 13–17, American Statistical Association, Alexandria, VA.

Goldenberg, K. L., Anderson, A. E., Willimack, D. K., Freedman, S. R., Rutchik, R. H., and Moy, L. M. (2002a), Experiences implementing establishment survey questionnaire development and testing at selected U.S. government agencies, paper presented at the International Conference on Questionnaire Development, Evaluation, and Testing Methods (QDET), Charleston, SC, Nov. 14–17.

Goldenberg, K. L., Willimack, D. K., Fisher, S. K., and Anderson, A. E. (2002b), Measuring key economic indicators in U.S. government establishment surveys, paper presented at the International Conference on Improving Surveys, Copenhagen, Aug. 25–28.

Gonzalez, M. E. (1988), *Measurement of Quality in Establishment Surveys*, Bureau of Labor Statistics, Washington DC.

Goodman, R., and Kish, L. (1950), Controlled selection – a technique in probability sampling, *Journal of the American Statistical Association* **45**(251): 350–372.

Gower, A. R. (1994), Questionnaire design for business surveys, *Survey Methodology* **20**: 125–136.

Gower, A. R., and Nargundkar, M. S. (1991), Cognitive aspects of questionnaire design: Business surveys versus household surveys, *Proceedings of the 1991 Annual Research Conference*, Arlington, March 17–20, US Bureau of the Census, Washington DC, pp. 299–312.

Granquist, L. (1991), Macro-editing—a review of some methods for rationalizing the editing of survey data, *Statistical Journal* **8**: 137–154.

Granquist, L. (1995), Improving the traditional editing process, in Cox, B. G., Binder, D. A., Chinnappa, B. N., Christianson, A., Colledge, M. J., and Kott, P. S., eds., *Business Survey Methods*, Wiley, New York, pp. 385–402.

Granquist, L., Kovar, J., and Nordbotten, S. (2006), Improving surveys – where does editing fit in? in *Statistical Data Editing Vol. 3, Impact of Data Quality*, United Nations, New York/ Geneva, pp. 355–361.

Gravem, D. F., Haraldsen, G., and Löfgren, T. (2011), Response burden trends and consequences, in Giesen, D., and Bavdaz, M., eds., *Proceedings of the BLUE-ETS Conference on Burden and Motivation in Official Business Surveys*, Statistics Netherlands, Heerlen, March 22–23, pp. 221–235.

Green, S., Skentelbery, R., and Viles, R. (2008), Process quality in the Office for National Statistics, paper presented at the 4th European Conference on Quality in Official Statistics, Rome, July 8–11.

Green, S., Jones, J., and Frost, J.-M. (2010a), User views on quality reporting, paper presented at the 5th European Conference on Quality in Official Statistics, Helsinki, May 4–6.

Green, S., Jones, J., and Williams, D. (2010b), Measuring respondent burden to statistical surveys, paper presented at the 5th European Conference on Quality in Official Statistics, Helsinki, May 4–6.

Gregory, A. (2011), *The Data Documentation Initiative (DDI): An Introduction for National Statistical Institutes*, Open Data Foundation, Tucson, AZ.

Grice, P. (1975), Logic and conversation, in Cole, P., and Morgan, J., eds., *Syntax and Semantics: Speech Acts*, Academic Press, Orlando, FL.

Grice, P. (1989), *Studies in the Way of Words*, Harvard University Press, London.

Groves, R. M. (1989), *Survey Errors and Survey Costs*, Wiley, New York.

Groves, R.M. (2011), Three eras of survey research, *Public Opinion Quarterly*, Special issue on the 75th anniversary of POQ, **75**(5): 861–871.

Groves, R. M., and Cialdini, R. B. (1991), Towards a useful theory of survey participation. *Proceedings of the Section on Survey Research Methods*, American Statistical Association, Alexandria, VA, pp. 88–97.

Groves, R. M., and Couper, M. P. (1998), *Nonresponse in Household Interview Surveys*, Wiley, New York.

Groves, R., and Heeringa, S. (2006), Responsive design for household surveys: Tools for actively controlling survey errors and costs, *Journal of the Royal Statistical Society, Series A* **169**(3): 439–457.

Groves, R. M., and McGonagle, K. A. (2001), A theory-guided interviewer training protocol regarding survey participation, *Journal of Official Statistics* **17**(2): 249–265 (available at `www.jos.nu`).

Groves, R. M., Cialdini, R. B., and Couper, M. P. (1992), Understanding the decision to participate in a survey, *Public Opinion Quarterly* **56**(4): 475–495.

Groves, R. M., Cantor, D., Couper, M., Levin, K., McGonagle, K., Singer, E., and Van Hoewyk, J. (1997), Research investigations in gaining participation from sample firms in the current employment statistics program, *Proceedings of the Section on Survey Research Methods*, American Statistical Association, Alexandria, VA, pp. 289–294.

Groves, R., Singer, E., and Corning, A. (2000), Levarage-saliency theory of survey participation. Description and illustration, *Public Opinion Quarterly* **64**(3): 299–308.

Groves, R. M., Dillman, D. A., Eltinge, J. L., and Little, R. J. A., eds. (2002), *Survey Nonresponse*, Wiley, New York.

Groves, R. M., Fowler, F. J., Jr., Couper, M. P., Lepkowski, J. M., Singer, E., and Tourangeau, R. (2004), *Survey Methodology*, Wiley, Hoboken, NJ.

Grubbs, F. E. (1969), Procedures for detecting outlying observations in samples, *Technometrics* **11**(1): 1–21.

Gruber, D. L. (2007), High quality data and collection systems through active communication with data providers, paper presented at the 56th Session of the International Statistical Institute (ISI), Lisbon, Aug. 22–29.

Guttman, L. (1950), The basis for scalogram analysis, in Stouffer, S., Guttman, L., Suchman, E., Lazarsfeld, P., Star, S., and Clausen, J., eds., *Measurement and Prediction,* Vol. IV, *The American Soldier*, Wiley, New York.

Gwartney, P. A. (2007), *The Telephone Interviewer's Handbook: How to Conduct Standardized Conversations*, Jossey-Bass, San Francisco.

Gwet, J.-P., and Lee, H. (2001), An evaluation of outlier-resistant procedures in establishment surveys, *Proceedings of the 2nd International Conference on Establishment Surveys (ICES-II)*, Buffalo, NY, June 17–21, American Statistical Association, Alexandria, VA, pp. 707–716.

Hak, T., and van Sebille, M. (2002), *The Response Process in Establishment Surveys*, internal report, Erasmus Research Institute of Management, Erasmus University/Statistics Netherlands, Rotterdam/Voorburg.

Hak, T., Willimack, D. K., and Anderson, A. E. (2003), Response process and burden in establishment surveys, *Proceedings of the Section on Government Statistics*, Joint Statistical Meetings, San Francisco, Aug. 3–7, American Statistical Association, Alexandria, VA, pp. 1724–1730.

Hales, J., and Webster, S. (2008), *Methodological Review of Research with Large Businesses,* Paper 2: *Making Contact and Response Issues*, Research Report 60, UK: Her Majesty's Revenue and Customs, London (available at `www.hmrc.gov.uk/research/paper2-contact.pdf`).

Hamilton, J. D. (1994), *Time Series Analysis*, Princeton University Press, Princeton NJ.

Hampton, P. (2005), *Reducing Administrative Burdens: Effective Inspection and Enforcement*, UK: Her Majesty's Treasury, London.

Haraldsen, G. (1999), *Data Collection Methodology: The Cookbook Approach* (in Norwegian: Sporreskjemametodikk, etter kokebokmethoden), ad Notam, Gyldendal, Norway.

Haraldsen, G. (2003), Searching for response burden in focus groups with business respondents, in Prüfer, P., Rexroth, M., and Fowler, F., eds., *QUEST 2003, Proceedings of the 4th Workshop on Questionnaire Evaluation Standards*, Mannheim, Germany, Oct. 21–23, ZUMA Nachrichten, Spezial Band 9, pp. 113–123.

Haraldsen, G. (2004), Identifying and reducing response burdens in Internet business surveys, *Journal of Official Statistics* **20**(2): 393–410 (available at `www.jos.nu`).

Haraldsen, G. (2009), Why don't all businesses report on web?, paper presented at the 4th International Workshop on Internet Survey Methodology, Bergamo, Italy, Sept. 17–19.

Haraldsen, G. (2010), Reflections about the impact business questionnaires have on the perceived response burden and the survey quality, paper presented at the European

SIMPLY Conference on Administrative Simplification in Official Statistics, Ghent, Belgium, Dec. 2–3.

Haraldsen, G., and Bergstrøm, Y. (2009), Turning grid questions into sequences in business web surveys: Why don't all businesses report on the web?, paper presented at the 4th International Workshop on Internet Survey Methodology, Bergamo, Italy, Sept. 17–19.

Haraldsen, G., and Jones, J. (2007), Paper and web questionnaires seen from the business respondent's perspective, *Proceedings of the 3rd International Conference on Establishment Surveys (ICES III)*, Montreal, June 18–21, American Statistical Association, Alexandria, VA, pp. 1040–1047.

Haraldsen, G., Kleven, Ø., and Stålnacke, M. (2006), Paradata indications of problems in Web Surveys, paper presented at the 3rd European Conference on Quality in Official Statistics, Cardiff, UK, April 24–26.

Haraldsen, G., Kleven, Ø., and Sundvoll, A. (2010), Quality indicators in data collection, paper presented at the 5th European Conference on Quality in Official Statistics, Helsinki, May 4–5.

Haraldsen, G., Snijkers, G., Roos, M., Sundvoll, A., Vik, T., and Stax, H.-P. (2011), Utilizing web technology in business data collection: Some Norwegian, Dutch and Danish experiences, paper presented at the 2011 European Conference on New Techniques and Technologies for Statistics (NTTS), Brussels, Feb. 22–24.

Harberger, A. C. (1971), Measuring the social opportunity cost of labour, *International Labor Review (June)*: 559.

Harley, M. D., Pressley, K. D., and Murphy, E. D. (2001), 2002 economic electronic style guide, paper presented at the 2001 International Statistics Canada Methodology Symposium: Achieving Data Quality in a Statistical Agency: A Methodological Perspective, Hull, Quebec, Oct. 18–19, CD-rom, Statistics Canada, Ottawa.

Harrell, L., Yu, H., and Rosen, R. (2007), Respondent acceptance of web and E-mail data reporting for an establishment survey, *Proceedings of the 3rd International Conference on Establishment Surveys (ICES-III)*, Montreal, June 18–21, American Statistical Association, Alexandria, VA, pp. 1442–1445.

Hartley, H. O., and Rao, J. N. K. (1962), Sampling with unequal probabilities and without replacement, *The Annals of Mathematical Statistics* **33**(2): 350–374.

Hatanaka, M. (1996), *Time-Series-Based Econometrics. Unit Roots and Cointegrations*, Oxford University Press, Oxford.

Haziza, D. (2009), Imputation and inference in the presence of missing data, in Pfeffermann, D., and Rao, C. R., eds., *Handbook of Statistics,* Vol. 29A, *Sample Surveys: Design, Methods and Applications*, North-Holland, Amsterdam, pp. 215–246.

Haziza, D., Chauvet, G., and Deville, J. C. (2010), Sampling and estimation in the presence of cut-off sampling, *Australian and New Zealand Journal of Statistics* **52**(3): 303–319.

Heath, R. (2011), Statistical implications of the global crisis: G-20 data initiative, paper presented at the Conference on Strengthening Sectoral Position and Flow Data in the Macroeconomic Accounts, Washington, DC, Feb. 28–March 2.

Hedeman, B., Vis van Heemst, G., and Fredriksz, H. (2006), *Project Management Based on Prince2*, Van Haren Publishing, Zaltbommel, The Netherlands.

Hedlin, D. (2003), Score functions to reduce business survey editing at the UK Office for National Statistics, *Journal of Official Statistics* **19**(2): 177–199 (available at `www.jos.nu`).

Hedlin, D. (2008), Small area estimation: A practitioner's appraisal, *Rivista Internazionale di Scienze Sociali* **116**(4): 407–417.

Hedlin, D., Dale, T., Haraldsen, G., and Jones, J. (2005), *Developing Methods for Assessing Perceived Response Burden*, Eurostat, Luxembourg.

Hedlin, D., Fenton, T., McDonald, J. W., Pont, M., and Wang, S. (2006), Estimating the undercoverage of a sampling frame due to reporting delays, *Journal of Official Statistics* **22**(1): 53–70 (available at www.jos.nu).

Heeringa, S. G., Brady, T., and Berglund, P. A. (2010), *Applied Survey Data Analysis*, Chapman & Hall, London.

Heerwegh, D. (2002), Describing response behaviour in web surveys using client side paradata, paper presented at the 1st International Workshop on Internet Survey Methodology, Mannheim, Germany, Oct. 17–19.

Heerwegh, D. (2003), Explaining response latencies and changing answers using client-side paradata from a web survey, *Social Science Computer Review* **21**(3): 360–373.

Hellerman, E. (1982), *Overview of the Hellerman I&O Coding System*, Internal report, US Census Bureau, Washington DC.

Herrmann, V., and Junker, C. (2008), Reduction of response burden and priority setting in the field of community statistics—initiatives at the European level, paper presented at the 4th European Conference on Quality in Official Statistics, Rome, July 8–11.

Hidiroglou, M. A. (1986), The construction of a self-representing sample of large units in a survey design, *The American Statistician* **40**(1): 27–31.

Hidiroglou, M. A., and Berthelot, J. M. (1986), Statistical editing and imputation for periodic business surveys, *Survey Methodology* **12**: 73–83.

Hidiroglou, M. A., and Smith, P. (2005), Developing small area estimates for business surveys at the ONS, *Statistics in Transition* **7**(3): 527–539.

Hidiroglou, M., and Srinath, K. P. (1981), Some estimators of a population total from simple random samples containing large units, *Journal of the American Statistical Association* **76**(375): 690–695.

Hidiroglou, M., and Srinath, K. P. (1993), Problems associated with designing subannual business surveys, *Journal of Business & Economic Statistics* **11**(4): 397–405.

Hill, T. P. (1988), Recent developments in index number theory and practice, *OECD Journal: Economic Studies* **10** (spring): 123–148.

Hix, D., and Hartson, H. R. (1993), *Developing User Interfaces: Ensuring Usability through Product & Process*, Wiley, New York.

HLG (2009), *European Commission High Level Group of Independent Stakeholders on Administrative Burdens: Opinion of the High Level Group. Subject: Priority Area Statistics*, Brussels, July 7.

HMRC (2010), *Large Business Methodology Review: Stage Two Report* (Ipsos MORI, Social Research Institute), Research report 98, UK: Her Majesty's Revenue and Customs, London (available at www.hmrc.gov.uk/research/research-report98.pdf).

HM Treasury (2004). *The Orange Book: Management of Risk – Principles and Concepts*, UK: Her Majesty's Treasury, London (available at www.hm-treasury.gov.uk/orange_book.htm).

Hoeken, H., Hornikx, J., and Hustinx, L. (2009), *Persuasive Texts: Research and Design* (in Dutch: Overtuigende teksten: Onderzoek en ontwerp), Uitgeverij Coutinho, Bussum, The Netherlands.

Hoekstra, M. (2007), *Analysis of Mode Effects in Business Surveys* (in Dutch: Analyse op mode effecten bij bedrijfsenquêtes), Internal report, Statistics Netherlands, Heerlen.

Hoffman, D. D. (1997), *Visual Intelligence*, Norton, New York.

Holmberg, A. (2004), Pre-printing effects in official statistics: An experimental study, *Journal of Official Statistics* **20**(2): 341–355 (available at www.jos.nu).

Holmberg, A., Lorenc, B., and Werner, P. (2010), Contact strategies to improve participation via the web in a mixed-mode mail and web survey, *Journal of Official Statistics* **26**(3): 465–480 (available at www.jos.nu).

Homans, G. C. (1961), *Social Behavior: Its Elementary Forms*, Harcourt, Brace & World, New York.

Hood, C. C., and Findley, D. F. (2003), Comparing direct and indirect seasonal adjustments of aggregate series, in Manna, M. and Peronaci, R., eds., *Seasonal Adjustment*, European Central Bank, Frankfurt, pp. 9–22.

Hoogendoorn, A. W. (2004), A questionnaire design for dependent interviewing that addresses the problem of cognitive satisficing, *Journal of Official Statistics* **20**(2): 219–232 (available at www.jos.nu).

Horgan, J. M. (2006), Stratification of skewed populations: A review, *International Statistical Review* **74**(1): 67–76.

Horvitz, D. G., and Thompson, D. J. (1953), A generalization of sampling without replacement from a finite universe, *Journal of the American Statistical Association* **47**(283): 663–685.

Houlihan, L. F. (2007), Using focus groups to improve response in monthly surveys, *Proceedings of the 3rd International Conference on Establishment Surveys (ICES-III)*, Montreal, June 18–21, American Statistical Association, Alexandria, VA, pp. 332–337.

Houston, G., and Bruce, A. G. (1993), gred: Interactive graphical editing for business surveys, *Journal of Official Statistics* **9**(1): 81–90 (available at www.jos.nu).

Hox, J. (1997), From theoretical concepts to survey questions, in Lyberg, L., Biemer, P., Collins, M., de Leeuw, E., Dippo, C., Schwarz, N., and Trewin, D., eds., *Survey Measurement and Process Quality*, Wiley, New York, pp. 47–69.

Hox, J. J., de Leeuw, E. D., and Dillman, D. A. (2008), The cornerstones of survey research, in de Leeuw, E. D., Hox, J. J., and Dillman, D. A., eds., *International Handbook of Survey Methodology*, Lawrence Erlbaum Associates, New York, pp. 1–17.

Huges, J. (2008), Prioritizing business respondents to target important non-response, paper presented at the 2008 International Statistics Canada Methodology Symposium: Data Collection: Challenges, Achievements and New Directions, Catalog 11-522-X, Statistics Canada, Ottawa, Oct. 28–31 (available at www.statcan.gc.ca/pub/11-522-x/2008000/article/10984-eng.pdf).

Hughes, P. J., McDermid, I., and Linacre, S. J. (1990), The use of graphical methods in editing, *Proceedings of the US Bureau of the 1990 Census Annual Research Conference*, US Bureau of the Census, Washington DC, pp. 538–550.

Hundepool, A. (2000), Statistical disclosure control software, *Proceedings of the 2nd International Conference on Establishment Surveys (ICES-II)*, Buffalo, NY, June 17–21, American Statistical Association, Alexandria, VA, pp. 897–903.

Hundepool, A., Domingo-Ferrer, J., Franconi, L., Giessing, S., Lenz, R., Naylor, J., Nordholt, E., Seri, G., and De Wolf, P. P. (2010), *Handbook on Statistical Disclosure Control*, version 1.2, ESSNet SDC (available at http://neon.vb.cbs.nl/casc/handbook.htm).

Hunter, L., and Carbonneau, J.-F. (2005), An active management approach to survey collection, paper presented at the 2005 International Statistics Canada Methodology Symposium: Methodological Challenges for Future Information Needs, Statistics Canada, Ottawa, Oct. 25–28.

Hutchins, E. (1995), *Cognition in the Wild*, MIT Press, Cambridge, MA.

Hyman, H. H. (1972), *Secondary Analysis of Sample Surveys*, Wiley, New York.

IGEN (Interagency Group on Establishment Nonresponse) (1998), Establishment nonresponse: Revisiting the issues and looking to the future, paper presented at the Council for Professional Association on Federal Statistics Conference, Nov.

Institute for Supply Management (2012), ISM's Manufacturing *Report on Business*®, March.

International Labour Organization (2003), Report III: Consumer price indices, paper presented at the 17th International Conference of Labour Statisticians, Geneva, Nov. 24–Dec. 3.

International Monetary Fund (2004), *Producer Price Index Manual Theory and Practice*, International Monetary Fund, Washington DC.

Ipsos MORI (2010), *Large Business Methodology Review: Stage Two Report* (Ipsos MORI, Social Research Institute), Research report 98, UK: Her Majesty's Revenue and Customs, London (available at www.hmrc.gov.uk/research/research-report98.pdf).

ISI (2010), *Declaration of Professional Ethics*, International Statistical Institute (ISI), The Hague, The Netherlands (available at www.isi-web.org/about/ethics-intro).

ISO (2006), *International Standard on Market, Opinion and Social Research*, ISO20252:2006, International Organization for Standardization, Geneva.

Jarrett, C., and Gaffney, G. (2009), *Forms that Work: Designing Web Forms for Usability*, Morgan Kaufmann Publishers, Amsterdam.

Jenkins, C. (1992), Questionnaire research in the schools and staffing survey: A cognitive approach, paper presented at the Joint Statistical Meetings, Boston, Aug. 9–13, American Statistical Association, Alexandria, VA.

Jeskanen-Sundström, H. (2002), Feedback reporting, personal email, Statistics Finland, Helsinki, June 14.

Jeskanen-Sundström, H. (2007), Contacts with organizations of industry, personal email, Statistics Finland, Helsinki, July 6.

Jobber, D., and O'Reilly, D. (1998), Industrial mail surveys, *Industrial Marketing Management* **27**(2): 95–107.

Jobber, D., Mirza, H., and Wee, K. (1991), Incentives and response rates to cross-national business surveys: A logit model analysis, *Journal of International Business Studies* **22**(4): 711–721.

Johnson, J. (2010), *Designing with the Mind in Mind: Simple Guide to Understanding*, Elsevier, Burlington, MA.

Johnson, L. (1996), *Choosing a Price Index Formula: A Survey of the Literature with an Application to Price Indexes for the Tradable and Nontradable Sectors*, Working Paper in Econometrics and Applied Statistics 96/1, Australian Bureau of Statistics, Canberra.

Jones, J. (2003), A Framework for Reviewing Data Collection Instruments in Business Surveys, *Survey Methodology Bulletin*, UK Office for National Statistics, London, **52**: 4–9.

Jones, J. (2011), Effects of different modes, especially mixed modes, on response rates, paper presented at the Workshop on Different Modes of Data Collection, Eurofond, Dublin, April 6–7.

Jones, J., Borgerson, H., Williams, G., Curzon, J., and Smith, A. (2004), Catalysts for change: The rationale for mixed mode data collection in the UK, paper presented at the 2nd European Conference on Quality and Methodology in Official Statistics, Mainz, Germany, May 24–26.

Jones, J., Rushbrooke, J., Haraldsen, G., Dale, T., and Hedlin, D. (2005), Conceptualising Total Business Survey Burden, *Survey Methodology Bulletin*, UK Office for National Statistics, London, **55**: 1–10.

Jones, J., Brodie, P., Williams, S., and Carter, J. (2007a), Improved questionnaire design yields better data: Experiences from the UK's annual survey of hours and earnings, paper presented at the 3rd International Conference on Establishment Surveys (ICES-III), Montreal, June 18–21, American Statistical Association, Alexandria, VA.

Jones, J., Haraldsen, G., and Dale, T. (2007b), Measuring and monitoring response burden in business surveys, paper presented at the 13th Meeting of the National Statistics Methodology Advisory Committee (MAC), UK Office for National Statistics.

Jones, J., Brodie, P., Williams, S., and Carter, J. (2008a), Improved Questionnaire Design Yields Better Data: Experiences from the U.K.'s Annual Survey of Hours and Earnings, *Survey Methodology Bulletin*, UK Office for National Statistics, London, **62**: 12–28.

Jones, J., Lewis, A., Woodland, S., Jones, G., and Byard, J. (2008b), Communicating with survey respondents at the UK Office for National Statistics, *Proceedings of the Survey Research Methods Section*, Joint Statistical Meetings, Denver, Aug. 3–7, American Statistical Association, Alexandria, VA, pp. 309–320.

Jones, J., Duff, H., and Doody, R. (2011), *Meeting the Needs of Users: Feedback from Users of the Index of Production and Index of Services*, UK Office for National Statistics, London.

Julius, D., and Butler, J. (1998), Inflation and growth in a service economy, *Bank of England Quarterly Bulletin* (Nov.) 338–346 (available at www.bankofengland.co.uk/publications/quarterlybulletin/service.pdf).

Juran, J. M., and Gryna, F. M., Jr. (1980), *Quality Planning and Analysis*, McGraw-Hill, New York.

Kahneman, D. (1973), *Attention and Effort*, Prentice-Hall, Englewood Cliffs, NJ.

Kalafatis, S., and Tsogas, M. (1994), Impact of the inclusion of an article as an incentive in industrial mail surveys, *Industrial Marketing Management* **23**(2): 137–143.

Kaplan, A., and White, G. (2002), Incentives in a business survey: A study in improving response rates, *Proceedings of the Section on Survey Research Methods*, Joint Statistical Meetings, New York, Aug. 11–15, American Statistical Association, Alexandria, VA, pp. 1756–1761.

Katz, D., and Kahn, R. L. (1978), *The Social Psychology of Organizations*, 2nd ed., Wiley, New York.

Kearney, A. T., and Kornbau, M. E. (2005), An automated industry coding application for new U.S. business establishments, *Proceedings of the Section on Business and Economic Statistics*, Joint Statistical Meetings, Minneapolis, Aug. 7–11, American Statistical Association, Alexandria, VA, pp. 867–874.

Keller, S., Hough, R. S., Falconer, T., and Curcio, K. (2011), Characteristics of large company respondents to an economic survey, paper presented at the Joint Statistical Meetings, Miami Beach, FL, July 30 – Aug. 4, American Statistical Association, Alexandria, VA.

Keller, W. J. (1996), EDI: Electronic Data Interchange for Statistical Data Collection and Dissemination, *Proceedings of the 1996 Annual Research Conference and Technology Interchange*, Arlington, VA, March 17–21, pp. 955–970.

Kennedy, J. M., Tarnai, J., and Wolf, J. G. (2010), Managing survey research projects, in Marsden, P. V., and Wright, J. D., eds., *Handbook of Survey Research*, 2nd ed., Emerald Group, Bingley, UK, pp. 575–590.

Keppel, G., and Wickens, T.D., 2013 (forthcoming), *Design and Analysis: A Researcher's Handbook*, 5th Edition, Pearson Prentice Hall, Upper Saddle River, NJ.

Keynes, J. M. (1936), *The General Theory of Employment, Interest and Money*, Atlantic Publishers and Distributors, New Delhi.

Kirk, R.E. (2013), *Experimental Design: Procedures for Behavioral Sciences*, 4th Edition, Sage, Thousand Oaks, CA.

Knaub, J. R. (2007), *Cutoff Sampling and Inference*, Interstat, Statistics on the Internet, April (available at http://interstat.statjournals.net/YEAR/2007/articles/0704006.pdf).

Knaub, J. R. (2011), Cut-off sampling and total survey error (letter to the editor), *Journal of Official Statistics* **27**(1): 135–138 (available at www.jos.nu).

Knottnerus, P. (2011), On the efficiency of randomized probability proportional to size sampling, *Survey Methodology* **37**(1): 95–102.

Kokic, P. N., and Bell, P. A. (1994), Optimal winsorizing cut-offs for a stratified finite population estimator, *Journal of Official Statistics* **10**: 419–435 (available at www.jos.nu).

Kornbau, M., Bouffard, J. and Vile, J. (2007), Steps to provide quality industrial coding for a business register, *Proceedings of the 3rd International Conference on Establishment Surveys (ICES-III)*, Montreal, June 18–21, American Statistical Association, Alexandria, VA, pp. 25–32.

Koop, G. (2009), *Analysis of Economic Data*, Wiley, Hoboken, NJ.

Kovar, J. G., and Whitridge, P. J. (1995), Imputation of business survey data, in Cox, B. G., Binder, D. A., Chinnappa, B. N., Christianson, A., Colledge, M. J., and Kott, P. S., eds., *Business Survey Methods*, Wiley, New York, pp. 403–424.

Kozak, M. (2004), Optimal stratification using random search method in agricultural surveys, *Statistics in Transition* **6**(5): 797–806.

Kreuter, F., ed. (2013), *Improving Surveys with Paradata: Analytic Uses of Process Information*, Wiley, Hoboken, NJ.

Kreuter, F., and Casas-Cordero, C. (2010), Paradata, Working paper, German Council for Social and Economic Data, Berlin (available at www.ratswd.de/en).

Kreuter, F., Couper, M., and Lyberg, L. (2010), The use of paradata to monitor and manage survey data collection, *Proceedings of the Survey Research Methods Section*, Joint Statistical Meetings, Vancouver, Canada, July 31 – Aug. 5, American Statistical Association, Alexandria, VA, pp. 282–296.

Kroese, A. H., and Renssen, R. H. (2000), New applications of old weighting techniques; constructing a consistent set of estimates based on data from different surveys, *Proceedings of the 2nd International Conference on Establishment Surveys (ICES-II)*, Buffalo, NY, June 17–21, American Statistical Association, Alexandria, VA, pp. 831–840.

Kroese, A. H., Renssen, R. H., and Trijssenaar, M. (2000), Weighting or imputation: Constructing a consistent set of estimates based on data from different sources, *Netherlands Official Statistics*, Special issue: Integrating administrative registers and household surveys **15**(Summer): 23–31.

Krosnick, J. A. (1991), Response strategies for coping with the cognitive demands fattitude measures in surveys, *Applied Cognitive Psychology*, Special issue: Cognition and survey measurement **5**(3): 213–236.

Krosnick, J. A., and Alwin, D. (1987), An evaluation of a cognitive theory of response-order effects in survey measurment, *Public Opinion Quarterly* **51**(2): 201–219.

Krosnick, J. A., and Presser, S. (2010), Question and questionnaire design, in Marsden, P. V., and Wright, J. D., eds., *Handbook of Survey Research*, 2nd ed., Emerald Group, Bingley, UK, pp. 263–313.

Krueger, R., and Casey, M. A. (2000), *Focus Groups: A Practical Guide for Applied Research*, 3rd ed., Sage, Thousand Oaks, CA.

Kydoniefs, L. (1993), The occupational safety and health survey, *Proceedings of the 1st International Conference on Establishment Surveys (ICES-I)*, Buffalo, NY, June 27–30, American Statistical Association, Alexandria, VA, pp. 99–106.

Kydoniefs, L., and Stinson, L. L. (1999), Standing on the outside, looking In: Tapping data users to compare and review surveys, *Proceedings of the Section on Survey Research Methods*, American Statistical Association, Alexandria, VA, pp. 968–972.

Labillois, T., and March, M. (2000), Cost-recovery business surveys–helping policy makers acquire information required to deal with newly arising issues, *Proceedings of the 2nd International Conference on Establishment Surveys (ICES-II)*, Buffalo, NY, June 17–21, American Statistical Association, Alexandria, VA, pp. 1250–1255.

Laffey, F. (2002), Business survey questionnaire review and testing at Statistics Canada, paper presented at the International Conference on Questionnaire Development, Evaluation, and Testing Methods (QDET), Charleston, SC, Nov. 14–17.

Laflamme, F. (2008), Data collection research using paradata at Statistics Canada, paper presented at the 2008 International Statistics Canada Methodology Symposium: Data Collection: Challenges, Achievements and New Directions, Catalog 11-522-X, Statistics Canada, Ottawa, Oct. 28–31.

Laflamme, F., and Karaganis, M. (2010), Implementation of responsive collection design for CATI surveys at Statistics Canada, paper presented at the 5th European Conference on Quality in Official Statistics, Helsinki, May 4–5.

Laflamme, F., Maydan, M., and Miller, A. (2008), Using paradata to actively manage data collection survey process, *Proceedings of the Survey Research Methods Section*, Joint Statistical Meetings, Denver, Aug. 3–7, American Statistical Association, Alexandria, VA, pp. 630–637.

Lane, J. (2010), Linking administrative and survey data, in Marsden, P. V., and Wright, J. D., eds., *Handbook of Survey Research*, 2nd ed., Emerald Group, Bingley, UK, pp. 659–680.

Latouche, M., and Berthelot, J. (1992), Use of a score function to prioritize and limit recontacts in editing business surveys, *Journal of Official Statistics* **8**(3): 389–400 (available at www.jos.nu).

Lavallée, P. (2002), Combining survey and administrative data: Discussion paper, *Proceedings of the 2nd International Conference on Establishment Surveys (ICES-II)*, Buffalo, NY, June 17–21, American Statistical Association, Alexandria, VA, pp. 841–844.

Lavallée, P., and Hidiroglou, M. A. (1988), On the stratification of skewed populations, *Survey Methodology* **14**: 33–43.

Lawrence, D., and McDavitt, C. (1994), Significance editing in the Australian survey of average weekly earnings, *Journal of Official Statistics* **10**(4): 437–447 (available at www.jos.nu).

Lawrence, C., and McKenzie, R. (2000), The general application of significance editing, *Journal of Official Statistics*, **16**(3): 243–253 (available at www.jos.nu).

Leach, N. L. (1999), Using cognitive research to redesign federal questionnaires for manufacturing establishments, paper presented at the Joint Statistical Meetings, Baltimore, Aug. 8–12, American Statistical Association, VA.

Lee, H. (1995), Outliers in business surveys, in Cox, B. G., Binder, D. A., Chinnappa, B. N., Christianson, A., Colledge, M. J., and Kott, P. S., eds., *Business Survey Methods*, Wiley, New York, pp. 503–526.

Leivo, J. (2010), Developing business data collection and measuring response burden, paper presented at the 5th European Conference on Quality in Official Statistics, Helsinki, May 4–6.

Lemaitre, G., and Dufour, J. (1987), An integrated method for weighting persons and families, *Survey Methodology* **13**: 199–207.

Lemmens, A., Croux, C., and Dekimpe, M. G. (2004), On the predictive content of production surveys: A pan-European study, in *ERIM Report Series Research in Management*, Erasmus Research Institute of Management, Rotterdam.

Leontief, W. (1936), Quantitative input and output relations in the economic system of the united states, *Review of Economics and Statistics* **18**(3): 105–125.

Lequiller, F., and Blades, D. (2006), *Understanding National Accounts*, OECD publication, Paris (available at www.oecd.org/std/na/38451313.pdf).

Lessler, J. T., and Forsyth, B. H. (1996), A coding system for appraising questionnaires, in Schwarz, N., and Sudman, S., eds., *Answering Questions, Methodology for Determining Cognitive and Communicative processes in Survey Research*, Jossey-Bass, San Francisco, pp. 259–291.

Lessler, J. T., and Kalsbeek, W. D. (1992), *Nonsampling Error in Surveys*, Wiley, New York.

Likért, R. (1932), *A Technique for the Measurement of Attitudes*, Archive of Psychology 140, Columbia University Press, New York.

Löfgren, T. (2010), Burden reduction by instrument design, in Giesen, D., ed., *Response Burden in Official Business Surveys: Measurement and Reduction Practices of National Statistical Institutes*, Deliverable 2. 2, BLUE-Enterprise and Trade Statistics Project (BLUE-ETS), European Commission, European Research Area, pp. 43–50 (available at www.blue-ets.istat.it/fileadmin/deliverables/Deliverable2.2.pdf).

Löfgren, T., Gravem, D. F., and Haraldsen, G. (2011), A glimse into the businesses' use of internal and external dats sources in decision-making processes, in Giesen, D., and Bavdaz, M., eds., *Proceedings of the BLUE-ETS Conference on Burden and Motivation in Official Business Surveys*, Statistics Netherlands, Heerlen, March 22–23, pp. 53–68.

Lohr, S. L. (1999), *Sampling: Design and Analysis*, Duxbury Press.

Lohr, S. (2007), Recent developments in multiple frame surveys, *Proceedings of the Survey Research Methods Section*, Joint Statistical Meetings, Salt Lake City, July 29 – Aug. 2, American Statistical Association, Alexandria, VA, pp. 3257–3264.

Lohr, S. (2010a), *Sampling: Design and Analysis*, 2nd ed., Brooks/Cole, Boston.

Lohr, S. L. (2010b), Dual frame surveys: Recent developments and challenges, paper presented at the 45th Scientific Meeting of the Italian Statistical Society, Padua, Italy, June 16–18 (available at http://homes.stat.unipd.it/mgri/SIS2010/Program/14-SSXIV_Mecatti/894-1542-1-RV%5B1%5D.pdf).

London, S., and Dommeyer, C. (1990), Increasing response to industrial mail surveys, *Industrial Marketing Management* **19**(3): 235–241.

Lorenc, B. (2006), Social distribution of the response process in a survey of schools, paper 2 in *Two Topics in Survey Methodology*, PhD thesis, Department of Statistics, Stockholm University.

Lorenc, B. (2007), Using the theory of socially distributed cognition to study the establishment survey response process, *Proceedings of the 3rd International Conference on Establishment Surveys (ICES-III)*, Montreal, June 18–21, American Statistical Association, Alexandria, VA, pp. 881–891.

Lowthian, P., Barton, J., and Bycroft, C. (2006), Standards and Guidance for Disclosure Control, *Survey Methodology Bulletin*, UK Office for National Statistics, London, **59**: 44–59.

Luppes, M. (1995), A content analysis of advance letters from expenditure surveys in seven countries, *Journal of Official Statistics* **11**(4): 461–480 (available at www.jos.nu).

Luzi, O., Di Zio, M., Guarnera, U., Manzari, A., de Waal, T., Pannekoek, J., Hoogland, J., Tempelman, C., Hulliger, B., and Kilchmann, D. (2007), *Recommended Practices for Editing and Imputation in Cross-Sectional Business Surveys in the ESS*, Eurostat, Luxembourg (available at http://epp.eurostat.ec.europa.eu/portal/page/portal/quality/documents/RPM_EDIMBUS.pdf).

Lyberg, L. (2011), The total survey error paradigm, Seminar at Statistics Norway, Oslo, Nov. 9–10, Lyberg Survey Quality Management Inc., Stockholm.

Lyberg, L., and Kasprzyk, D. (1997), Some aspects of post-survey processing, in Lyberg, L., Biemer, P., Collins, M., de Leeuw, E., Dippo, C., Schwarz, N., and Trewin, D., eds., *Survey Measurement and Process Quality*, Wiley, New York, pp. 353–370.

Lyberg, L., Biemer, P., Collins, M., de Leeuw, E., Dippo, C., Schwarz, N., and Trewin, D., eds. (1997), *Survey Measurement and Process Quality*, Wiley, New York.

Lynn, P., and Nicolaas, G. (2010), Making good use of survey paradata, *Survey Practice* (April) (available at www.surveypractice.org).

Lynn, P., and Sala, M. (2004), *The Contact and Response Process in Business Surveys: Lessons Learned from a Multimode Survey of Employers in the UK*, Working papers of the Institute for Social and Economic Research, paper 2004-12, University of Essex, Colchester, UK.

Macauly, F. R. (1931), *The Smoothing of Time Series*, National Bureau of Economic Research, New York.

Macintyre, C. (2011), *Findings from the first 100 Assessment Reports*, Monitoring Brief 2/11, UK Statistics Authority, London.

Mahajan, S. (2007), Development, compilation and use of input-output supply and use tables in the United Kingdom national accounts, paper presented at the 16th International Input-Output Conference, Istanbul, July 2–6 (available at http://unstats.un.org/unsd/EconStatKB/Attachment107.aspx).

Manfreda, K. L., Vehovar, V., and Batagelj, Z. (2001), Web versus mail questionnaire for an institutional surveys, *Proceedings of the ASC International Conference: The Challenge of the Internet*, Chesham, UK, May 11–12, Association for Survey Computing, Chesham, pp. 79–90 (available at www.asc.org.uk/publications/proceedings/ASC2001Proceedings.pdf).

Manning, A., and Atkinson, D. (2009), Toward a comprehensive editing and imputation structure for NASS – integrating the parts, paper presented at the Conference of European Statisticians, Work Session on Statistical Data Editing, Neuchâtel, Switzerland, Oct. 5–7.

Marske, R., and Stempowski, D. M. (2008), Company-centric communication approaches for business survey respondent management, paper presented at the 2008 International Statistics Canada Methodology Symposium: Data Collection: Challenges, Achievements and New Directions, Catalog 11-522-X, Statistics Canada, Ottawa, Oct. 28–31.

Marske, R., Torene, L., and Hartz, M. (2007), Company-centric communication approaches for business survey response, *Proceedings of the 3rd International Conference on Establishment Surveys (ICES-III)*, Montreal, Montreal, June 18–21, American Statistical Association, Alexandria, VA, pp. 941–952.

Martelli, B. M. (2005), Part 1—relationship between response rates and data collection methods, Task Force on Improvement of response Rates and Minimisation of Respondent Load, paper presented at the Joint European Commission–OECD Workshop on International Development of Business and Consumer Tendency Surveys, Brussels, Nov. 14–15.

Massell, P. B. (2004), Comparing statistical disclosure control methods for tables: Identifying the key factors, paper presented at the Joint Statistical Meetings, Toronto, Canada, Aug. 8–12, American Statistical Association, Alexandria, VA.

Massell, P. B. and Funk, J. M. (2007), Protecting the confidentiality of tables by adding noise to the underlying microdata, *Proceedings of the 3rd International Conference on Establishment Surveys (ICES-III)*, Montreal, June 18–21, American Statistical Association, Alexandria, VA, pp. 1009–1020.

Massey, J. T. (1988), An overview of telephone coverage, in Groves, R. M., Biemer, P. P., Lybergy, L. E., Massey, J. T., NichollsII, W. L., and Waksberg, J., eds., *Telephone Survey Methodology*, Wiley, New York, pp. 3–8.

Mathiowetz, N. A., and McGonagle, K. A. (2000), An assessment of the current state of dependent interviewing in household surveys, *Journal of Official Statistics* **16**(4): 401–418 (available at www.jos.nu).

McAfee, R. P., and Johnson, J. S. (2006), *Introduction to Economic Analysis*, version 2.0, California Institute of Technology, Pasadena.

McCarthy, J. S. (2001), Using respondent requests for help to develop quality data collection instruments: The 2000 Census of Agriculture Content Test, paper presented at the 56th Annual Conference of the American Association for Public Opinion Research (AAPOR), Montreal, May 17–20, also in *Proceedings of the Joint Statistical Meetings*, American Statistical Association, Alexandria, VA, CD-rom.

McCarthy, J. S. (2007), Like, but oh, how different: The effect of different questionnaire formats in the 2005 Census of Agriculture Content Test, *Proceedings of the 3rd International Conference on Establishment Surveys (ICES-III)*, Montreal, June 18–21, American Statistical Association, Alexandria, VA, pp. 1194–1199.

McCarthy, J. S. (2008), Using data mining techniques to analyze survey reporting errors, paper presented at the 2nd International Workshop on Business Data Collection Methodology (BDCM), Ottawa, Oct. 22–24.

McCarthy, J., and Buysse, D. (2010), Bento box questionnaire testing: Multi-method questionnaire testing for the 2012 Census of Agriculture, paper presented at the 65th Annual Conference of the American Association for Public Opinion Research (AAPOR), Chicago, May 13–16.

McCarthy, J. S., Johnson, J., and Ott, K. (1999), Exploring the relationship between survey participation and survey sponsorship: What do respondents and non-respondents think of us?, paper presented at the International Conference on Survey Nonresponse, Portland, OR, Oct. 28–31.

McCarthy, J. S., Beckler, D. G., and Qualey, S. M. (2006), An analysis of the relationship between survey burden and nonresponse: If we bother them more, are they less cooperative?, *Journal of Official Statistics*, **22**(1): 97–112 (available at `www.jos.nu`).

McCarthy, J. S., Thomas, J., and Atkinson, D. (2009), Innovative uses of data mining techniques in the production of official statistics, paper presented at the Annual Federal Committee of Statistical Methodology (FCSM) Research Conference, Washington, DC, Nov 2–4.

McConnell, S. (1996), *Rapid Development: Taming Wild Software Schedules*, Microsoft Press, Redmond, WA.

McGregor, D. (1985), *The Human Side of Enterprise: 25th Anniversary Printing*, McGraw-Hill, New York.

McGuckin, R. H. (1995), Establishment microdata for economic research and policy analysis: Looking beyond the aggregates, *Journal of Business and Economic Statistics* **13**(1): 121–126.

McKenzie, R. (2005), Assessing and minimising the impact of non-response on survey estimates, Task Force on Improvement of Response Rates and Minimisation of Respondent Load, paper presented at Joint European Commission–OECD Workshop on International Development of Business and Consumer Tendency Surveys.

McLuhan, M. 1964. *Understanding Media: The Extensions of Man*, Mentor, New York.

McQuail, D. (2010), *McQuail's Mass Communication Theory*, Sage, London.

Meier, J. D., Farre, C., Prashant, B. Barber, S., and Rea, D. (2007), *Performance Testing Guidance for Web Applications*, Microsoft online library (available at `http://msdn.microsoft.com/en-us/library/bb924375.aspx`).

Merrington, R., Torrey, B., and van Heerden, L. (2009), Measurement of respondent load at statistics New Zealand, paper presented at the 57th Session of the International Statistical Institute (ISI), Durban, South Africa, Aug. 16–22.

Miles, M. B., and Huberman, A. M. (1994), *Qualitative Data Analysis: An Expanded Sourcebook:*, 2nd ed., Sage, Thousand Oaks, CA.

Miller, J. E. (2004), *The Chicago Guide to Writing about Numbers*, University of Chicago Press, Chicago.

Mills, K., and Palmer, K. (2006), Using qualitative methods in feasibility studies: investigating the availability of local unit data for business surveys, paper presented at the 3rd European Conference on Quality in Official Statistics, Cardiff, UK, April 24–26; also published in *Survey Methodology Bulletin*, UK Office for National Statistics, London, 60: 1–8.

Minton, G. (1972), Verification error in single sampling inspection plans for processing survey data, *Journal of the American Statistical Association* **67**(337): 46–54.

Mitchell, B. R. (1988), *British Historical Statistics*, Cambridge University Press, Cambridge.

Mohl, C. (2007), The continuing evolution of generalized systems at Statistics Canada for business survey processing, *Proceedings of the 3rd International Conference on Establishment Surveys (ICES-III)*, Montreal, June 18–21, American Statistical Association, Alexandria, VA, pp. 758–768.

Monsour, N. J., ed. (1985), Evaluation of the 1977 economic censuses of the United States, *Journal of Official Statistics* **1**(3): 331–350 (available at www.jos.nu).

Monsour, N. J., and Wolter, K. M. (1989), Evaluation of economic censuses at the United States Bureau of the Census, *Proceedings of the International Statistical Association*, pp. 517–535.

Moore, D., and Ollinger, M. (2007), Effectiveness of monetary incentives and other stimuli experimentally carried out across industry populations for establishment surveys, *Proceedings of the 3rd International Conference on Establishment Surveys (ICES-III)*, Montreal, June 17–21, American Statistical Association, Alexandria, VA, pp. 525–532.

Morgan, D. L. (1997), *Focus Groups as Qualitative Research*, 2nd ed., Qualitative Research Methods Series 16, Sage, Thousand Oaks, CA.

Morren, M. (2008), *The 2006 Structural Business Survey: Response Analysis* (in Dutch: De PS-2006: Een evaluatie van de respons), Research report, Statistics Netherlands, Heerlen.

Morren, M., and Snijkers, G. (2011), *Audit Trails Used to Measure Response Burden* (in Dutch: Audit trails voor het meten van de Enquêtedruk), Internal report, Statistics Netherlands, Heerlen.

Morrison, R. L., and Anderson, A. E. (2005), The effect of data collection software on the cognitive survey response process, paper presented at the Joint Statistical Meetings, Minneapolis, Aug. 7–11, American Statistical Association, VA.

Morrison, R. L., Stettler, K., and Anderson, A. E. (2004), Using vignettes in cognitive research on establishment surveys, *Journal of Official Statistics* **20**(2): 319–340 (available at www.jos.nu).

Morrison, R. L., Stokes, S. L., Burton, J., Caruso, A., Edwards, K. K., Harley, D., Hough, C., Hough, R., Lazirko, B. A., and Proudfoot, S. (2008), *Economic Directorate Guidelines on Questionnaire Design*, US Census Bureau, Washington, DC (available at www.census.gov/srd/Economic_Directorate_Guidelines_on_Questionnaire_Design.pdf).

Morrison, R. L., Dillman, D. A., and Christian, L. M. (2010), Questionnaire design guidelines for establishment surveys, *Journal of Official Statistics* **26**(1): 43–85 (available at www.jos.nu).

Morton-Williams, J. (1993), *Interviewer Approaches*, Dartmouth Publishing Company, Aldershot, UK.

Mowen, J. C., and Cialdini, R. B. (1980), On implementing the door-in-the-face compliance technique in a business context, *Journal of Marketing Research* **17**: 253–258.

Moy, L., and Stinson, L. (1999), Two sides of a single coin? Dimensions of change suggested in different settings, *Proceedings of the Section on Survey Research Methods*, Baltimore, Aug. 8–12, American Statistical Association, Alexandria, VA, pp. 44–53.

Mudryk, W., Joyce, B., and Xie, H. (2004), Generalized quality control approach for ICR data capture in Statistics Canada's centralized operations, paper presented at the 2nd

European Conference on Quality and Methodology in Official Statistics, Mainz, Germany, May 24–26.

Mueller, C., and Phillips, M. A. (2000), The genesis of an establishment survey: Research and development for the job openings and labor turnover survey at the BLS, *Proceedings of the Survey Research Methods Section*, Indianapolis, Aug. 13–17, American Statistical Association, Alexandria, VA, pp. 366–370.

Mueller, C., and Wohlford, J. (2000), Developing a new business survey: Job openings and labor turnover survey at the Bureau of Labor Statistics, paper presented at the 2nd International Conference on Establishment Surveys (ICES-II), Buffalo, NY, June 17–21, also in *Proceedings of the Survey Research Methods Section*, Indianapolis, Aug. 13–17, American Statistical Association, Alexandria, VA, pp. 360–365.

Mulrow, J. M. (2008), Getting an establishment survey to the right person in the organization, *Proceedings of the Government Statistics Section*, Joint Statistical Meetings, Denver, Aug. 3–7, American Statistical Association, Alexandria, VA, pp. 1702–1709.

Mulrow, J. M., Carlson, L., Jankowski, J., Shackelford, B., and Wolfe, R. (2007a), A face-lift or reconstructive surgery? What does it take to renew a 53-year old survey?*Proceedings of the 3rd International Conference on Establishment Surveys (ICES-III)*, Montreal, June 18–21, American Statistical Association, Alexandria, VA, pp. 208–213.

Mulrow, J. M., Freedman, S., and Rutchik, R. (2007b), Record keeping studies—love 'em or leave 'em, *Proceedings of the 3rd International Conference on Establishment Surveys (ICES-III)*, Montreal, June 18–21, American Statistical Association, Alexandia, VA, pp. 56–62.

Murphy, E., Marquis, K., Hoffman, R., Saner, L., Tedesco, H., Harris, C., and Roske-Hofstrand, R. (2000), *Improving Electronic data collection and Dissemination through Usability Testing*, Working Papers in Survey Methodology, US Census Bureau, Washington, DC.

Murphy, E. D., Nichols, E. M., Anderson, A. E., Harley, M. D., and Pressley, K. D. (2001), Building usability into electronic data-collection forms for economic censuses and surveys, *Proceedings of the Annual Federal Committee on Statistical Methodology (FCSM) Research Conference*, Arlington, VA, Nov. 14–16, pp. 113–122.

National Statistician Guidance (2009), *Presentation and Publication of Official Statistics*, UK Office for National Statistics, London.

Neter, J., and Waksberg, J. (1964), A study of response error in expenditures data from household interviews, *Journal of the American Statistical Association* **59**(305): 18–55.

Newby, R., Watson, J., and Woodliff, D. (2003), SME survey methodology: Response rates, data quality, and cost effectiveness, *Entrepreneurship Theory and Practice* **28**(2): 163–172.

Neyman, J. (1934), On the two different aspects of the representative method: The method of stratified sampling and the method of purposive selection, *Journal of the Royal Statistical Society* **97**(4): 558–625.

Nguyen, L., Davies, K., Oddy, C., St-Jean, H., and Melki, L. (2008), Generalized quality control for optical data capture at Statistics Canada, paper presented at the 2008 International Statistics Canada Methodology Symposium: Data Collection: Challenges, Achievements and New Directions, Statistics Canada, Ottawa, Oct. 28–31.

Nicholls, II, W. L., Baker, R. P., and Martin, J. (1997), The effect of new data collection technologies on survey data quality, in Lyberg, L., Biemer, P., Collins, M., de Leeuw, E., Dippo, C., Schwarz, N., and Trewin, D., eds., *Survey Measurement and Process Quality*, Wiley, New York, pp. 221–248.

Nicholls, II, W. L., Mesenbourg, Jr., T. L., Andrews, S. H., and de Leeuw, E. (2000), Use of new data collection methods in establishment surveys, *Proceedings of the 2nd International Conference on Establishment Surveys (ICES-II)*, Buffalo, NY, June 17–21, American Statistical Association, Alexandria, VA, pp. 373–382.

Nichols, E., and Willimack, D. (2001), Balancing confidentiality and burden concerns in censuses and surveys of large businesses, paper presented at the Annual Federal Committee on Statistical Methodology (FCSM) Research Conference Arlington, VA, Nov. 14–16 (available at `http://www.fcsm.gov/01papers/Nichols.htm`).

Nichols, E., Tedesco, H., and King, R. (1998), *Results from Usability Testing of Possible Electronic Questionnaires for the 1998 Library Media Center Public School Questionnaire Field Test*, Human-Computer Interaction Memorandum Series 20, US Census Bureau, Statistical Research Division, Washington, DC.

Nichols, E. M., Willimack, D. K., and Sudman, S. (1999), Who are the reporters: A study of government data providers in large, multi-unit companies, paper presented at the Joint Statistical Meetings, Baltimore, Aug. 8–12, American Statistical Association, Alexandria, VA.

Nichols, E., Murphy, E., and Anderson, A. E. (2001a), *Results from Cognitive and Usability Testing of Edit Messages for the 2002 Economic Census (First Round)*, Human-Computer Interaction Report Series 39, US Census Bureau, Statistical Research Division, Washington, DC.

Nichols, E., Murphy, E., and Anderson, A. E. (2001b), *Usability Testing Results of the 2002 Economic Census Prototype RT-44401*, Human-Computer Interaction Report Series 49, US Census Bureau, Statistical Research Division, Washington, DC.

Nielsen, J. (1993), *Usability Engineering*, Morgan Kaufmann, San Diego, CA.

Nielsen, J., and Loranger, H. (2006), *Prioritizing Web Usability*, New Riders, Berkeley, CA.

Nielsen, J., and Mack, R. L., eds. (1994), *Usability Inspection Methods*, Wiley, New York.

Nielson, J., and Pernice, K. (2010), *Eyetracking Web Usability*, New Riders, Berkeley, CA.

Nordbotten, S. (1963), Automatic Editing of Individual Statistical Observations, paper presented at the 2nd Conference of European Statisticians, Statistical Standards and Studies, UN Statistical Commission and Economic Commission for Europe, New York (available at `http://nordbotten.com/articles/ece1963.pdf`).

Norman, D. A. (2004), *Emotional Design: Why We Love (or Hate) Everyday Things*, Basic Books, Cambridge, MA.

Notstrand, N., and Bolin, E. (2008), Measurement and follow-up of the response burden from enterprises by the register of data providers concerning enterprises and organizations at Statistics Sweden, paper presented at the 2008 International Statistics Canada Methodology Symposium: Data Collection: Challenges, Achievements and New Directions, Ottawa, Oct. 28–31.

Oakland (City of) (2004), *Oakland on Quality Management*, Elsevier Butterworth-Heinemann, Amsterdam.

O'Brien, E. M. (1997), Redesigning economic surveys of establishments, paper presented at the 52nd Annual Conference of the American Association for Public Opinion Research (AAPOR), Norfolk, VA, May.

O'Brien, E. M. (2000a), *A Cognitive Appraisal Methodology for Establishment Survey Questionnaires*, Statistical Policy Working Paper 30, US Census Bureau, Washinton DC; also published in the *1999 Federal Committee on Statistical Methodology (FCSM) Research Conference: Complete Proceedings* (Part 1 of 2), Office of Management and Budget, Washington, DC, Nov. 15–17, pp. 307–316.

O'Brien, E. M. (2000b), Respondent role as a factor in establishment survey response, *Proceedings of the 2nd International Conference on Establishment Surveys (ICES-II)*, American Statistical Association, Buffalo, NY, June 17–21, Alexandria, VA, pp. 1462–1467.

O'Brien, E., Fisher, S., Goldenberg, K., and Rosen, R. (2001), Application of cognitive methods to an establishment survey: A demonstration using the current employment statistics survey, paper presented at the 56th Annual Conference of the American Association for Public Opinion Research (AAPOR), Montreal, May 17–20, also in *Proceedings of the Joint Statistical Meetings*, American Statistical Association, Alexandria, VA, CD-rom.

OECD (2003), *Business Tendency Surveys. A Handbook*, Organization for Economic Cooperation and Development (OECD), Paris.

OECD (2006), *Structural and Demographic Business Statistics*, Organization for Economic Cooperation and Development (OECD), Paris (available at http://unstats.un.org/unsd/trade/s_geneva2011/refdocs/IADs/Structural%20and%20Demographic%20Business%20Statistics%20(OECD%20-%202006).pdf).

OECD (2009), *Completion of Changes to the Main Economic Indicators Paper Publication*, Organization for Economic Cooperation and Development (OECD), Paris, Nov. (available at www.oecd.org/std/44026920.pdf).

OECD (2012), *Measuring Regulatory Performance. A Practitioner's Guide to Perception Surveys*, Organization for Economic Cooperation and Development (OECD), Paris.

Office for National Statistics, UK (2007), *Guidelines for Measuring Statistical Quality*, UK Office for National Statistics, London.

Office for National Statistics, UK (2011a), *GSS Quality Good Practice 2011*, UK Office for National Statistics, London (available at www.ons.gov.uk/ons/guide-method/best-practice/gss-best-practice/index.html).

Office for National Statistics, UK (2011b), *Retail Sales Statistical Bulletin*, UK Office for National Statistics, London, Aug.

Office for National Statistics, UK (2011c). *History of and Differences between the Consumer Prices Index and Retail Prices Index*, UK Office for National Statistics, London.

Office for National Statistics, UK (2012), *Retail Sales Statistical Bulletin*, UK, Office for National Statistics, London, January.

Ohlsson, E. (1995), Co-ordination of samples using permanent random numbers, in Cox, B. G., Binder, D. A., Chinnappa, B. N., Christianson, A., Colledge, M. J., and Kott, P. S., eds., *Business Survey Methods*, Wiley, New York, pp. 153–169.

Ohlsson, E. (1998), Sequential Poisson sampling, *Journal of Official Statistics* **14**(2): 149–162 (available at www.jos.nu).

Öller, L. E. (1990), Forecasting the business cycle using survey data, *International Journal of Forecasting* **6**(4): 453–461.

O'Neill, G. E. (2007), Improving the effectiveness of interviewer administered surveys through refusal avoidance training, *Proceedings of the 3rd International Conference on Establishment Surveys (ICES-III)*, Montreal, June 18–21, American Statistical Association, Alexandria, VA, pp. 233–239.

Oomens, P., and Timmermans, G. (2008), The Dutch approach to reducing the real and perceived administrative burden on enterprises caused by statistics, paper presented at the 94th DGINS (Directors-General of the National Statistical Institutes) Conference: Reduction of administrative burden through official statistics, Vilnius, Lithuania, Sept. 25–26.

Oppeneer, M., and Luppes, M. (1998), Improving communication with providers of statistical information in business surveys, paper presented at the GSS Methodology Conference, Government Statistical Service, London, June 29.

Orchard, C. B., Coyle, B., Jones, J., and Green, S. (2009), Cost-benefit analysis of proposed new data requirements, paper presented at the 16th Meeting of the Government Statistical Service Methodology Advisory Committee (GSS MAC), London, May 19.

O'Reilly, J. (2009), Paradata and Blaise: A review of recent applications and research, paper presented at the 12th International Blaise Users Conference (IBUC), Riga, Latvia, June 2–4.

Ortega, M. (2007), Requesting voluntary data from non-financial corporations. The experience of the Banco de España CBSO, paper presented at the 56th Session of the International Statistical Institute (ISI), Lisbon, Aug. 22–29.

Osmotherly, E., Graham, T., and Pepper, M. (1996), *Osmotherly Report: Statistical Surveys: Easing the Burden on Business*, a report by the Osmotherly Steering Group, UK Office for National Statistics, London.

Ouwehand, P., and Schouten, B. (2011), *Representativity of VAT and Survey Data for Short Term Business Statistics*, Discussion paper 201106, Statistics Netherlands, The Hague/Heerlen.

Paben, S. (1998), The reinterview program for the BLS compensation surveys, paper presented at the Joint Statistical Meetings, Dallas, Aug. 9–13, American Statistical Association, Alexandria, VA.

Pafford, B. V. (1988), *The Influence of Using Previous Data in the 1986 April ISO Grain Stock Survey*, National Agricultural Statistical Service, Washington DC.

Palmer, S. E. (1999), *Vision Science: Photons to Phenomenology*, Bradford Book, London.

Palmisano, M. (1988), The application of cognitive survey methodology to an establishment survey field test, *Proceedings of the Survey Research Methods Section*, American Statistical Association, Alexandria, VA, pp. 179–184.

Pannekoek, J. (2009), *Research on Edit and Imputation Methodology: The Throughput Programme*, discussion paper 09022, Statistics Netherlands, The Hague.

Parsons, T. (1956), Suggestions for a sociological approach to the theory of organizations, I, *Administrative Science Quarterly*, **1**(1): 63–85.

Paxson, M. C., Dillman, D. A., and Tarnai, J. (1995), Improving response to business mail surveys, in Cox, B. G., Binder, D. A., Chinnappa, B. N., Christianson, A., Colledge, M. J., and Kott, P. S., eds., *Business Survey Methods*, Wiley, New York, pp. 303–317.

Peck, M., and Maclean, E. (2008), Telephone Data Entry—Enhancing the System, *Survey Methodology Bulletin*, UK Office for National Statistics, London, **63**: 52–60.

Pereira, H. J. (2011), Simplified business information (IES)—is coordination between public entities really possible?, in Giesen, D., and Bavdaz, M., eds., *Proceedings of the BLUE-ETS Conference on Burden and Motivation in Official Business Surveys*, Statistics Netherlands, Heerlen, March 22–23, pp. 177–188.

Perloff, R. M. (2010), *The Dynamics of Persuasion: Communication and Attitudes in the 21st Century*, Routledge, New York.

Perry, J. (2007), Implementing coding tools for a new classification, *Proceedings of the 3rd International Conference of Establishment Surveys (ICES-III)*, Montreal, June 18–21, American Statistical Association, Alexandria, VA, pp. 33–41.

Peternelj, P. M., and Bavdaz, M. (2011), Response burden measurement, in Giesen D., and Raymond-Blaess, V., eds., *Inventory of Published Research: Response Burden Measurement and Reduction in Official Business Statistics*, Deliverable 2.1, BLUE-Enterprise and Trade Statistics Project (BLUE-ETS), European Commission, European Research Area, pp. 13–26 (available at www.blue-ets.istat.it/fileadmin/deliverables/Deliverable2.1.pdf).

Petroni, R., Sigman, R., Willimack, D., Cohen, S., and Tucker, C. (2004a), Response rates and nonresponse in establishment surveys – BLS and Census Bureau, paper presented to Federal Economic Statistics Advisory Committee (FESAC), Washington DC, Dec. 14.

Petroni, R., Sigman, R., Willimack, D., Cohen, S., and Tucker, C. (2004b), Response rates and nonresponse in BLS and Census Bureau establishment surveys, *Proceedings of the Survey Research Methods Section*, Joint Statistical Meetings, Toronto, Canada, Aug. 8–12, American Statistical Association, Alexandria, VA.

Pfeiffer, M., and Walter, P. (2007), The OeNB's experience in cooperating with information providers for Austria's new balance of payments system, paper presented at the 56th Session of the International Statistical Institute (ISI), Lisbon, Aug. 22–29.

Phillips, J. M., Mitra, A., Knapp, G., Simon, A., Temperly, S., and Lakner, E. (1994), The determinants of acquiscence to preprinted information on survey instruments, paper presented at the 54th Annual Conference of the American Associaton for Public Opinion Research Conference (AAPOR), St. petersburg Beach, FL, May 13–16.

Phipps, P. A. (1990), Applying cognitive theory to an establishment mail survey, *Proceedings of the Section on Survey Research Methods*, American Statistical Association, Alexandria, VA, pp. 608–612.

Phipps, P. A., Butani, S. J., and Chun, Y. I. (1995), Research on establishment—survey questionnaire design, *Journal of Business and Economic Statistics* **13**(3): 337–346.

Pickford, S., Cunningham, J., Lynch, R., Radice, J., and White, G. (1988), *Government Economic Statistics: A Scrutiny Report*, UK Her Majesty's Stationery Office (HMSO), London.

Pierce, J. R., and Schott, P. K. (2012), Concording U.S. harmonized system codes over time, *Journal of Official Statistics* **28**(1): 53–68 (available at www.jos.nu).

Pierzchala, M. (1995), Editing systems and software, in Cox, B. G., Binder, D. A., Chinnappa, B. N., Christianson, A., Colledge, M. J., and Kott, P. S., eds., *Business Survey Methods*, Wiley, New York, pp. 425–442.

Pietsch, L. (1995), Profiling large businesses to define frame units, in Cox, B. G., Binder, D. A., Chinnappa, B. N., Christianson, A., Colledge, M., and Kott, P. S., eds., *Business Survey Methods*, Wiley, New York, pp. 101–114.

Planas, C. (1997), *The Analysis of Seasonality in Economic Statistics*, Eurostat Working Group Document, Eurostat, Luxembourg.

Pol, B., Swankhuisen, C., and van Vendeloo, P. (2007), *A New Approach in Government Communication* (in Dutch: Nieuwe aanpak in overheidscommunicatie), Coutinho, Bussum, The Netherlands.

Porter, L.W., Lawler, E. E., III, and Hackman, J. R. (1975), *Behavior in Organizations*, McGraw-Hill, New York.

Prest, A. R., and Turvey, R. (1965), Cost-benefit analysis: A survey, *The Economic Journal*, **75**(300): 683–735.

Proctor, M. (1994), Analysing survey data, in Gilbert, N., ed., *Researching Social Life*, Sage, London, pp. 239–254.

Pustjens, H., and Wieser, M. (2011), A consistency unit at Statistics Netherlands: Reducing asymmetries in national accounts and related statistics, in *The Impact of Globalization on National Accounts*, United Nations Economic Commission for Europe (UNECE), New York/Geneva, pp. 23–24.

Rahim, M. A., and Currie, S. (1993), Optimizing sample allocation with multiple response variables. *Proceedings of the Survey Research Methods Section*, Joint Statistical Meetings, San Francisco, Aug. 8–12, American Statistical Association, Alexandria, VA.

Rainer, N. (2004), Measuring response burden: The response burden barometer of Statistics Austria, paper presented at the 2nd European Conference on Quality and Methodology in Official Statistics, Mainz, Germany, May 24–26.

Ramanayake, A., and Zayatz, L. (2010), *Balancing Disclosure Risk with Data Quality*, Research Report Series, US Census Bureau, Washington, DC.

Ramirez, C., (1997), Effects of precontacting on response and cost in self-administered establishment surveys, *Proceedings of the Survey Research Methods Section*, Joint Statistical Meetings, Anaheim, CA, Aug. 10–14, American Statistical Association, Alexandria, VA, pp. 1000–1005.

Ramirez, C. (2002), Strategies for subject matter expert review in questionnaire design, paper presented at the International Conference on Questionnaire Development, Evaluation, and Testing Methods (QDET), Charleston, SC, Nov. 14–17.

Ramirez, C., and McCarthy, J. (2007), Non-response reduction methods in establishment surveys, Introductory Overview Lecture presented at the 3rd International Conference on Establishment Surveys (ICES-III), Montreal, 18–21, American Statistical Association, Alexandria, VA.

Rao, J. N. K. (2003), *Small Area Estimation*, Wiley, Hoboken, NJ.

Redline, C. (2011), *Clarifying survey questions*. Dissertation, Faculty of the Graduate School of the University of Maryland, Collegepart, MD.

Redline, C., Dillman, D. A., Carley-Baxter, L., and Creecy, R. H. (2004), Factors that influence reading and comprehension of branching instructions in self-administered questionnaires, *Allgemeines Statistisches Archiv* **89**(1): 21–38.

Regan, C., and Green, S. (2010), *Process Quality Measurement Guide*, Report, UK Office for National Statistics, London.

Renssen, R. H., Camstra, A., Huegen, C., Hacking, W. J. G., and Huigen, R. D. (1998), *A Model for Evaluating the Quality of Business Surveys*, Research report, Statistics Netherlands, Heerlen.

Revilla, P. (2008), Editing strategy and Administrative burden: Services for reporting units, presentation at the joint meeting of Statistics Spain and Statistics Netherlands, Statistics Spain, Madrid, Nov. 6–7.

Ritzen, J. (2007) Statistical business register: Content, place and role in economic statistics, Introductory Overview Lecture, *Proceedings of the 3rd International Conference on Establishment Surveys (ICES-III)*, Montreal, June 18–21, American Statistical Association, Alexandria, VA, pp. 179–191.

Ritzen, J. (2012), Business register profiling of multinational enterprises, paper presented at the 23rd meeting of the Wiesbaden Group on Business Registers, Washington DC, Sept. 17–20 (available at `www.census.gov/epcd/wiesbaden/agenda/papers.html`).

Rivest, L.-P. (2002), A generalization of the Lavallée and Hidiroglou algorithm for stratification in business surveys, *Survey Methodology* **28**(2): 191–198.

Riviére, P. (2002), What makes business statistics special? *International Statistical Review* **70**(1): 145–159.

Robbins, N. B. (2005), *Creating More Effective Graphs*, Wiley, Hoboken, NJ.

Robbins, S. P., Judge, T. A., and Campbell, T. T. (2010), *Organizational Behaviour*, Financial Times Prentice-Hall, Harlow, UK.

Roels, J. (2010), *The Effect of Enforcement in the International Trade Survey and Structural Business Survey* (in Dutch: Effectmeting Handhaving Internationale Handel, Productie Statistiek en Prodcom), Internal report, Statistics Netherlands, Heerlen.

Roos, M. (2010), Using XBRL in a statistical context. The case of the Dutch taxonomy project, *Journal of Official Statistics* **26**(3): 559–575 (available at `www.jos.nu`).

Roos, M. (2011), Combining web surveys and XBRL, paper presented at the 5th International Workshop on Internet Survey Methodology, Statistics Netherlands, The Hague, Aug. 29–31.

Rosen, R. J. (2007), Multi-mode data collection: Why, when, how?, paper presented at the International Conference on Establishment Surveys (ICES-III), Montreal, June 18–21, American Statistical Association, Alexandria, VA.

Rosen, R. J., Clayton, R. L., and Rubino, T. B. (1991), Controlling nonresponse in an establishment survey, *Proceedings of the Survey Research Methods Section*, Joint Statistical Meetings, Atlanta, Aug. 18–22, American Statistical Association, Alexandria, VA, pp. 587–591.

Rosen, R., Kapani, V., and Clancy P. (2005), Data capture using FAX and intelligent character and optical character recognition (ICR/OCR) in the current employment statistics survey (CES), paper presented at the Annual Federal Committee on Statistical Methodology (FCSM) Research Conference, Arlington, VA, Nov. 14–16.

Rowlands, O., Eldridge, J., and Williams, S. (2002), Expert review followed by interviews with editing staff – effective first steps in the testing process for business surveys, paper presented at the International Conference on Questionnaire Development, Evaluation, and Testing Methods (QDET), Charleston, SC, Nov. 14–17.

Roy, D. (2001), Market research and the evolution of Internet services at Statistics Canada, paper presented at the 53rd Session of the International Statistical Institute (ISI), Seoul, South Korea, Aug. 22–29.

Rutchik, R. H., and Freedman, S. R. (2002), Establishments as respondents: Is conventional cognitive interviewing enough?, paper presented at the International Conference on

Questionnaire Development, Evaluation, and Testing Methods (QDET), Charleston, SC, Nov. 14–17.

Saint-Pierre, E., and Bricault, M. (2011), The common editing strategy and the data processing of business statistics surveys, paper presented at Conference of European Statisticians, Work Session on Statistical Data Editing, Ljubljana, Slovenia, May 9–11.

Sala, M., and Lynn, P. (2005), *The Impact of a Mixed-mode Data Collection Design on Nonresponse Bias on a Business Survey*, Working papers of the Institute for Social and Economic Research, paper 2005-16, University of Essex, Colchester, UK.

Saner, L. D., and Pressley, K. D. (2000), Assessing the usability of an electronic establishment survey instrument in a natural use context, *Proceedings of the 2nd International Conference on Establishment Surveys (ICES-II)*, Buffalo, NY, June 17–21, American Statistical Association, Alexandria, VA, pp. 1634–1639.

Särndal, C.-E., Swensson, B., and Wretman, J. (1992), *Model Assisted Survey Sampling*, Springer, New York.

Schaalma, H, and Kok, G. (2007), Intervention mapping (in Dutch: Interventieontwikkeling), in Brug, J., van Assema, P., and Lechner, L., eds., *Health Education and Behaviour Change: A Planned Approach* (in Dutch: Gezondheidsvoorlichting en gedragsverandering: Een planmatige aanpak), Van Gorcum, Assen, The Netherlands.

Schaeffer, N. C., and Presser, S. (2003), The science of asking questions, *Annual Review of Sociology* **29**: 65–88.

Schechter, S., Stinson, L. L., and Moy, L. (1999), Developing and Testing Aggregate Reporting Forms for Data on Race and Ethnicity, *Statistical Policy Working Paper 30, Federal Committee on Statistical Methodology (FCSM) Research Conference: Complete Proceedings* (Part 1 of 2), Office of Management and Budget, Washington, DC, Nov. 15–17, pp. 337–346.

Scheuren, F. (2000), *Macro and Micro Paradata for Survey Assessment*, unpublished report, Urban Institute, Washington, DC.

Scheuren, F. (2004), *What is a Survey*, American Statistical Association, Alexandria, VA (available at www.whatisasurvey.info or www.amstat.org/sections/srms/pamphlet.pdf).

Schober, M. F. (1998), Different kinds of conversational perspective-taking, in Russel, S. R., and Kreuz, R. J., eds., *Social and Cognitive Psychological Approaches to Interpersonal Communication*, Lawrence Erlbaum, Mahwah, NJ, pp. 145–174.

Schouten, B., Cobben, F., and Bethlehem, J. (2009), Indicators for the representativeness of survey response, *Survey Methodology* **35**(1): 101–113.

Schubert, P., Guiver, T., MacDonald, R., and Yu, F. (2006), Using quality measures to manage statistical risk in business surveys, paper presented at the 3rd European Conference on Quality in Official Statistics, Cardiff, UK, April 24–26.

Schwarz, N. (1996), *Cognition and communication: Judgmental biases, research methods, and the logic of conversation*, Memorial lecture series, Lawrence Erlbaum, Hillsdale, NJ.

Schwarz, N., Knäuper, B., Hippler, H. J., Noelle-Neumann, E., and Clark, F. (1991), Rating scales: Numeric values may change the meaning of scale labels, *Public Opinion Quarterly* **55**(4): 570–582.

Schwarz, N., Knäuper, B., Oyserman, D., and Stich, C. (2008), The psychology of asking questions, in de Leeuw, E. D., Hox, J. J., and Dillman, D. A., eds., *International Handbook of Survey Methodology*, Lawrence Erlbaum Associates, New York, pp. 28–34.

Schwede, L., and Ellis, Y. (1997), *Children in Custody Questionnaire Redesign Project Final Report: Results of the Split-Panel and Feasibility Tests,* Report prepared by the US Census Bureau for the Department of Justice, Washington DC.

Schwede, L., and Gallagher, C. (1996), *Children in Custody Questionnaire Redesign Project: Results from Phase 3 Cognitive Testing of the Roster Questionnaire*, Report prepared by the US Census Bureau for the Department of Justice, Washington DC.

Schwede, L., and Moyer, L. (1996), *Children in Custody Questionnaire Redesign Project: Results from Phase 2 Questionnaire Development and Testing*, Report prepared by the US Census Bureau for the Department of Justice, Washington DC.

Schwede, L., and Ott, K. (1995), *Children in Custody Questionnaire Redesign Project: Results from Phase 1*, report prepared by the US Census Bureau for the Department of Justice, Washington DC.

Sear, J. (2011), Response burden measurement and motivation at Statistics Canada, in Giesen, D., and Bavdaz, M., eds., *Proceedings of the BLUE-ETS Conference on Burden and Motivation in Official Business Surveys*, Statistics Netherlands, Heerlen, March 22–23, pp. 151–160.

Sear, J., Hughes, J., Vinette, I., and Bozzato, W. (2007), Holistic response management of business surveys at Statistics Canada, *Proceedings of the 3rd International Conference on Establishment Surveys (ICES-III)*, Montreal, June 18–21, American Statistical Association, Alexandria, VA, pp. 953–958.

Searls, D. T. (1966), An estimator which reduces large true observations, *Journal of the American Statistical Association* **61**(316): 1200–1204.

Seens, D. (2010), *Analysis of Regulatory Compliance Costs:* Part II, *Paperwork Time Burden, Costs of Paperwork Compliance and Paperwork Simplification*, December 2010 briefing, Government of Canada, Publishing and Depository Services, Ottawa (available at `www.reducingpaperburden.gc.ca/Survey`).

Segers, J. H. G. (1977), *Sociological Research Methods* (in Dutch: Sociologische onderzoeksmethoden), Van Gorcum, Assen/Amsterdam.

Shackelford, B., and Helig, S. (2011), Effect of form design on the reporting of complex concepts in a survey of businesses: Evidence from the business R&D and innovation survey, paper presented at the Joint Statistical Meetings, Miami Beach, Fl, July 30 – Aug. 4, American Statistical Association Alexandria, VA.

Shao, J., and Steel, P. (1999), Variance estimation for survey data with composite imputation and nonnegligible sampling fractions, *Journal of the American Statistical Association* **94**(445): 254–265.

Shaw, W. D. (1992), Searching for the opportunity cost of an individual's time, *Land Economics* **68**(1): 107–115.

Shiskin, J., Young, A. H., and Musgrove, J. C. (1967), *The X-11 Variant of the Census Method II Seasonal Adjustment Program*, Technical Paper 15, US Bureau of the Census, Washington DC (available at `www.census.gov./ts/papers/ShiskinYoungMusgrave1967.pdf`).

Sieber, J. E., and Bernard Tolich, M. (2012), *Planning Ethically Responsible Research*, Applied Social Research Methods Series, Sage, Thousand Oaks, CA.

SIGMA (2008), *Modern statistics for modern society, SIGMA, The Bulletin of European Statistics*, Vol. 2, Eurostat, Luxembourg.

Signore, M., Carbini, R., D'Orazio, M., Brancato, G., and Simeoni, G. (2010), Assessing quality through auditing and self-assessment, paper presented at the 5th European Conference on Quality in Official Statistics, Helsinki, May 4–5.

Simon, H. A. (1997), *Administrative Behavior: A Study of Decision-Making Processes in Administrative Organizations*, 4th ed., The Free Press, New York.

Sinclair, J. (1837), *Memoirs of the Life and Works of the Late Right Honourable Sir John Sinclair, Bart*, Vol. II, William Blackwood and Sons, Edinburgh.

Singer, E. (2002), The use of incentives to reduce nonresponse in household surveys, in Groves, R. M., Dillman, D. A., Eltinge J. L., and Little, R. J. A., eds., *Survey Nonresponse*, Wiley, Hoboken, NJ, pp. 163–177.

Singer, E. (2008), Ethical issues in surveys, in de Leeuw, E. D., Hox, J. J., and Dillman, D. A., eds., *International Handbook of Survey Methodology*, Lawrence Erlbaum Associates, New York, pp. 78–96.

Singh, M. P., Hidiroglou, M. A., Gambino, J. G., and Kovačević, M. S. (2001), Estimation methods and related systems at Statistics Canada, *International Statistical Review* **69**(3): 461–485.

Skentelbery, R., Finselbach, H., and Dobbins, C. (2011), Improving the efficiency of editing for ONS business surveys, paper presented at the United Nations, Economic Commission for Europe and Conference of European Statisticians, Working Session on Statistical Data Editing, Ljubljana, May 9–11.

Skinner, C. (2009), *Statistical Disclosure Control For Survey Data*, Working Paper M09/03, University of Southampton, Southampton.

Skinner, C. J., Holmes, D. J. and Smith, T. M. F. (1986), The effect of sample design on principal component analysis, *Journal of the American Statistical Association*, **81**(395): 789–798.

Skinner, C. J., Holmes, D. J., and Holt, D. (1994), Multiple frame sampling for multivariate stratification, *International Statistical Review* **62**(3): 333–347.

Slanta, J., and Krenzke, T. (1994), Applying the Lavallée and Hidiroglou method to obtain stratification boundaries for the Census Bureau's annual capital expenditures survey, *Proceedings of the Survey Research Methods Section*, Joint Statistical Meetings, Toronto, Canada, Aug. 13–18, American Statistical Association, Alexandria, VA, pp. 693–698.

Smeets, M. (2006), *The Effect of Enforcement* (in Dutch: Effectmeting handhaving), Internal report, Statistics Netherlands, Heerlen.

Smith, P. (2001), Multivariate allocation for commodity surveys, *Proceedings of the 2nd International Conference on Establishment Surveys (ICES-II)*, Buffalo, NY, June 17–21, American Statistical Association, Alexandria, VA, pp. 137–146.

Smith, P. (2004), Perspectives on response rates and nonresponse in establishment surveys, paper presented to the Federal Statistics Advisory Committee (FESAC), Washington DC, Dec. 14.

Smith, P., and Penneck, S. (2009), *100 Years of the Census of Production in the UK*, GSS Methodological Series 38, UK Office for National Statistics, Newport.

Smith, P., and Perry, J. (2001), Surveys of business register quality in Central European countries, *Proceedings of the 2nd International Conference on Establishment Surveys (ICES-II)*, Buffalo, NY, June 17–21, American Statistical Association, Alexandria, VA, pp. 1105–1110.

Smith, P., and Perry, J. (2005), A review of strategies for surveying rare and difficult to reach populations in ONS's establishment surveys, paper presented at the 2004 International Statistics Canada Symposium: Innovative Methods for Surveying Difficult-to-Reach Populations, Statistics Canada, Ottawa.

Smith, P., Highland, F., Taylor, D., George, T., and Wolfgang, G. (2007), A comprehensive view of capture data quality, paper presented at the Conference on Innovative Methodologies for Censuses in the New Millennium, Aug. 31 – Sept. 2, Lisbon.

Smith, T. M. F. (1978), Principles and problems in the analysis of repeated surveys, in Namboodiri, N. K., ed., *Survey Sampling and Measurement*, Academic Press, New York, pp. 201–216.

Snijkers, G. (1992), Computer assisted interviewing: Telephone or personal?, in Westlake, A., Banks, R., Payne, C., and Orchard, T., eds., *Survey and Statistical Computing*, North-Holland, Amsterdam, pp. 137–146.

Snijkers, G. (2002), *Cognitive Laboratory Experiences: On Pre-Testing Computerized Questionnaires and Data Quality*, PhD thesis Utrecht University, Statistics Netherlands, Heerlen.

Snijkers, G. (2008), Getting data for business statistics: A response model, paper presented at the 4th European Conference on Quality in Official Statistics, Rome, July 8–11.

Snijkers, G., and Bavdaz, M. (2011), Business surveys, in Lovric, M., ed., *International Encyclopedia of Statistical Science*, Springer Verlag, Berlin, pp. 191–194.

Snijkers, G., and Giesen, D. (2010), NSI's practices concerning business response burden and motivation: Results from BLUE-ETS WP-2, introductory presentation to the opening session of the European SIMPLY Conference on Administrative Simplification in Official Statistics, Ghent, Belgium, Dec. 2–3.

Snijkers, G., and Lammers, L. (2007), The annual structural business survey: Audit trails and completion behaviour, paper presented at the 3rd International Workshop on Internet Survey Methodology, Lillehammer, Norway, Sept. 17–19.

Snijkers, G., and Luppes M. (2000), The best of two worlds: Total design method and new Kontiv design. An operational model to improve respondent cooperation, *Proceedings of the 2nd International Conference on Establishment Surveys (ICES-II)*, Buffalo, NY, June 17–21, American Statistical Association, Alexandria, VA, pp. 361–372.

Snijkers, G., and Morren, M. (2010), Improving web and electronic questionnaires: The case of the audit trail, paper presented at the 5th European Conference on Quality in Official Statistics, Helsinki, May 4–5.

Snijkers, G., and Willimack, D. K. (2011), The missing link: From concepts to questions in economic surveys, paper presented at the 2nd European Establishment Statistics Workshop (EESW11), Neuchâtel, Switzerland, Sept. 12–14.

Snijkers, G., Berkenbosch, B., and Luppes, M. (2007a), Understanding the decision to participate in a business survey, *Proceedings of the 3rd International Conference on Establishment Surveys (ICES-III)*, Montreal, June 18–21, American Statistical Association, Alexandria, VA, pp. 1048–1059.

Snijkers, G., Onat, E., and Vis-Visschers, R. (2007b), The annual structural business survey: Developing and testing an electronic form, *Proceedings of the 3rd International Conference on Establishment Surveys (ICES-III)*, Montreal, June 18–21, American Statistical Association, Alexandria, VA, pp. 456–463.

Snijkers, G., Göttgens, R., and Hermans, H. (2011a), Data collection and data sharing at Statistics Netherlands: Yesterday, today, tomorrow, paper presented at the 59th Plenary Session of the Conference of European Statisticians (CES), United Nations Economic Commission for Europe (UNECE), Geneva, June 14–16 (available at www.unece.org/fileadmin/DAM/stats/documents/ece/ces/2011/20.e.pdf).

Snijkers, G., Haraldsen, G., Sundvoll, A., Vik, T., and Stax, H. P. (2011b), Utilizing web technology in business data collection: Some Norwegian, Dutch and Danish experiences, paper presented at the 2011 European Conference on New Techniques and Technologies for Statistics (NTTS), Brussels, Feb. 22–24.

Sperry, S., Edwards, B., and Dulaney, R. (1998), Evaluating interviewer use of CAPI navigation features, in Couper, M. P., Baker, R. P., Bethlehem, J., Clark, C. Z. F., Martin, J., Nicholls, W. L., II, and O'Reilly, J. M., eds., *Computer Assisted Survey Information Collection*, Wiley, New York, pp. 351–365.

Squire, P. (1988), Why the 1936 Literary Digest poll failed, *Public Opinion Quarterly* **52**(1): 125–133.

Stäglin, R., Gnoss, R., and Sturm, R. (2006), The Measurement of response burden caused by official statistics in Germany, *Statistika, Journal for Economy and Statistics* **2006**(2): 170–174 (available at http://panda.hyperlink.cz/cestapdf/pdf06c2/staeglin.pdf).

Stanley, J., and Safer, M. (1997), Last time you had 78, how many do you have now? The effect of providing previous reports on current reports of cattle inventories, paper presented at the 52nd Annual Conference of the American Association for Public Opinion Research (AAPOR), Norfolk, May 15–18, also in *Proceedings of the Section on Survey Research Methods*, American Statistical Association, Alexandria, VA, pp. 875–880.

Statistics Canada (2002), *Statistics Canada's Quality Assurance Framework*, Catalog 12-586-XIE, Statistics Canada, Ottawa (available at http://unstats.un.org/unsd/industry/meetings/eg2008/AC158-11.PDF).

Statistics Canada (2005), Multinational Enterprise (MNE) project final report – Phase 1, paper presented at the 53rd Plenary Session of the Conference of European Statisticians (CES), United Nations Economic Commission for Europe (UNECE), Geneva, June 13–15 (available at www.unece.org/fileadmin/DAM/stats/documents/ces/2005/wp.3.e.pdf).

Statistics Canada (2009), *Statistics Canada Quality Guidelines*, Catalog 12-539-X, Statistics Canada, Ottawa (available at http://unstats.un.org/unsd/dnss/docs-nqaf/Canada-12-539-x2009001-eng.pdf).

Statistics Canada (2010), *Survey Methods and Practices*, Catalog 12-587-X, Statistics Canada, Ottawa (available at www.statcan.gc.ca/pub/12-587-x/12-587-x2003001-eng.pdf).

Statistics Netherlands (2008/2012), *Statistics Netherlands' Quality Declaration*, Statistics Netherlands, The Hague/Heerlen (available at www.cbs.nl/en-GB/menu/organisatie/kwaliteitsverklaring/default.htm).

Statistics Norway (2007), *Strategy for Data Collection*, Statistics Norway, Oslo (available at www.ssb.no/english/about_ssb/strategy/strategy2007.pdf).

Statistics South Africa (2010), *South African Statistical Quality Assessment Framework. Operational Standards and Guidelines*, Statistics South Africa, Pretoria (available at www.statssa.gov.za/inside_statssa/standardisation/SASQAF_OpsGuidelines_Edition_1.pdf).

Stern, M. J., (2008), The use of client-side paradata in analyzing the effects of visual layout on changing responses in web surveys, *Field Methods* **20**(4): 377–398.

Stettler, K., and Featherston, F. (2010), Early stage scoping: Building high quality survey instruments without high costs, paper presented at the 65th Annual Conference of the American Association for Public Opinion Research (AAPOR), Chicago, May 13–16.

Stettler, K. J., and Willimack, D. K. (2001), Designing a questionnaire on the confidentiality perceptions of business respondents, paper presented at the 2001 International Statistics Canada Methodology Symposium: Achieving Data Quality in a Statistical Agency: A Methodological Perspective Hull, Quebec, Oct. 18–19, CD-rom, Statistics Canada, Ottawa.

Stettler, K. J., Morrison, R. L., and Anderson, A. E. (2000), Results of cognitive interviews studying alternative formats for economic census forms, *Proceedings of the 2nd International Conference on Establishment Surveys (ICES-II)*, Buffalo, NY, June 17–21, American Statistical Association, Alexandria, VA, pp. 1646–1651.

Stettler, K. J., Willimack, D. K., and Anderson, A. E. (2001), Adapting cognitive interviewing methodologies to compensate for unique characteristics of establishments, paper presented at the 56th Annual Conference of the American Association for Public Opinion Research (AAPOR), Montreal, May 17–20, also in Proceedings of the the Joint Statistical Meetings, American Statistical Association, Alexandria, VA, CD-rom.

Steve, K., and Sharp, J. (2010), Cognitive interviews in establishment surveys: Getting quality feedback in a diverse setting of U.S. ferry operators, paper presented at the 65th Annual Conference of the American Association for Public Opinion Research (AAPOR), Chicago, May 13–16.

Stevens, S. S. (1946), On the theory of scales of measurement, *Science* (*new series*), **103**(2684): 677–680.

Struijs, P., and Willeboordse, A. J. (1995), Changes in populations of statistical units, in Cox, B. G., Binder, D. A., Chinnappa, B. N., Christianson, A., Colledge, M. J., and Kott, P. S., eds., *Business Survey Methods*, Wiley, New York, pp. 65–84.

Studds, S. (2008), iCADE, the data capture system of the U.S. Census Bureau, paper presented at the 2008 International Statistics Canada Methodology Symposium: Data Collection: Challenges, Achievements and New Directions, Catalog 11-522-X, Statistics Canada, Ottawa, Oct. 28–31.

Sudman, S., and. Bradburn, N. M. (1982), *Asking Questions: A Practical Guide to Questionnaire Design*, Jossey-Bass, San Francisco.

Sudman, S., Bradburn, N., and Schwarz, N. (1996), *Thinking about Answers: The Application of Cognitive Processes to Survey Methodology*, Jossey-Bass, San Francisco.

Sudman, S., Willimack, D. K., Nichols, E., and Mesenbourg, T. L. (2000), Exploratory research at the U.S. Census Bureau on the survey response process in large companies, *Proceedings of the 2nd International Conference on Establishment Surveys (ICES-II)*, Buffalo, NY, June 17–21, American Statistical Association, Alexandria, VA, pp. 327–337.

Swatman, P. M. C., and Swatman, P. A. (1991), Electronic data interchange: Organisational opportunity not technical problem, paper presented at the 2nd Australian Conference on Database and Information Systems, University of New South Wales, Sydney, Feb.

Sweet, E., and Ramos, M. (1995), *Evaluation Results from a Pilot Test of a Computerized Self-Administered Questionniare (CSAQ) for the 1994 Industrial Research and Development (R&D) Survey*, Working Paper in Survey Methodology SM95/22, US Census Bureau, Statistical Research Division, Washington, DC.

Sweet, E., Marquis, K., Sedevi, B., and Nash, F. (1997), *Results of Expert Review of Two Internet R&D Questionnaires*, Human Computer Interaction Report Series 1, US Department of Commerce, US Census Bureau, Statistical Research Division, Washington, DC.

Swick, R., and Bathgate, D. (2007), US services producer price indices, *Proceedings of the 3rd International Conference on Establishment Surveys (ICES-III)*, Montreal, June 18–21, American Statistical Association, Alexandria, VA, pp. 305–316.

Sykes, W. (1997), *Quality Assurance Study on ONS Employment Questions: Qualitative Research*, UK Office for National Statistics, Social Survey Division, London.

Thomas, M. (2002), Standard outcome codes and methods of calculating response rates in business surveys at the Office for National Statistics, paper presented at the Government Statistical Service (GSS) Conference 2002, UK Office for National Statistics, Newport.

Thomas, P. (2002a). *Data Collection Initiatives and Business Surveys in the Office for National Statistics*, UK Office for National Statistics, London.

Thomas, P. (2002b), *Using Telephone Data Entry as a Data Collection Mode for Business Surveys*, UK Office for National Statistics, London.

Thomas, P. (2007), Using telephone data entry as a data collection mode for business surveys, paper presented at the 56th Session of the International Statistical Institute (ISI), Lisbon, Aug. 22–29.

Thurstone, L. L. (1928), Attitudes can be measured, *American Journal of Sociology* **33**: 529–544.

Tillé, Y. (2006), *Sampling Algorithms*, Springer, New York.

Tillé, Y., and Matei, A. (2011), *Package "Sampling"*, version 2.4 (available at `http://cran.r-project.org/web/packages/sampling/index.html`).

Tomaskovic-Devey, D., Leiter, J., and Thompson, S. (1994), Organizational survey nonresponse, *Administrative Science Quarterly* **39**: 439–457.

Torres van Grinsven, V., and Snijkers, G. (2013), Sentiments of Business Survey Respondents on Social Media, *Proceedings of the 2013 European Conference on New Techniques and Technologies for Statistics (NTTS)*, Brussels, March 5–7, pp. 399–408.

Torres van Grinsven, V., Bolko, I., Bavdaz, M., and Biffignandi, S. (2011), Motivation in business surveys, in Giesen, D., and Bavdaz, M., eds., *Proceedings of the BLUE-ETS Conference on Burden and Motivation in Official Business Surveys*, Statistics Netherlands, Heerlen, March 22–23, pp. 7–22.

Tourangeau, R. (1984), Cognitive science and survey methods, in Jabine, T., Straf, M., Tanur, J. M., and Tourangeau, R., eds., *Cognitive Aspects of Survey Design. Building a Bridge between Disciplines*, National Accademy Press, Washington, DC, pp. 73–100.

Tourangeau, R., and Yan, T. (2007), Sensitive questions in surveys, *Psychological Bulletin* **133**(5): 859–883.

Tourangeau, R., Conrad, F. G., Arens, S. F., Lee, S., and Smith, E. (2006), Everyday concepts and classification errors: Judgments of disability and residence, *Journal of Official Statistics* **22**(3): 385–418 (available at `www.jos.nu`).

Tourangeau, R., Rips, L. J., and Rasinski, K. (2000), *The Psychology of Survey Response*, Cambridge University Press, New York.

Trivellato, U. (1999), Issues in the design and analysis of panel studies: A cursory review, *Quality & Quantity* **33**(3): 339–352.

Trochim, W. M. (2009), *The Research Methods Knowledge Base*, Atomic Dog Publishing, Cincinnati.

Tufte, E. R. (2006), *Beautiful Evidence*, Graphics Press, Cheshire, CT.

Tuttle, A. D. (2008), *Establishment Respondents as Survey Managers: Evaluating and Applying an Organizational Model of Survey Response*, poster presented at the 63rd Annual Conference of the American Association for Public Opinion Research (AAPOR), New Orleans, May 13–15.

Tuttle, A. D. (2009), Establishment respondents as survey managers: Using survey design features to empower respondents to find and recruit knowledgeable company personnel for assistance, paper presented at the Annual Federal Committee of Statistical Methodology (FCSM) Research Conference, Washington, DC, Nov. 2–4.

Tuttle, A. D., and Willimack, D. K. (2005), Privacy principles and data sharing: Implications of CIPSEA for economic survey respondents, paper presented at the Annual Federal Committee of Statistics (FCSM) Research Conference, Washington, DC, Nov. 14–16.

Tuttle, A. D., and Willimack, D. K. (2012), Social network analysis as a tool for assessing respondent burden and measurement error in establishment surveys, paper presented at the 4th International Conference on Establishment Surveys (ICES-IV), Montreal, June 11–14, American Statistical Association, Alexandria, VA.

Tuttle, A. D., Morrison, R. L., and Willimack, D. K. (2010a), From start to pilot: A multi-method approach to the comprehensive redesign of an economic survey questionnaire, *Journal of Official Statistics* **26**(1): 87–103 (available at www.jos.nu).

Tuttle, A. D., Pick, K., Hough, R. S., and Mulrow, J. M. (2010b), Experimenting with pre-contact strategies for reducing nonresponse in an economic survey, *Proceedings of the Survey Research Methods Section*, Joint Statistical Meetings, Vancouver, Canada, July 31 – Aug. 5, American Statistical Association, Alexandria, VA, CD-rom.

UK ONS (2010), *Office for National Statistics Compliance Plan 2010: Reducing the Administrative Burden on Business Resulting from ONS Surveys*, UK Office for National Statistics, London.

Underwood, C., Small, C., and Thomas, P. (2000), Improving the efficiency of data validation and editing activities for business surveys, *Proceedings of the 2nd International Conference on Establishment Surveys (ICES-II)*, Buffalo, NY, June 17–21, American Statistical Association, Alexandria, VA, pp. 1262–1267.

UNECE (2007), *Register-Based Statistics in the Nordic Countries – Review of Best Practices with Focus on Population and Social Statistics*, United Nations Economic Commission for Europe, Geneva.

UNECE (2011), *The Impact of Globalization on National Accounts*, United Nations Economic Commission for Europe, New York/Geneva.

Unger, R., and Chandler, C. (2009), *The Project Guide to UX Design: For User Experience Designers in the Field or in the Making*, New Riders, Berkeley, CA.

United Nations (1947), *Measurement of National Income and the Construction of Social Accounts*, Studies and Reports on Statistical Methods 7, United Nations, Geneva.

United Nations (1953), *A System of National Accounts and Supporting Tables,* Studies in Methods, series F, no. 2, rev. 1, United Nations, New York.

United Nations (1968), *A System of National Accounts and Supporting Tables,* Studies in Methods, series F, no. 2, rev. 3, United Nations, New York.

United Nations (1999), *A Systems Approach to National Accounts Compilation,* Studies in Methods, *Handbook of National Accounting*, series F, no. 77, United Nations, New York.

United Nations (2001), *Guidelines on the Application of New Technology to Population Data Collection and Capture*, United Nations Economic and Social Commission for Asia and the Pacific (ESCAP), Bangkok, Thailand (available at www.unescap.org/stat/pop-it/pop-guide/index.asp#guide-capture).

United Nations (2003), *National Accounts: A Practical Introduction*, United Nations, New York (available at http://unstats.un.org/unsd/publication/SeriesF/seriesF_85.pdf).

United Nations (2007), *Statistical Units*, United Nations, Department of Economic and Social Affairs, Statistics Division, New York (available at http://unstats.un.org/unsd/ISDTS/docs/StatisticalUnits.pdf).

United Nations (2008), *System of National Accounts*, United Nations New York.

United Nations Economic Commission for Europe (2007), *Managing Statistical Confidentiality & Microdata Access: Principles and Guidelines for Good Practice*, United Nations Economic Commission for Europe, Geneva.

United Nations Economic Commission for Europe (2009a), *Making Data Meaningful*. Part 1: *A Guide to Writing Stories about Numbers*, UNECE, Geneva.

United Nations Economic Commission for Europe (2009b), *Making Data Meaningful*. Part 2: *A Guide to Presenting Statistics*, UNECE, Geneva.

United Nations Economic Commission for Europe (2009c), *Statistical Metadata in a Corporate Context: A Guide for Managers*, part A, UNECE, Geneva.

United Nations Statistical Commission (1994), *Fundamental Principles of Official Statistics*, Economic and Social Council Supplement 9, United Nations, New York.

US Census Bureau (1987), *1982 Economic Censuses and Census of Governments Evaluation Studies*, US Government Printing Office, Washington, DC.

US Census Bureau (1989), *Recordkeeping Practices Survey*, US Census Bureau, Economic Census Staff, Washington, DC.

US Census Bureau (2011a), *Statistical Quality Standards*, US Census Bureau, Washington, DC (available at www.census.gov/quality/standards).

US Census Bureau (2011b), *Statistical Quality Standard A2: Developing Data Collection Instruments and Supporting Materials*, US Census Bureau, Washington DC (available at http://www.census.gov/quality/standards/standarda2.html).

US Paperwork Reduction Act of 1995 (1980), Title 44, chapter 35, United States Code (44 U.S.C. 3501 et seq.), US Government Printing Office, Washington, D.C. (accessed at www.gpo.gov/fdsys/pkg/USCODE-2011-title44/pdf/USCODE-2011-title44-chap35.pdf or www.archives.gov/federal-register/laws/paperwork-reduction).

Vale, S. (2009), Generic statistical business process model, paper presented at the Joint UNECE/Eurostat/OECD Work Session on Statistical Metadata, April 2009 (available at http://www1.unece.org/stat/platform/display/metis/Generic+Statistical+Business+Process+Model+Paper).

Vale, S. (2010), Exploring the relationship between DDI, SDMX and the generic statistical business process model, paper presented at the 2nd Annual European DDI Users Group Meeting, Utrecht, Dec. 8–9.

Vale, S. (2010), The generic statistical business process model, next steps, paper presented at the Metadata Working Group, Luxembourg, June 17–18.

Valliant, R., Dorfman, A. H., and Royall, R. M. (2000), *Finite Population Sampling and Inference: A Prediction Approach*, Wiley, New York.

Valliant, R., and Gentle, J. E. (1997), An application of mathematical programming to sample allocation, *Computational Statistics and Data Analysis* **25**(3): 337–360.

van den Brakel, J. A. (2010), Sampling and estimation techniques for the implementation of new classification systems: The change-over from NACE Rev. 1.1 to NACE Rev. 2 in business surveys, *Survey Research Methods* **4**(2): 103–119.

van den Brakel, J. A., Smith, P. A., and Compton, S. (2008), Quality procedures for survey transitions—experiments, time series and discontinuities, *Survey Research Methods* **2**(3): 123–141.

van den Ende, J. (1994), The number factory: Punched-card machines at the Dutch Central Bureau of Statistics, *IEEE Annals of the History of Computing* **16**(3): 15–24.

van der Geest, T. M. (2001), *Web Site Design is Communication Design*, John Benjamins Publishing, Amsterdam.

van Nederpelt, P. W. M. (2009), *Checklist Quality of Statistical Output*, Report, Statistics Netherlands, The Hague/Heerlen.

van Nederpelt, P. W. M. (2012), *Object-Oriented Quality and Risk Management (OQRM): A Practical, Scalable and Generic Method to Manage Quality and Risk*, Lulu Press, Raleigh, NC.

Van Nest, J. G. (1987), Content evaluation pilot study, in *1982 Economic Censuses and Census of Governments: Evaluation Studies*, US Census Bureau, Washington, DC.

Vardigan, M., Heus, P., and Thomas, W. (2008), Data documentation initiative: Towards a standard for the social sciences, *The International Journal of Digital Curation* **3**(1): 107–113.

Vehovar, V., and Manfreda, K. L. (2000), Costs and errors of web surveys in establishment surveys, *Proceedings of the 2nd International Conference on Establishment Surveys (ICES-II)*, Buffalo, NY, June 17–21, American Statistical Association, Alexandria, VA, pp. 449–453.

Vendrig, J. P. (2005), *The Burden of Administrative Obligations: What Does Irritate Entrepreneurs the Most?* (in Dutch: De (over)last van Administratieve Lasten: Waar ergeren MKB-ondernemers zich het meest aan?), EIM Business & Policy Research, Zoetermeer, The Netherlands.

Vennix, B. (2009), *Incentives for Businesses: An Empirical Study*, master's thesis, Utrecht University, The Netherlands.

Vik, T. (2007), Automated data collection from internal enterprise production systems, paper presented at the 3th Workshop on Internet Survey Methodology, Lillehammer, Norway, Sept. 17–19.

Voineagu, V., Dumitrescu, I., and Ştefănescu, D. (2008), Ways to change the negative perception of statistics, paper presented at the 94th DGINS (Directors-General of the National Statistical Institutes) Conference: Reduction of administrative burden through official statistics, Vilnius, Lithuania, Sept. 25–26.

Vonk, R. (2009), *Social Psychology* (in Dutch: Sociale psychologie), Noordhoff Uitgevers, Groningen/Houten, The Netherlands.

Wagner, J. (2010), The fraction of missing information as a tool for monitoring the quality of survey data, *Public Opinion Quarterly* **74**(2): 223–243.

Wainer, H. (1984), How to display data badly?, *The American Statistician* **38**(2): 137–147.

Walker, G. (2011), Developing financial statistics for policy, *Economic and Labour Market Review* **5**(1): 46–58 (UK Office for National Statistics).

Wallgren, A., and Wallgren, B. (2007), *Register-Based Statistics – Administrative Data for Statistical Purposes*, Wiley, Hoboken, NJ.

Wang, J. H. (2004), Non-response in the Norwegian business tendency survey, in *Documents*, Statistics Norway, Oslo.

Ward, R., and Doggett, T. (1991), *Keeping Score: The First Fifty Years of the Central Statistical Office*, Her Majesty's Stationary Office, London.

Ware, C. (2004), *Information Visualization: Perception for Design*, Morgan Kaufmann, San Francisco.

Ware-Martin, A. J. (1999), *Introducing and Implementing Cognitive Interviewing Techniques at the Energy Information Administration: An Establishment Survey Environment*, Statistical Policy Working Paper 28, Seminar on Interagency Coordination and Cooperation, Office of Management and Budget, Federal Committee on Statistical Methodology (FCSM), Washington, DC.

Ware-Martin, A., Adler, R. K., and Leach, N. L. (2000), Assessing the impact of the redesign of the manufacturing energy consumption survey, *Proceedings of the 2nd International Conference on Establishment Surveys* (*ICES-II*), Buffalo, NY, June 17–21 American Statistical Association, Alexandria, VA, pp. 1488–1492.

Weeks, W. (2008), Register building and updating in developing countries, *Proceedings of the International Conference on Establishment Surveys* (*ICES-III*), Montreal, June 18–21 American Statistical Association, Alexandria, VA, pp. 1257–1265.

Weisberg, H. F. (2005), *The Total Survey Error Approach: A Guide to the New Science of Survey Research*, University of Chicago Press, Chicago.

Wenemark, M., Persson, A., Noorlind, H., Svensson, T., and Kristenson, M. (2011), Applying motivation theory to achieve increased response rates, respondent satisfaction, and data quality, *Journal of Official Statistics* **27**(2): 393–414 (available at www.jos.nu).

Wenzowski, M. J. (1988), ACTR – a generalised automated coding system, *Survey Methodology* **14**: 299–308.

Wenzowski, M. J. (1995), Advances in automated and computer assisted coding software at Statistics Canada, paper presented at the 3rd International Blaise Users Conference (IBUC), Helsinki, Sept. 18–20.

West, B. (2011), Paradata in survey research: Examples, utility, quality, and future directions, *Survey Practice* (Aug.) (available at www.surveypractice.org).

White House (2006), *Questions and answers when designing surveys for information collections*, Office of Information and Regulatory Affairs, Office and Management and Budget, Washington, DC (available at www.whitehouse.gov/sites/default/files/omb/assets/omb/inforeg/pmc_survey_guidance_2006.pdf).

Willeboordse, A. (1997/1998), *Handbook on the Design and Implementation of Business Surveys*, Eurostat, Luxembourg.

Willenborg, L., and de Waal, T. (2001), *Elements of Statistical Disclosure Control*, Lecture Notes in Statistics 155, Springer, New York.

Williams, N. (2006), ACTR/IDBR Test Evaluation Report, *Survey Methodology Bulletin*, UK Office for National Statistics, London, **57**: 33–38.

Willimack, D. K. (2007), Considering the establishment survey response process in the context of the administrative sciences, *Proceedings of the 3rd International Conference on Establishment Surveys* (*ICES-III*), Montreal, June 18–21, American Statistical Association, Alexandria, VA, pp. 892–903.

Willimack, D. K. (2008), Issues in the design and testing of business survey questionnaires: What we know now that we didn't know then—and what we still don't know, paper presented at the Conference on Reshaping Official Statistics, International Association on Official Statistics (IAOS), Shanghai, Oct. 14–16.

Willimack, D. K., and Dalzell, J. L. (2006), An examination of non-contacts as a source of non-response in a business survey, paper presented at the 3rd European Conference on Quality in Official Statistics, Cardiff, UK, April 24–26.

Willimack, D. K., and Gibson, P. (2010), Collaborating for quality: A cross-discipline approach to questionnaire content evaluation in business surveys, paper presented at the 5th European Conference on Quality in Official Statistics, Helsinki, May 4–6.

Willimack, D. K., and Nichols, E. (2001), Building an alternative response process model for business surveys, paper presented at the 56th Annual Conference of the American Association for Public Opinion Research (AAPOR), Montreal, May 17–20, also in *Proceedings of the Joint Statistical Meetings*, American Statistical Association, Alexandria, VA, CD-rom.

Willimack, D., and Nichols, E. (2010), A hybrid response process model for business surveys, *Journal of Official Statistics* **26**(1): 3–24 (available at www.jos.nu).

Willimack D. K., Nichols, E., and Sudman, S. (2002), Understanding unit and item non-response in business surveys, in Groves, R. M., Dillman, D. A., Eltinge, J. L., and Little, R. J. A., eds., *Survey Nonresponse*, Wiley, Hoboken, NJ pp. 213–227.

Willimack, D. K., Lyberg, L., Martin, J., Japec, L., and Whitridge, P. (2004), Evolution and adaptation of questionnaire development, evaluation, and testing methods for establishment surveys, in Presser, S., Rothgeb, J. M., Couper, M. P., Lessler, J. T., Martin, E., Martin, J., and Singer, E., eds., *Methods for Testing and Evaluating Survey Questionnaires*, Wiley, Hoboken, NJ, pp. 385–407.

Willis, G. B. (2005), *Cognitive Interviewing. A Tool for Improving Questionnaire Design*, Sage, London.

Wilmot, A., Jones, J., Dewar, A., Betts, P., Harper, R., and Simmons, E. (2005), *Public Confidence in Official Statistics: A Qualitative Study on Behalf of the Office for National Statistics and the Statistics Commission*, UK Office for National Statistics, London.

Wirth, H. (2011), Services for EU-SILC and EU-LFS German Microdata Lab at GESIS, paper presented at the European Workshop to Introduce the EU-SILC and the EU-LFS Data, Manchester, UK, Aug. 4–5.

Wolter, K., and Monsour, N. (1986), Conclusions from economic census evaluation studies, *Proceedings of the 2nd US Census Bureau Annual Research Conference*, pp. 41–53.

Wood, J., Šova, M. G., and Richardson, I. (2007), Assessing the quality of SPPIs, *Proceedings of the 3rd International Conference on Establishment Surveys* (*ICES-III*), Montreal, June 18–21, American Statistical Association, Alexandria, VA, pp. 317–325.

Wroblewski, L. (2008), *Web Form Design: Filling in the Blanks*, Rosenfeld Media, New York.

Yan, T. (2005), *Gricean Effects In Self-Adminstered Surveys*, Dissertation, Faculty of the Graduate School of the University of Maryland, College Park, MD.

Yu, F. (2007), Steps to provide quality industrial coding for a business register, *Proceedings of the 3rd International Conference on Establishment Surveys (ICES-III)*, Montreal, June 18–21, American Statistical Association, Alexandria, VA, pp. 23–24.

Zayatz, L., Evans, T., and Slanta, J. (2000), Using noise for disclosure limitation of establishment tabular data, *Proceedings of the 2nd International Conference on Establishment Surveys (ICES-II)*, Buffalo, NY, June 17–21, American Statistical Association, Alexandria, VA, pp. 877–886.

Zayatz, L. (2007), Disclosure avoidance practices and research at the U.S. Census Bureau: An update, *Journal of Official Statistics* **23**(2): 253–265 (available at www.jos.nu).

Zhang, L.-C. (2008), On some common practices of systematic sampling, *Journal of Official Statistics* **24**(4): 557–569 (available at www.jos.nu).

Zhang, L.-C. (2012), Topics of statistical theory for register-based statistics and data integration, *Statistica Neerlandica* **66**(1): 41–63.

Zissis, D., Lekkas, D., and Koutsabasis, P. (2011), Cryptographic dysfunctionality – a survey on user perceptions of digital certificates, paper presented at the 7th ICGS3 4th e-Democracy Joint Conferences, Thessaloniki, Greece, Aug. 24–26.

Zukerberg, A., and Lee, M. (1997), *Better Formatting for Lower Response Burden*, Working Paper in Survey Methodology SM97/02, US Census Bureau, Statistical Research Division, Washington, DC.

Zukerberg, A., Nichols, E., and Tedesco, H. (1999), Designing surveys for the next millennium: Internet questionnaire design issues, paper presented at the 54th Annual Conference of the American Association for Public Opinion Research (AAPOR), St. Petersburg Beach, FL, May 13–16.

Index

Designing and Conducting Business Surveys, First Edition.
Ger Snijkers, Gustav Haraldsen, Jacqui Jones, and Diane K. Willimack.

WILEY SERIES IN SURVEY METHODOLOGY

Established in Part by WALTER A. SHEWHART AND SAMUEL S. WILKS

Editors: *Mick P. Couper, Graham Kalton, J. N. K. Rao, Norbert Schwarz, Christopher Skinner*

Editor Emeritus: *Robert M. Groves*

The ***Wiley Series in Survey Methodology*** covers topics of current research and practical interests in survey methodology and sampling. While the emphasis is on application, theoretical discussion is encouraged when it supports a broader understanding of the subject matter.

The authors are leading academics and researchers in survey methodology and sampling. The readership includes professionals in, and students of, the fields of applied statistics, biostatistics, public policy, and government and corporate enterprises.

ALWIN · Margins of Error: A Study of Reliability in Survey Measurement
BETHLEHEM · Applied Survey Methods: A Statistical Perspective
BETHLEHEM, COBBEN, and SCHOUTEN · Handbook of Nonresponse in Household Surveys
BIEMER · Latent Class Analysis of Survey Error
*BIEMER, GROVES, LYBERG, MATHIOWETZ, and SUDMAN · Measurement Errors in Surveys
BIEMER and LYBERG · Introduction to Survey Quality
BIEMER · Latent Class Analysis of Survey Error
BRADBURN, SUDMAN, and WANSINK ·Asking Questions: The Definitive Guide to Questionnaire Design—For Market Research, Political Polls, and Social Health Questionnaires, *Revised Edition*
BRAVERMAN and SLATER · Advances in Survey Research: New Directions for Evaluation, No. 70
CHAMBERS and SKINNER (editors) · Analysis of Survey Data
COCHRAN · Sampling Techniques, *Third Edition*
CONRAD and SCHOBER · Envisioning the Survey Interview of the Future
COUPER, BAKER, BETHLEHEM, CLARK, MARTIN, NICHOLLS, and O'REILLY (editors) · Computer Assisted Survey Information Collection
COX, BINDER, CHINNAPPA, CHRISTIANSON, COLLEDGE, and KOTT (editors) · Business Survey Methods
*DEMING · Sample Design in Business Research
DILLMAN · Mail and Internet Surveys: The Tailored Design Method
FULLER · Sampling Statistics
GROVES and COUPER · Nonresponse in Household Interview Surveys
GROVES · Survey Errors and Survey Costs
GROVES, DILLMAN, ELTINGE, and LITTLE · Survey Nonresponse
GROVES, BIEMER, LYBERG, MASSEY, NICHOLLS, and WAKSBERG · Telephone Survey Methodology
GROVES, FOWLER, COUPER, LEPKOWSKI, SINGER, and TOURANGEAU · Survey Methodology, *Second Edition*
*HANSEN, HURWITZ, and MADOW · Sample Survey Methods and Theory, Volume 1: Methods and Applications

HARKNESS, VAN DE VIJVER, and MOHLER (editors) · Cross-Cultural Survey Methods
KALTON and HEERINGA · Leslie Kish Selected Papers
KISH · Statistical Design for Research
*KISH · Survey Sampling
KORN and GRAUBARD · Analysis of Health Surveys
KREUTER (editor) · Improving Surveys with Paradata: Analytic Uses of Process Information
LEPKOWSKI, TUCKER, BRICK, DE LEEUW, JAPEC, LAVRAKAS, LINK, and SANGSTER (editors) · Advances in Telephone Survey Methodology
LESSLER and KALSBEEK · Nonsampling Error in Surveys
LEVY and LEMESHOW · Sampling of Populations: Methods and Applications, *Fourth Edition*
LUMLEY · Complex Surveys: A Guide to Analysis Using R
LYBERG, BIEMER, COLLINS, de LEEUW, DIPPO, SCHWARZ, TREWIN (editors) · Survey Measurement and Process Quality
LYNN · Methodology of Longitudinal Surveys
MADANS, MILLER, and MAITLAND (editors) · Question Evaluation Methods: Contributing to the Science of Data Quality
MAYNARD, HOUTKOOP-STEENSTRA, SCHAEFFER, and VAN DER ZOUWEN · Standardization and Tacit Knowledge: Interaction and Practice in the Survey Interview
PORTER (editor) · Overcoming Survey Research Problems: New Directions for Institutional Research, No. 121
PRESSER, ROTHGEB, COUPER, LESSLER, MARTIN, MARTIN, and SINGER (editors) · Methods for Testing and Evaluating Survey Questionnaires
RAO · Small Area Estimation
REA and PARKER · Designing and Conducting Survey Research: A Comprehensive Guide, *Third Edition*
SARIS and GALLHOFER · Design, Evaluation, and Analysis of Questionnaires for Survey Research
SÄRNDAL and LUNDSTRÖM · Estimation in Surveys with Nonresponse
SCHWARZ and SUDMAN (editors) · Answering Questions: Methodology for Determining Cognitive and Communicative Processes in Survey Research
SIRKEN, HERRMANN, SCHECHTER, SCHWARZ, TANUR, and TOURANGEAU (editors) · Cognition and Survey Research
SNIJKERS, HARALDSEN, JONES, and WILLIMACK · Designing and Conducting Business Surveys
STOOP, BILLIET, KOCH and FITZGERALD · Improving Survey Response: Lessons Learned from the European Social Survey
SUDMAN, BRADBURN, and SCHWARZ · Thinking about Answers: The Application of Cognitive Processes to Survey Methodology
UMBACH (editor) · Survey Research Emerging Issues: New Directions for Institutional Research No. 127
VALLIANT, DORFMAN, and ROYALL · Finite Population Sampling and Inference: A Prediction Approach
WALLGREN and WALLGREN · Register-based Statistics: Administrative Data for Statistical Purposes

Printed in the United States
By Bookmasters